THE URBAN GEOGRAPHY READER

Drawing on a rich diversity of theoretical approaches and analytical strategies, urban geographers have been at the forefront of understanding the global and local processes shaping cities and of making sense of the urban experiences of a wide variety of social groups. Through their links with those working in the fields of urban policy and design, urban geographers have also played an important role in the analysis of the economic and social problems confronting cities.

The Urban Geography Reader captures this diversity of scholarship by presenting a stimulating selection of articles and excerpts by leading figures in their fields organized around seven themes. The themes address the changing economic, political, socio-cultural, and technological conditions of contemporary urbanization and the range of individual and collective responses. The *Reader* provides an unparalleled resource for geography and urban studies by offering a selection of original source material on a range of issues related to contemporary cities. Designed to aid understanding, the *Reader* features extensive editorial input in the form of general, section, and individual extract introductions.

Bringing together in one volume "classic" and contemporary pieces of urban geography, studies undertaken in cities in different parts of the world, and examples of theoretical and applied research, *The Urban Geography Reader* will prove invaluable for those studying the complex geographies of urban areas.

Nicholas R. Fyfe is Reader in Human Geography at the University of Dundee.

Judith T. Kenny is Associate Professor in Geography at the University of Wisconsin-Milwaukee.

THE ROUTLEDGE URBAN READER SERIES

Series editors

Richard T. LeGates

Professor of Urban Studies, San Francisco State University

Frederic Stout

Lecturer in Urban Studies, Stanford University

The Routledge Urban Reader Series responds to the need for comprehensive coverage of the classic and essential texts that form the basis of intellectual work in the various academic disciplines and professional fields concerned with cities.

The readers focus on the key topics encountered by undergraduates, graduates and scholars in urban studies and allied fields. They discuss the contributions of major theoreticians and practitioners and other individuals, groups, and organizations that study the city or practise in a field that directly affects the city.

As well as drawing together the best of classic and contemporary writings on the city, each reader features extensive general, section and selection introductions prepared by the volume editors to place the selections in context, illustrate relations among topics, provide information on the author and point readers towards additional related bibliographic material.

Each reader will contain:

- Approximately thirty-six *selections* divided into approximately six sections. Almost all of the selections will be previously published works that have appeared as journal articles or portions of books.
- A *general introduction* describing the nature and purpose of the reader.
- Two- to three-page *section introductions* for each section of the reader to place the readings in context.
- A one-page *selection introduction* for each selection describing the author, the intellectual background of the selection, competing views of the subject matter of the selection and bibliographic references to other readings by the same author and other readings related to the topic.
- A plate section with twelve to fifteen plates and illustrations at the beginning of each section.
- An index.

The types of readers and forthcoming titles are as follows:

THE CITY READER

The City Reader: third edition – an interdisciplinary urban reader aimed at urban studies, urban planning, urban geography and urban sociology courses – will be the *anchor urban reader*. Routledge published a first edition of *The City Reader* in 1996 and a second edition in 2000. *The City Reader* has become one of the most widely used anthologies in urban studies, urban geography, urban sociology and urban planning courses in the world.

URBAN DISCIPLINARY READERS

The series will contain *urban disciplinary readers* organized around social science disciplines. The urban disciplinary readers will include both classic writings and recent, cutting-edge contributions to the respective disciplines. They will be lively, high-quality, competitively priced readers which faculty can adopt as course texts and which will also appeal to a wide audience.

TOPICAL URBAN ANTHOLOGIES

The urban series will also include *topical urban readers* intended both as primary and supplemental course texts and for the trade and professional market.

INTERDISCIPLINARY ANCHOR TITLE

The City Reader: third edition
Richard T. LeGates and Frederic Stout (eds)

URBAN DISCIPLINARY READERS

The Urban Geography Reader
Nicholas R. Fyfe and Judith T. Kenny (eds)

The Urban Sociology Reader
Jan Lin and Christopher Mele (eds)

Forthcoming:

The Urban Politics Reader
Elizabeth Strom and John Mollenkopf (eds)

The Urban and Regional Planning Reader
Eugenie Birch (ed.)

The Urban Design Reader
Michael Larice and Elizabeth Macdonald (eds)

TOPICAL URBAN READERS

The City Cultures Reader: second edition
Malcolm Miles and Tim Hall, with Iain Borden (eds)

The Cybercities Reader
Stephen Graham (ed.)

The Sustainable Urban Development Reader
Stephen M. Wheeler and Timothy Beatley (eds)

Forthcoming:

The Global Cities Reader
Neil Brenner and Roger Keil (eds)

For further information on The Routledge Urban Reader Series
please visit our website:

www.geographyarena.com/geographyarena/urbanreaderseries

or contact:

Andrew Mould
Routledge
Haines House
21 John St
London WC1N 2BP
UK
andrew.mould@tandf.co.uk

Richard T. LeGates
Urban Studies Program
San Francisco State University
1600 Holloway Avenue
San Francisco, California 94132
(415) 338-2875
dlegates@sfsu.edu

Frederic Stout
Urban Studies Program
Stanford University
Stanford, California 94305-6050
(650) 725-6321
fstout@stanford.edu

The Urban Geography Reader

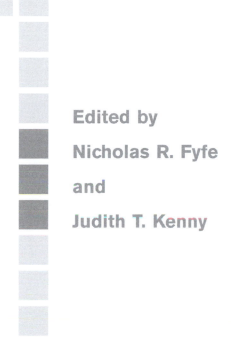

Edited by

Nicholas R. Fyfe

and

Judith T. Kenny

Routledge
Taylor & Francis Group

LONDON AND NEW YORK

First published 2005
by Routledge
2 Park Square, Milton Park, Abingdon, Oxon OX14 4RN

Simultaneously published in the USA and Canada
by Routledge
270 Madison Ave, New York, NY 10016

Routledge is an imprint of the Taylor & Francis Group

Typeset in Amasis MT Lt and Akzidenz Grotesk by Graphicraft Limited, Hong Kong
Printed and bound in Great Britain by Bell & Bain Ltd, Glasgow

British Library Cataloguing in Publication Data
A catalogue record for this book is available from the British Library

Library of Congress Cataloging in Publication Data
A catalog record for this book has been requested

ISBN 0–415–30701–5 (hbk)
ISBN 0–415–30702–3 (pbk)

To Gillian, Alexander and Christopher Fyfe
and
To James and Judith Keyes Kenny

Contents

Acknowledgments

Richard LeGates suggested this project to us, generously offered advice as we began, and then offered problem-solving insight along the way. We thank him for his trust and support. At Routledge we would like to thank our Commissioning Editor Andrew Mould, and Editorial Assistants Melanie Attridge and Anna Somerville, all of whom have shown remarkable patience in the face of missed deadlines. In Dundee, Sarah Reid and Jim Ford provided invaluable assistance scanning articles and diagrams; and we owe a special thank you to Pat Michie who worked so efficiently and skillfully to prepare the final manuscript to a very tight deadline. In Milwaukee, Cordella Jones and Chuanrong Zhang offered assistance in putting the text in order, and Danlin Yu provided invaluable technical expertise with the illustrations. Whenever possible Amy Lewis improved the quality of the prose with her editing assistance. Friends and colleagues generously offered their photographs to enrich the Reader's illustrations, including Rina Ghose, Thomas Harvey, Scott Purl, Elvin Wyly, and Jeffrey Zimmerman. For inspiring us with both their insight and confusion, we wish to thank the geography and urban studies students in our classes at the University of Dundee, University of Wisconsin-Milwaukee, and University of Canterbury, New Zealand. A special acknowledgment of appreciation goes to Deanna Benson for her help as research assistant.

Our families also deserve special mention. Nick would like to thank Gillian, Alexander, and Christopher for providing much help and encouragement to get the book finished, as well as plenty of reminders of when it was time to do something completely different. Judith would like to thank her extended family for cheerfully overlooking the distraction of the book during reunions in Oregon and Maine; Tom Hubka for providing the daily support, essential encouragement, and appropriate distractions to keep her working; and her parents, James and Judith K. Kenny, for understanding and inspiration.

INTRODUCTION

As this book's cover image illustrates, the towers of Daimler-Chrysler House and Sony Center soar over Berlin's Potsdamer Platz today. They punctuate the material and symbolic transformation of the city since the fall of the Berlin Wall in 1989. Europe's busiest plaza during the roaring twenties, bombed during World War II, and then a vacant monument to Cold War tensions for nearly forty-five years, this impressive place in Berlin's history now communicates a dramatic turn from its divided past and emphasizes a commitment to integration within the global economy. Befitting the importance of its landscape to Berlin's residents and residents of Germany as a whole, lively debate surrounds the changes in Potsdamer Platz. Critics question the quality of public life supported by the new corporate, mixed-use developments and the missed opportunity to otherwise define the future cultural and economic capital of Europe. Indisputedly, however, Berliners, other Europeans, and international visitors stream through Potsdamer Platz, the city's transportation hub, surrounded by the bustle of a new Berlin (Marshall and Young 2000).

This one example of the remarkable transformations associated with today's urban space and social life encapsulates an array of questions that urban geographers are well positioned to address. How has the influx of capital and labor to Berlin restructured its local economy and defined its place in the global economy? What forms of partnership are being negotiated between government and private interests to construct a new urban landscape? Whose vision of the future is being produced; and what memories of the past are being preserved? As a city once closed off opens up to a changing population, how will members of that new population deal with social and cultural change? What might individuals of various backgrounds expect to experience on the streets of Berlin?

Berlin's phoenix-like "rise from the ashes" offers a singularly impressive case study for the analyses of urban change at the turn of the millennium (see Till 2005; related Smith 1999). Yet there is no disputing that remarkable change worthy of examination is going on elsewhere as well. Paraphrasing urban geographer Paul Knox, cities are not what they were a few years ago (1993). This simple statement sums up complex demographic and spatial change. Never before have the majority of people resided in cities as they do today, and the urban environment that they are experiencing is decisively different as a result of new economic, social, cultural, and political dynamics.

Urban geographers are making important contributions to the analysis of the material and symbolic aspects of these late twentieth- and early twenty-first-century spatial transformations. *The Urban Geography Reader* highlights such work by considering several themes that researchers actively engage in their work, such as: globalization and transnationalism; economic restructuring; governance, urban politics, and inequality; social differentiation of urban space; spatial form and symbolism; and the influence of technologies. Although the majority of the readings that we have selected for inclusion might be described as reflecting the recent past, we begin the collection with a separate section of articles generally viewed as works that inspired and organized urban geography in its development.

Before launching into the selection of readings, however, we offer a brief history of urban geography to acknowledge the contributions of a variety of theoretical perspectives found within the subfield and to place contemporary geographic literature in context. Narrating the changes in urban geography since its organization in the 1950s generally involves an acknowledgment of the changing theoretical influences on a decade-by-decade basis. It would be misleading, however, to think that each successive decade's new discoveries in theory and methodology wiped out what came before. Instead, new perspectives joined established approaches over time and consequently produced a rather eclectic subfield but nevertheless one with impressive breadth and a membership that includes scholars, practitioners, and policy-makers. Given the expansive literature associated with urban geography, difficult decisions had to be made about what to include; that is, what themes and which authors. The final section of this Introduction will clarify the themes that organize the Reader and will offer suggestions on how to read this edited anthology as well.

Founding urban geography

More than one author has observed that geographers did not initiate the study of urban space (LeGates and Stout 2003). The credit for that is generally given to members of the Chicago School of Sociology. In the language of science, Chicago sociologists developed the study of human ecology by collecting extensive ethnographies of various social and ethnic groups to assess qualities of community and individual deviance, and by mapping out social difference in the city to assess urban growth and social change. Acknowledgment of that influence along with other early stepping-stones in the geographic study of urbanization is located in Part 1 of the Reader, Foundations. Geographers began to systematize the subfield of urban geography in the mid-1950s by borrowing from the modernist social sciences of sociology and neo-classical economics, and building from there (Harris 1990; Mayer 1990; Taaffe 1990). This is not meant to suggest that there were no urban geography courses or research projects prior to these outside influences but historical geographies of cities dominated what had been produced earlier. Moving beyond "mere description," founding members emphasized a research agenda reflecting its status as a science focused on spatial generalizations. The urban economy was central to the work of the early urban geographers; using quantitative empirical methods, rational theories were "translated into policy prescriptions for a new and improved urban reality" (Barnes 2003: 483). Barnes notes that "in every way urban geography upheld the ideal of modernity." Harold Mayer, a significant early leader in urban geography, described himself as a futurist and spoke immediately prior to his death in 1994 with great enthusiasm of the benefits that new technologies allow for improving knowledge in a field of study with "limitless possibilities" (Mayer 1994). This initial vision of urban geography's scientific potential, quantitative methodology, and application for planning and policy has defined the discipline for many to the present time.

The features and themes that defined the subfield in *Readings in Urban Geography* (1959) lent themselves to geography's experience of the quantitative revolution in the late 1950s and 1960s. Arguably an evolution rather than a revolution, during this period geographical analysis relied on inferential statistics and theoretical models such as central place theory, industrial location theory, urban factorial ecology, and the rank-size rule. This spatial science defined the curriculum in geography for an extended period and still characterizes the approach valued in many university departments in the United States, and in A level and GCSE work in Britain (Lees 2002). Ronald J. Johnston commented with some regret that the quantitative revolution is rapidly receding from human geography's institutional memory (cited in Barnes 2001). Trevor Barnes concurred with his assessment and shared his own sense of regret given the pivotal moment that the quantitative revolution represented for human geography, establishing it as a social science, and molding it theoretically, methodologically, and sociologically for decades. We recommend reading Barnes' interpretation of the quantitative revolution (e.g. 2001). It offers impressive insight into the discipline from the late 1950s, as he considers the context in which it unfolded and the sociology of knowledge within the discipline.

Marxist and humanistic critiques

The "grip" of the quantitative revolution began to loosen against the background of the 1960s and debates dealing with poverty, civil rights, gender and racial inequality, and war. In David Livingstone's history of geographic thought, he titled his discussion of this period "Statistics don't bleed" (1992). A dispassionate science of spatial geometry did not bleed; nor did it appear to effectively address the difficult issues of the day. Richard Morrill's (1965) study of Seattle's ghetto expansion and Harold Rose's (1970) analysis of Milwaukee objectively analyzed issues of race using diffusion models, which did suggest to one of their contemporaries hope for "a new era of geographic research" (Deskin cited in Peake and Schein 2000). Analysis of spatial patterns of segregation did not explain the social processes that produced it, however. A critical search for an alternative to the theoretical framework and methodology of urban geography's dominant spatial analytic approach was under way.

Inspiration for alternative theoretical frameworks came from two directions by the early 1970s. The rediscovery of the writings of Marx and Engels introduced a radical political economy approach to urban geography, with one of the earliest leaders in this new direction being David Harvey. Harvey's personal progress is considerable considering his major contribution to the spatial science of geography in *Explanations in Geography* (1969). Others revisited the Chicago School's work, this time to emphasize its ethnographic methodology in research focused on the individual and social groups. At the risk of oversimplifying the development of the humanistic critique (see Ley 2000), David Ley's development as a geographer serves as an example of this response to spatial science and its contribution to urban geography. These two critiques evolved through the 1970s to offer well-established epistemological and methodological programs a decade later. Perhaps ironically, in the 1980s, proponents of Marxist geography and humanistic geography spent considerable time debating each other's positions, ignoring for the most part the (by now) "traditional" geography of the quantitative revolution.

The establishment of *Antipode: A Radical Journal of Geography* in 1969 by graduate students and faculty at Clark University suggests a means of dating the strength of radical scholarship in geography. Another landmark is associated with David Harvey's personal and perceptible progress from liberal to radical geographer in his collection of essays, *Social Justice and the City* (1973). This influential book, focused on inequality in the housing market, provides evidence of Harvey's working through an array of liberal assumptions about the city, and ultimately rejecting them for a systematic Marxist analysis. Harvey's article, "The Urban Process under Capitalism" (1978) included in Part 3 of this volume, details the outcome of political and economic processes that produce the geography of capitalism. The political economy approach within Marxist geography (see Smith 2000) continues to contribute insight in our consideration of the consequences of globalization and the processes of suburbanization and gentrification as demonstrated in the work of Richard Walker (p. 121), Neil Smith (p. 128), Paul Knox (p. 281) and others.

In contrast to the critique from the early radical geographers who emphasized social *structures* "beneath the visible and conscious designs of active human subjects" in order to expose the logic of capitalism, concern for the active role of human *agency* motivated a number of their contemporaries who resisted the definition of geography as a spatial science. A classic in geography, Kevin Lynch's *The Image of the City* (1960), addresses individuals' perceptions of the built environment in an attempt to correct the effects of modernist development in America's cities by making them more "legible," more "readable," and consequently more "liveable." According to Lynch, people read the city as a text based on their own experience of space and understanding of behavioral cues in the landscape. His technique to analyze city dwellers' mental maps continues to be influential in architects' and planners' efforts to make cities more legible for residents and visitors. Although categorized as behavioral geography, Lynch's work shares with the humanist critique that followed a similar focus on values, meanings, and individuals' experience of space.

In the title of his book, *Black Inner City as Frontier Outpost: Images and Behavior of a Philadelphia Neighborhood* (1974), David Ley reveals his background and early interest in behavioral geography. Through his lengthy field research in that Philadelphia neighborhood, however, Ley began his shift from behavioral geography to a social geography influenced by the humanistic critique. Rather eclectic in the variety of humanistic philosophies associated with it, for our purposes, humanistic geography might be best understood in terms of its methodological approach. As was the case in Ley's research in Philadelphia, a battery of qualitative methods (such as participant observation, in-depth interviewing, analysis of documents) are associated with an effort to uncover meaning and understand the "taken-for-granted" qualities of everyday life in a specific place. As key concepts in humanistic geography, place and "sense of place" re-entered geographic vocabulary as once again acceptable and now worthy of theorizing.

During the 1980s several cross-currents brought closer together the divide that had existed between political economy and humanistic geography perspectives. Increasing attention was paid to place by political and economic geographers who sought to theorize the concept as manifesting specificity within the context of general processes. *Geography Matters!* (Massey and Allen 1984) and *Place and Politics* (Agnew 1987) provide some evidence that place had gained theoretical credibility beyond humanistic geography. An interest in the historical geography of class also brought humanist Marxists and humanistic geographers closer together in their shared interest in the struggle of people. Finally, drawing on the structuration theory of sociologist Anthony Giddens (1979), the possibility of bridging the great divide between the Marxists' privileging of social systems and structures and the humanistic geographers' privileging of knowledgeable human agents could be considered (see Gregory 2000).

Feminism and postmodernism

In the 1980s, two distinct movements – albeit with a degree of overlap – introduced new critiques aimed at urban theory that had paid insufficient attention to various categories of difference. Feminism and postmodernism challenged modernism and its privileging of knowledge based on changeless, foundational relationships, thus overlooking the contingencies of space and time. Both explore categories of difference as socially constructed (rather than inherent and foundational) and recognize that, as a consequence of constructed difference, knowledge is necessarily partial based on an individual's position within society. With this intellectual challenge, the "rational man" of spatial science and even the "human" agency of humanistic geography could no longer go unexamined, while differences other than class came into discussions of urban space.

In examining the specific contributions of each theoretical perspective, we will begin with feminism since it takes some historical precedent in the discipline. As the impact of feminism was felt in society in general during the rise of the women's movement of the 1970s, its influence was felt within geography as both a political movement and a theoretical perspective. Reflecting its arrival on geographers' agenda, we might cite the Gender and Geography Research Group organized in the United Kingdom in 1979, and Janice Monk and Susan Hanson's (1982) exhortation to American geographers "On not excluding half of the humans in human geography." By bringing women explicitly into the study of urban geographies, researchers questioned how cities' spatial organization affected women's lives and how urban development itself reflected and reinforced society's assumptions about women (e.g. McDowell 1983). In important research that focused on employment, Linda McDowell and Doreen Massey (1984) were among the first to examine the spatially and temporally contingent nature of women's labor by considering the movement of capital in late nineteenth- and early twentieth-century Britain as capital was shifted to localities where women's "docile and cheap" labor could be exploited.

Since the earliest of the feminist studies in geography, variations among feminists within the discipline have occurred (see Monk 1994). However, we can define in general terms three strands of feminist geography, including: (1) the geography of women, which focuses on the effects of gender inequality; (2) socialist feminist geography, which explores the relationship between capitalism and patriarchy with particular attention to the spatial separation of women and men; and (3) feminist geographies of difference, which focus on the construction of gendered identities (see Pratt 2000). This last category, feminist geographies of difference, indicates a move away from a primary focus on gender and class systems to acknowledge the various parameters around which identity is formed. Such a move reflects the impact of the "cultural turn" in geography, a turn that will be considered more closely below, and the influence of psychoanalytic theories. Postmodernist ideas brought attention to difference(s) in the construction of gender relations across races, nationalities, ages, and sexualities – to name just a few categories of difference. The "cultural turn" also placed (pun intended) arguments about particularity and specificity in gender identities at the center of a comparative feminist research agenda. In this volume, for instance, the empirical studies by feminist geographers Geraldine Pratt and Susan Hanson (1987) and Liz Bondi (1999) (Part 5) underscore the extent to which local attachments and specific local cultures influence decisions related to the location of work and home – with significance for both the construction of local labor markets and the gentrification process for different neighborhoods.

To sort out distinctions between feminism and postmodernism requires some attention before proceeding to a discussion of postmodernism and what has been most vividly described as a "vast and sprawling" literature (Barnes 2003). Linda McDowell offers a means of distinguishing feminist scholarship from postmodernist by acknowledging that both feminism and postmodernism break down the simple binary distinction between women and men. She states: "the insistence on changing relationships of power and inequality based on a gendered division is what distinguishes feminist from postmodern analyses of difference." McDowell then concludes: "In this sense, feminist scholarship remains a modernist project with political and progressive aims" (1999: 228). Nonetheless, forms of feminism tend to be identified with the large tent of postmodernism and we will return to geographies of gender and sexuality in the discussion which follows.

David Ley asked in 1993, "What is there left to say about postmodernism?" Ten years later, in a return to the topic, he noted that "this unusually promiscuous theoretical programme has already burrowed its way across every discipline in the humanities and social sciences" (2003: 537). Reference to postmodernism has become so commonplace that apparently interest is beginning to diminish. Ley suggests that the declining interest creates an opportunity to reassess the trajectory of this intellectual movement and his review provides interesting insight into the disciplinary debate. That postmodernism's critique has "burrowed its way across" disciplines complicates understanding of the term given its multiple expressions. That it has effectively burrowed "in," however, warrants the effort to understand its intellectual history in the discipline.

Among the first geographers to engage in debates surrounding postmodernism, Michael Dear instructed geographers on how to untangle its three component parts: style, epoch, and method (1986). Postmodernism's critique of style, particularly architecture and planning design, has been brought quite explicitly into urban geographic studies, such as in articles in this volume by Michael Dear (p. 138), Paul Knox (p. 281), Jon Goss (p. 293), and David Ley (p. 304). By emphasizing the ravages of modernist planning and responses by a range of social movements in the early stages of postmodernism, Ley is one of the few geographers to suggest that perhaps there could be a critical postmodernism if both form and intentions are taken into consideration (1993, 2003). Jon Goss cautiously considers the possibility in his analyses while elaborating on the manipulation of symbols and meanings in the built environment experienced by today's shoppers touring shopping malls and festival marketplaces. David Harvey's argument in *Conditions of Postmodernity* (1989), however, has been quite influential in its rejection of postmodernism as a style of surfaces that masks basic changes in the political economy. Paul Knox concurs, stating that the inventive stage of the postmodernist style has passed and we are left with conformity in the chosen style of the times.

Harvey's use of postmodernity in the title of his book underscores the extent to which postmodernism is tied to an epoch. Whether it is a phase of the modern or a break from it has been a matter of dispute. Harvey concludes that the condition of postmodernity must be understood in terms of the changes in the capitalist global economy since 1973 and the social and cultural forms that have flowed from them. The fragmentation of the mass market into market niches is associated with post-Fordism, and changing management practices reinforce the strength of the culture industries as they create not only product identity but also the consumer's identification with products. Accompanying restructuring and globalization, class identity is replaced by the multiplication of class fractions defined, it is argued, by appropriate consumption patterns. The culture of consumption becomes a distinguishing feature of the historical period. All of these influences consequently shape landscapes. And with this we loop around to style once again.

Postmodernism as method draws together the influence of post-colonialism and post-structuralism as well as feminist theory with the common thread being, as mentioned above, a challenge to foundational knowledge and an embrace of multiple voices. Post-colonial theory, as demonstrated in Kay Anderson's article in this volume (p. 219), focuses on the impact of colonialism on the colonized and the colonizers historically and into the present by considering the reproduction and transformation of colonial representations, relations, and practices. Post-structuralist theory, actually a body of work associated with a diverse group of French theorists that dates from the 1970s, may generally be described as a questioning of the transparent relationship between material reality and the languages we use to represent them. Foucault's studies of the relations between knowledge, discourses, representations, and power have had a particular influence on research within geography – with discourse referring to "frameworks that embrace particular combinations of narratives, concepts, ideologies, and signifying practices, each relevant to a particular realm of social action" (Barnes and Duncan 1992: 8). In the 1990s, the use of discourse had become standard vocabulary for many urban geographers (see e.g. the article by Judith Kenny in this volume (p. 315).

Marxist geographer Neil Smith wryly offered his view of postmodernism, when he wrote: "The Enlightenment is dead, Marxism is dead, the working class movement is dead . . . and the author does not feel very well either" (cited in Harvey 1989: 325). The author in this instance does not refer to Smith but to the loss of authority of the author. Postmodernism's emphasis on the constructed nature of meaning focuses on the interpreter's reading rather than the author's intentions. Thus reading replaced writing as the site of meaning at any particular time. In the extremes to which the postmodern critique might be taken, this lack of authority undermines an ability to establish "truth claims" and, thus, a foundation for political action. Just as Linda McDowell chose to distinguish between feminism and postmodernism, others have taken the critique and stopped short of the radical relativism that would undermine political positioning by recognizing both the materiality of the world which serves to anchor meaning, and the ability to share meaning based within a particular context of place and time.

In the area of gender and sexuality research, however, the influence of identity politics combines with post-structural theories in Queer Theory. Queer Theory, an intellectual movement developed in the 1990s, challenges the presumption that heterosexuality is the normal sexuality, and rejects notions of fixed sexuality and gender characteristics. The exchange between Queer Theory and geography has been two-way, as exemplified in the work by Gill Valentine (p. 263, this volume). She discusses the heterosexist notions that construct geographies of public space by limiting behavior that falls outside acceptable norms and demonstrates how that space can be politicized. Lesbian behavior becomes a political act when performed where it disrupts dominant expectations of gender and sexuality. We might note that public space has become an important research theme in the last decade for two reasons: (1) privatization of public space appears to be increasing given new landscape elements and practices of urban governance; and (2) attention to categories of difference and the geographies associated with them raises issues with the inclusiveness of "public" and community (e.g. Mitchell 1996; cf. Amin 2002).

(Re)examining urban geography

With this outline of the changing theoretical frameworks that influence research in urban geography, we end with several observations. Debates at various times in the subdiscipline have marked intractable differences of perspective that divide urban geographic research into separate camps. At other times, the differences are minimized as the influences of particular arguments are absorbed. For instance, the recent history of the discipline has been discussed in terms of the "cultural turn." Once dealt with as a matter to be embraced or decried, researchers from either side of the culture/economy divide are increasingly integrating or at least acknowledging the need to deal with both the cultural and economic transformation of the city. Trevor Barnes offers an important analysis of urban geography's intellectual history, noting the important exchange between cultural and economic perspectives and, though "things now seem less straightforward, messier, mixed up, and contaminated," our analyses better reflect reality (2003: 488). We note the calls for "rematerializing" geography as another expression of concern that research themes have drifted from the questions and concerns which urban geographers might most constructively contribute (Lees 2002).

There have also been calls to "reimagine" the city – that is, to engage, in a much more sustained and systematic way than in the past, with the complexity of the city. Reacting to the tendency within urban research to over-generalize from a handful of very specific cities, such as Los Angeles or the global financial centers of London, New York, and Tokyo, Amin and Graham (1997: 418) argue that it is important to capture a sense of "The contemporary city [as] a variegated and multiplex entity." In particular, they distil several elements of the "multiplexity of urban life" which they believe need to be taken seriously in contemporary urban research. These include the ways in which cities combine intense face-to-face interaction as well as technologically mediated communication with the world beyond the city; and the increasingly complex institutional base of cities as established patterns of urban govern-ment give way to more complex networks of urban governance. This emphasis on "multiplexity" is developed further in Amin and Thrift's (2002) *Cities: Reimagining the Urban*. Although we typically think of particular sites or moments when imagining a city – "Paris as cafe life, New York as Manhattan, Calcutta as the noise of traffic" – cities are now "extraordinarily intricate, and for this, difficult to generalize" (2002: 1). For Amin and Thrift, this means understanding cities as being porous, "as spatially open and cross-cut by many different kinds of mobilities, from flows of people to commodities and information . . . a recognition that urban life is the irreducible product of mixture" (2002: 3).

These are the kinds of debate that will stimulate urban geography and shape its agenda in the near future. The prospects for the subfield are impressive as progress is made with important issues and theoretical discussions (see Aitken *et al.* 2003). The claim that "geography matters" no longer requires the same degree of explanation that it once did, and certainly the majority of urban, social, and economic theorists acknowledge that space and place are profoundly important in the constitution of social life.

Section themes and reading guidelines

Compiling a disciplinary reader to join The Routledge Urban Reader Series provided us with an oppor-tunity to reflect upon representative writings on essential themes in a dynamic discipline. Despite the distance between our two departments, we held similar views on what constituted the relevant issues to be addressed in *The Urban Geography Reader*. This consensus reflects the rich cross-fertilization of Anglo-American geography that produces similar research agendas as well as the extent to which common themes of urban life exist.

Nonetheless, the organization of the selections in this Reader into seven themes was not easy because all of the themes overlap and interconnect in a variety of different ways. "Globalization," for example, involves complex processes of economic "restructuring" in which new information and communications "technologies" and new forms of political "governance" play crucial roles. Our attempts to place particular articles in specific sections should therefore be viewed as one of several possible permutations in rela-tion to the organization of the readings. Indeed, we have indicated in our commentaries on individual articles in each section the need to cross-reference articles found in other sections. Nevertheless, the thematic structure adopted in the Reader does offer a coherent and comprehensive way of engaging with the range and richness of research in urban geography, as well as providing important insights into changes in the urban landscape and the differentiated character of the urban experience. The particular concerns of each of the thematic sections are spelt out in more detail below.

Foundations

Urban geographers immediately recognize references to Burgess' concentric zone model, Hoyt's sector model, Harris and Ullman's multi-nuclei model and central place theory. This section contains the four articles that introduced these analyses of urban space and, thus, provides an opportunity to consider the original arguments that proved influential in the development of the subfield and for an extended period after. We deal with cities today that have changed significantly since the models were developed, and yet they are frequently cited in discussions of urban form even if to dismiss their continued relevance. Consequently, they are a part of urban geography's basic vocabulary, and provide an opportunity to consider the development of geographic thought and the sociology of knowledge.

Globalization

A term which is hotly contested and debated, globalization is, according to the UK Commission on Social Sciences, one of the "big issues of our times" (quoted in Martin 2004: 147). For urban geographers in particular, globalization is important, because of claims that it is leading to greater homogenization and a decline in local difference. In challenging such claims, the selections in this section reveal the uneven impact of the processes of globalization on urban areas, showing that geographical differences between cities are now more, not less, important. In addition, this section also challenges the economistic tendencies of the globalization literature, and demonstrates the significant political and socio-cultural dimensions of globalization.

Restructuring

This section focuses mainly on the economic transformations that underpin the making of urban land-scapes, providing an opportunity to explore the importance of the radical, Marxist tradition within urban geography. As was indicated earlier in this Introduction, in the 1970s urban geography became a key entry point for radical ideas into human geography and evidence of this is provided in this section in studies of the changing configuration of urban space associated with processes such as suburbaniza-tion and gentrification. More recently, however, and amidst claims that a new form of urbanism – the postmodern city – has emerged, many urban geographers now insist on a more pluralistic under-standing of "restructuring," using the term in the context of a range of political, socio-cultural as well as economic changes that characterize advanced capitalist societies.

Politics, governance, and inequality

This section provides an opportunity to focus on the city as a site of political power. Urban geographers have long been interested in how cities are administered and governed as well as in the struggles and conflicts that surround the unequal distribution of resources and services in urban areas. In the past, attention would have focused largely on the role of urban government in these matters. Today, as this section explores, formal, hierachical government is increasingly giving way to much more complex "poly-centric, non-hierarchical and non-directive" forms of governance (Parker 2004) characterized by the increasing involvement of private sector and not-for-profit interests in the management of the city.

Difference

Historically, the diversity associated with modern cities has been both a source of celebration and concern. At the same time, social scientists have tended to poorly theorize concepts of difference with the exception of class. That lacuna is being filled by the contributions of feminist theory as well as post-structural theory and its attention to the social construction of various categories of difference such as race and ethnicity, gender, and sexuality. The entries in this section explore the contributions of geographers who examine these three categories of difference and the spatial construction of urban life. Difference in its various forms is considered as a dynamic force in place-making in these urban studies whether at the scale of the neighborhood or the street.

Form and symbolism

References to the postmodern city have become commonplace and yet a degree of confusion continues to surround the term. Are we referring to an epoch, an architectural style, or an intellectual critique of modernism? While "postmodern" may not be specifically addressed in each, the articles in this section assess various theoretical and methodological approaches to the examination of the built environment as they explore the relationship between its symbolic meaning and material form. The selection also provides insight into research themes associated with the contemporary Western city, such as: the privatization of public space; landscapes of consumption; and the politics of meaning as manifested in the built environment.

Technologies

From street lights to sewers, trams to telecommunications, different technologies have a crucial functional as well as symbolic role in the development of cities. Although often overlooked (literally in the case of the networked infrastructures that lie beneath city streets) or taken for granted, technology is an intrinsic feature of urban life. As Amin and Thrift observe, the city is a "mechanosphere, a set of constantly evolving systems or networks, mechanic assemblages which intermix categories like the biological, technical, social, economic, and so on" (2002: 78). This section explores this intermixing, focusing on the importance of water and telecommunications networks, surveillance technologies, and geographical information systems, in the making of urban landscapes. Eschewing a naive technological determinism, the contributors to this section all share a common agenda in viewing the city as a "sociotechnical process" (Graham and Marvin 2001).

Once satisfied with these as representative themes in contemporary urban geography, the challenge came in selecting only thirty entries to address them. In the editing process, we became very aware of what was being left out of a volume of essential readings in urban geography. The post-socialist city and cities in developing countries are absent for the most part. Although regrettable, we hope that these omissions have been rectified to a degree by our offering recommendations for other readings and including a critique of the dominantly Western bias of global theory (Chakravorty, p. 84).

The necessary narrowing of the focus for this volume still leaves us with an impressive array of articles from which to choose, a process made both difficult and rewarding by the vibrancy of the work produced by urban geographers over the past decade or more, the breadth of this subdiscipline, and the significance of today's changing urban experience. To manage this "embarrassment of riches," we cover additional ground in the literature by offering recommendations for further reading at the end of each section introduction and at the end of the introductions to every selection. To extend our reach, we also decided to avoid duplicating entries provided in *The City Reader*. Thus you are encouraged to look there for work by Kevin Lynch, Edward Soja, Peter Hall, and an additional selection by David Harvey. In addition, given the interdisciplinary demands of urban study, it is difficult to contemplate urban geography without acknowledging the work of urban specialists outside the discipline that either inspire or provoke us. Again we recommend *The City Reader* for a selection of interdisciplinary readings that include Engels, Manuel Castells, Mike Davis, Saskia Sassen, John Mollenkopf, Sharon Zukin, and Andreas Duany and Elizabeth Plater-Zyberk, as well as others.

As to the articles that make up this collection, we should note that few of the selections in the original are of a length suitable for an edited anthology. As editors, we made choices on what portion of the original should remain. We have confidence in the integrity of the arguments made in each, but on more than one occasion regret at what was left out. If a particular article suits your interest in a topic, return to the original, because these readings are necessarily partial in order to introduce as many important ideas as possible. We hope, in fact, that these readings will inspire you to go beyond the pages that you find here.

References

Agnew, J. (1987) *Place and Politics: The Geographical Mediation of State and Society*. Boston: Allen and Unwin.

Aitken, S., Mitchell, D. and Staeheli, L. (2003) "Urban geography." In G. Gaile and C. Willmott (eds), *Geography at the Dawn of the 21st Century*. Oxford: Oxford University Press.

Amin, A. (2002) "Ethnicity and the multicultural city: living with diversity," *Environment and Planning A* 34: 959–980.

Amin, A. and Graham, S. (1997) "The ordinary city," *Transactions of the Institute of British Geographers* 22: 411–429.

Amin, A. and Thrift, N. (2002) *Cities: Reimagining the Urban*. Cambridge: Polity Press.

Barnes, T. (2001) "Lives lived and lives told: biographies of geography's quantitative revolution," *Environment and Planning D: Society and Space* 19: 409–429.

Barnes, T. (2003) "The '90s show: culture leaves the farm and hits the streets," *Urban Geography* 24(6): 479–492.

Barnes, T. and Duncan, J. (1992) *Writing Worlds: Discourse, Text and Metaphor in the Representation of Landscape*. New York: Routledge.

Dear, M. (1986) "Postmodernism and planning," *Environment and Planning D: Society and Space* 4: 367–384.

Fincher, R. and Jacobs, J. (1998) *Cities of Difference*. New York: Guilford Press.

Giddens, A. (1979) *Central Problems in Social Theory: Action, Structure and Contradiction in Social Analysis*. London: Macmillan.

Graham, S. and Marvin, S. (2001) *Splintering Urbanism: Networked Infrastructures, Technological Mobilities, and the Urban Condition*. London: Routledge.

Gregory, D. (2000) "Humanistic geography" and "Structuration theory." In R. Johnston *et al.*, *The Dictionary of Human Geography*. Oxford: Blackwell.

Harris, C. (1990) "Urban geography in the United States: my experience of the formative years," *Urban Geography* 11(4): 403–417.

Harvey, D. (1969) *Explanations in Geography*. London: Edward Arnold.

Harvey, D. (1973) *Social Justice and the City*. London: Edward Arnold.

Harvey, D. (1989) *The Condition of Postmodernity*. Oxford: Blackwell.

Knox, P. (1993) *The Restless Urban Landscape*. Englewood Cliffs, NJ: Prentice Hall.

Lees, L. (2002) "Rematerializing geography: the 'new' urban geography," *Progress in Human Geography* 26(1): 101–112.

LeGates, R. and Stout, F. (2003) *The City Reader*. London: Routledge.

Ley, D. (1974) *Black Inner City as Frontier Outpost: Images and Behavior of a Philadelphia Neighborhood.* Washington, DC: Association of American Geographers.

Ley, D. (1993) "Postmodernism, or the cultural logic of advanced intellectual capital," *Tijdschrift voor Economische en Sociale Geografie* 84: 171–174.

Ley, D. (2000) "Humanistic geography." In R. Johnston, D. Gregory, G. Pratt and M. Watts (eds), *The Dictionary of Human Geography.* Oxford: Blackwell.

Ley, D. (2003) "Forgetting postmodernism? Recuperating a social history of local knowledge," *Progress in Human Geography* 27(5): 537–560.

Livingstone, D. (1992) *The Geographical Tradition: Episodes in the History of a Contested Enterprise.* Oxford: Blackwell.

Lynch, K. (1960) *The Image of the City.* Boston, MA: MIT Press.

McDowell, L. (1983) "Towards an understanding of gender division of urban space," *Environment and Planning D: Society and Space* 1: 15–30.

McDowell, L. (1999) *Gender, Identity and Place: Understanding Feminist Geographies.* Minneapolis: University of Minnesota Press.

McDowell, L. and Massey, D. (1984) "A woman's place?" In D. Massey and J. Allen (eds), *Geography Matters!: A Reader.* New York: Cambridge University Press.

Marshall, A. and Young, S. (2000) "A new city rises from Berlin's no-man's land," *Boston Globe*, 5 November.

Martin, R. (2004) "Geography: making a difference in a globalizing world," *Transactions of the Institute of British Geographers* 29: 147–150.

Massey, D. and Allen, J. (eds) (1984) *Geography Matters!: A Reader.* New York: Cambridge University Press.

Mayer, H. (1990) "A half-century of urban geography in the United States," *Urban Geography* 11(4): 418–421.

Mayer, H. (1994) Interview by Judith Kenny, Department of Geography, University of Wisconsin-Milwaukee, Milwaukee Wisconsin. Unpublished.

Mayer, H. and Kohn, C. (eds) (1959) *Readings in Urban Geography.* Chicago, IL: Chicago University Press.

Mitchell, D. (1996) "Public Space and the City: Special Issue," *Urban Geography* 17(2): 127–131.

Monk, J. (1994) *Contextualizing Feminism: International Perspectives.* IGU Working Paper 27, Commission on Gender and Geography, Washington, DC.

Monk, J. and Hanson, S. (1982) "On not excluding half of the human in human geography," *The Professional Geographer* 34: 11–23.

Morrill, R. (1965) "The Negro ghetto: problems and alternatives," *Geographical Review* 55: 339–361.

Parker, S. (2004) *Urban Theory and the Urban Experience: Encountering the City.* London: Routledge.

Peake, L. and Schein, R. (2000) "Racing geography into the new millennium: studies of 'race' and North American geographies," *Social and Cultural Geography* 1(2): 133–142.

Pratt, G. (2000) "Feminist geography." In R. Johnston *et al.*, *The Dictionary of Human Geography.* Oxford: Blackwell.

Rose, H. (1970) "The development of an urban subsystem: the case of the Negro ghetto," *Annals of the Association of American Geographers* 60(1): 1–17.

Smith, F.M. (1999) "The neighbourhood as site for contesting German reunification." In J. Sharp, P. Routledge, C. Philo and R. Paddison (eds), *Entanglements of Power: Geographies of Domination/Resistance.* London: Routledge.

Smith, N. (2000) "Marxist geography." In R. Johnston *et al.*, *The Dictionary of Human Geography.* Oxford: Blackwell.

Taaffe, E. (1990) "Some thoughts on the development of urban geography in the United States during the 1950s and 1960s," *Urban Geography* 11(4): 422–431.

Till, K. (2005) *The New Berlin: Memory, Politics, Place.* Minneapolis: University of Minnesota Press.

PART ONE

Foundations

German immigrants in front of their boarding house in Milwaukee, Wisconsin (late 1880s). (Courtesy of Judith Kenny)

INTRODUCTION TO PART ONE

Why consider urban scholarship that pre-dates the end of World War II? Students generally acknowledge the need to pay some attention to their discipline's academic history. In the case of the following
four articles, however, such attention goes beyond "paying one's (disciplinary) dues" since, in various ways,
each has influenced urban geography for over six decades. Although only two were written by geographers,
reference to all four may be found in the first textbooks devoted to the subdiscipline and in the majority
of today's introductory texts. For their continued visibility, we might title this section "the pioneers" in
recognition of the scholars who produced them – or "the classics" in deference to the texts' perennial presence. Instead, the section is titled "Foundations" to acknowledge that during the mid-part of the twentieth
century there was a point at which geographers deliberately constructed a new subfield devoted to the
analysis of urban areas, and those urban geographers viewed these articles as worthy of study.

Although their impact on the development of urban geography cannot be disputed, consideration of the
continued significance of these four texts requires individual examination. Even after such an examination,
given the array of themes and approaches contained within urban geography today, a lively debate would
follow as to their place in the discipline. Those debates are reflected in the themes of the later sections. While
urban geographers may be committed to a progressive discipline, they continue to engage in the literature
of the past whether: to extend urban theory associated with the earlier writings; or more likely, to contrast
contemporary conditions with historical ones, including the assumptions of earlier schools of thought; and/or
to acknowledge the consequence of geographic thought on contemporary and current policy and the
construction of knowledge. We engage this literature for various reasons, but certainly they have contributed basic vocabulary and metaphors to our ongoing discussion of urban patterns and process.

Placing these influential articles in context requires recognition of the particular time and specific place
associated with the emergence of the subfield. Urban geography developed in the late 1940s from a
timely mix of traditional geography, the influence of the Chicago School of Sociology, and contemporary
responses to city planning, and emerged as a systematic area of study in the 1950s with the modernist
goals of contemporary social sciences (Taaffe 1990: 422). For this first generation of American and
British urban geographers, *Readings in Urban Geography* (Mayer and Kohn 1959) is the text that defined
the new subfield of urban geography during their student careers. In their Introduction to this pivotal text,
editors Harold Mayer (1916–1994) and Clyde Kohn (1911–1989) describe the field as "now at a stage
at which some of its concepts and generalizations have been clearly formulated and a large number of
hypotheses [are] stated as bases for further investigation" (1). Furthermore, they announce, they are
emphasizing concepts and theories instead of the descriptive approaches of the past. The Mayer and Kohn
Reader pulled together several strands of work into a dominantly spatial framework, thus outlining the
basic structure of urban geography that persisted in the university curriculum well into the 1980s.

Indicative of the "accelerating rate at which the field of urban geography is advancing," Mayer and Kohn
claim that much of the scholarship generated prior to 1945 had been superseded by more recent contributions drawing on new theoretical formulations, more sophisticated methods, and newly developed
modern techniques. Yet the earlier work of land economist Homer Hoyt, and that of geographers Chauncy
Harris and Edward Ullman, met their standard of evaluation and were included in the volume, while sociologist Ernest Burgess' influence was present through a borrowing of his concentric zone model.

Borrowing, as we shall see, is a process that greatly contributed to the development of the new subdiscipline. The fact that this is the only section in *The Urban Geography Reader* that contains work by non-geographers underscores the significance of the field's net borrowing at its inception. Although the four authors included here did not share the same disciplinary background, they did share the same geographic location. Each of these pioneers of urban spatial theory is associated with the University of Chicago which undoubtedly facilitated borrowing, and raises interesting questions about both the sociology of knowledge and efforts to define disciplinary boundaries (see Entriken 1980).

Early twentieth-century industrial Chicago figured prominently in the development of urban theory, serving as the iconic North American model for urbanization and the site of fieldwork for members of the Chicago School of Sociology. Burgess' "The Growth of the City," the first essay in this section, exemplifies the seminal work on the social structure of the city conducted by members of the Chicago School. Not only did "The Growth of the City" produce one of the most famous models in social science, but the volume in which it was originally published – *The City* (1925) edited by Chicago School members Robert Park, Ernest Burgess and Roderick McKenzie – also had enormous influence. As geographer Michael Dear recently observed, *The City* still "retains a tremendous vitality far beyond its interest as a historical document" (2003: 500). This vitality corresponds, Dear argues, to the book's expression of the modernist analytical paradigm applied to the urban condition, which remained popular for most of the twentieth century and, arguably, well into the twenty-first. The assumptions that underpin this paradigm include: the linear progress of society that evolves from traditional to modern and with that progression, a changing construction of social relations; an agency-based interpretation of urban process with individuals' choices, rather than structural constraints, explaining the urban condition; and the concept of the city as a unified whole in which the center organizes the hinterland.

Burgess' essay incorporates each of these assumptions into his explanation of both the pattern and process of growth in a concentric zone model. The appealing simplicity of Burgess' model may explain the fascination that has gripped generations of geographers and other social scientists who continue to contemplate concentric zones of land use despite numerous critiques of Burgess' evaluation of urban pattern and process. It is an interesting comment about the history of the subdiscipline that in one of the first urban geography textbooks, Robert Dickinson (1947) wrote that the pattern applied specifically to early twentieth-century, industrial Chicago. By the 1950s, however, the systematic approach of successive urban geography texts generalized its significance beyond that of Chicago.

Arguably, Burgess' concentric zone model has heuristic value in examining portions of late nineteenth- and early twentieth-century North American cities. Even when considering portions of Los Angeles, viewed by many as the urban opposite of Burgess' model, Ed Soja (1989) uses the vocabulary of zones and sectors to describe certain, older sections of the city. For the most part, however, the Chicago School and the concentric ring model has become the foil for which contemporary geographers critique urban theory related to the early twentieth-century city. Dear and Flusty (p. 138), as well as Soja, discuss the Chicago School's conception of the city to underscore how a postmodern view of urban process shifts away significantly from modernist perspectives on the city. Others, such as Walker and Lewis (p. 121), underscore the failings of the Chicago School model of urban growth in explaining even early twentieth-century metropolitan areas, by noting that its emphasis on the city's social geography obscures the operation of the American city's political economy and the early suburbanization of industry.

In 1939, economist Homer Hoyt developed a revised version of Burgess' model theorizing urban growth as a star-shaped, rather than concentric, pattern of development as land use radiated out from the urban center along transportation corridors. His sector theory, based on his attention to the residential real estate market, emphasized that once variations in land values arise, similar land use patterns persist as a city expands. The Federal Housing Administration commissioned this study, the results of which were published in *The Structure and Growth of Residential Neighborhoods in American Cities* (1939). The second essay that follows in this section is drawn from his 1939 study because it describes both the concept of "filtering" in the housing market as wealthier property owners sought the novelty of new development, and the relationship between socio-economic characteristics of residential populations and the stability of neighborhoods. As discussions of gentrification and urban form in later sections remind us (see Smith, p. 128, Bondi, p. 251, and Ley, p. 304), Hoyt's model reflected contemporary trends rather than a universal principle for the housing market. Reflecting the influence of human ecology on British social geography, however, attention to urban residential patterns was sustained through the 1960s in Britain. Mann adjusted Hoyt's and Burgess' models to correspond to the British city in 1965, while the primary focus of urban geographic research in the United States shifted to a neoclassical economic inspired location theory (see Johnston 1971).

Although Hoyt's sector model was viewed as progress in theorizing the internal structure of the city, the impact of his work was experienced more widely because of its influence on the interpretation of changing housing markets. As the Federal Housing Administration's transmittal letter stated, it was deemed a useful "guide to the development of housing and the creation of a sound mortgage market" (iii). Perhaps the most influential land economist of his era, Hoyt's study had profound impacts on American housing assessment practice and thus on the geography of metropolitan areas for an extended period, since it favored new construction and inscribed social and racial prejudice in lending practices. Published research reflects contemporary priorities and perspectives for society, and Hoyt's work powerfully underscores the potential immediate and long-term impact of research on policies and programs that shape the urban environment – with all of its consequences (see Jackson 1985).

The empirical requirements of Hoyt's theorizing presented employment opportunities for a number of Chicago geographers during the late Depression years. Hoyt was appointed Director of Research for the Chicago Plan Commission in 1939, which employed Harold Mayer, among others, to compile a land use survey. Mayer, later editor of the first urban geography reader, was then a graduate student at the University of Chicago along with fellow students Chauncy Harris and Edward Ullman. As Mayer reminisced in 1990, the relationship between urban geography and urban planning at this time was particularly close and would further shape the nature of the new subdiscipline in its applied and economic-oriented direction.

Edward Ullman, author of the third essay in this section and co-author of the fourth, was one of the most influential geographers working during the early development of urban geography. In "A Theory of Location for Cities," Ullman brought the work of German geographer Walter Christaller to the attention of English-speaking geographers. Christaller's central place theory introduced concern for the geographic structure among cities rather than the structure within a single city. Ullman was a doctoral student reflecting upon the regular spatial pattern he had observed in the cities of Iowa when he "discovered" Christaller's book on the system of service centers in southern Germany. Ullman credited his discovery of Christaller to the German land economist August Losch (Harris 1977: 597). Losch not only facilitated the diffusion of Christaller's work but also produced a comparable spatial theory oriented to the distribution of manufacturing centers soon after (1940).

Ullman ascribed to Christaller's central place theory significance for an interpretation of settlement distribution comparable to the concentric zone theory's contribution to the interpretation of land use within cities. Such high praise reflected the desire to develop generalizations for a discipline focused on location theory. Christaller himself had sought to complement Von Thunen's location theory on agricultural land use around market centers (1826) and Alfred Weber's theory of industrial location (1909). As he built on a tradition within the German social sciences, however, Christaller rejected traditional geographical methods of inquiry focused on empirical evidence and emphasized instead the development of theory based on logic that could then be "confronted with reality" (Berry and Harris 1970: 116). This vision of the discipline presaged the development of geography as a spatial science during the 1950s. Despite Ullman's early efforts to set a new direction for geographic research, however, Christaller's research was not fully recognized among English-speaking geographers until the rise of the "Washington School" in the 1960s (see Johnston 1997: 66–70).

In the final essay of this section, Chauncy Harris and Edward Ullman not only address the nature of cities but they outline a framework for urban geography itself by summarizing two general categories of patterns and relationships for analysis – systems of cities and the system within a city. The relationships that they are interested in are primarily economic and, as they explicitly and confidently state, the purpose of that knowledge is to assist in planning the future of cities.

Despite this optimism, the multi-nuclei model that they offer as a reflection of a growing fragmentation of the metropolitan area suggests greater unpredictability than is found in either Burgess' or Hoyt's models of urban growth. In part, this might be explained by their attention to a range of economic activities rather than the singular residential focus of the two earlier models. They summarized the pattern by noting that specialized cells of activity would develop according to specific requirements of certain activities, different rent-paying abilities, and the tendency for some kinds of economic activity to cluster together. Harris explains that he observed multiple nuclei in the traditional urban development of various German cities as well as in London (1990: 411). However, the majority of geographers interpret Harris' and Ullman's spatial typology as a forecast of the future of the decentralizing Western city.

Their framework for urban geography, along with the introduction of the multi-nuclei model of urban form, distinguishes "The Nature of Cities" as a landmark in the subdiscipline's development. Harris and Ullman assert that there is both an ordered spatial pattern in a city's layout and identifiable hierarchies among cities – and more than a half-century later these general categories may still be discerned in the discipline. Today, however, questions related to globalization have replaced central place theory in our understanding of the hierarchy among cities, and models of urban form have become increasingly decentralized. As Robert Lake notes in a fiftieth-year commemoration of "The Nature of Cities," Harris and Ullman's article "established a benchmark against which we can trace the cumulative progress of the field" (1997: 2). Thus, while it suggests certain genealogical links, it also underscores dramatic change in both our cities and our discipline.

References and suggestions for further reading

Berry, B. and Harris, C. (1970) "Walter Christaller: an appreciation," *Geographical Review* 60: 116–19.

Dear, M. (2003) "The Los Angeles School of Urbanism: an intellectual history," *Urban Geography* 24, 6: 493–509.

Dickinson, R. (1947) *City Region and Regionalism: A Geographical Contribution to Human Ecology*, London: Routledge & Kegan Paul.

Entriken, N. (1980) "Robert Park's human ecology and human geography," *Annals of the Association of American Geographers* 70: 43–58.

Harris, C. (1977) "Edward Louis Ullman, 1912–1976," *Annals of the Association of American Geographers* 67, 4: 595–600.

Harris, C. (1990) "Urban geography in the United States: my experience of the formative years," *Urban Geography* 11, 4: 403–17.

Jackson, K. (1985) *The Crabgrass Frontier: The Suburbanization of the United States*. Oxford: Oxford University Press.

Johnston, R.J. (1971) *Urban Residential Patterns: An Introductory Review*, London: G. Bell and Sons.

Johnston, R.J. (1997) *Geography and Geographers: Anglo-American Human Geography since 1945*. London: Arnold.

Lake, R. (1997) "Chauncy Harris and Edward Ullman, 'The nature of cities': a fiftieth year commemoration," *Urban Geography* 18, 1: 1–3.

Losch, A. (1940; trans. 1954) *The Economics of Location*, New Haven, CT: Yale University Press.

Mayer, H. (1990) "A half-century of urban geography in the United States," *Urban Geography* 11, 4: 418–21.

Mayer, H. and Kohn, C. (1959) *Readings in Urban Geography*, Chicago, IL: University of Chicago Press.

Park, R., Burgess, E. and McKenzie, R. (1925) *The City*, Chicago, IL: University of Chicago Press.

Soja, E. (1989) *Postmodern Geographies: The Reassertion of Space in Critical Social Theory*. London: Verso.

Taaffe, E. (1990) "Some thoughts on the development of urban geography in the United States during the 1950s and 1960s," *Urban Geography* 11, 4: 422–31.

Von Thunen, J. (1826; trans. 1966) *Von Thunen's Isolated State*, Oxford: Pergamon Press.

Weber, A. (1909; trans. 1966) *Theory of the Location of Industry*, Chicago, IL: University of Chicago Press.

"The Growth of the City: An Introduction to a Research Project"

from Robert Park *et al., The City* (1925)

Ernest W. Burgess

Editors' Introduction

Years after writing his groundbreaking text, Ernest Burgess (1886–1966) described the period from 1916 to 1923 as the "discovering the physical pattern of the city" phase for the Chicago School of Sociology (1964: 6). Scores of graduate student researchers, directed by Burgess and his colleagues, went out into Chicago to develop extensive and detailed case studies of the city's various neighborhoods. Attempting to find patterns in the patchwork of neighborhoods, Burgess drew upon these case studies and the mapping exercises associated with them to create an urban model that reflected social and economic distinctions. The concentric zone model, radiating out from the central business district, represented increasing degrees of cultural assimilation as well as greater economic and social status with each successive residential zone. Although Burgess subtitled his essay "An Introduction to a Research Project" and acknowledged his debt to the extensive collection of data for this one city, the zonal model became a significant and long-standing representation of *the* North American city – particularly in regard to the correlation of social and spatial distance for various members of society. One author underscores this impact by calling the concentric zone model the most famous diagram in social science (cited in Duncan 1996: 256).

An examination of this seminal work on the internal structure of the city highlights contemporary definitions of social problems in terms of immigration and the critical issue of immigrant assimilation, as well as the generalized spatial arrangement of the early twentieth-century North American industrial city. In the Chicago School of Sociology research agenda, the contemporary concern for inner city slums became part of a broader set of anxieties about the specialization and segmentation of urban society as a whole. Chicago, with its rapid expansion and social complexity, exemplified the segmented urban world and became the prototype of the urban future.

Burgess asked, "In what way are individuals incorporated into the life of a city?" and then answered with a discussion of social organization and disorganization. The force of the consequent changes was "measured in the physical growth and expansion of cities." Despite a vocabulary suggesting social dysfunction, his model presented an image of progressive and inevitable improvement as inner city residents moved outward from the zone of transition (Zone II) to successive zones of better environmental conditions. The inner city neighborhoods might contain areas of blight, but they were viewed as temporary residences as the recent immigrants shared in the process of spatial and social mobility.

The consensual imagery of this mobility fits more easily with the "melting-pot" metaphor of American popular culture than perhaps it does with the Chicago School's language of urban ecology. The urban ecologists' metaphors represented the city as a social organism governed by the struggle for existence. Certainly the experiences of individuals in the zone of transition involved struggle. The pressure of low-status in-migrants arriving in the inner city fueled the dynamic of the model with this pressure characterized as invasion that led to succession in waves outward. Burgess' model of ecological change and urban growth assumes that the negative environmental effect of low-status residents influences wealthier residents' choice to move farther out, while within each zone, segregation of smaller areas developed on the basis of language, culture, and race. These "natural areas" – comparable to neighborhoods – reflected a symbiotic relationship expressed in the residents' shared sentiments, traditions, and local interests. The dynamic growth model suggested eventual improvement and – with restrictions on immigration – a possible "natural" emptying out of Zone I.

After 1930 Burgess refocused on his primary career interest in the family and left others to critique his model. Numerous critics took up the challenge. Hoyt (p. 28) reformulated the residential model by changing the zones to sectors and arguing that the desire of the wealthy for new homes fueled urban growth and development. Geographers Harris and Ullman (p. 46) observed that nodes other than the central business district influenced socio-spatial organization and offered their multi-nuclei model.

Burgess' concentric zone model and ecological approach continued to offer a compelling approach for many geographers based in part on its promise of universal laws founded on fundamental regularities. Factorial ecology, geography's adaptation of human ecology, developed during the 1960s as the language of science appealed to members of the quantitative revolution and computers offered greater opportunity to manage large amounts of data. Berry and Rees' (1969) work on a factorial ecology of Calcutta demonstrates one effort to apply this approach to explain urban form, growth and change in the developing world. This article also distinguishes factorial ecology from forms of social area analysis that preceded it by analyzing a whole battery of socio-economic, demographic, and other characteristics to see what structural dimensions result.

Successive generations of urban geography students become acquainted with Burgess' concentric zone model, and through that with the early twentieth-century work of the Chicago School of Sociology. As a consequence, in each passing decade, there are efforts in the discipline to engage with this work. For example, Peter Jackson (1984) discusses the ethnographic research associated with the Chicago School, and "the pivotal concepts" of social disorganization and moral order. Barney Warf (1990) argues that despite the justified critique of urban ecology, the Chicago School offers insight into the formation of urban residential geographies – and recommends a reconstituted ecological analysis by "graft(ing on) a conception of the division of labor . . . informed by structuration theory." Most recently, Michael Dear (2002) launched a new research agenda and a new school, the LA School of urbanism, by situating it in contrast to *The City*. In *From Chicago to L.A.: Making Sense of Urban Theory* (2002), formal notice is given of the "resistible rise of the L.A. School" by using the original chapters of *The City* as "points of departure" (3, ix). Whether to embrace or reject, such gestures define a piece of literature as a classic that requires attention.

The outstanding fact of modern society is the growth of great cities. Nowhere else have the enormous changes which the machine industry has made in our social life registered themselves with such obviousness as in the cities. In the United States the transition from a rural to an urban civilization, though beginning later than in Europe, has taken place, if not more rapidly and completely, at any rate more logically in its most characteristic forms.

All the manifestations of modern life which are peculiarly urban – the skyscraper, the subway, the department store, the daily newspaper, and social work – are characteristically American. The more subtle changes in our social life, which in their cruder manifestations are termed "social problems," problems that alarm and bewilder us, such as divorce, delinquency, and social unrest, are to be found in their most acute forms in our largest American cities. The profound and "subversive" forces which have wrought these changes are measured in the physical growth and expansion of cities. That is the significance of the comparative statistics of Weber, Bucher, and other students.

These statistical studies, although dealing mainly with the effects of urban growth, brought out into clear relief certain distinctive characteristics of urban as compared with rural populations. The larger proportion of women to men in the cities than in the open country, the greater percentage of youth and middle-aged, the higher ratio of the foreign-born, the increased heterogeneity of occupation increase with the growth of the city and profoundly alter its social structure. These variations in the composition of population are indicative of all the changes going on, in the social organization of the community. In fact, these changes are a part of the growth of the city and suggest the nature of the processes of growth.

The only aspect of growth adequately described by Bucher and Weber was the rather obvious process of the *aggregation* of urban population. Almost as overt a process, that of *expansion*, has been investigated from a different and very practical point of view by groups interested in city planning, zoning, and regional surveys. Even more significant than the increasing density of urban population is its correlative tendency to overflow, and so to extend over wider areas, and to incorporate these areas, into a larger communal life. This paper, therefore, will treat first of the expansion of the city, and then of the less-known processes of urban metabolism and mobility which are closely related to expansion.

EXPANSION AS PHYSICAL GROWTH

The expansion of the city from the standpoint of the city plan, zoning, and regional surveys is thought of almost wholly in terms of its physical growth. Traction studies have dealt with the development of transportation in its relation to the distribution of population throughout the city. The surveys made by the Bell Telephone Company and other public utilities have attempted to forecast the direction and the rate of growth of the city in order to anticipate the future demands for the extension of their services. In the city plan the location of parks and boulevards, the widening of traffic streets, the provision for a civic center, are all in the interest of the future control of the physical development of the city.

This expansion in area of our largest cities is now being brought forcibly to our attention by the Plan for the Study of New York and Its Environs, and

by the formation of the Chicago Regional Planning Association, which extends the metropolitan district of the city to a radius of 50 miles, embracing 4,000 square miles of territory. Both are attempting to measure expansion in order to deal with the changes that accompany city growth. In England, where more than one-half of the inhabitants live in cities having a population of 100,000 and over, the lively appreciation of the bearing of urban expansion on social organization is thus expressed by C.B. Fawcett:

> One of the most important and striking developments in the growth of the urban populations of the more advanced peoples of the world during the last few decades has been the appearance of a number of vast urban aggregates, or conurbations, far larger and more numerous than the great cities of any preceding age. These have usually been formed by the simultaneous expansion of a number of neighboring towns, which have grown out toward each other until they have reached a practical coalescence in one continuous urban area. Each such conurbation still has within it many nuclei of denser town growth, most of which represent the central areas of the various towns from which it has grown, and these nuclear patches are connected by the less densely urbanized areas which began as suburbs of these towns. The latter are still usually rather less continuously occupied by buildings, and often have many open spaces.
>
> These great aggregates of town dwellers are a new feature in the distribution of man over the earth. At the present day there are from thirty to forty of them, each containing more than a million people, whereas only a hundred years ago there were, outside the great centers of population on the waterways of China, not more than two or three. Such aggregations of people are phenomena of great geographical and social importance; they give rise to new problems in the organization of the life and well-being of their inhabitants and in their varied activities. Few of them have yet developed a social consciousness at all proportionate to their magnitude, or fully realized themselves as definite groupings of people with many common interests, emotions and thoughts. ('British Conurbations in 1921,' *Sociological Review* XIV, April 1922: 111–12)

In Europe and America the tendency of the great city to expand has been recognized in the term "the metropolitan area of the city," which far over-runs its political limits, and in the case of New York and Chicago, even state lines. The metropolitan area may be taken to include urban territory that is physically contiguous, but it is coming to be defined by that facility of transportation that enables a business man to live in a suburb of Chicago and to work in the loop, and his wife to shop at Marshall Field's and attend grand opera in the Auditorium.

EXPANSION AS A PROCESS

No study of expansion as a process has yet been made, although the materials for such a study and intimations of different aspects of the process are contained in city planning, zoning, and regional surveys. The typical processes of the expansion of the city can best be illustrated, perhaps, by a series of concentric circles, which may be numbered to designate both the successive zones of urban extension and the types of areas differentiated in the process of expansion.

Figure 1 represents an ideal construction of the tendencies of any town or city to expand radially from its central business district – on the map "the Loop" (I). Encircling the downtown area there is normally an area in transition, which is being invaded by business and light manufacture (II). A third area (III) is inhabited by the workers in industries who have escaped from the area of deterioration (II) but who desire to live within easy access of their work. Beyond this zone is the "residential area" (IV) of high-class apartment buildings or of exclusive "restricted" districts of single family dwellings. Still farther, out beyond the city limits, is the commuters' zone: suburban areas, or satellite cities, within a thirty- to sixty-minute ride of the central business district.

This chart brings out clearly the main fact of expansion, namely, the tendency of each inner zone to extend its area by the invasion of the

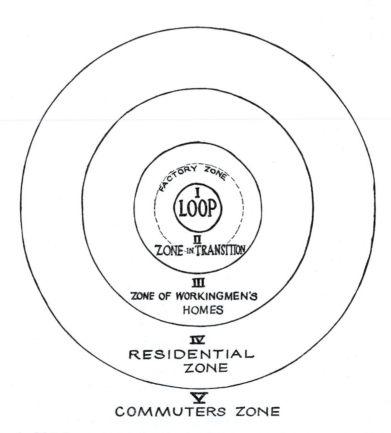

Figure 1 The growth of the city.

next outer zone. This aspect of expansion may be called *succession*, a process which has been studied in detail in plant ecology. If this chart is applied to Chicago, all four of these zones were in its early history included in the circumference of the inner zone, the present business district. The present boundaries of the area of deterioration were not many years ago those of the zone now inhabited by independent wage-earners, and within the memories of thousands of Chicagoans contained the residences of the "best families." It hardly needs to be added that neither Chicago nor any other city fits perfectly into this ideal scheme. Complications are introduced by the lake front, the Chicago River, railroad lines, historical factors in the location of industry, the relative degree of the resistance of communities to invasion, etc.

Besides extension and succession, the general process of expansion in urban growth involves the antagonistic and yet complementary processes of concentration and decentralization. In all cities there is the natural tendency for local and outside transportation to converge in the central business district. In the downtown section of every large city we expect to find the department stores, the skyscraper office buildings, the railroad stations, the great hotels, the theaters, the art museum, and the city hall. Quite naturally, almost inevitably, the economic, cultural, and political life centers here. The relation of centralization to the other processes of city life may be roughly gauged by the fact that over half a million people daily enter and leave Chicago's "loop." More recently sub-business centers have grown up in outlying zones. These "satellite loops" do not, it seems, represent the "hoped for" revival of the neighborhood, but rather a telescoping of several local communities into a larger economic unity. The Chicago of yesterday, an agglomeration of country towns and immigrant colonies, is undergoing a process of reorganization into a centralized decentralized system of local communities coalescing into sub-business areas visibly or invisibly dominated by the central business district. The actual processes of what may be called centralized decentralization are now being studied in the development of the chain store, which is only one illustration of the change in the basis of the urban organization.

Expansion, as we have seen, deals with the physical growth of the city, and with the extension of the technical services that have made city life not only livable, but comfortable, even luxurious. Certain of these basic necessities of urban life are possible only through tremendous development of communal existence. Three millions of people in Chicago are dependent upon one unified water system, one giant gas company, and one huge electric light plant. Yet, like most of the other aspects of our communal urban life, this economic cooperation is an example of cooperation without a shred of what the "spirit of cooperation" is commonly thought to signify. The great public utilities are a part of the mechanization of life in great cities, and have little or no other meaning for social organization.

Yet the processes of expansion, and especially the rate of expansion, may be studied not only in the physical growth and business development, but also in the consequent changes in the social organization and in personality types. How far is the growth of the city, in its physical and technical aspects, matched by a natural but adequate readjustment in the social organization? What, for a city, is a normal rate of expansion, a rate of expansion with which controlled changes in the social organization might successfully keep pace?

SOCIAL ORGANIZATION AND DISORGANIZATION AS PROCESSES OF METABOLISM

These questions may best be answered, perhaps, by thinking of urban growth as a resultant of organization and disorganization analogous to the anabolic and katabolic processes of metabolism in the body. In what way are individuals incorporated into the life of a city? By what process does a person become an organic part of his society? The natural process of acquiring culture is by birth. A person is born into a family already adjusted to a social environment – in this case the modern city. The natural rate of increase of population most favorable for assimilation may then be taken as the excess of the birth-rate over the death-rate, but is this the normal rate of city growth? Certainly, modern cities have increased and are increasing in population at a far higher rate. However, the natural rate of growth may be used to measure the disturbances of metabolism caused by any excessive increase, as those which followed the great influx of southern Negroes to northern cities since the war. In a similar way all cities show deviations in composition by age and sex from a standard population such as that of Sweden, unaffected in recent years by any great emigration or immigration. Here again, marked variations, as any great excess of males over females, or of females over males, or in the proportion of children, or of grown men or women, are symptomatic of abnormalities in social metabolism.

Normally the processes of disorganization and organization may be thought of as in reciprocal relationship to each other, and as co-operating in a moving equilibrium of social order toward an end vaguely or definitely regarded as progressive. So far as disorganization points to reorganization and makes for more efficient adjustment, disorganization must be conceived not as pathological, but as normal. Disorganization as preliminary to reorganization of attitudes and conduct is almost invariably the lot of the newcomer to the city, and the discarding of the habitual, and often of what has been to him the moral, is not infrequently accompanied by sharp mental conflict and sense of personal loss. Oftener, perhaps, the change gives sooner or later a feeling of emancipation and an urge toward new goals.

In the expansion of the city a process of distribution takes place which sifts and sorts and relocates individuals and groups by residence and occupation. The resulting differentiation of the cosmopolitan American city into areas is typically all from one pattern, with only interesting minor modifications (see Figure 2). Within the central business district or on an adjoining street is the "main stem" of "hobohemia," the teeming Rialto of the homeless migratory man of the Middle West. In the zone of deterioration encircling the central business section are always to be found the so-called "slums" and "bad lands," with their submerged regions of poverty, degradation, and disease, and

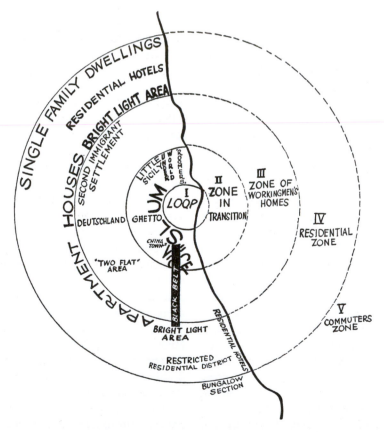

Figure 2 Urban areas.

their underworlds of crime and vice. Within a deteriorating area are rooming-house districts, the purgatory of "lost souls." Nearby is the Latin Quarter, where creative and rebellious spirits resort. The slums are also crowded to over flowing with immigrant colonies – the Ghetto, Little Sicily, Greek-town, Chinatown – fascinatingly combining old world heritages and American adaptations. Wedging out from here is the Black Belt with its free and disorderly life. The area of deterioration, while essentially one of decay, of stationary or declining population, is also one of regeneration, as witness the mission, the settlement, the artists' colony, radical centers – all obsessed with the vision of a new and better world.

The next zone is also inhabited predominatingly by factory and shop workers, but skilled and thrifty. This is an area of second immigrant settlement, generally of the second generation. It is the region of escape from the slum, the *Deutschland* of the aspiring Ghetto family. For *Deutschland* (literally "Germany") is the name given, half in envy, half in derision, to that region beyond the Ghetto where successful neighbors appear to be imitating German Jewish standards of living. But the inhabitant of this area in turn looks to the "Promised Land" beyond, to its residential hotels, its apartment-house region, its "satellite loops," and its "bright light" areas.

This differentiation into natural economic and cultural groupings gives form and character to the city. For segregation offers the group, and thereby the individuals who compose the group, a place and a role in the total organization of city life. Segregation limits development in certain directions, but releases it in others. These areas tend to accentuate certain traits, to attract and develop their kind of individuals, and so to become further differentiated.

The division of labor in the city likewise illustrates disorganization, reorganization and increasing differentiation. The immigrant from rural communities in Europe and America seldom brings with him economic skill of any great value in our industrial, commercial, or professional life. Yet interesting occupational selection has taken place by nationality, explainable more by racial temperament or circumstance than by old-world economic background as Irish policemen, Greek ice-cream parlors, Chinese laundries, Negro porters, Belgian janitors, etc.

The facts that in Chicago one million (996,589) individuals gainfully employed reported 509 occupations, and that over 1,000 men and women in *Who's Who* gave 116 different vocations give some notion of how in the city the minute differentiation of occupation "analyzes and sifts the population, separating and classifying the diverse elements" (sic). These figures also afford some intimation of the complexity and complication of the modern industrial mechanism and the intricate segregation and isolation of divergent economic groups. Inter-related with this economic division of labor is a corresponding division into social classes and into cultural and recreational groups. From this multiplicity of groups, with their different patterns of life, the person finds his congenial social world and – what is not feasible in the narrow confines of a village – may move and live in widely separated, and perchance conflicting, worlds. Personal disorganization may be but the failure to harmonize the canons of conduct of two divergent groups.

If the phenomena of expansion and metabolism indicate that a moderate degree of disorganization may and does facilitate social organization, they indicate as well that rapid urban expansion is accompanied by excessive increases in disease, crime, disorder, vice, insanity and suicide, rough indexes of social disorganization. But what are the indexes of the causes, rather than of the effects, of the disordered social metabolism of the city? The excess of the actual over the natural increase of population has already been suggested as a criterion. The significance of this increase consists in the immigration into a metropolitan city like New York and Chicago of tens of thousands of persons annually. Their invasion of the city has the effect of a tidal wave inundating first the immigrant colonies, the ports of first entry, dislodging thousands of inhabitants who overflow into the next zone, and so on and on until the momentum of the wave has spent its force on the last urban zone. The whole effect is to speed up expansion, to speed up industry, to speed up the "junking" process in the area of deterioration (II). These internal movements of the population become the more significant for study. What movement is going on in the city, and how may this movement be measured? It is easier, of course, to classify movement within the city than to measure it. There is the movement from residence to residence, change of occupation, labor turnover, movement to and from work, movement for recreation and adventure. This leads to the question: what is the significant aspect of movement for the study of the changes in city life? The answer to this question leads directly to the important distinction between movement and mobility.

MOBILITY AS THE PULSE OF THE COMMUNITY

Movement, per se, is not an evidence of change or of growth. In fact, movement may be a fixed and unchanging order of motion, designed to control a constant situation, as in routine movement. Movement that is significant for growth implies a change of movement in response to a new stimulus or situation. Change of movement of this type is called *mobility*. Movement of the nature of routine finds its typical expression in work. Change of movement, or mobility, is characteristically expressed in adventure. The great city, with its "bright lights," its emporiums of novelties and bargains, its palaces of amusement, its underworld of vice and crime, its risks of life and property from accident, robbery, and homicide, has become the region of the most intense degree of adventure and danger, excitement and thrill.

Mobility, it is evident, involves change, new experience, stimulation. Stimulation induces a response of the person to those objects in his environment which afford expression for his wishes. For the person, as for the physical organism, stimulation is essential to growth. Response to stimulation is wholesome so long as it is a correlated *integral* reaction of the entire personality. When the reaction is *segmental*, that is, detached from, and uncontrolled by, the organization of personality, it tends to become disorganizing or pathological. That is why stimulation for the sake of stimulation, as in the restless pursuit of pleasure, partakes of the nature of vice.

The mobility of city life, with its increase in the number and intensity of stimulations, tends inevitably to confuse and to demoralize the person. For an essential element in the mores and in personal morality is consistency, consistency of the type that is natural in the social control of the primary group. Where mobility is the greatest, and

where in consequence primary controls break down completely, as in the zone of deterioration in the modern city, there develop areas of demoralization, of promiscuity, and of vice.

In our studies of the city it is found that areas of mobility are also the regions in which are found juvenile delinquency, boys' gangs, crime, poverty, wife desertion, divorce, abandoned infants, vice.

These concrete situations show why mobility is perhaps the best index of the state of metabolism of the city. Mobility may be thought of, in more than a fanciful sense, as the "pulse of the community." Like the pulse of the human body, it is a process which reflects and is indicative of all the changes that are taking place in the community, and which is susceptible of analysis into elements which may be stated numerically.

The elements entering into mobility may be classified under two main heads: (1) the state of mutability of the person, and (2) the number and kind of contacts or stimulations in his environment. The mutability of city populations varies with sex and age composition, and the degree of detachment of the person from the family and from other groups. All these factors may be expressed numerically. The new stimulations to which a population responds can be measured in terms of change of movement or of increasing contacts. Statistics on the movement of urban population may only measure routine, but an increase at a higher ratio than the increase of population measures mobility. In 1860 the horse-car lines of New York City carried about 50,000,000 passengers; in 1890 the trolley cars (and a few surviving horse-cars) transported about 500,000,000; in 1921, the elevated, subway, surface, and electric and steam suburban lines carried a total of more than 2,500,000,000 passengers. In Chicago the total annual rides per capita on the surface and elevated lines were 164 in 1890; 215 in 1900; 320 in 1910; and 338 in 1921. In addition, the rides per capita on steam and electric suburban lines almost doubled between 1916 (23) and 1921 (41), and the increasing use of the automobile must not be overlooked. For example, the number of automobiles in Illinois increased from 131,140 in 1915 to 833,920 in 1923.

Mobility may be measured not only by these changes of movement, but also by increase of contacts. While the increase of population of Chicago in 1912–22 was less than 25 per cent (23.6 per cent), the increase of letters delivered to Chicagoans was double that (49.6 per cent) – (from 693,048,196 to 1,038,007,854). In 1912 New York had 8.8 telephones; in 1922, 16.9 per 100 inhabitants. Boston had, in 1912, 10.1 telephones; ten years later, 19.5 telephones per 100 inhabitants. In the same decade the figures for Chicago increased from 12.3 to 21.6 per 100 population. But increase of the use of the telephone is probably more significant than increase in the number of telephones. The number of telephone calls in Chicago increased from 606,131,928 in 1914 to 944,010,586 in 1922, an increase of 55.7 per cent, while the population increased only 13.4 per cent.

Land values, since they reflect movement, afford one of the most sensitive indexes of mobility. The highest land values in Chicago are at the point of greatest mobility in the city, at the corner of State and Madison streets, in the Loop. A traffic count showed that at the rush period 31,000 people an hour, or 210,000 men and women in sixteen and one-half hours, passed the southwest corner. For over ten years land values in the Loop have been stationary but in the same time they have doubled, quadrupled and even sextupled in the strategic corners of the "satellite loops," an accurate index of the changes which have occurred. Our investigations so far seem to indicate that variations in land values, especially where correlated with differences in rents, offer perhaps the best single measure of mobility, and so of all the changes taking place in the expansion and growth of the city.

In general outline, I have attempted to present the point of view and methods of investigation which the department of sociology is employing in its studies in the growth of the city, namely, to describe urban expansion in terms of extension, succession, and concentration; to determine how expansion disturbs metabolism when disorganization is in excess of organization; and, finally, to define mobility and to propose it as a measure both of expansion and metabolism, susceptible to precise quantitative formulation, so that it may be regarded almost literally as the pulse of the community. In a way, this statement might serve as an introduction to any one of five or six research projects under way in the department. The project, however, in which I am directly engaged is an attempt to apply these methods of investigation to a cross-section of the city – to put this area, as it were, under the microscope, and so to study in more detail and with greater control and precision the processes which have been described here in the large. For this purpose the West Side Jewish community has been selected. This community includes the so-called "Ghetto," or area of first settlement, and Lawndale, the so-called "Deutschland," or area of second settlement. This area has certain obvious advantages for this study, from the standpoint of expansion, metabolism, and mobility. It exemplifies the tendency to expansion radially from the business center of the city. It is now relatively a homogeneous cultural group. Lawndale is itself an area in flux, with the tide of migrants still flowing in from the Ghetto and a constant egress to more desirable regions of the residential zone. In this area, too, it is also possible to study how the expected outcome of this high rate of mobility in social and personal disorganization is counteracted in large measure by the efficient communal organization of the Jewish community.

Editors' references and suggestions for further reading

Berry, B. and Rees, P. (1969) "Factorial ecology of Calcutta," *American Journal of Sociology* 74: 445–491.

Burgess, E. (1964) *Contributions to Urban Sociology*, Chicago, IL: University of Chicago Press.

Dear, M. (2002) *From Chicago to L.A.: Making Sense of Urban Theory*, Thousand Oaks, CA: Sage.

Duncan, J.S. (1996) "Me(trope)olis: or Hayden White among the urbanists," *Re-Presenting the City*, New York: New York University Press.

Jackson, P. (1984) "Social disorganisation and moral order in the city," *Transactions of the Institute of British Geographers* 9: 168–180.

Warf, B. (1990) "The reconstruction of social ecology and neighborhood change in Brooklyn," *Environment and Planning D: Society and Space* 8: 73–96.

"The Pattern of Movement of Residential Rental Neighborhoods"

from *The Structure and Growth of Residential Neighborhoods in American Cities* (1939)

Homer Hoyt

Editors' Introduction

While working for the Federal Housing Administration in the last years of the Depression, Homer Hoyt (1895–1984) undertook a study of patterns of rental values in American cities. The study's sponsor specifically sought information that would predict mortgage lending risks in different types of neighborhoods, while Hoyt, a land economist, also valued this opportunity to assess the overall structure of cities and the dynamics of land use change. Data from over 140 cities, drawn from the real properties inventory of the Civil Works Administration and physical inventories completed by the Public Works Administration, informed his analysis. Although this inductive model was not based on a single city, its origins may be traced to his doctoral work on Chicago, which evaluated one hundred years of land values in that metropolitan area (1933).

Hoyt's analysis engaged directly with Burgess' concentric zone model (p. 19) as he sought to amplify and correct aspects of the model's representation of the internal structure of the city and the explanation of its growth. Both Burgess' zonal and Hoyt's sector models were developed to fit the "facts" for the early twentieth-century North American city, those being the significance of the private enterprise economy, the assumed dominance of the central business district, and a still growing mass transportation system. Thus it is not surprising that Hoyt's sectors share a great deal with Burgess' concentric zone pattern. Their primary concerns were different however; and, thus, the processes of growth and change they emphasized relied on different factors. Burgess focused on the inner city, low-status in-migrants, and issues of social organization. Hoyt's primary concern was with suburban locations, the affluent, and housing quality.

In the first part of *The Structure and Growth of Residential Neighborhoods in American Cities*, Hoyt's critique of the zonal model emphasizes the ribbons of development that extended along commercial streets leading out of the central business district, and the tendency of industry to follow railroad lines and rivers. Furthermore, he notes the presence of poor and middle-income housing developments along with affluent residential areas on the urban periphery, and estimates in fact that the highest rental areas occupied little more than 25 percent of the periphery (75). Hoyt concludes his description of the city's general structure by noting that residential development in American cities was more *axial* than concentric, conforming to a pattern of sectors rather than concentric circles. High-rent sections, he noted, were located only on the edge of one or more sectors of the city and that in "some quadrants the area of low-rent houses extended like the cut of pie from the center to the periphery of the city."

Claiming significance for his model, Hoyt stated that "[t]his sector theory of rent areas is of fundamental importance in analyzing neighborhoods in American cities for the purpose of locating markets for retail sales or for determining the risk in residential mortgages" (76). He concluded that declining property values were a function of aging housing stocks and the low status of certain residents. In "The Pattern of Movement of Residential Rental Neighborhoods," the second part of the book, Hoyt underscores the significance of the movement of residential rental neighborhoods as a matter of concern for homeowners or investors in residential mortgages. Why place emphasis on rental values and the behavior of affluent households? His attention to rental properties reflects conditions of the late 1930s when nearly 60 percent of American urban households relied on the rental market (Knox 1994: 241). Hoyt theorized that the dynamic of the housing market depended on the transfer of obsolete housing from higher-income households, who vacate in order to occupy more fashionable dwellings, to middle-income households. Thus he contributed the term "filtering" to housing market vocabulary, arguing that status and the desire for novelty drives growth, resulting in the filtering of housing stock down the social scale.

Countering the emphasis on mobility involved in the work of Hoyt and others influenced by the School of Human Ecology, Walter Firey (1945) offered a contemporary critique arguing that excessive emphasis had been placed on economic interpretations of locational decisions. By focusing on three Boston neighborhoods – the wealthy Beacon Hill neighborhood, the public space called the Commons, and the North End neighborhood – largely settled by Italian immigrants, Firey demonstrated that space in certain central city locations becomes a symbol for cultural values, and thus sentiment more than economics influences the locational process.

Although Hoyt defended his sectoral model throughout his writings, in 1941 he observed that the automobile was decentralizing the streetcar city and changing its shape. He predicted this decentralizing pattern of growth in the post-war era and elaborated on his filtering theory in an article published in 1943. However, it was left to Chauncy Harris and Edward Ullman (p. 46) two years later to offer both a schematic of Hoyt's sector model and their view of the decentralized urban area in the form of the multi-nuclei model.

The significance and longevity of Hoyt's sectoral model are matched or perhaps exceeded by the impact of his work on real estate assessment practices in the United States. As explained by Kenneth Jackson (1985), the predictive aspects of his urban growth theory became a self-fulfilling reality particularly through his legitimization of the use of race and ethnicity in real estate appraisals and his influence on such federal agencies as the Home Owners Loan Corporation (HOLC). The "Residential Security Maps" of the HOLC incorporated Hoyt's evaluation of property values and mortgage risk. Following HOLC guidelines, a red pencil line circling a neighborhood, reflecting the perceived risk associated with certain racial or ethnic groups and/or the age of the housing stock, marked the area as unworthy for investment. This practice introduced the term "red-lining" to American real estate vocabulary. Not until the late 1960s did federal policy change to address the disinvestment in central city neighborhoods produced by these assessment practices.

Of the various shifts that take place in the internal structure of a city as a result of population growth, the movement of the residential rental neighborhoods most vitally concerns the home owner or the investor in residential mortgages. This monograph is primarily a study of residential areas; the other types of land uses are considered because of their influence upon the home sections of the city.

. . . [F]rom the high rental areas that are frequently located on the periphery of one or more sectors of American cities, there is a downward gradation of rents until one reaches the low rent areas near the business center. The low rent areas are usually large and may extend from this center to the periphery on one side of the urban community. The high, low, and intermediate rental neighborhoods, however, did not always occupy these locations on the urban site. Their present positions are the points reached in the course of a movement taking place over a period of time. It is not a movement of buildings but a shifting and a change in the character of occupants that produces neighborhood change. New patterns of rent areas are formed as the city grows and adds new structures by both vertical and lateral expansion.

[. . .]

The high rent neighborhoods of a city do not skip about at random in the process of movement – they follow a definite path in one or more sectors of the city.

Apparently there is a tendency for neighborhoods within a city to shift in accordance with what may be called the sector theory of neighborhood change. The understanding of the framework within which this principle operates will be facilitated by considering the entire city as a circle and various neighborhoods as falling into sectors radiating out from the center of that circle. No city conforms exactly to this ideal pattern, of course, but the general figure is useful inasmuch as in our American cities the different types of residential areas tend to grow outward along rather distinct radii, and new growth on the arc of a given sector tends to take on the character of the initial growth in that sector.

Thus if one sector of a city first develops as a low rent residential area, it will tend to retain that character for long distances as the sector is extended through process of the city's growth. On the other hand, if a high rent area becomes established in another sector of the city, it will tend to grow or expand within that sector, and new high grade areas will tend to establish themselves in the sector's outward extension. This tendency is portrayed by the shifts in the location of the fashionable residential areas in six American cities between 1900 and 1936 (Figure 1). Generally speaking, different sectors of a city present different characters according to the original types of the neighborhoods within them.

In considering the growth of a city, the movement of the high rent area is in a certain sense the most important because it tends to pull the growth of the entire city in the same direction. The homes of the leaders of society are located at some point in the high rent area. This location is the point of highest rents or the high rent pole. Residential rents grade downward from this pole

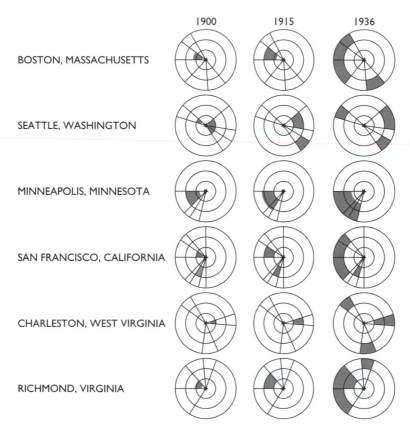

Figure 1 Shift in location of fashionable residential areas in six US cities, 1900 to 1936. Fashionable residential areas indicated by solid black.

as lesser income groups seek to get as close to it as possible. This high rent pole tends to move outward from the center of the city along a certain avenue or lateral line. The new houses constructed for the occupancy of the higher rental groups are situated on the outward edges of the high rent area. As these areas grow outward, the lower and intermediate rental groups filter into the homes given up by the higher income groups. In New York City the movement was up Fifth Avenue, starting at Washington Square and proceeding finally to Ninety-sixth Street in the course of a century. In Chicago, there were three high rental areas, moving southward along Michigan and Wabash Avenues, westward in the band between Jackson and Washington Streets, and northward along La Salle and Dearborn Streets to the Lake Shore Drive.

Sometimes the high rent pole jumps to new areas on the periphery of the city, as in the case of the development of Shaker Heights in Cleveland, Ohio, and Coral Gables in Miami, Fla., but usually these new areas are in the line of growth of the high rent areas. In Charleston, W. Va., the high grade neighborhood moved from the center of the city along Kanawha Street until it reached the river, and then the new high grade area jumped to new locations in the hills in the south and north. In Seattle, Wash., the high grade neighborhood started near the center of the city and moved northeast in one sector of the city – the location along the lake on the periphery. At the same time the high grade development sprang up to the northwest, jumping intervening low grade areas.

In Minneapolis, Minn., there was a movement of the high grade neighborhood to the southwest, starting at the center of the city and repeating the same type of growth until it reached the outer edge of the city in a lake region. In Richmond, Va., the sector of the city containing Monument Avenue first developed as a high grade area. The movement of the high grade neighborhood continued out along the line of Monument Avenue until it reached the city limits and then it expanded fan shape in a sector to the north and west. At the same time a high grade development started to the north in a sector which was bisected by Chamberlayne Street.

[. . .]

High rent or high grade residential neighborhoods must almost necessarily move outward toward the periphery of the city. The wealthy seldom reverse their steps and move backward into the obsolete houses which they are giving up. On each side of them is usually an intermediate rental area, so they cannot move sideways. As they represent the highest income group, there are no houses above them abandoned by another group. They must build new houses on vacant land. Usually this vacant land lies available just ahead of the line of march of the area because, anticipating the trend of fashionable growth, land promoters have either restricted it to high grade use or speculators have placed a value on the land that is too high for the low rent or intermediate rental group. Hence the natural trend of the high rent area is outward, toward the periphery of the city in the very sector in which the high rent area started. The exception to this outward movement is the development of de luxe apartment areas in old residential areas. This will be treated more fully on a following page.

[. . .]

In all of the cities studied, the high grade residential area had its point of origin near the retail and office center. This is where the higher income groups work, and is the point that is the farthest removed from the side of the city that has industries or warehouses. In each city, the direction and pattern of its future growth then tends to be governed by some combination of the following considerations:

(1) *High grade residential growth tends to proceed from the given point of origin, along established lines of travel or toward another existing nucleus of buildings or trading centers.* – This principle is illustrated by the movement of the high grade residential neighborhood of Chicago along the main axes of the roads like Cottage Grove Avenue, leading south around the bend of Lake Michigan to the east, of main roads like Madison Street leading westward, and of roads following the lake northward to Milwaukee. . . .

(2) *The zone of high rent areas tends to progress toward high ground which is free from the risk of floods and to spread along lake, bay, river, and ocean fronts, where such water fronts are not used for industry.* – The movement of high grade residential neighborhoods away from river bottoms to higher ground or to wooded

hills is illustrated by numerous examples. In San Francisco, Calif., the wealthy moved from the lowland along the bay to Knob Hill which was less subject to fogs and smoke. In Washington, D.C., the high grade neighborhoods moved from the mud flats along the Potomac in the southeast quadrant and from the lowland in the southwest quadrant, to the higher land in the northwest section. . . .

[I]n cities located on relatively flat land near rivers, bays, lakes, or oceans, the high grade residential neighborhood tends to expand in long lines along the water front that is not used for industrial purposes. Thus in Chicago, the lake front on the north side is the front yard of the city and is preempted for high grade residential use for a distance of nearly 30 miles north of the business center. In New York City, a high grade residential area grew northward along the Hudson River on Riverside Drive from 72nd Street to Riverdale in the West Bronx. . . . [T]hus, where such lakes, rivers, bays, or ocean fronts exist and offer the attractions of bathing, yachting, cool breezes in summer, and a wide expanse of water with its uninterrupted view, rent areas tend to follow the contour of the water front in long, narrow lines of growth.

(3) *High rent residential districts tend to grow toward the section of the city which has free, open country beyond the edges and away from "dead end" sections which are limited by natural or artificial barriers to expansion.* – The lure of open fields, golf courses, country clubs, and country estates acts as a magnet to pull high grade residential areas to sections that have free, open country beyond their borders and away from areas that run into "dead ends." Thus the high grade neighborhood of Washington, D.C., grows northwest toward expanding open country and estates. Thus, the expansion of high grade neighborhoods to the north of Baltimore, Md., to the south of Kansas City, Mo., and to the north of New York City in Westchester County is into areas with a wide expanse of country beyond them.

(4) *The higher priced residential neighborhood tends to grow toward the homes of the leaders of the community.* – In Washington, D.C., the White House, in New York, the homes of the Astors

and the Vanderbilts were the magnets that pulled the members of society in their direction. One fashionable home, an outpost on the prairie, standing near Sixteenth Street and Prairie Avenue in Chicago in 1836, gave prestige to the section and caused other leaders of fashion to locate near the same spot.

(5) *Trends of movement of office buildings, banks, and stores, pull the higher priced residential neighborhoods in the same general direction.* – The stores, offices, and banks in the central business district usually move in the direction of the high rent area, but follow rather than lead the movement of the high rent neighborhood. Sometimes, however, when an office building center becomes established at a certain point, it facilitates the growth of a high rent area in sections that are conveniently accessible to it. Thus the office building center in the Grand Central District in New York City has aided the growth of the de luxe apartment area in Park Avenue and also the exclusive suburban towns in Westchester that are served by fast express trains entering the Grand Central Station. . . . [I]n Washington, D.C., the northwestward trend of the office buildings, while the result of the pull of the high grade areas to the northwest, also favored the further growth of the northwest area because it made those areas more accessible to offices. Similarly, the trend of office buildings on North Michigan Avenue in Chicago favored the northward growth of the de luxe apartment area.

(6) *High grade residential areas tend to develop along the fastest existing transportation lines.* – The high grade residential areas in Chicago grew along the main plank road, horse car, cable car, and suburban railroad routes. In New York City, the elevated lines and subways paralleled Fifth Avenue. Fast commuters' trains connect New York City with the high grade suburban homes in Montclair, the Oranges, and Maplewood in New Jersey, in Scarsdale, Pelham, and Bronxville in Westchester, and in Forest Hills, Kew Gardens, Flushing, and Hempstead in Long Island. . . . [I]n Washington, D.C., the best areas are on the main transportation arteries – Connecticut Avenue, Massachusetts Avenue, and Sixteenth Street leading directly to the White House.

(7) *The growth of high rent neighborhoods continues in the same direction for a long period of time.* – In New York City, the march of the fashionable areas continued up Fifth Avenue from Washington Square to Central Park for over a century. The high grade neighborhoods in Chicago moved south, west, and north from their starting points in or near the present "Loop" to present locations – 7 to 20 miles distant – in the course of a century. In the century after the Revolutionary War, the high grade area of Washington, D.C., moved from the Capitol to the Naval Observatory.

... [I]n Miami, Fla., Minneapolis, Minn., Seattle, Wash., Charleston, W. Va., Salt Lake City, Utah, and many other cities, this same continuous outward movement of high rent areas has been maintained for long periods of time. Except under the unusual conditions now to be described, there have been no reversals of this long continued trend.

(8) *De luxe high rent apartment areas tend to be established near the business center in old residential areas.* – One apparent exception to the rule that high rent neighborhoods do not reverse their trend of growth is found in the case of de luxe apartment areas like Streeterville in Chicago and Park Avenue in New York City. This exception is a very special case, however, and applies only to intensive high grade apartment developments in a few metropolitan centers. When the high rent single family home areas have moved far out on the periphery of the city, some wealthy families desire to live in a colony of luxurious apartments close to the business center. Because of both the intensive use of the land by use of multiple family structures and the high rents charged it pays to wreck existing improvements.

Such apartments can rise even in the midst of a poor area because the tall building itself, rising from humble surroundings like a feudal castle above the mud huts of the villeins, is a barrier against intrusion. Thus, when the railroad tracks were depressed under Park Avenue in New York City and the railroads were electrified, that street, originally lined with shanties, became the fashionable apartment avenue of New York City. In Chicago, the wall of apartments on the sands where Captain Streeter once had his shack is now occupied by the most exclusive social set. In both cases, there was a renaissance of an old neighborhood. It is only where intensive apartment uses occupy the land that such an apparent reversal of trend occurs.

(9) *Real estate promoters may bend the direction of high grade residential growth.* – While it is almost impossible for real estate developers to reverse the natural trend of growth of high grade neighborhoods, even by the expenditure of large sums of money and great promotional effort, it is possible for them to accelerate a natural trend or to bend a natural trend of growth.

... [S]o ... [t]he developers of Roland Park in Baltimore, Shaker Heights near Cleveland, and the Country Club District of Kansas City take large areas in the line of growth and establish high grade communities by means of building restrictions, architectural control, community planning, and other barriers against invasion. ...

[I]n all these cases, the high rent area was in the general path of growth; but ... [t]he favored area became the fashionable center ... [based] on the promotional skill and the money expended by individual promoters.

As a result of some or all of these forces, high rent neighborhoods thus become established in one sector of the city, and they tend to move out in that sector to the periphery of the city. Even if the sector in which the high rent growth begins does not possess all of the advantages, it is difficult for the high rent neighborhood to change its direction suddenly or to move to a new quarter of the city. For as the high rent neighborhood grows and expands, the low and intermediate areas are likewise growing and expanding, and they are taking up and utilizing land alongside the high rent area as well as in other sectors of the city. When these other areas have acquired a low rent character, it is very difficult to change that character except for intensive apartment use. Hence, while in the beginning of the growth of the city, high rent neighborhoods may have a considerable choice of direction in which to move, that range of choice is narrowed as the city grows and begins to be filled up on one or more sides by low rent structures.

It is possible for high rent neighborhoods to take over sections which are marred by a few shacks. These are swept aside or submerged by the tide of growth. Negro houses have even been bought up and moved away in some southern cities to make way for a high grade development. This possibility exists where the houses are flimsy or scattered, where the land is cheap, where it is held by one owner, or where the residents are under the domination of others. It is extremely difficult otherwise. The cost of acquiring and tearing down substantial buildings and the practical impossibility of acquiring large areas from scattered owners, usually prevent high grade areas from taking over land once it has been fairly well occupied by middle or low grade residential uses.

Now that the radius of the settled area of cities has been greatly extended by the automobile, however, there is little difficulty in securing land for the expansion of high rent areas; for the high rent sector of the city expands with an ever widening arc as one proceeds from the business center.

The next vital question to be considered is how the various types of high rent areas are affected by the process of dynamic growth of the city and how the various types are related to each other in historical sequence. The first type of high rent development was the axial type with high grade homes in a long avenue or avenues leading directly to the business center. The avenue was a social bourse, communication being maintained by a stream of fashionable carriages, the occupants of which nodded to their acquaintances in other passing carriages or to other friends on the porches of the fine residences along the way. Such avenues were lined with beautiful shade trees and led to a park or parks through a series of connecting boulevards. Examples of this type of development, in the decades from 1870 to 1900, are Prairie and South Michigan Avenues, Washington and Jackson Streets and the Lake Shore Drive in Chicago, Fifth Avenue in New York City . . . [a]nd Summit Street in St. Paul, Minn. The fashionable area in this type of development expanded in a long string in a radial line from the business center. There was usually an abrupt transition within a short distance on either side of the high grade street.

The axial type of high rent area rapidly became obsolete with the growth of the automobile. When the avenues became automobile speedways, dangerous to children, noisy, and filled with gasoline fumes, they ceased to be attractive as home sites for the well-to-do. No longer restricted to the upper classes, who alone could maintain prancing steeds and glittering broughams, but filled with *hoi polloi* jostling the limousines with their flivvers, the old avenues lost social caste. The rich then desired seclusion away from the "madding crowd" whizzing by and honking their horns. Mansions were then built in wooded areas, screened by trees. The very height of privacy is now attained by some millionaires whose homes are so protected from the public view by trees that they can be seen from outside only from an airplane.

The well-to-do who occupy most of the houses in the high rent brackets have done likewise in segregated garden communities. The new type of high grade area was thus not in the form of a long axial line but in the form of a rectangular area, turning its back on the outside world, with winding streets, woods, and its own community centers. Such new square or rectangular areas are usually located along the line of the old axial high grade areas. The once proud mansions still serve as a favorable approach to the new secluded spots. As some of the old axial type high rent areas still maintain a waning prestige and may still be classed as high rent areas, the new high rent area takes a fan-shaped or funnel form expanding from a central stem as it reaches the periphery of the city. The old stringlike development of high rent areas still asserts itself, however, in the cases of expansion of high rent areas along water fronts like Lake Michigan, Miami Beach, and the New Jersey coast. The automobile, however, has made accessible hilly and wooded tracts on which houses are built on the crest of hills along winding roads.

The fashionable suburban town, which had its origin even before the Civil War, has remained a continuous type of high grade area. Old fashionable towns like Evanston, Oak Park, and Lake Forest near Chicago, have maintained their original character and expanded their growth. Other new high grade suburban towns have been established. The de luxe apartment area has been a comparatively recent development, coming after 1900, when the wealthy ceased to desire to maintain elaborate town houses and when the high grade single-family home areas began to be located far from the business center. A group of wealthy people, desiring to live near the business center and to avoid the expense and trouble of maintaining a retinue of servants, sought the convenience of tall elevator apartments.

. . . [I]ntermediate rental groups tend to occupy the sectors in each city that are adjacent to the high rent area. Those in the intermediate rental group have incomes sufficient to pay for new houses with modern sanitary facilities. Hence, the new growth of these middle-class areas takes place on the periphery of the city near high grade areas or sometimes at points beyond the edge of older middle-class areas.

Occupants of houses in the low rent categories tend to move out in bands from the center of the city mainly by filtering up into houses left behind by the high income groups, or by erecting shacks on the periphery of the city. They live in either second-hand houses in which the percentage needing major repairs is relatively high or in newly constructed shacks on the periphery of the city. These shacks frequently lack modern plumbing facilities and are on unpaved streets. The shack fringe of the city is usually in the extension of a low rent section.

Within the low rent area itself there are movements of racial and national groups. Until only comparatively recently, the immigrants poured from Europe into the oldest and cheapest quarters on the lower East Side of New York and on the West Side of Chicago. The earlier immigrants moved out toward the periphery of the city. These foreign groups moved in bands or straight lines out from the railroad stations near the central business district. . . . [W]ith the decline of immigration after the World War, new immigrants ceased to fill the old houses in the downtown area and this outward progression of foreign groups slackened. Many of the tenements in the lower east side were boarded up, and some of the oldest quarters near the central business district of Chicago were demolished. During the World War and after, however, there was a great influx of Negroes into the northern cities to take the place of European immigration. The Negro neighborhood in Harlem, New York, expanded in concentric circles. In Chicago, the Negroes burst the bounds of their old area along State Street and the Rock Island tracks, Twenty-second and Thirty-ninth Street. . . . [I]n this

movement in Chicago, they spread into an area formerly occupied by middle-class and some high income families. The area, however, was becoming obsolete and did not offer vigorous resistance to the incoming of other racial groups.

Thus, in the framework of the city there is a constant dynamic shifting of rental areas. There is a constant outward movement of neighborhoods because as neighborhoods become older they tend to be less desirable.

Forces constantly and steadily at work are causing a deterioration in existing neighborhoods. A neighborhood composed of new houses in the latest modern style, all owned by young-married couples with children, is at its apex. At this period of its vigorous youth, the neighborhood has the vitality to fight off the disease of blight. The owners will strenuously resist the encroachment of inharmonious forces because of their pride in their homes and their desire to maintain a favorable environment for their children. The houses, being in the newest and most popular style, do not suffer from the competition of any superior house in the same price range, and they are marketable at approximately their reproduction cost under normal conditions.

Both the buildings and the people are always growing older. Physical depreciation of structures and the aging of families constantly are lessening the vital powers of the neighborhood. Children grow up and move away. Houses with increasing age are faced with higher repair bills. This steady process of deterioration is hastened by obsolescence; a new and more modern type of structure relegates these structures to the second rank. The older residents do not fight so strenuously to keep out inharmonious forces. A lower income class succeeds the original occupants. Owner occupancy declines as the first owners sell out or move away or lose their homes by foreclosure. There is often a sudden decline in value due to a sharp transition in the character of the neighborhood or to a period of depression in the real estate cycle.

These internal changes due to depreciation and obsolescence in themselves cause shifts in the locations of neighborhoods. When, in addition, there is poured into the center of the urban organism a stream of immigrants or members of other racial groups, these forces also cause dislocations in the existing neighborhood pattern.

The effects of these changes vary according to the type of neighborhood and can best be described by discussing each one in turn. The highest grade neighborhood, occupied by the mansions of the rich, is subject to an extraordinary rate of obsolescence. The large scale house, modeled after a feudal castle or a palace, has lost favor even with the rich. When the wealthy residents seek new locations, there is no class of a slightly lower income which will buy the huge structures because no one but wealthy persons can afford to furnish and maintain them. There is no class filtering up to occupy them for single family use. Consequently, they can only be converted into boarding houses, offices, clubs, or light industrial plants, for which they were not designed. Their attraction of these types of uses causes a deterioration of the neighborhood and a further decline in value. These mansions frequently become white elephants like those on Arden Park and East Boston Boulevard in Detroit, Mich.

On the other hand, houses in intermediate rental neighborhoods designed for small families can be handed down to a slightly lower income group as they lose some of their original desirability because of age and obsolescence. There is a loss of value when a transition to a lower income group occurs, but the house is still used for the essential purpose for which it was designed; and the loss of value is not so great. There is always a class filtration to occupy the houses in the intermediate rental neighborhoods. Hence, a certain stability of value is assured.

Since the buildings in low rent areas are occupied by the poorest unskilled or casual workers, collection losses and vacancy ratios are highest. The worst buildings are condemned or removed by demolition to save taxes. Formerly these worst quarters in the old law tenements of New York or the West Side of Chicago were occupied by newly arrived immigrants. With the decline of immigration, this submarginal fringe of housing is being wrecked or boarded up as the residents filter up to better houses. Thus, intermediate rental neighborhoods tend to preserve their stability better than either the highest or lowest rental areas.

The erection of new dwellings on the periphery of a city, made accessible by new circulatory systems, sets in motion forces tending to draw population from the older houses and to cause all groups to move up a step leaving the oldest and cheapest houses to be occupied by the poorest families or to be vacated. The constant competition of new areas is itself a cause of neighborhood shifts. Every building boom, with its new crop of structures equipped with the latest modern devices, pushes all existing structures a notch down in the scale of desirability.

Editors' references and suggestions for further reading

Firey, W. (1945) "Sentiments and symbolism as ecological variables," *American Sociological Review* 10, 2: 140–148.

Hoyt, H. (1943) "The structure of American cities in the post-war era," *American Journal of Sociology* 48, 4: 475–481.

Jackson, K. (1985) *Crabgrass Frontier: The Suburbanization of the United States*, New York: Oxford University Press.

Knox, P. (1994) *Urbanization: An Introduction to Urban Geography*, Englewood Cliffs, NJ: Prentice Hall.

"A Theory of Location for Cities"

from *American Journal of Sociology* (1941)

Edward L. Ullman

Editors' Introduction

As exemplified in Burgess' (p. 19) and Hoyt's (p. 28) writings, during the early twentieth century, urban residential patterns held an increasingly important place on the research agenda of North American geographers, sociologists and other urban scholars, while the economic base of cities received little attention. Geographer Edward Ullman (1912–1976) sought to remedy this oversight by urging his colleagues to take up work on a theory of location for cities, arguing that beyond its important theoretical significance the knowledge gained would provide practical applications in regional planning. In the following article, Ullman proposes the foundation for this research agenda by asking the question – "What are the causes for the existence, present size, and character of a city?" – and then arguing that the German geographer Walter Christaller (1893–1969) supplies the theoretical framework for such study with his central-place theory. Indeed, Ullman offers the endorsement that: "central-place theory probably provides as valid an interpretation of settlement distribution over the land as the concentric-zone theory does for land within cities."

Christaller's *Central Places in Southern Germany* (1933), a study of the size and spatial arrangement of retail activities among urban centers, reached a North American audience through Ullman's article. The striking regularities that Christaller noted in the size and spacing of settlements in southern Germany inspired and offered a test of the model's logic. Two elementary principles serve as the foundation of central-place theory defining the location and hierarchy of the settlements – range and threshold. Range defines the distance that consumers would travel for particular types of goods, and threshold is the minimum number of consumers required to sustain a business with particular goods or services. Assuming that consumers are utility-maximizing decision-makers, Christaller assumes further that accessibility is a direct function of distance and, thus, shops locate (or may be found) centrally within their hinterlands. Showing what might appear in certain idealized conditions, this normative theory projects a hexagonal network of shop locations in central places since a hexagon is the most efficient geometrical shape to exhaust the market territory without overlap.

Ullman's introduction of central-place theory to American urban scholars launched a classic in the literature of urban geography and, as a consequence, the hexagon pattern of central-place theory is as recognizable as Burgess' concentric zones. Christaller's 1933 publication had made little impact on his contemporaries in Germany but when Ullman, as a doctoral student, drew upon it to contemplate the regular spatial patterns of cities in Iowa, he set in motion a research agenda focused on locational theory that directed geographic research in the 1950s and 1960s. Moving to the University of Washington in 1951, Ullman helped establish the "Washington School," which became known for its theoretical testing, empirical work, and contributions to planning. The work of a Washington graduate, Brian Berry's *Geography of Market Centers and Retail Distribution* (1967), is perhaps the best-known analysis of settlement patterns that applies central-place theory. Ullman maintained a reputation throughout his career as a pioneer in urban and transportation geography with such work as "The Nature of Cities" (p. 46) and "The Role of Transportation and the Bases for Interaction" (1956).

Today we recognize the limitations of central-place theory, since our contemporary concern for the external structure of the city suggests less the dependence relationship of the city and its hinterland than the links among cities. This theoretical development is addressed in Part 2 with the discussion of "global cities" and world-systems theory.

To gain insight into the history of an idea and the scholars associated with it, read Brian Berry and Chauncy Harris' memorial for Christaller (1970). Harris also published a tribute to Ullman which addresses the important influence of Christaller's work on his early career (1977). Read as well Richard Symanski and John Agnew's *Order and Skepticism: Human Geography and the Dialectic of Science* (1981) to appreciate the extreme to which fascination with a spatial theory might be taken.

■ ■ ■ ■ ■ ■

Periodically in the past century the location and distribution of cities and settlements have been studied. Important contributions have been made by individuals in many disciplines. Partly because of the diversity and uncoordinated (sic) nature of the attack and partly because of the complexities and variables involved, a systematic theory has been slow to evolve, in contrast to the advances in the field of industrial location.[1]

The first theoretical statement of modern importance was von Thunen's *Der isolierte Staat*, initially published in 1826, wherein he postulated an entirely uniform land surface and showed that under ideal conditions a city would develop in the center of this land area and concentric rings of land use would develop around the central city. In 1841 Kohl investigated the relation between cities and the natural and cultural environment, paying particular attention to the effect of transport routes on the location of urban centers.[2] In 1894 Cooley admirably demonstrated the channelizing influence that transportation routes, particularly rail, would have on the location and development of trade centers.[3] He also called attention to break in transportation as a city-builder just as Ratzel had earlier. In 1926 Haig sought to determine why there was such a large concentration of population and manufacturing in the largest cities.[4] Since concentration occurs where assembly of material is cheapest, all business functions, except extraction and transportation, ideally should be located in cities where transportation is least costly. Exceptions are provided by the processing of perishable goods, as in sugar centrals, and of large weight-losing commodities, as in smelters. Haig's theoretical treatment is of a different type from those just cited but should be included as an excellent example of a "concentration" study.

In 1927 Bobeck[5] showed that German geographers since 1899, following Schluter and others, had concerned themselves largely with the internal geography of cities, with the pattern of land use and forms within the urban limits, in contrast to the problem of location and support of cities. Such preoccupation with internal urban structure has also characterized the recent work of geographers in America and other countries. Bobeck insisted with reason that such studies, valuable though they were, constituted only half the field of urban geography and that there remained unanswered the fundamental geographical question: "What are the causes for the existence, present size, and character of a city?" Since the publication of this article, a number of urban studies in Germany and some in other countries have dealt with such questions as the relations between city and country.[6]

A theoretical framework for study of the distribution of settlements is provided by the work of Walter Christaller.[7] The essence of the theory is that a certain amount of productive land supports an urban center. The center exists because essential services must be performed for the surrounding land. Thus the primary factor explaining Chicago is the productivity of the Middle West; location at the southern end of Lake Michigan is a secondary factor. If there were no Lake Michigan, the urban population of the Middle West would in all probability be just as large as it is now. Ideally, the city should be in the center of a productive area.[8] The similarity of this concept to von Thunen's original proposition is evident.

Apparently many scholars have approached the scheme in their thinking.[9] Bobeck claims he presented the rudiments of such an explanation in 1927. The work of a number of American rural sociologists shows appreciation for some of Christaller's preliminary assumptions, even though done before or without knowledge of Christaller's work and performed with a different end in view. Galpin's epochal study of trade areas in Walworth County, Wisconsin, published in 1915, was the first contribution. Since then important studies bearing on the problem have been made by others.[10] These studies are confined primarily to smaller trade centers but give a wealth of information on distribution of settlements which independently substantiates many of Christaller's basic premises.

As a working hypothesis one assumes that normally the larger the city, the larger its tributary area. Thus there should be cities of varying size ranging from a small hamlet performing a few simple functions, such as providing a limited shopping and market center for a small contiguous area, up to a large city with a large tributary area composed of the service areas of many smaller towns and providing more complex services, such as wholesaling, large-scale banking, specialized retailing, and the like. Services performed purely for a surrounding area are termed "central" functions by Christaller, and the settlements performing them "central" places. An industry using raw materials imported from outside the local region and shipping its products out of the local area would not constitute a central service.

Ideally, each central place would have a circular tributary area, as in von Thunen's proposition, and the city would be in the center. However, if three or more tangent circles are inscribed in an area, unserved spaces will exist; the best theoretical shapes are hexagons, the closest geometrical figures to circles which will completely fill an area (Figure 1).[11]

Christaller has recognized typical-size settlements, computed their average population, their distance apart, and the size and population of their tributary areas in accordance with his hexagonal theory. . . . He also states that the number of central

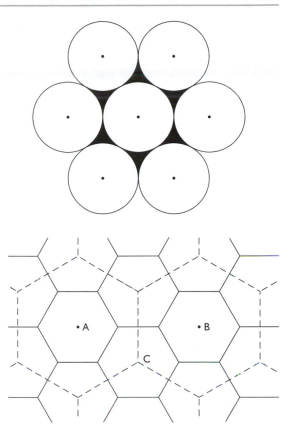

Figure 1 Theoretical shapes of tributary areas. Circles leave unserved spaces, hexagons do not. Small hexagons represent service areas for smaller places, large hexagons (dotted lines) represent service areas for next higher rank central places.

places follows a norm from largest to smallest in the following order: 1:2:6:18:54, etc.[12]

All these figures are computed on the basis of South Germany, but Christaller claims them to be typical for most of Germany and western Europe. The settlements are classified on the basis of spacing each larger unit in a hexagon of next-order size. . . . [T]he initial distance figure of 7 km. between the smallest centers is chosen because 4–5 km., approximately the distance one can walk in one hour, appears to be a normal service-area limit for the smallest centers. Thus, in a hexagonal scheme, these centers are about 1 km. apart. Christaller's maps indicate that such centers are spaced close to this norm in South Germany. In the larger categories the norms for distance apart and size of centers appear to be true averages; but variations from the norm are the rule, although wide discrepancies are not common in the eastern portion of South Germany, which is less highly industrialized than the Rhine-Ruhr areas in the west. The number of central places of each rank varies rather widely from the normal order of expectancy.

The theoretical ideal appears to be most nearly approached in poor, thinly settled farm districts – areas which are most nearly self-contained. In some other sections of Germany industrial concentration seems to be a more important explanation, although elements of the central-place type of distribution are present. Christaller points out that Cologne is really the commercial center for the Ruhr industrial district even though it is outside the Ruhr area. Even in mountain areas centrality is a more important factor than topography in fixing the distribution of settlements. Christaller states that one cannot claim that a certain city is where it is because of a certain river – that would be tantamount to saying that if there were no rivers there would be no cities.

Population alone is not a true measure of the central importance of a city; a large mining, industrial, or other specialized function town might have a small tributary area and exercise few central functions. In addition to population, therefore, Christaller uses an index based on number of telephones in proportion to the average number per thousand inhabitants in South Germany, weighted further by the telephone density of the local sub-region. A rich area such as the Palatinate supports more telephones in proportion to population than a poor area in the Bavarian Alps; therefore, the same number of telephones in a Palatinate town would not give it the same central significance as in the Alps. He claims that telephones, since they are used for business, are a reliable index of centrality. Such a thesis would not be valid for most of the United States, where telephones are as common in homes as in commercial and professional quarters.

Some better measures of centrality could be devised, even if only the number of out-of-town telephone calls per town. Better still would be some measure of actual central services performed. It would be tedious and difficult to compute the amount, or percentage, of business in each town drawn from outside the city, but some short cuts might be devised. If one knew the average number of customers required to support certain specialized

functions in various regions, then the excess of these functions over the normal required for the urban population would be an index of centrality.[13] In several states rural sociologists and others have computed the average number of certain functions for towns of a given size. With one or two exceptions only small towns have been analyzed. Retail trade has received most attention, but professional and other services have also been examined. These studies do not tell us actually what population supports each service, since the services are supported both by town and by surrounding rural population, but they do provide norms of function expectancy which would be just as useful.[14]

A suggestive indicator of centrality is provided by the maps which Dickinson has made for per capita wholesale sales of cities in the United States.[15] On this basis centers are distributed rather evenly in accordance with regional population density. Schlier has computed the centrality of cities in Germany on the basis of census returns for "central" occupations.[16] Refinement of some of our census returns is desirable before this can be done entirely satisfactorily in the United States, but the method is probably the most promising in prospect.

Another measure of centrality would be the number of automobiles entering a town, making sure that suburban movements were not included. Figures could be secured if the state-wide highway planning surveys in forty-six states were extended to gather such statistics.

The central-place scheme may be distorted by local factors, primarily industrial concentration or main transport routes. Christaller notes that transportation is not an areally operating principle, as the supplying of central goods implies, but is a linearly working factor. In many cases central places are strung at short intervals along an important transport route, and their tributary areas do not approximate the ideal circular or hexagonal shape but are elongated at right angles to the main transport line.[17] In some areas the reverse of this normal expectancy is true. In most of Illinois, maps depicting tributary areas show them to be elongated parallel to the main transport routes, not at right angles to them.[18] The combination of nearly uniform land and competitive railways peculiar to the state results in main railways running nearly parallel and close to one another between major centers.

In highly industrialized areas the central-place scheme is generally so distorted by industrial concentration in response to resources and transportation that it may be said to have little significance as an explanation for urban location and distribution, although some features of a central-place scheme may be present, as in the case of Cologne and the Ruhr.

In addition to distortion, the type of scheme prevailing in various regions is susceptible to many influences. Productivity of the soil,[19] type of agriculture and intensity of cultivation, topography, governmental organization, are all obvious modifiers. In the United States, for example, what is the effect on distribution of settlements caused by the sectional layout of the land and the regular size of counties in many states? In parts of Latin America many centers are known as "Sunday towns"; their chief functions appear to be purely social, to act as religious and recreational centers for holidays – hence the name "Sunday town."[20] Here social rather than economic services are the primary support of towns, and we should accordingly expect a system of central places with fewer and smaller centers, because fewer functions are performed and people can travel farther more readily than commodities. These underlying differences do not destroy the value of the theory; rather they provide variations of interest to study for themselves and for purposes of comparison with other regions.

The system of central places is not static or fixed; rather it is subject to change and development with changing conditions.[21] Improvements in transportation have had noticeable effects. The provision of good automobile roads alters buying and marketing practices, appears to make the smallest centers smaller and the larger centers larger, and generally alters trade areas.[22] Since good roads are spread more uniformly over the land than railways, their provision seems to make the distribution of centers correspond more closely to the normal scheme.[23]

Christaller may be guilty of claiming too great an application of his scheme. His criteria for determining typical-size settlements and their normal number apparently do not fit actual frequency counts of settlements in many almost uniform regions as well as some less rigidly deductive norms.[24]

Bobeck in a later article claims that Christaller's proof is unsatisfactory.[25] He states that two-thirds of the population of Germany and England live in cities and that only one-third of these cities in Germany are real central places. The bulk are primarily industrial towns or villages inhabited solely by farmers. He also declares that exceptions in the rest of the world are common, such as the purely rural districts of the Tonkin Delta of Indo-China, cities based on energetic entrepreneurial activity, as some Italian cities, and world commercial ports such as London, Rotterdam, and Singapore. Many of these objections are valid; one wishes that Christaller had better quantitative data and were less vague in places. Bobeck admits, however, that the central-place theory has value and applies in some areas.

The central-place theory probably provides as valid an interpretation of settlement distribution over the land as the concentric-zone theory does for land use within cities. Neither theory is to be thought of as a rigid framework fitting all location facts at a given moment. Some, expecting too much, jettison the concentric-zone theory; others, realizing that it is an investigative hypothesis of merit, regard it as a useful tool for comparative analysis.

Even in the closely articulated national economy of the United States there are strong forces at work to produce a central-place distribution of settlements. It is true that products under our national economy are characteristically shipped from producing areas through local shipping points directly to consuming centers which are often remote. However, the distribution of goods or imports brought into an area is characteristically carried on through brokerage, wholesale, and retail channels in central cities.[26] This graduated division of functions supports a central-place framework of settlements. Many non-industrial regions of relatively uniform land surface have cities distributed so evenly over the land that some sort of central-place theory appears to be the prime explanation.[27] It should be worth while to study this distribution and compare it with other areas.[28] In New England, on the other hand, where cities are primarily industrial centers based on distant raw materials and extra-regional markets, instead of the land's supporting the city the reverse is more nearly true: the city supports the countryside by providing a market for farm products, and thus infertile rural areas are kept from being even more deserted than they are now.

The forces making for concentration at certain places and the inevitable rise of cities at these favored places have been emphasized by geographers and other scholars. The phenomenal growth of industry and world trade in the last hundred years and the concomitant growth of cities justify this emphasis but have perhaps unintentionally caused the intimate connection between a city and its surrounding area partially to be overlooked. Explanation in terms of concentration is most important for industrial districts but does not provide a complete areal theory for distribution of settlements. Furthermore, there is evidence that "of late . . . the rapid growth of the larger cities has reflected their increasing importance as commercial and service centers rather than as industrial centers."[29] Some form of the central-place theory should provide the most realistic key to the distribution of settlements where there is no marked concentration – in agricultural areas where explanation has been most difficult in the past. For all areas the system may well furnish a theoretical norm from which deviations may be measured.[30] It might also be an aid in planning the development of new areas. If the theory is kept in mind by workers in academic and planning fields as more studies are made, its validity may be tested and its structure refined in accordance with regional differences.

NOTES

1 Cf. Tord Palander, *Beitriige sur Standortstheorie* (Uppsala, Sweden, 1935), or E.M. Hoover, Jr., *Location Theory and the Shoe and Leather Industries* (Cambridge, Mass., 1937).

2 J.G. Kohl, *Der Verkehr und die Ansiedlungen der Menschen in ihrer Abangikeit von der Gestaltung der Erdoberflache* (2nd edn; Leipzig, 1850).

3 C.H. Cooley, "The Theory of Transportation," *Publications of the American Economic Association*, IX (May, 1894), 1–148.

4 R.M. Haig, "Toward an Understanding of the Metropolis: Some Speculations Regarding the Economic Basis of Urban Concentration," *Quarterly Journal of Economics*, XL (1926), 179–208.

5 Hans Bobeck, "Grundfragen der Stadt Geographie," *Geographischer Anzeiger*, XXVIII (1927), 213–24.

6 A section of the International Geographical Congress at Amsterdam in 1938 dealt with "Functional Relations between City and Country." The papers are published in Vol. II of the *Comptes rendus* (Leiden: E.J. Brill, 1938). A recent American study is C.D. Harris, "Salt Lake City: A Regional Capital" (Ph.D. diss., University of Chicago, 1940). Pertinent also is R.E. Dickinson, "The Metropolitan Regions of the United States," *Geographical Review*, XXIV (1934), 278–91.

7 *Die zentralen Orte in Suddeutschland* (Jena, 1935); also a paper (no title) in *Comptes rendus du Congres internationale de geographie Amsterdam* (1938), II, 123–37.

8 This does not deny the importance of "gateway" centers such as Omaha and Kansas City, cities located between contrasting areas in order to secure exchange benefits. The logical growth of cities at such locations does not destroy the theory to be presented (cf. R.D. McKenzie's excellent discussion in *The Metropolitan Community* [New York, 1933], pp. 4ff.).

9 Cf. Petrie's statement about ancient Egypt and Mesopotamia: "It has been noticed before how remarkably similar the distances are between the early nome [*sic*] capitals of the Delta (twenty-one miles on an average) and the early cities of Mesopotamia (averaging twenty miles apart). Some physical cause seems to limit the primitive rule in this way. Is it not the limit of central storage of grain, which is the essential form of early capital? Supplies could be centralised up to ten miles away; beyond that the cost of transport made it better worth while to have a nearer centre" (W.M. Flinders Petrie, *Social Life in Ancient Egypt* [London, 1923; reissued, 1932], pp. 3–4).

10 C.J. Galpin, *Social Anatomy of an Agricultural Community* (University of Wisconsin Agricultural Experiment Station Research Bull. 34 [1915]), and the restudy by J.H. Kolb and R.A. Polson, *Trends in Town–Country Relations* (University of Wisconsin Agricultural Experiment Station Research Bull. 117 [1933]); B.L. Melvin, *Village Service Agencies of New York State, 1925* (Cornell University Agricultural Experiment Station Bull.

493 [1929]), and *Rural Population of New York, 1855–1925* (Cornell University Agricultural Experiment Station Memoir 116 [1928]); Dwight Sanderson, *The Rural Community* (New York, 1932), esp. pp. 488–514, which contains references to many studies by Sanderson and his associates; Carle C. Zimmerman, *Farm Trade Centers in Minnesota, 1905–29* (University of Minnesota Agricultural Experiment Station Bull. 269 [1930]); T. Lynn Smith, *Farm Trade Centers in Louisiana 1905 to 1931* (Louisiana State University Bull. 234. [1933]); Paul H. Landis, *South Dakota Town–Country Trade Relations, 1901–1931* (South Dakota Agricultural Experiment Station Bull. 274 [1932]), and *The Growth and Decline of South Dakota Trade Centers, 1901–1933* (Bull. 279[1938]), and *Washington Farm Trade Centers, 1900–1935* (State College of Washington Agricultural Experiment Station Bull. 360 [1938]). Other studies are listed in subsequent footnotes.

11 See August Losch, "The Nature of the Economic Regions," *Southern Economic Journal*, V (1938), 73. Galpin (*op. cit.*) thought in terms of six tributary-area circles around each center. See also Kolb and Polson, *op. cit.*, 30–41.

12 Barnes and Robinson present some interesting maps showing the average distance apart of farmhouses in the driftless area of the Middle West and in southern Ontario. Farmhouses might well be regarded as the smallest settlement units in a central place scheme, although they might not be in the same numbered sequence (James A. Barnes and Arthur H. Robinson, "A New Method for the Representation of Dispersed Rural Population," *Geographical Review*, XXX [1940], 134–37).

13 In Iowa, for example, almost all towns of more than 450 inhabitants have banks, half of the towns of 250–300, and 20 per cent of the towns of 100–150 (according to calculations made by the author from population estimates in *Rand McNally's Commercial Atlas* for 1937).

14 See particularly the thorough study by B.L. Melvin, *Village Service Agencies, New York State 1925*; C.R. Hoffer, *A Study of Town–Country Relationships* (Michigan Agricultural Experiment Station Special Bull. 181 [1928]) (data on number of retail stores and professions per town); H.B. Price and C.R. Hoffer, *Services of Rural Trade Centers in Distribution of Farm Supplies* (Minnesota Agricultural Experiment Station Bull. 249 [1938]); William J. Reilly, *Methods for the Study of Retail Relationships* ("Bureau of Business Research Monographs," No. 4, University of Texas Bull. 2944 [1929]), p. 26; J.H. Kolb, *Service institutions of Town and Country* (Wisconsin Agricultural Experiment Station Research Bull. 66 [1925] (town size in relation to support of institutions); Smith, *op. cit.*, pp. 32–40; Paul H. Landis, *South Dakota Town–Country Trade Relations, 1901–1931*, p. 20 (population per business enterprise), and pp. 24–25 (functions per town size); Zimmerman, *op. cit.*, pp. 16 and 51ff.

15 *Op. cit.*, pp. 280–81.

16 Otto Schlier, "Die zentralen Orte des Deutschen Reichs," *Zeitschrift der Gesellschaft fur Erdkunde zu Berlin* (1937), pp. 161–70. See also map constructed from Schlier's figures in R.E. Dickinson's valuable article, "The Economic Regions of Germany," *Geographical Review*, XXVIII (1938), 619. For use of census figures in the United States see Harris, *op. cit.*, 3–12.

17 For an illustration of this type of tributary area in the ridge and valley section of east Tennessee see H.V. Miller, "Effects of Reservoir Construction on Local Economic Units," *Economic Geography*, XV (1939), 242–49.

18 See, e.g., *Marketing Atlas of the United States* (New York: International Magazine Co., Inc.) or *A Study of Natural Areas of Trade in the United States* (Washington, D.C.: U.S. National Recovery Administration, 1935).

19 Cf. the emphasis of Sombart, Adam Smith, and other economists on the necessity of surplus produce of land in order to support cities. Fertile land ordinarily produces more surplus and consequently more urban population, although the town "may not always derive its whole subsistence from the country in its neighborhood" (Adam Smith, *The Wealth of Nations* ["Modern Library" edition; New York, 1937], p. 357; Werner Sombart, *Der moderne Kapitalismus* [2nd rev. edn; Munich and Leipzig, 1916], 1, 130–31).

20 For an account of such settlements in Brazil see Pierre Deffontaines, "Rapports fonctionnés entre les agglomerations urbaines et rurales: un example en pays de colonisation, le Bresil,"

Comptes rendus du Congres internationale de geographie Amsterdam (1938), II, 139–44.

21 The effects of booms, droughts, and other factors on trade-center distribution by decades are brought out in Landis' studies for South Dakota and Washington. Zimmerman and Smith also show the changing character of trade-center distribution (see n. 10 of this paper for references). Melvin calls attention to a "village population shift lag"; in periods of depressed agriculture villages in New York declined in population approximately a decade after the surrounding rural population had decreased (B.L. Melvin, *Rural Population of New York*, 1855–1925, 120).

22 Most studies indicate that only the very smallest hamlets (under 250 population) and crossroads stores have declined in size or number. The larger small places have held their own (see Landis for Washington, *op. cit.*, 37, and his *South Dakota Town–Country Trade Relations 1901–1931*, 34–36). Zimmerman in 1930 (*op. cit.*, 41) notes that crossroads stores are disappearing and are being replaced by small villages. He states further: "It is evident that claims of substantial correlation between the appearance and growth of the larger trading center and the disappearance of the primary center are more or less unfounded. Although there are minor relationships, the main change has been a division of labor between the two types of centers rather than the complete obliteration of the smaller in favor of the larger" (32). For further evidences of effect of automobiles on small centers see R.V. Mitchell, *Trends in Rural Retailing in Illinois 1926 to 1938* (University of Illinois Bureau of Business Research Bull., Ser. 59 [1939]), 31ff., and Sanderson, *op. cit.*, 564, as well as other studies cited above.

23 Smith (*op. cit.*, p. 54) states: "There has been a tendency for centers of various sizes to distribute themselves more uniformly with regard to the area, population, and resources of the state." Most of the changes seem to be in the direction of a more efficient pattern of rural organization. This redistribution of centers in conjunction with improved methods of communication and transportation has placed each family in frequent contact with several trade centers.

24 This statement is made on the basis of frequency counts by the author for several midwestern states (see also Schlier, *op. cit.*, 165–69, for Germany).

25 Hans Bobeck, "Uber einige functionelle Stadttypen und ihre Beziehungen zum Lande," *Comptes rendus du Congres internationale de geographie Amsterdam* (1938), II, 88. Many would jettison the concentric-zone theory; others, realizing that it is an investigative hypothesis of merit, regard it as a useful tool for comparative analysis.

26 Harris, *op. cit.*, 87.

27 For a confirmation of this see the column diagram on p. 73 of Losch (*op. cit.*), which shows the minimum distances between towns in Iowa of three different size classes. The maps of trade-center distribution in the works of Zimmerman, Smith, and Landis (cited earlier) also show an even spacing of centers.

28 The following table [not reproduced] gives the average community area for 140 villages in the United States in 1930. In the table notice throughout that (1) the larger the village, the larger its tributary area in each region and (2) the sparser the rural population density, the larger the village tributary area for each size class (contrast mid-Atlantic with Far West, etc.).

Although 140 is only a sample of the number of villages in the country, the figures are significant because the service areas were carefully and uniformly delimited in the field for all villages (E. deS. Brunner and J.D. Kolb, *Rural Social Trends* [New York, 1933], p. 95; see also E. deS. Brunner, G.S. Hughes, and M. Patten, *American Agricultural Villages* [New York, 1927], ch. Ii).

In New York 26 sq. mi. was found to be the average area per village in 1920. "Village" refers to any settlement under 2,500 population. Nearness to cities, type of agriculture, and routes of travel are cited as the three most important factors influencing density of villages. Since areas near cities are suburbanized in some cases, as around New York City, the village-density in these districts is correspondingly high. Some urban counties with smaller cities (Rochester, Syracuse, and Niagara Falls) have few suburbs, and consequently the villages are farther apart than in many agricultural counties (B.L. Melvin,

Rural Population of New York, 1925–1955, pp. 88–89; the table on p. 89 shows number of sq. mi. per village in each New York county).

In sample areas of New York State the average distance from a village of 250 or under to another of the same size or larger is about 3 miles; for the 250–749 class it is 3–5 miles; for the 750–1,249 class, 5–7 miles (B.L. Melvin, *Village Service Agencies, New York*, 1925, p. 102; in the table on p. 103 the distance averages cited above are shown to be very near the modes).

Kolb makes some interesting suggestions as to the distances between centers. He shows that spacing is closer in central Wisconsin than in Kansas, which is more sparsely settled (J.H. Kolb, *Service Relations of Town and Country* [Wisconsin Agricultural Experimental Station Research Bull. 58 (1923)]; see pp. 7–8 for theoretical graphs).

In Iowa, "the dominant factor determining the *size* of convenience-goods areas is distance" (*Second State Iowa Planning Board Report* [Des Moines, April, 1935], p. 198). This report contains fertile suggestions on trade areas for Iowa towns. Valuable detailed reports on retail trade areas for some Iowa counties have also been made by the same agency.

29 U.S. National Resources Committee, *Our Cities – Their Role in the National Economy: Report of the Urbanism Committee* (Washington: Government Printing Office, 1937), p. 37.

30 Some form of the central-place concept might well be used to advantage in interpreting the distribution of outlying business districts in cities (d. Malcolm J. Proudfoot, "The Selection of a Business Site," *Journal of Land and Public Utility Economics*, XIV [1938], esp. 373ff.).

Editors' references and suggestions for further reading

Berry, B. (1967) *Geography of Market Centers and Retail Distribution*, Englewood Cliffs, NJ: Prentice Hall.

Berry, B. and Harris, C. (1970) "Walter Christaller: an appreciation," *The Geographical Review* 60: 116–119.

Christaller, W. (1933; 1966) *Central Places in Southern Germany*, Englewood Cliffs, NJ: Prentice-Hall.

Harris, C. (1977) "Edward Louis Ullman, 1912–1976," *Annals of AAG* 67, 4: 595–600.

Symanski, R. and Agnew, J. (1981) *Order and Skepticism: Human Geography and the Dialectic of Science*, Washington, DC: Association of American Geographers.

Ullman, E. (1956) "The nature of cities," in W. Thomas (ed.) *Man's Role in Changing the Face of the Earth*, Chicago, IL: University of Chicago Press.

ONE

"The Nature of Cities"

from *Annals of the American Academy of Political and Social Science* (1945)

Chauncy D. Harris and Edward L. Ullman

Editors' Introduction

Several months after the conclusion of World War II, Chauncy Harris (1914–2003) and Edward Ullman's (1912–1976) classic article on the economic base and internal structure of the city appeared in a special issue of the *Annals of the American Academy of Political and Social Science* dedicated to "Building the Future City." "The Nature of Cities" brought together previously developed work of both authors, including Ullman's publication on central-place theory (p. 37) and Harris' classifications of various types of cities and suburbs (1943a, 1943b). In that regard, one might have expected the article to be less influential. Instead, it instilled a conceptual vocabulary and organizational approach in the subdiscipline of urban geography that until fairly recently was so ingrained as to be almost "second nature" (Agnew 1997: 5).

In their article, Harris and Ullman laid out a two-part research agenda that joined an interest in the economic foundations and external system of cities with an examination of the internal structure of cities. Their programmatic statement underscored a particular research interest in locational theory and its application in land use planning. The combination of intellectual optimism, social commitment, and belief in the efficacy of city planning reflected in the following article corresponded to Americans' optimism and the particular tone of the modern social sciences at the conclusion of World War II. Western geographers, once focused primarily on differences among cities based on a cultural historical tradition, were called upon to develop generalizations to "build the future city in such a manner that the advantages of urban concentration can be preserved for the benefit of man and the disadvantages minimized."

While Harris and Ullman's summary of urban geographic knowledge relied on the import of European locational theories of inter-urban systems, they engaged directly in a discussion of intra-urban structure that was almost exclusively conducted by American urbanists (see Lichtenberger 1997: 9). They noted Burgess' concentric zone theory (p. 19), created the schematic for Hoyt's sector theory (p. 28), and then collaboratively developed the Multiple Nuclei model. The Multiple Nuclei model departed significantly from the two earlier models by representing a metropolitan area that was not defined by distances from the central business district, but was instead based on patterns of usage in surrounding land. Unlike Burgess and Hoyt's models, the center no longer molded the periphery.

Concerned with more than the social differentiation of residential areas, the most important aspect of Harris and Ullman's new model focused on the development of special purpose districts. These districts were nodes of economic activity that required specialized facilities or benefited from the clustering of uses. With the continued expansion of the urban areas, the absorption of former independent settlements or the development of new nodes around outlying business districts created a metropolitan area with multiple nuclei. By constructing this model, they acknowledged a new logic of transportation and location as the social, economic, and political forces that had been unleashed in the 1920s by the increased use of the automobile, which changed the form of the city. An appreciation of their insight increases when one considers that nearly two-thirds of the United States' population lived in central cities during the early 1940s. Yet, a new decentralized metropolitan form had begun to take physical shape and to suggest the direction of the future city.

Ullman and Harris' article proved very influential in the multidisciplinary urban studies literature, due in large part to the clarity and simplicity of its classification schemes and models. Yet were they too simplistic? Today few would challenge the heuristic value of the Multiple Nuclei model as it applied to the American city for the period in which it was developed. Geographers continue to find heuristic value in models, as demonstrated by the undifferentiated, decentralized grid-like "Keno capitalism" model that Michael Dear and Steven Flusty offer as a representation of the structure of the postmodern city (p. 138). Harris and Ullman's contemporary emphasis on "the" nature of the city, however, requires comment. As it generalized the conditions of the mid-twentieth-century American city, the article lacked historical and geographical awareness of its particular context (Agnew 1997). Even at the time, the schematic mappings were difficult to apply directly to other cultural and political systems (Lichtenberger 1997). Yet few would dispute the article's lasting and influential impact on urban geography. In a recent commemoration of this classic article, Robert Lake recommends that "The Nature of Cities" be evaluated in terms of the genealogical ties that link our intellectual history. If it indeed "established a benchmark against which we can trace the cumulative progress of the field over the ensuing half century," this seminal work provides an important perspective on contemporary urban geography (Lake 1997: 2; Harris 1997).

■ ■ ■ ■ ■ ■

Cities are the focal points in the occupation and utilization of the earth by man. Both a product of and an influence on surrounding regions, they develop in definite patterns in response to economic and social needs.

Cities are also paradoxes. Their rapid growth and large size testify to their superiority as a technique for the exploitation of the earth, yet by their very success and consequent large size they often provide a poor local environment for man. The problem is to build the future city in such a manner that the advantages of urban concentration can be preserved for the benefit of man and the disadvantages minimized.

Each city is unique in detail but resembles others in function and pattern. What is learned about one helps in studying another. Location types and internal structure are repeated so often that broad and suggestive generalizations are valid, especially if limited to cities of similar size, function, and regional setting. This paper will be limited to a discussion of two basic aspects of the nature of cities – their support and their internal structure. Such important topics as the rise and extent of urbanism, urban sites, culture of cities, social and economic characteristics of the urban population, and critical problems will receive only passing mention.

THE SUPPORT OF CITIES

As one approaches a city and notices tall buildings rising above the surrounding land and as one continues into the city and observes the crowds of people hurrying to and fro past stores, theaters, banks, and other establishments, one naturally is struck by the contrast with the rural countryside. What supports this phenomenon? What do the people of the city do for a living?

The support of a city depends on the services it performs not for itself but for a tributary area. Many activities serve merely the population of the city itself. Barbers, dry cleaners, shoe repairers, grocerymen, bakers, and movie operators serve others who are engaged in the principal activity of the city, which may be mining, manufacturing, trade, or some other activity.

The service by which the city earns its livelihood depends on the nature of the economy and of the hinterland. Cities are small or rare in areas either of primitive, self-sufficient economy or of meager resources. As Adam Smith stated, the land must produce a surplus in order to support cities. This does not mean that all cities must be surrounded by productive land, since strategic location with reference to cheap ocean highways may enable a city to support itself on the specialized surplus of distant lands. Nor does it mean that cities are parasites living off the land. Modern mechanization, transport, and a complex interdependent economy enable much of the economic activity of mankind to be centered in cities. Many of the people engaged even in food production are actually in cities in the manufacture of agricultural machinery.

The support of cities as suppliers of urban services for the earth can be summarized in three categories, each of which presents a factor of urban causation:[1]

1. Cities as central places performing comprehensive services for a surrounding area. Such cities tend to be evenly spaced throughout productive territory. For the moment this may be considered the "norm" subject to variation primarily in response to the ensuing factors.
2. Transport cities performing break-of-bulk and allied services along transport routes, supported by areas which may be remote in distance but close in connection because of the city's strategic location on transport channels. Such cities tend to be arranged in linear patterns along rail lines or at coasts.
3. Specialized-function cities performing one service such as mining, manufacturing, or recreation for large areas, including the general tributary areas of hosts of other cities. Since the principal localizing factor is often a particular resource such as coal, water power, or a beach, such cities may occur singly or in clusters.

Most cities represent a combination of the three factors, the relative importance of each varying from city to city.

Cities as central places

Cities as central places serve as trade and social centers for a tributary area. If the land base is homogeneous these centers are uniformly spaced, as in many parts of the agricultural Middle West. In areas of uneven resource distribution, the distribution of cities is uneven. The centers are of varying sizes, ranging from small hamlets closely spaced with one or two stores serving a local tributary area through larger villages, towns, and cities more

widely spaced with more special services for larger tributary areas, up to the great metropolis such as New York or Chicago offering many specialized services for a large tributary area composed of a whole hierarchy of tributary areas of smaller places. Such a net of tributary areas and centers forms a pattern somewhat like a fish net spread over a beach, the network regular and symmetrical where the sand is smooth, but warped and distorted where the net is caught in rocks.

The central-place type of city or town is widespread through the world, particularly in non-industrial regions. In the United States it is best represented by the numerous retail and wholesale trade centers of the agricultural Middle West, Southwest, and West. Such cities have imposing shopping centers or wholesale districts in proportion to their size; the stores are supported by the trade of the surrounding area. This contrasts with many cities of the industrial East, where the centers are so close together that each has little trade support beyond its own population.

Not only trade but social and religious functions may support central places. In some instances these other functions may be the main support of the town. In parts of Latin America, for example, where there is little trade, settlements are scattered at relatively uniform intervals through the land as social and religious centers. In contrast to most cities, their busiest day is Sunday, when the surrounding populace attend (sic) church and engage in holiday recreation, thus giving rise to the name "Sunday town."

Most large central cities and towns are also political centers. The county seat is an example. London and Paris are the political as well as trade centers of their countries. In the United States, however, Washington and many state capitals are specialized political centers. In many of these cases the political capital was initially chosen as a centrally located point in the political area and was deliberately separated from the major urban center.

Cities as transport foci and break-of-bulk points

All cities are dependent on transportation in order to utilize the surplus of the land for their support. This dependence on transportation destroys the symmetry of the central-place arrangement, inasmuch as cities develop at foci or breaks of transportation, and transport routes are distributed unevenly over the land because of relief or other limitations. City organizations recognize the importance of efficient transportation, as witness their constant concern with freight-rate regulation and with the construction of new highways, port facilities, airfields, and the like.

Mere focusing of transport routes does not produce a city, but according to Cooley, if break of bulk occurs, the focus becomes a good place to process goods. Where the form of transport changes, as transferring from water to rail, break of bulk is inevitable. Ports originating merely to transship cargo tend to develop auxiliary services such as repacking, storing, and sorting. An example of simple break-of-bulk and storage ports is Port Arthur-Fort William, the twin port and wheat-storage cities at the head of Lake Superior; surrounded by unproductive land, they have arisen at the break-of-bulk points on the cheapest route from the wheat-producing Prairie Provinces to the markets of the East. Some ports develop as entrepots, such as Hong Kong and Copenhagen, supported by transshipment of goods from small to large boats or vice versa. Servicing points or minor changes in transport tend to encourage growth of cities as establishment of division points for changing locomotives on American railroads.

Transport centers can be centrally located places or can serve as gateways between contrasting regions with contrasting needs. Kansas City, Omaha, and Minneapolis-St. Paul serve as gateways to the West as well as central places for productive agricultural regions, and are important wholesale centers. The ports of New Orleans, Mobile, Savannah, Charleston, Norfolk and others served as traditional gateways to the Cotton Belt with its specialized production. Likewise, northern border metropolises such as Baltimore, Washington, Cincinnati, and Louisville served as gateways to the South, with St. Louis a gateway to the Southwest. In recent years the South has been developing its own central places, supplanting some of the monopoly once held by the border gateways. Atlanta, Memphis, and Dallas are examples of the new southern central places and transport foci.

Changes in transportation are reflected in the pattern of city distribution. Thus the development of railroads resulted in a railroad alignment of cities which still persists. The rapid growth of automobiles and widespread development of highways in recent decades, however, has changed the trend toward a more even distribution of towns. Studies in such diverse localities as New York and Louisiana have shown a shift of centers away from exclusive alignment along rail routes. Airways may reinforce this trend or stimulate still different patterns of distribution for the future city.

Cities as concentration points for specialized services

A specialized city or cluster of cities performing a specialized function for a large area may develop at a highly localized resource. The resort city of Miami, for example, developed in response to a favorable climate and beach. Scranton, Wilkes-Barre, and dozens of nearby towns are specialized coal-mining centers developed on anthracite coal deposits to serve a large segment of the northeastern United States. Pittsburgh and its suburbs and satellites form a nationally significant iron-and-steel manufacturing cluster favored by good location for the assembly of coal and iron ore and for the sale of steel to industries on the coal fields.

Equally important with physical resources in many cities are the advantages of mass production and ancillary services. Once started, a specialized city acts as a nucleus for similar or related activities, and functions tend to pyramid, whether the city is a seaside resort such as Miami or Atlantic City, or, more important, a manufacturing center such as Pittsburgh or Detroit. Concentration of industry in a city means that there will be a concentration of satellite services and industries – supply houses, machine shops, expert consultants, other industries using local industrial by-products or waste, still other industries making specialized parts for other plants in the city, marketing channels, specialized transport facilities, skilled labor, and a host of other facilities; either directly or indirectly, these benefit industry and cause it to expand in size and numbers in a concentrated place or district. Local personnel with the know-how in a given industry also may decide to start a new plant producing similar or like products in the same city. Furthermore, the advantages of mass production itself often tend to concentrate production in a few large factories and cities. Examples of localization of specific manufacturing industries are clothing in New York City, furniture in Grand Rapids, automobiles in the Detroit area, pottery in Stoke-on-Trent in England, and even such a specialty as tennis rackets in Pawtucket, Rhode Island.

Such concentration continues until opposing forces of high labor costs and congestion balance the concentrating forces. Labor costs may be lower in small towns and in industrially new districts; thus some factories are moving from the great metropolises to small towns; much of the cotton textile industry has moved from the old industrial areas of New England to the newer areas of the Carolinas in the South. The tremendous concentration of population and structures in large cities exacts a high cost in the form of congestion, high land costs, high taxes, and restrictive legislation.

Not all industries tend to concentrate in specialized industrial cities; many types of manufacturing partake more of central-place characteristics. These types are those that are tied to the market because the manufacturing process results in an increase in bulk or perishability. Bakeries, ice cream establishments, ice houses, breweries, soft-drink plants, and various types of assembly plants are examples. Even such industries, however, tend to be more developed in the manufacturing belt because the density of population and hence the market is greater there.

The greatest concentration of industrial cities in America is in the manufacturing belt of northeastern United States and contiguous Canada, north of the Ohio and east of the Mississippi. Some factors in this concentration are: large reserves of fuel and power (particularly coal), raw materials such as iron ore via the Great Lakes, cheap ocean transportation on the eastern seaboard, productive agriculture (particularly in the west), early settlement, later immigration concentrated in its cities, and an early start with consequent development of skilled labor, industrial know-how, transportation facilities, and prestige.

The interdependent nature of most of the industries acts as a powerful force to maintain this area as the primary home of industrial cities in the United States. Before the war, the typical industrial city outside the main manufacturing belt had only

a single industry of the raw-material type, such as lumber mills, food canneries, or smelters (Longview, Washington; San Jose, California; Anaconda, Montana). Because of the need for producing huge quantities of ships and airplanes for a two-ocean war, however, many cities along the Gulf and Pacific coasts have grown rapidly during recent years as centers of industry.

Application of the three types of urban support

Although examples can be cited illustrating each of the three types of urban support, most American cities partake in varying proportions of all three types. New York City, for example, as the greatest American port is a break-of-bulk point; as the principal center of wholesaling and retailing it is a central-place type; and as the major American center of manufacturing it is a specialized type. The actual distribution and functional classification of cities in the United States, more complex than the simple sum of the three types, has been mapped and described elsewhere in different terms.[2]

The three basic types therefore should not be considered as a rigid framework excluding all accidental establishment, although even fortuitous development of a city becomes part of the general urban-supporting environment. Nor should the urban setting be regarded as static; cities are constantly changing and exhibit characteristic lag in adjusting to new conditions.

Ample opportunity exists for use of initiative in strengthening the supporting base of the future city, particularly if account is taken of the basic factors of urban support. Thus a city should examine: (1) its surrounding area to take advantage of changes such as newly discovered resources or crops, (2) its transport in order to adjust properly to new or changed facilities, and (3) its industries in order to benefit from technological advances.

INTERNAL STUCTURE OF CITIES

Any effective plans for the improvement or rearrangement of the future city must take account of the present pattern of land use within the city, of the factors which have produced this pattern, and of the facilities required by activities localized within particular districts.

Although the internal pattern of each city is unique in its particular combination of details, most American cities have business, industrial, and residential districts. The forces underlying the pattern of land use can be appreciated if attention is focused on three generalizations of arrangement – by concentric zones, sectors, and multiple nuclei.

Concentric zones

According to the concentric-zone theory, the pattern of growth of the city can best be understood in terms of five concentric zones (Figure 1).[3]

1. *The central business district.* – This is the focus of commercial, social, and civic life, and of transportation. In it is the downtown retail district with its department stores, smart shops, office buildings, clubs, banks, hotels, theaters, museums, and organization headquarters. Encircling the downtown retail district is the wholesale business district.
2. *The zone in transition.* – Encircling the downtown area is a zone of residential deterioration. Business and light manufacturing encroach on residential areas characterized particularly by rooming houses. In this zone are the principal slums, with their submerged regions of poverty, degradation, and disease, and their underworlds of vice. In many American cities it has been inhabited largely by colonies of recent immigrants.
3. *The zone of independent workingmen's homes.* – This is inhabited by industrial workers who have escaped from the zone in transition but who desire to live within easy access of their work. In many American cities second-generation immigrants are important segments of the population in this area.
4. *The zone of better residences.* – This is made up of single-family dwellings, of exclusive "restricted districts," and of high-class apartment buildings.
5. *The commuters' zone.* – Often beyond the city limits in suburban areas or in satellite cities, this is a zone of spotty development of high-class residences along lines of rapid travel.

Sectors

The theory of axial development, according to which growth takes place along main transportation routes or along lines of least resistance to form a star-shaped city, is refined by Homer Hoyt in his sector theory, which states that growth along a particular axis of transportation usually consists of similar types of land use (Figure 1).[4] The entire city is considered as a circle and the various areas as sectors radiating out from the center of that circle; similar types of land use originate near the center of the circle and migrate outward toward the periphery. Thus a high-rent residential area in the eastern quadrant of the city would tend to migrate outward, keeping always in the eastern quadrant. A low-quality housing area, if located in the southern quadrant, would tend to extend outward to the very margin of the city in that sector. The migration of high-class residential areas outward along established lines of travel is particularly pronounced on high ground, toward open country, to homes of community leaders, along lines of fastest transportation, and to existing nuclei of buildings or trading centers.

Multiple nuclei

In many cities the land-use pattern is built not around a single center but around several discrete nuclei (Figure 1). In some cities these nuclei have existed from the very origins of the city; in others they have developed as the growth of the city stimulated migration and specialization. An example of the first type is Metropolitan London, in which "The City" and Westminster originated as separate points separated by open country, one at the center of finance and commerce, the other at the center of political life. An example of the second type is Chicago, in which heavy industry, at first localized along the Chicago River in the heart of the city, migrated to the Calumet District, where it acted as a nucleus for extensive new urban development.

The initial nucleus of the city may be the retail district in a central-place city, the port or rail facilities in a break-of-bulk city, or the factory, mine, or beach in a specialized-function city.

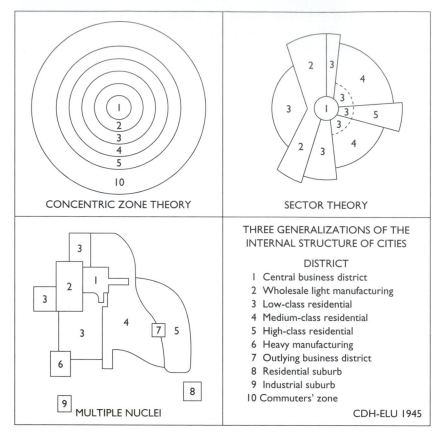

CONCENTRIC ZONE THEORY

SECTOR THEORY

MULTIPLE NUCLEI

THREE GENERALIZATIONS OF THE
INTERNAL STRUCTURE OF CITIES

DISTRICT
1 Central business district
2 Wholesale light manufacturing
3 Low-class residential
4 Medium-class residential
5 High-class residential
6 Heavy manufacturing
7 Outlying business district
8 Residential suburb
9 Industrial suburb
10 Commuters' zone

CDH-ELU 1945

Figure 1 The pattern of growth of the city.

The rise of separate nuclei and differentiated districts reflects a combination of the following four factors:

1. Certain activities require specialized facilities. The retail district, for example, is attached to the point of greatest intracity accessibility, the port district to suitable water front, manufacturing districts to large blocks of land and water or rail connection, and so on.
2. Certain like activities group together because they profit from cohesion.[5] The clustering of industrial cities has already been noted above under "Cities as concentration points for specialized services." Retail districts benefit from grouping which increases the concentration of potential customers and makes possible comparison shopping. Financial and office-building districts depend upon facility of communication among offices within the district. The Merchandise Mart of Chicago is an example of wholesale clustering.
3. Certain unlike activities are detrimental to each other. The antagonism between factory development and high-class residential development is well known. The heavy concentrations of pedestrians, automobiles, and streetcars in the retail district are antagonistic both to the railroad facilities and the street loading required in the wholesale district and to the rail facilities and space needed by large industrial districts, and vice versa.
4. Certain activities are unable to afford the high rents of the most desirable sites. This factor works in conjunction with the foregoing. Examples are bulk wholesaling and storage activities requiring much room, or low-class housing unable to afford the luxury of high land with a view.

The number of nuclei which result from historical development and the operation of localization forces varies greatly from city to city. The larger the city, the more numerous and specialized are the nuclei. The following districts, however, have developed around nuclei in most large American cities.

The central business district

This district is at the focus of intracity transportation facilities by sidewalk, private car, bus, streetcar, subway, and elevated. Because of asymmetrical growth of most large cities, it is generally not now in the areal center of the city but actually near one edge, as in the case of lake-front, riverside, or even inland cities; examples are Chicago, St. Louis, and Salt Lake City. Because established internal transportation lines converge on it, however, it is the point of most convenient access from all parts of the city, and the point of highest land values. The retail district, at the point of maximum accessibility, is attached to the sidewalk; only pedestrian or mass-transportation movement can concentrate the large numbers of customers necessary to support department stores, variety stores, and clothing shops, which are characteristic of the district. In small cities financial institutions and office buildings are intermingled with retail shops, but in large cities the financial district is separate, near but not at the point of greatest intracity facility. Its point of attachment is the elevator, which permits three-dimensional access among offices, whose most important locational factor is accessibility to other offices rather than to the city as a whole. Government buildings also are commonly near but not in the center of the retail district. In most cities a separate "automobile row" has arisen on the edge of the central business district, in cheaper rent areas along one or more major highways; its attachment is to the highway itself.

The wholesale and light-manufacturing district

This district is conveniently within the city but near the focus of extra city transportation facilities. Wholesale houses, while deriving some support from the city itself, serve principally a tributary region reached by railroad and motor truck. They are, therefore, concentrated along railroad lines, usually adjacent to (but not surrounding) the central business district. Many types of light manufacturing which do not require specialized buildings are attracted by the facilities of this district or similar districts: good rail and road transportation, available loft buildings, and proximity to the markets and labor of the city itself.

The heavy industrial district

This is near the present or former outer edge of the city. Heavy industries require large tracts of space, often beyond any available in sections already subdivided into blocks and streets. They also require good transportation, either rail or water. With the development of belt lines and switching yards, sites on the edge of the city may have better transportation service than those near the center. In Chicago about a hundred industries are in a belt three miles long, adjacent to the Clearing freight yards on the southwestern edge of the city. Furthermore, the noise of boiler works, the odors of stockyards, the waste disposal problems of smelters and iron and steel mills, the fire hazards of petroleum refineries, and the space and transportation needs which interrupt streets and accessibility – all these favor the growth of heavy industry away from the main center of the large city. The Calumet District of Chicago, the New Jersey marshes near New York City, the Lea marshes near London, and the St. Denis district of Paris are examples of such districts. The stockyards of Chicago, in spite of their odors and size, have been engulfed by urban growth and are now far from the edge of the city. They form a nucleus of heavy industry within the city but not near the center, which has blighted the adjacent residential area, the "back-of-the-yards" district.

The residential district

In general, high-class districts are likely to be on well-drained, high land and away from nuisances such as noise, odors, smoke, and railroad lines. Low-class districts are likely to arise near factories and railroad districts, wherever located in the city.

Because of the obsolescence of structures, the older inner margins of residential districts are fertile fields for invasion by groups unable to pay high rents. Residential neighborhoods have some measure of cohesiveness. Extreme cases are the ethnically segregated groups, which cluster together although including members in many economic groups; Harlem is an example.

Minor nuclei

These include cultural centers, parks, outlying business districts, and small industrial centers. A university may form a nucleus for a quasi-independent community; examples are the University of Chicago, the University of California, and Harvard University. Parks and recreation areas occupying former wasteland too rugged or wet for housing may form nuclei for high-class residential areas; examples are Rock Creek Park in Washington and Hyde Park in London. Outlying business districts may in time become major centers. Many small institutions and individual light manufacturing plants, such as bakeries, dispersed throughout the city may never become nuclei of differentiated districts.

Suburbs or satellites

Suburbs, either residential or industrial, are characteristic of most of the larger American cities.[6] The rise of the automobile and the improvement of certain suburban commuter rail lines in a few of the largest cities have stimulated suburbanization. Satellites differ from suburbs in that they are separated from the central city by many miles and in general have little daily commuting to or from the central city, although economic activities of the satellite are closely geared to those of the central city. Thus Gary may be considered a suburb but Elgin and Joliet are satellites of Chicago.

Appraisal of land-use patterns

Most cities exhibit not only a combination of the three types of urban support, but also aspects of the three generalizations of the land-use pattern. An understanding of both is useful in appraising the future prospects of the whole city and the arrangement of its parts.

As a general picture subject to modification because of topography, transportation, and previous land use, the concentric-zone aspect has merit. It is not a rigid pattern, inasmuch as growth or arrangement often reflects expansion within sectors or development around separate nuclei.

The sector aspect has been applied particularly to the outward movement of residential districts. Both the concentric-zone theory and the sector theory emphasize the general tendency of central residential areas to decline in value as new construction takes place on the outer edges; the sector theory is, however, more discriminating in its analysis of that movement.

Both the concentric zone, as a general pattern, and the sector aspect, as applied primarily to residential patterns, assume (although not explicitly) that there is but a single urban core around which land use is arranged symmetrically in either concentric or radial patterns. In broad theoretical terms such an assumption may be valid, inasmuch as the handicap of distance alone would favor as much concentration as possible in a small central core. Because of the actual physical impossibility of such concentration and the existence of separating factors, however, separate nuclei arise. The specific separating factors are not only high rent in the core, which can be afforded by few activities, but also the natural attachment of certain activities to extra-urban transport, space, or other facilities, and the advantages of the separation of unlike activities and the concentration of like functions.

The constantly changing pattern of land use poses many problems. Near the core, land is kept vacant or retained in antisocial slum structures in anticipation of expansion of higher-rent activities. The hidden costs of slums to the city in poor environment for future citizens and excessive police, fire, and sanitary protection underlie the argument for a subsidy to remove the blight. The transition zone is not everywhere a zone of deterioration with slums, however, as witness the rise of high-class apartment development near the urban core in the Gold Coast of Chicago or Park Avenue in New York City. On the fringe of the city, over-ambitious subdividing results in unused land to be crossed by urban services such as sewers and transportation. Separate political status of many suburbs results in a lack of civic responsibility for the problems and expenses of the city in which the suburbanites work.

NOTES

1 For references see Edward Ullman, "A Theory of Location for Cities," *American Journal of Sociology*, XLVI (May, 1941), 853–64.

2 Chauncy D. Harris, "A Functional Classification of Cities in the United States," *Geographical Review*, XXXIII (January, 1943), 85–99.

3 Ernest W. Burgess, "The Growth of the City," in *The City*, ed. Robert E. Park, Ernest W. Burgess, and Roderick D. McKenzie (Chicago: University of Chicago Press, 1925), pp. 47–62; and Ernest W. Burgess, "Urban Areas," in *Chicago: An Experiment in Social Science Research*, ed. T.V. Smith and Leonard D. White (Chicago: University of Chicago Press, 1929), pp. 113–38.

4 Homer Hoyt, "City Growth and Mortgage Risk," *Insured Mortgage Portfolio*, Vol. I, Nos. 6–10 (December, 1936–April, 1937), *passim*; and *idem.* (U.S. Federal Housing Administration), *The Structure and Growth of Residential Neighborhoods in American Cities* (Washington: Government Printing Office, 1939), *passim*.

5 Exceptions are service-type establishments such as some grocery stores, dry cleaners, and gasoline stations.

6 Chauncy D. Harris, "Suburbs," *American Journal of Sociology*, XLIX (July, 1943), p. 6.

Editors' references and suggestions for further reading

Agnew, J. (1997) "Commemoration and criticism: fifty years after the publication of Harris and Ullman's 'The Nature of Cities'," *Urban Geography* 18: 4–6.

Harris, C. (1943a) "A functional classification of cities in the United States," *Geographical Review* 33: 86–99.

Harris, C. (1943b) "Suburbs," *The American Journal of Sociology* 49: 1–13.

Harris, C. (1997) "The nature of cities and urban geography in the last half century," *Urban Geography* 18: 15–35.

Lake, R. (1997) "Chauncy Harris and Edward Ullman, 'The Nature of Cities': a fiftieth year commemoration," *Urban Geography* 18: 1–3.

Lichtenberger, E. (1997) "Harris and Ullman's 'The Nature of Cities': the paper's historical context and its impact on further research," *Urban Geography* 18: 7–14.

PART TWO

Globalization

The towers of Paris' La Defense office center. (Courtesy of Harold Mayer Collection, American Geographical Society Library)

INTRODUCTION TO PART TWO

Describing a walk through her local shopping street in London, the UK geographer Doreen Massey (1994) is intrigued by the rich diversity of cultures she routinely encounters. There is a shop selling Indian saris, a newspaper stand selling papers from every county of the Irish Republic, a convenience store run by a Muslim, while overhead there is the constant drone of airplanes making their way into Heathrow, one of the world's major international airports. What Massey finds striking about this mix of people and activities is the way they evoke a "global sense of place," an awareness of the global interconnectedness between a specific locality and the wider world. This notion of global interconnectedness is central to the concept of globalization. The term conjures up an image of a world of linkages and networks which "is shaking up our existing ways of life, no matter where we happen to be" (Giddens 1999: 19).

Although twenty years ago the term was hardly used, globalization now seems to be ever present in the speeches of politicians and the pronouncement of business leaders. Indeed, it has become something of a catch-all term, "used by many to bundle together virtually all the 'goods' and 'bads' of contemporary society" (Dicken 2004: 5). But the very pervasiveness of the term means its precise meaning is increasingly difficult to pin down. In most discussions of globalization, however, two features stand out. First, globalization indicates a specific scale of social, economic or political activity that is world-wide in scope where "events, decisions and activities in one part of the world can come to have significant consequences for individuals and communities in quite distant parts of the globe" (McGrew 1992: 23). Second, globalization is constituted by "trans-state" processes, processes that do not simply cross borders, but "operate as if borders were not there" (Taylor *et al.* 2002: 3). Examples of such "trans-state" processes include everything from global financial transactions, to global warming and the activities of international terrorists. Beyond these basic characteristics, however, there is much debate about the nature and importance of globalization. On the one hand, there are the "sceptics" who view globalization as nothing new, emphasizing the long history of economic exchange and political relationships between different parts of the globe. On the other hand, there are the "radicals" who claim we now live in a "global village," a borderless world in which nation-states are irrelevant and cultures homogenized (see Giddens 1999). For most geographers, the significance of globalization lies somewhere between these extreme sceptical and radical positions. There is something qualitatively different about the global economy today compared with that in the past in terms of intensity of interconnections and the crucial role played by transnational corporations, yet despite globalization (or even because of it) political spaces, like nation-states, as well as local cultures, continue to matter.

Against this general background, what are the more specific ways in which globalization has impacted on urban geography? In making the selections for this section, four themes stand out. First, although urban development is a truly global phenomena (as of 2000 more people now live in urban than rural areas in the world (Clark 2003)), some cities are clearly more important in global terms than others. In the past, cities like London, Amsterdam and Venice were very significant in the organization of international trade and the exercise of imperial power. Today, a new range of cities have emerged as the command and control centers of global capitalism. These include London, New York, Tokyo and Paris, cities "distinguished not by their size or their status as capital cities of large countries, but by the range and extent of their economic power" (Clark 2003: 156). These so-called world or global cities are the product of some fundamental geoeconomic, geopolitical and technological changes (Knox 2002). The emergence of a new international division of labor and the increasing importance of transnational corporations means that high-level management, development, and design functions are based increasingly in the major cities of the world's core economies. Political changes have also reinforced the development of world cities as a result of new modes of (de)regulation that have created fiscal environments attractive to international businesses. And advances in telecommunications, resulting in so-called time-space compression, have further added to the advantages of major cities as the nodal points in a new "informational economy." The product of these complex changes are world cities, "nodal points that function as control centers for the interdependent skein of material, financial, and cultural flows which, together, support and sustain globalization" (Knox 2002: 332). In the first reading in this section, Beaverstock, Smith and Taylor examine this world city network and, by focusing on the interconnections between world cities, provide important insights into some of the specific, material *processes* that constitute globalization.

While it is clearly important to analyze the "complex syndrome of processes" that make up globalization, it is also vital to recognize that such processes "generate geographically specific, highly uneven, concrete *outcomes*" (Dicken 2004: 16). A concern with mapping such outcomes has emerged as a second important theme in urban geographical studies of globalization. At one level, this is reflected in an interest in the growing cultural heterogeneity of cities. Edward Soja, for example, has vividly described how more than one-third of the nine million residents of Los Angeles County are now foreign-born and that the region contains some of the largest enclaves outside their home country of a wide variety of cultures, ranging from Mexican to Filipino, Salvadoran to Samoan (Soja 1995: 130). But cultural diversity is not the only hallmark of globalization; so too is growing socio-economic inequality. In contrast to claims that globalization generates far more "winners" than "losers," there is compelling evidence that the landscape of globalization is one of "staggeringly high peaks of affluence and deep troughs of deprivation" (Dicken 2004: 17). Nowhere is this more evident, some would argue, than in world/global cities. According to Sassen (1991, 2001), one of the most influential analysts of global cities, the processes of economic change in such cities are leading to a growing polarization of their occupational and income structures. Yet, although such arguments might hold true for certain cities, such as New York and Los Angeles, there is evidence, discussed by Hamnett in the second reading, that trying to extend the "global reach" of this social polarization thesis to all world/global cities might be problematic. In particular, Hamnett's research, focused on the Randstadt, a major urban region in the Netherlands, indicates that the outcomes of globalization are crucially mediated by national, place-specific contexts.

Although world/global cities have provided considerable scope for urban geographers to contribute to wider debates about the processes and outcomes of globalization, there are also significant limitations with this perspective. Not only does a focus on world cities mean that "much of the theorizing and empirical research is based on the experiences of the United States, West Europe, and other countries in the core of the world economy" (Grant and Nijman 2002: 320), it also consigns many of the cities in the world to a position of "structural irrelevance" (Robinson 2002: 536). In countering this highly ethnocentric perspective on globalization, many urban geographers have turned their attention to the impacts of globalization on cities which are "off the world cities map" (Grant and Nijman 2002; Robinson 2002). This is the third theme of this section and provides the rationale for the selection of Chakravorty's study of Calcutta. Like other geographers, Chakravorty is concerned that the globalization debate is not nearly as "global as it should be" (Grant and Nijman 2002: 320) and, in particular, that the development of so-called Third World cities is typically viewed separately from the development of Western, "First World cities." The example of Calcutta shows how misleading such a perspective is. The historical and geographical development of this mega-city (its current population is over eleven million) is inextricably linked to its colonial and postcolonial connections to the wider world economy. Indeed, Chakravorty's analysis exemplifies the view that "all cities are world cities" (King 1990) in the sense that urban development in one part of the globe can never be fully understood without reference to urban development elsewhere in the world.

In addition to challenging the ethnocentrism of globalization debates, urban geographers have also called into question the narrowly economistic perspective which informs many accounts of globalization. This is the fourth theme of this section. While the economic dimensions of globalization are of immense importance, it is also vital to consider the ways in which these overlap and intertwine with political and cultural aspects of globalization. As terms such as "Coca-colonization" and "McDonaldization" illustrate, for example, the economic power of global businesses is often inextricably bound up with issues of, in this case, American cultural hegemony (Short and Kim 1999: 4). Indeed, there have been claims that "economic globalization inevitably leads to cultural homogenization where local differences are simply obliterated" (Jackson 2002: 294). However, as Nijman's reading shows, such pessimism is perhaps premature. Drawing on work in anthropology which has highlighted how global flows of migrants and tourists are relocating cultures and creating so-called ethnoscapes, Nijman shows how the presence of international tourists in Amsterdam is influencing the cultural landscapes of the city's historic core. Infamous across the world for being a zone of tolerance in relation to the sex industry and "soft" drug-taking, Nijman argues that the cultural landscape of central Amsterdam is now characterized by a complex mix of "authenticity and artificiality" as it responds to the pressures of global tourism.

In sum, this section clearly demonstrates that urban geographers are not only responding to the wider globalization research agenda through their analyses of the processes shaping world/global cities and by mapping the outcomes of such processes. They are also actively shaping that agenda by focusing attention on aspects of urban development that challenge the ethnocentric and economistic character of many globalization debates.

References and suggestions for further reading

Clark, D. (2003) *Urban World/Global City*, London: Routledge.

Dicken, P. (2004) "Geographers and 'globalization': (yet) another missed boat?" *Transactions of the Institute of British Geographers* 29: 5–26.

Giddens, A. (1999) *Runaway World: How Globalization is Reshaping Our Lives*, London: Profile Books.

Grant, R. and Nijman, J. (2002) "Globalization and the corporate geography of cities in the less-developed world," *Annals of the Association of American Geographers* 92, 2: 320–40.

Jackson, P. (2002) "Consumption in a globalizing world," in R.J. Johnston, P.J. Taylor and M.J. Watts (eds) *Geographies of Global Change: Remapping the World*, Oxford: Blackwell, pp. 283–95.

King, A. (1990) *Urbanism, Colonialism and the World-economy*, London: Routledge.

Knox, P. (2002) "World cities and the organization of global space," in R.J. Johnston, P.J. Taylor and M.J. Watts (eds) *Geographies of Global Change: Remapping the World*, Oxford: Blackwell, pp. 328–39.

McGrew, A.G. (1992) "Conceptualizing global politics," in A.G. McGrew and P.G. Lewis (eds) *Global Politics: Globalization and the Nation-state*, Cambridge: Polity Press, pp. 1–28.

Massey, D. (1994) "A global sense of place," in D. Massey *Space, Place and Gender*, Cambridge: Polity Press, pp. 146–56.

Robinson, J. (2002) "Global and world cities: a view from off the map," *International Journal of Urban and Regional Research* 26, 3: 531–54.

Sassen, S. (1991) *Global City: New York, London, Tokyo*, Princeton, NJ: Princeton University Press.

Sassen, S. (2001) *Global City: New York, London, Tokyo* (2nd edn), Princeton, NJ: Princeton University Press.

Short, J.R. and Kim, T-H. (1999) *Globalization and the City*, Harlow: Longman.

Soja, E. (1995) "Postmodern urbanization: the six restructuring of Los Angeles," in S. Watson and K. Gibson (eds) *Postmodern Cities and Spaces*, Oxford: Blackwell, pp. 125–37.

Taylor, P.J., Watts, M. and Johnston, R.J. (2002) "Geography/globalization," in R.J. Johnston, P.J. Taylor and M.J. Watts (eds) *Geographies of Global Change: Remapping the World*, Oxford: Blackwell, pp. 1–17.

"World-City Network: A New Metageography?"

from *Annals of the Association of American Geographers* (2000)

Jonathan V. Beaverstock, Richard G. Smith, and Peter J. Taylor

Editors' Introduction

Ever since the term "world cities" was first coined by the town planner Patrick Geddes in 1915, there has been a fascination with those places where, according to Geddes, "the world's business is done." In 1966 Peter Hall's *The World Cities* used a range of criteria relating to political power, trade, finance and communications, to identify London, New York, Moscow, Tokyo, and the urban complexes of the Randstad in Holland and the Rhine-Ruhr in Germany, as the places at the top of the global urban hierarchy. Twenty years later, Friedmann's (1986) "World city hypothesis" used the clustering of the headquarters of transnational corporations to single out London, Paris, Tokyo, New York, Los Angeles and Chicago as the "commanding heights" of the global economy. More recently still, Sassen's definition of global cities as strategic sites for the management of the global economy and the production of the most advanced services and financial operations (Sassen 1991) has yielded the now familiar triad of London, New York and Tokyo as the cities at the apex of the world's urban system. Yet, important and influential as all these studies are, their focus on the attributes and characteristics of world cities, rather than the relationships and connections between them, is viewed increasingly as a critical weakness of world city research. As Taylor *et al.* (2002) have noted, the focus on world cities as "nodes" rather than as parts of "networks" is rather paradoxical given that the essence of world cities is their relations one to another. In this reading, Beaverstock, Smith and Taylor tackle this paradox head-on by focusing on the inter-city relations of world cities.

The theoretical foundations of Beaverstock *et al.*'s analysis of the world-city network derive from the work of the urban sociologist Manuel Castells. Ever since his examination of collective consumption in *The Urban Question: A Marxist Approach* (1977) and urban protest movements in *The City and the Grass Roots: A Cross-cultural Theory of Urban Social Movements* (1983), Castells has been an important source of conceptual inspiration for urban geographers. Here Beaverstock *et al.* draw on his more recent writings on "the information age," and, specifically, *The Rise of the Network Society* (1996), to underpin their argument that world cities should be conceptualized as a process located in a networked "space of flows" rather than as specific places (Allen 1999). From this neo-Marxist perspective, world cities are important because of what flows through them (in terms of information, money, ideas, people and so on) rather than what they statically contain. Castells, however, offers little empirical material with which to develop this theoretical framework, so Beaverstock *et al.* draw on their own innovative dataset, comprising the global location strategies of advanced producer-service firms, for exploring inter-city relations. This quantitative information on the global office networks of firms involved in activities such as banking, commercial law and accountancy provides the raw materials for examining the contours of world-city networks in a variety of different ways. Using presence/absence data, for example, Beaverstock *et al.* are able to map out "shared presences" (indicating the numbers of firms with offices in pairs of world cities) and the "probabilities of connection" (indicating how likely or unlikely it is that a firm based in one world city will have an office in another).

Conceptually and methodologically, Beaverstock *et al.*'s analysis is of considerable importance. Conceptually, it marks a decisive shift away from analysing world cities in terms of their internal similarities and differences or ranking them according to their "world city-ness" on the basis of the number of transnational corporate headquarters they contain. Instead, the focus is very much on inter-city relations and on the flows between world cities. Methodologically, the paper is significant because the study of world-city networks requires new kinds of data. Existing datasets are typically "state-centric" (i.e. they contain data collected for states and that are about states), and although this allows international comparisons it does not facilitate trans-state analysis. Beaverstock *et al.*'s dataset of the distribution of the offices of producer-service firms between cities provides an innovative way to address this data deficiency problem and thus allow trans-state, inter-city analysis (see also Short *et al.* 1996). Despite these strengths, however, concerns have been raised about the theoretical foundations that underpin Beaverstock *et al.*'s analysis of world-city networks. Indeed, in an intriguing "auto critique," one of the co-authors of this selection, Richard Smith (2003a: 31; see also 2003b), has likened their research on world cities to "counting door knockers" when "What is needed are new approaches that help us to go beyond counting – to go through these doors to find out precisely how networks work and are maintained over long distances." For Smith this means dispensing with Castells' neo-Marxist thinking, which is viewed as too abstract and lacking in any strong sense of human agency, and instead turning to actor-network and non-representational theories which would bring into focus how networks are constantly made by both human and non-human actors.

Although many geographers have contributed to the study of world cities (see e.g. D. Clark's *Urban World/Global City* (London: Routledge, 2003) and J. Short and Y-H. Kim's *Globalization and the City* (Harlow: Longman, 1999)), arguably the most sustained and innovative analyses have emerged from the Globalization and World Cities (GaWC) Study Group and Network based in the Department of Geography at Loughborough University, of which Beaverstock, Smith and Taylor are key members. Jonathan Beaverstock is Professor of Economic Geography at Loughborough and has particular research interests in the strategic management, organizational networks and spatialities of financial and professional services in economic globalization. Peter Taylor is Professor of Human Geography at Loughborough and was founding editor of the journals *Political Geography* and *Review of International Political Economy*. His many books include *Political Geography: World-economy, Nation State and Locality* (Harlow: Longman; the most recent edition appeared in 1999 and is co-authored with Colin Flint); *The Way the Modern World Works: World Hegemony to World Impasse* (Chichester: Wiley, 1996); and *Modernities: A Geohistorical Interpretation* (Cambridge: Polity Press, 1999). With Paul Knox he is co-editor of *World Cities in a World-system* (Cambridge: Cambridge University Press, 1995) and with David Slater he is co-editor of *The American Century: Consensus and Coercion in the Projection of American Power* (Oxford: Blackwell, 1999). Richard Smith is a Lecturer in Geography at Leicester University and has research interests in globalization and cities and in poststructuralism.

■ ■ ■ ■ ■ ■

You cannot have a geography of anything that is unconnected. No connections, no geography.

(Gould 1991: 4)

Our inability to measure and compare the flows of information between global command centres is a major problem for research on the global urban hierarchy.

(Short and Kim 1999: 38)

During the Apollo space flights, it was reported that one of the astronauts, looking back to Earth, expressed amazement that he could see no boundaries. This new view of our world as the "blue planet," famously captured in NASA photo 22727 of the whole and unshadowed globe (Cosgrove 1991: 127), contradicted the taken-for-granted, state-centric Ptolemaic model or image of world-space that most modern people carry around in their heads: a world of grids, graticule, and territorial boundaries. . . . As a further jolt to the arrogance of modernity, it was soon accepted as a truism that the only "man-made" artifact visible from space was the ancient Great Wall of China. Interestingly, however, the Great Wall is not the only visible feature: at night, modern settlements are clearly visible as pinpricks of electric light on a black canvas. . . . The globality of modern society is clear for all to see in the photo prints, communicated back to Earth, of lights delimiting a global pattern of cities, consisting of a broad swathe girdling the mid-latitudes of the northern hemisphere plus many oases of light elsewhere.

The fact that these "outside views" of Earth identified a world-space of settlements rather than the more familiar world-space of countries has contributed to the growth of contemporary, "One World" rhetoric (also "Spaceship Earth" or "Whole Earth"), which has culminated in "borderless world" theories of globalization. Of course, geographies do not depend solely upon visibility or metaphors. The fact that state boundaries are missing from space-flight photographs tells us nothing, therefore, about the current power of states to affect world geography. The photographs can, however, influence "metageography," or the "spatial structures through which people order their knowledge of the world" (Lewis and Wigen 1997: ix). In the modern world, this has been notably Eurocentric and state-based in character. It is this mosaic spatial structure that the night-time photographs challenge since, first and foremost, people live in settlements. In this paper, we consider the largest pinpricks of light, "the world cities" whose transnational functions materially challenge states and their territories. These cities exist in a world of flows, linkages, connections, and relations. World cities represent an alternative metageography, one of networks rather than the mosaic of states.

Historically, cities have always existed in environments of linkages, both material flows and information transfers. They have acted as centers from where their hinterlands are serviced and connected to wider realms. This is reflected in how economic geographers have treated economic sectors: primary and secondary activities are typically mapped as formal agricultural or industrial regions, tertiary activities as functional regions, epitomized by central-place theory. Why is our concern for contemporary cities in a world of flows any different from this previous tertiary activity and its study? First, the twentieth century has witnessed a remarkable sectoral turnabout in advanced economies:

originally defined by their manufacturing industry, economic growth has become increasingly dependent on service industries. Second, this trend has been massively augmented by more recent developments in information technology that has enabled service and control to operate not only more rapidly and effectively, but crucially on a global scale. Contemporary world cities are an outcome of these economic changes. The large electric pinpricks of light on space photos are actually connected by massive electronic flows of information, a new functional space that will be crucial to geographical understanding in the new millennium.

This paper . . . reports preliminary research on the empirical groundwork required for describing the new metageography of relations between world cities. Such a modest goal is made necessary because of a critical empirical deficit within the world-city literature on intercity relations. . . .

ATTRIBUTES WITHOUT RELATIONS: RESEARCH CLUSTERS IN WORLD-CITY STUDIES

Studies of world cities are generally full of information that facilitates evaluations of individual cities and comparative analyses of several cities. Yet the data upon which these analyses are based has been overwhelmingly derived from measures of city attributes (Taylor 1999). Such information is useful for estimating the general importance of cities and for studying intracity processes, but it tells us nothing directly about relations between cities. Hence cities can be ranked by attributes, but a hierarchical ordering aimed at uncovering flows or networks requires a different type of data based upon measures of relations between cities (Taylor 1997). It is the dearth of relational data that is the "dirty little secret" (Short et al. 1996) of this research area. In other words, we know about the nodes but not the links in this new metageography. Of course, a proper understanding requires an integrated knowledge of both nodes and links. Hence, our brief review of the main clusters of world-city research has two purposes: first, to illustrate the pervasive nature of the Achilles heel, and second, to find world-city formation processes that can direct our search for information on world-city network-formation processes.

Early studies: from cosmopolitanism to corporate economy

Peter Hall (1966) initiated the modern study of world cities with a very comprehensive study of the attributes – politics, trade, communication facilities, finance, culture, technology, and higher education – that placed London, Paris, Randstad-Holland, Rhine-Ruhr, Moscow, New York, and Tokyo at the top of the world urban hierarchy. Stephen Hymer initiated the "economic turn" in world-city studies that has continued to dominate to the present. In an emerging global economy, he argued, corporate control mechanisms were crucial, and hence multinational corporation headquarters tend to be concentrated in the "world's major cities – New York, London, Paris, Bonn and Tokyo...along with Moscow and Peking" (1972: 50). Using the distribution of headquarters to rank cities has since become commonplace...but although such attribute data can define the relative importance of cities, it cannot specify a hierarchy within a network.

Command centers and basing points: The New International Division of Labor and World-City Hypothesis

Most studies of world urban hierarchies have drawn inspiration from John Friedmann's (Friedmann and Wolff 1982; Friedmann 1986) seminal world-city hypothesis. Following Cohen's (1981) new hierarchy of predominant (New York, Tokyo, London) and secondary-level (Osaka, Rhine-Ruhr, Chicago, Paris, Frankfurt, and Zurich) world cities, Friedmann drew his ideas from the organizational implications for capital in Frobel et al.'s (1980) New International Division of Labor thesis. The restructuring of industrial production in the 1970s posed new problems for capital that world cities helped solve by becoming both command centers and basing points for capital in its perennial movement around the globe. Friedmann's hierarchy is frequently cited for its pedagogic and heuristic value, but it is based upon a limited attributive survey whose key variables and heights were difficult to measure and calibrate. These included the concentration of finance, multinational corporation headquarters, business services, manufacturing activity, transportation, and population (Friedmann 1986: 72).

Many have now developed elaborate critiques of Friedmann's original hierarchy (e.g., Beaverstock et al. 2000; Korff 1987; Taylor 1997), and Friedmann himself (1995) has noted some of its limitations. Yet subsequent research has allocated cities to hierarchies based upon their command-and-control criteria, measuring attributes of world cities and then ranking them in order of magnitude....

International financial centers

The rise of global financial markets has been one of the most noted elements of economic globalization, and their integration into international financial centers has stimulated a particular strand of world-city research. The pioneering work of Reed (1981) produced the first major quantitative analysis of world cities. Using a multivariate analysis of more than fifty financial, cultural, economic, geographical, and political attributes across seventy-six cities between 1900–1980, Reed produced an evolving hierarchy of international financial centers. Like Friedmann's (1986), this hierarchy has been widely discussed for its pedagogic value ..., but its principal weakness is that it neglects relations between financial centers. No matter how sophisticated an analysis of attribute data, such rankings can only produce a hierarchy, and world-city financial relations can only be inferred. Furthermore, consideration of the functions of financial intermediaries, with no inter-financial center data analysis, as advocated by Meyer, does not constitute a "world city hierarchy of financial centres" (1998: 428).

The producer-service complex and the "triad" of global cities

If international financial centers represent an "unpacking" of the world-city concept, Sassen's (1991) concept of global city is the beginning of a repacking. In addition to financial services, she identifies the production of other advanced producer services (e.g., accountancy, advertising, insurance, commercial law, etc.) in the creation of global city complexes, as epitomized by London, New York, and Tokyo, with concentrations of theoretical and practical knowledges. The servicing of global capital in these localized complexes creates the concentration of functions we know as world cities. Yet while Sassen interprets this triad of cities as the apex of a global urban hierarchy, her analyses are likewise wholly dependent upon attribute data. The result of her many studies (Sassen 1994a, 1994b, 1995) is a rich knowledge of the triad in comparative terms but with no direct evidence of relations between the three cities or between them and other cities.

The Los Angeles School

A part of the postmodern and cultural "turns" in geography and urban planning has been the rise of the so-called "Los Angeles School," which highlights the conceptual shift from the largely positivist Chicago School of the early twentieth century.... By focusing on a single city, Los Angeles, as the archetypal, paradigmatic, or "celebrity city" of contemporary world-city processes (the place where it "all comes together"; Soja 1989: 8), however, relations between cities have largely been neglected. One strand within this school, Michael Storper's "social organization of economic reflexivity" (1997: 244), is focused upon the proximity of places serving as vital innovation centers for capital. The city is treated by Storper (1997: 222) as a "privileged site" for reflexivity, because of its embedded knowledge and learning structures. Going beyond producer services to specialized manufacturing, with Hollywood as the classic "privileged site," this work nevertheless fails to treat the wider spatial role of cities, such that their connections in an economy of flows remains invisible. While reference is made to the "society of cities" (Storper 1997: 222), intercity relations are precisely what is absent from this work. Economic reflexivity need not be strictly local and territorial; we agree with Amin and Thrift (1992), that there is critical reflexivity embedded in and reproduced through global corporate networks: or, as we would like to put it, through the relational network of world cities.

In summary, the world-cities literature is seriously unbalanced: it has a surfeit of interesting theoretical concepts for treating the nodes of the world-city network, but these exist alongside a deficit in empirical concern for measuring relations between the nodes.

WORLD CITIES IN A NETWORK SOCIETY

One author, in particular, has attempted to advance theoretical knowledge of the world-city network. Castells (1996) conceptualizes the contemporary informational economy as operating through a "space of flows" that constitute a network society. This operates at several levels, one of which is the world-city network. Thus, instead of the static world-city concepts considered above – centers, points, complexes, sites – Castells conceptualizes world cities as processes "by which centers of production and consumption of advanced services, and their ancillary local societies, are connected in a global network" (1996: 380). Hence, cities accumulate and retain wealth, control, and power because of what flows through them, rather than what they statically contain, as is typically measured with attribute data.

It was not part of Castell's (1996) brief to engage in new data generation, and therefore, despite his theoretical contribution, his work reflects the prevailing use of attribute data (Taylor 1999). His chief use of data to specify space of flows is a broad resolution (one origin, nine destinations) set of data from Federal Express originally analyzed by Michelson and Wheeler (1994: 382–83). Castells does not offer an empirical advance on the world-city network, but with the other theoretical studies of nodes, he provides a framework for our empirical work on world-city network formation in a space of relations. World cities are produced by relations of corporate networking activities and connectivity between cities based upon knowledge complexes and economic reflexivity. These fruitful concepts notwithstanding, the key to unlocking the "spaces of relations" of world cities is new data collection (Smith and Timberlake 1995a, 1995b).

GLOBAL OFFICE LOCATION STRATEGIES

The only published data available for studying relations between cities at a global scale are international airline-passenger statistics. Not surprisingly, therefore, empirical studies that present networks of world cities have focused upon this source (Keeling 1995; Kunzmann 1998; Rimmer 1998). There are, however, serious limitations to these statistics as descriptions of relations between world cities (Taylor 1999): first, the information includes much more than trips associated with world-city processes (e.g., tourism), and second, important intercity trips within countries are not recorded in international data (e.g., New York–Toronto does feature in the data, New York–Los Angeles does not). While the latter can be overcome by augmenting the data with domestic flight statistics, the particularities of hub-and-spoke systems operated by airlines create another important caveat to using this data to describe the world-city network.

Studying the global location strategies of advanced producer-service firms is an alternative approach for describing world-city networks, one which overcomes these problems. Firms that provide business services on a global scale have to decide on the distribution of their practitioners and professionals across world cities. Setting up an office is an expensive undertaking, but a necessary investment if the firm believes that a particular city is a place where it must locate in order to fulfill its corporate goals. Hence the office geographies of advanced producer firms provide a strategic insight into world-city processes by interpreting intrafirm office networks as intercity relations. In this argument, world-city network formation consists of the aggregate of the global location strategies of major, advanced producer-service firms.

Information on the office networks of firms can be obtained by investigating a variety of sources, such as company web sites, internal directories, handbooks for customers, and trade publications. We have collected data on the distributions of offices for 74 companies (covering accountancy, advertising, banking/finance, and commercial law) in 263 cities. An initial analysis of this data identified the 143 major office centers in these cities, and 55 of these were designated world cities on the basis of the number, size, and importance of their offices (for details of this classification exercise, see Beaverstock et al. 1999a). No other such roster of world cities exists; it is used here as the basic framework for studying the world-city network.

AN INTERCITY GLOBAL NETWORK

The roster of 55 world cities is divided into three levels of service provision comprising 10 Alpha cities, 10 Beta cities, and 35 Gamma cities. Only

the Alpha cities – Chicago, Frankfurt, Hong Kong, London, Los Angeles, Milan, New York, Paris, Singapore, and Tokyo – are used in this section to illustrate how office geographies can define intercity relations. Note the geographical spread of these top 10 world cities; they are distributed relatively evenly across three regions we have previously identified as the major "globalization arenas": the U.S., western Europe, and Pacific Asia (Beaverstock et al. 1999b). World-city network patterns are constructed for these Alpha world cities, using simple presence/absence data for the largest 46 firms in the data (all of these firms have offices in 15 or more different cities).

Shared presences are shown in Table 1. Each cell in this intercity matrix indicates the number of firms with offices in both cities. Thus, London and New York "share" 45 of the 46 firms; only one firm in the data does not have offices in both of these cities. Obviously these two cities are the places to be for a corporate-service firm with serious global pretensions. This finding is not, of course, at all surprising; interest comes when lower levels of intercity relations are explored. In Figure 1a, the highest 20 shared presences are depicted at two levels of relation. The higher level picks out Sassen's (1991) trio of global cities – London, New York, and Tokyo – as a triangular relationship (but note that, in addition, Hong Kong has such a relationship with London and New York). Bringing in the lower level of relations, London and New York have shared presences with eight other cities in all, but note again the high Pacific Asia profile in this data: Singapore joins with Tokyo and Hong Kong in showing relations with five other cities, the same level as Paris. This contrasts with the U.S. world cities below New York; Los Angeles is in the next-to-bottom class of shared presences with Frankfurt and Milan, and Chicago stands alone, with no intercity relations at the minimum level for inclusion in the diagram. This pattern can be interpreted in terms of the different degrees of political fragmentation in the three major globalization arenas. In the most fragmented, Pacific Asia, there

Figure 1 (opposite) World-city network. (a) Shared presences among Alpha world cities; (b) primary vectors (probabilities of links) among Alpha world cities; (c) secondary vectors (probabilities of links) among Alpha world cities.

Table 1 Relations between Alpha world cities: shared firm presences

	CH	FF	HK	LN	LA	ML	NY	PA	SG	TK
				Number of firms with offices in both cities						
Chicago										
Frankfurt	21									
Hong Kong	21	30								
London	23	32	38							
Los Angeles	21	23	29	33						
Milan	19	28	29	32	22					
New York	23	32	38	45	32	32				
Paris	21	30	32	35	27	28	34			
Singapore	20	30	34	35	26	29	35	32		
Tokyo	23	30	34	37	30	29	37	32	32	

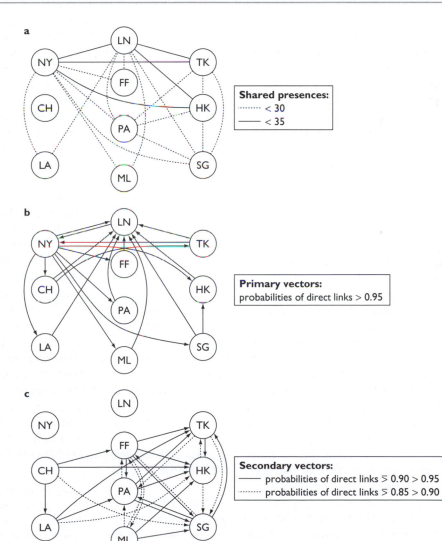

a

Shared presences:
········· < 30
——— < 35

b

Primary vectors:
probabilities of direct links > 0.95

c

Secondary vectors:
——— probabilities of direct links ⩾ 0.90 > 0.95
········· probabilities of direct links ⩾ 0.85 > 0.90

Table 2 Matrix of office-presence linkage indices for alpha world cities

Linkage from	Linkage to									
	CH	FF	HK	LN	LA	ML	NY	PA	SG	TK
Chicago		89	89	100	91	79	100	89	83	100
Frankfurt	67		93	100	72	87	100	95	94	95
Hong Kong	60	82		100	80	80	100	85	92	90
London	59	77	87		78	78	98	83	83	86
Los Angeles	67	73	89	100		70	97	84	81	89
Milan	59	88	93	100	67		100	88	91	93
New York	59	77	87	98	77	77		79	83	85
Paris	64	85	90	100	80	81	97		90	90
Singapore	60	87	98	100	78	83	100	92		95
Tokyo	64	84	93	100	83	81	100	87	88	

is no dominant world city, so that presences are needed in at least three cities to cover the region: Hong Kong for China, Singapore for southeast Asia, and Tokyo for Japan (Taylor 2000). In contrast, the U.S. consists of a single state such that one city can suffice for a presence in that market. The result is that New York throws a shadow effect over other U.S. cities. In between, western Europe is becoming more unified politically, but numerous national markets remain so that London does not dominate its regional hinterland to the same degree as New York.

Shared presences define a symmetric matrix that shows sizes, but not the direction, of intercity relations. By contrast, Table 2 is an asymmetric matrix showing probabilities of connections. Each cell contains the probability that a firm in city A will have an office in city B. Thus, Table 2 shows that if you do business with a Chicago-based firm, then there is a 0.91 probability that that firm will also have an office in Frankfurt. On the other hand, go to a Frankfurt-based firm, and the probability of it having an office in Chicago is only 0.66. Such asymmetry is represented by vectors in Figures 1b and 1c. Primary vectors are defined by probabilities above 0.95. Note that all cities connect to London and New York at this level (Figure 1b). As in Figure 1a, only Tokyo and Hong Kong reach this highest category of connection, but each with only one link. Again, it is also interesting to look at the lower level relations, and these are shown in Figure 1c. This diagram reinforces the

interpretation concerning the three globalization arenas presented above: Chicago and Los Angeles have no inward vectors from the other arenas in what is largely a Eurasian pattern of connections. Vectors to the Pacific Asian cities dominate, but Frankfurt and Paris also have a reasonable number of inward vectors.

This is the first time intercity relations on a global scale have been studied in this way. As expected of such initial research, several opportunities for further investigations are suggested, not least using more cities and more sophisticated network analysis to tease out further features of the contemporary world-city network. But the most urgent task is to go beyond this cross-sectional analysis and study changes over time in order to delineate the evolution of world-city network formation. Only in this way will we be able to make informed assessments of how the network will develop in the new millennium and how this will affect different cities. For instance, is the New York shadow effect growing or declining? We simply do not know.

[. . .]

CONCLUSION: METAGEOGRAPHIC DYSTOPIA?

Riccardo Petrella, sometimes referred to as the "official futurist of the European Union" (1995: 21), has warned of the rise of a "wealthy archipelago of city regions . . . surrounded by an impoverished lumpenplanet" (p. 21). He envisages a scenario in which the 30 most powerful city regions (the CR-30) will replace the G-7 (the seven most powerful states), presiding over a new global governance by 2025. Such a scenario is given credence by the fact that contemporary world cities are implicated in the current polarization of wealth and wages accompanying economic globalization. World-city practitioners and professionals operating in a global labor market have demanded and received "global wages" (largely in the form of bonuses) to create a new income category of the "waged rich"; with reference to London, they have been called the new "Super Class" (Adonis and Pollard 1997).

Petrella sets out his global apartheid dystopia as a warning about current trends so as to alert us to the dangers ahead. But cities do not have to play the bete noire role of the future. It is within cosmopolitan cities that cultural tensions can be best managed and creatively developed. Certainly modern states, in their ambition to be nation-states, have an appalling record in dealing with matters of cultural difference. But the key point is that this is not a simple matter of cities versus states (Taylor 2000). World cities are not eliminating the power of states, they are part of a global restructuring which is "rescaling" power relations, in which states will change and adapt as they have done many times in previous restructurings (Brenner 1998). The "renegotiations" going on between London's world role and the nation's economy, between New York's world role and the U.S. economy, and with all other world cities and their encompassing territorial "home" economies, are part of a broader change affecting the balance between networks and territories in the global space-economy. In this paper, we have illustrated how empirical analysis of city economic networks might be undertaken to complement traditional economic geography's concern for comparative advantage between states. Our one firm conclusion is that in the new millennium, we cannot afford to ignore this new metageography, the world-city network.

REFERENCES FROM THE READING

Adonis, A. and Pollard, S. (1997) *A Class Act: The Myth of Britain's Classless Society*, London: Hamish Hamilton.

Amin, A. and Thrift, N. (1992) "Neo-Marshallian Nodes in Global Networks," *International Journal of Urban and Regional Research* 16: 571–87.

Beaverstock, J.Y.; Smith, R.G. and Taylor, P.J. (1999a) "A Roster of World Cities," *Cities* 16.

Beaverstock, J.Y.; Smith, R.G.; and Taylor, P.J. (1999b) "The Long Arm of the Law: London's Law Firms in a Globalizing World-Economy," *Environment and Planning A* 31: 187–92.

Beaverstock, J.Y.; Smith, R.G.; Taylor, P.J.; Walker, D.R.E.; and Lorimer, H. (2000) "Relational Studies of Globalization and World Cities: Three Measurement Methodologies," *Applied Geography* 20.

Brenner, N. (1998) "Global Cities, Glocal States: Global City Formation and State Territorial Restructuring in Contemporary Europe," *Review of International Political Economy* 5: 1–37.

Castells, M. (1996) *The Rise of the Network Society*, Oxford: Blackwell.

Cohen, R.B. (1981) "The New International Division of Labour, Multinational Corporations and Urban Hierarchy," in M. Dear and A. Scott (eds) *Urbanisation and Urban Planning in Capitalist Society*, London: Methuen, pp. 287–315.

Cosgrove, D. (1991) "New World Orders," in C. Philo (ed.) *New Words, New Worlds: Reconceptualising Social and Cultural Geography*, Aberystwyth, Wales, UK: Cambrian Publishers, pp. 125–30.

Cosgrove, D. and Rogers, A. (1991) "Territory, Locality and Place," in C. Philo (ed.) *New Words, New Worlds: Reconceptualising Social and Cultural Geography*, Aberystwyth, Wales, UK: Cambrian Publishers, pp. 36–38.

Friedmann, J. (1986) "The World City Hypothesis," *Development and Change* 17: 69–83.

Friedmann, J. (1995) "Where We Stand: A Decade of World City Research," in P.L. Knox and P.J. Taylor (eds) *World Cities in a World System*, Cambridge: Cambridge University Press, pp. 21–47.

Friedmann, J. and Wolff, G. (1982) "World City Formation: An Agenda for Research and Action," *International Journal of Urban and Regional Research* 3: 309–44.

Frobel, E.; Heinrichs, J.; and Kreye, O. (1980) *The New International Division of Labour*, Cambridge: Cambridge University Press.

Gould, P. (1991) "Dynamic Structures of Geographic Space," in S.D. Brunn and T.R. Leinbach (eds) *Collapsing Space and Time: Geographic Aspects of*

Communication and Information, London: Harper-Collins, pp. 3–30.

Hall, P. (1966) *The World Cities*, London: Heinemann.

Hymer, S. (1972) "The Multinational Corporation and the Law of Uneven Development," in J. Bhagwati (ed.) *Economics and World Order from the 1970s to the 1990s*, London: Collier-Macmillan, pp. 113–40.

Keeling, D.J. (1995) "Transportation and the World City Paradigm," in P.L. Knox and P.J. Taylor (eds) *World Cities in a World System*, Cambridge: Cambridge University Press, pp. 115–31.

Korff, R. (1987) "The World City Hypothesis: A Critique," *Development and Change* 17: 483–95.

Kunzmann, K.R. (1998) "World City Regions in Europe: Structural Change and Future Challenges," in Fu-chen Lo and Yue-Man Yeung (eds) *Globalization and the World of Large Cities*, Tokyo: United Nations University Press, pp. 37–75.

Lewis, M.W. and Wigen, K.E. (1997) *The Myth of Continents. A Critique of Metageography*, Berkeley: University of California Press.

Meyer, D.R. (1998) "World Cities as International Financial Centres," in Fu-Chen Lo and Yue-Man Yeung (eds) *Globalization and the World of Large Cities*, Tokyo: United Nations University Press, pp. 410–32.

Michelson, R.L. and Wheeler, J.O. (1994) "The Flow of Information in a Global Economy: The Role of the American Urban System in 1990," *Annals of the Association of American Geographers* 84: 87–107.

Petrella, R. (1995) "A Global Agora vs. Gated City-Regions," *New Perspectives Quarterly*, winter: 21–22.

Reed, H.C. (1981) *The Pre-eminence of International Financial Centres*, New York: Praeger.

Rimmer, P.J. (1998) "Transport and Telecommunications among World Cities," in Fu-Chen Lo and Yue-Man Yeung (eds) *Globalization and the World of Large Cities*, Tokyo: United Nations University Press, pp. 433–70.

Rogers, R. (1997) *Cities for a Small Planet*, London: Faber and Faber.

Sassen, S. (1991) *The Global City*, Princeton. NJ: Princeton University Press.

Sassen, S. (1994a) *Cities in a World Economy*, London: Pine Forge Press.

Sassen, S. (1994b) "The Urban Complex in a World Economy," *International Social Science Journal* 46: c43–62.

Sassen, S. (1995) "On Concentration and Centrality in the Global City," in P.L. Knox and P.J. Taylor (eds) *World Cities in a World System*, Cambridge: Cambridge University Press, pp. 63–78.

Sassen, S. (1999) "Global Financial Centers," *Foreign Affairs* 78: 75–87.

Short, J.R. and Kim, Y.H. (1999) *Globalization and the City*, Harlow, UK: Longman.

Short, J.R.; Kim, Y.; Kuss, M.; and Wells, H. (1996) "The Dirty Little Secret of World Cities Research," *International Journal of Urban and Regional Research* 20: 697–717.

Smith, D.A. and Timberlake, M. (1995a) "Conceptualising and Mapping the Structure of the World Systems City System," *Urban Studies* 32: 287–302.

Smith, D.A. and Timberlake, M. (1995b) "Cities in Global Matrices: Toward Mapping the World-System's City System," in P.L. Knox and P.J. Taylor (eds) *World Cities in a World System*, Cambridge: Cambridge University Press, pp. 79–97.

Soja, E. (1989) *Postmodern Geographies: The Reassertion of Space in Critical Social Theory*, London: Verso.

Storper, M. (1997) *The Regional World*, New York: Guilford Press.

Taylor, P.J. (1997) "Hierarchical Tendencies amongst World Cities: A Global Research Proposal," *Cities* 14: 323–32.

Taylor. P.J. (1999) "The So-called World Cities: The Evidential Structure of a Literature," *Environment and Planning A* 31: 1901–04.

Taylor, P.J. (2000) "World Cities and Territorial States under Conditions of Contemporary Globalization," *Political Geography* 19: 5–32.

Editors' references and suggestions for further reading

Allen, J. (1999) "Cities of power and influence: settled formations," in J. Allen, D. Massey and M. Pryke (eds) *Unsettling Cities: Movement/Settlement*, London: Routledge, pp. 181–228.

Castells, M. (1977) *The Urban Question: A Marxist Approach*, London: Edward Arnold.

Castells, M. (1983) *The City and the Grass Roots: A Cross-cultural Theory of Urban Social Movements*, Berkeley: University of California Press.

Castells, M. (1996) *The Rise of the Network Society*, Oxford: Blackwell.

Friedmann, J. (1986) "The world city hypothesis," *Development and Change* 17: 69–83.

Hall, P. (1966) *The World Cities*, London: Weidenfeld & Nicolson.

Sassen, S. (1991) *Global City: New York, London, Tokyo*, Princeton, NJ: Princeton University Press.

Short, J., Kim. Y., Kuus, M. and Wells, H. (1996) "The dirty little secret of world cities research: data problems in comparative analysis," *International Journal of Urban and Regional Research* 20: 697–717.

Smith, R.G. (2003a) "World city actor-networks," *Progress in Human Geography* 27, 1: 25–44.

Smith, R.G. (2003b) "World city topologies," *Progress in Human Geography* 27, 5: 561–82.

Taylor, P., Walker, D. and Beaverstock, J. (2002) "Firms and their global service networks," in S. Sassen (ed.) *Global Networks, Linked Cities*, New York: Routledge, pp. 93–115.

T W O

"Social Polarisation in Global Cities: Theory and Evidence"

from *Urban Studies* (1994)

Chris Hamnett

Editors' Introduction

The previous reading examined the nature of relationships and networks between world or global cities. In this reading the spatial focus shifts from the networks to the nodes in order to consider what is happening *within* such cities and, in particular, claims that they are characterized by a polarized or dual socio-economic structure. The main proponent of this dual city thesis is Sassen (1991/2001), whose studies of the global cities of New York, London and Tokyo suggest they have a socio-economic profile that increasingly resembles an hourglass: absolute growth at the top and bottom ends of the occupational and income distributions and a decline in the middle of the distribution. The key driver of this phenomenon, Sassen argues, is economic restructuring which has involved a decline in manufacturing employment and a substantial growth both in highly skilled, well-paid jobs in the headquarters of transnational corporations and the advanced services sector (such as management consultancy, advertising, and legal services), and in very poorly paid, low-skilled service jobs, such as secretarial staff and office cleaners. Sassen identifies a high degree of socio-economic symbiosis between these extremes of rich and poor in the global city. Not only do corporate headquarters and advanced producer services generate low-wage jobs directly through the structure of the work process but also indirectly through the high-income lifestyles of those employed in these areas.

In this reading by Hamnett, Sassen's social polarization thesis is subjected to critical scrutiny. Beginning with the conceptual foundations of the thesis Hamnett detects some important weaknesses. First, the notion of polarization deployed by Sassen is riddled with ambiguity, partly because it is used to refer to both occupational and income polarization and partly because it fails to distinguish between absolute and relative polarization. The latter point is particularly important because Sassen focuses on absolute polarization in terms of the growth in jobs at the high- and low-income extremes of the labour market and ignores the way in which polarization might also result from relative or differential shrinkage. The latter may occur when there are declining numbers of low-paid, unskilled jobs while at the same time there are increasing levels of professionalization based around post-industrial forms of employment requiring professional, managerial and technical skills. Hamnett also takes issue with Sassen's contention that an expansion in low-wage service jobs is driving large-scale ethnic immigration in global cities. Hamnett argues that the inverse is also possible, with growth at the bottom end of the occupational distribution being contingent on large-scale immigration and that it is the existence of a large supply of cheap labour that enables a substantial low-wage service sector to exist. For Hamnett this problematic aspect of Sassen's social polarization thesis stems largely from her attempts to generalize from the experiences of New York and Los Angeles, cities which are unusual in terms of the scale of immigration and thus the supply of low-wage workers. Using empirical evidence from the Randstad, a major urban area of Holland incorporating the cities of Amsterdam, the Hague, Rotterdam and Utrecht, Hamnett further sharpens his critique of Sassen's social polarization thesis by demonstrating the difficulties of generalizing her arguments in a Western European context. In particular, he shows that it is professionalization rather than polarization which is the dominant feature of the changing socio-economic structure of the Randstad.

By introducing greater conceptual precision into the global cities debate and by highlighting the difficulties of uncritically transferring ideas from one socio-economic context to another, Hamnett's article makes a very important contribution to discussions of the impact of globalization on cities. His research also touches on another key question: are global cities "dual cities" characterized by a new spatial order in terms of strengthening patterns of segregation between the richest and poorest socio-economic groups (Marcuse and van Kempen, 2000)? In the Randstad, Hamnett's evidence indicates that the answer is an unequivocal "No," largely because of government income and housing policies which have ensured that low-income groups are not disproportionately concentrated in social housing. In other global (and globalizing) cities, too, there is little evidence to support the notion that these have become "dual cities." There are, of course, areas of extreme affluence which are often in close proximity to areas of extreme poverty (see readings by Smith and Short, this volume), but global cities are much more complex and diverse than the term dual city suggests. As Fainstein and Campbell (1996: 8) observe, "As well as containing rich and poor, they encompass groups of upwardly mobile immigrants and aspiring artists, masses of unionized government employees, large student populations, and vast numbers of middle-level white collar workers."

Sassen's global cities thesis, the culmination of earlier work on immigration and the mobility of labour and capital, is set out most clearly in *The Global City: New York, London, Tokyo* (Princeton NJ: Princeton University Press, 1991; 2nd edn 2001). She is also author of *Cities in a World Economy* (California, Pine Forge Press, 1998). Detailed case studies of the impact of globalization on the internal geographies of cities may be found in P. Marcuse and R. van Kempen (eds), *Globalizing Cities: A New Spatial Order?* (Oxford: Blackwell, 2000), while D. Clark's *Urban World/Global City* (London: Routledge, 2003) also provides a useful introductory overview of work on globalization and cities.

Chris Hamnett is Professor of Geography at Kings College, University of London. With interests in urban social geography and social change, he is editor of *Social Geography: A Reader* (London: Arnold, 1996), the author of *Winners and Losers: The Home Ownership Market in Britain* (London: UCL Press, 1998), and co-author with H. Feigenbaum and J. Henig of *Shrinking the State: The Political Underpinnings of Privatization* (Cambridge: Cambridge University Press, 1998). Most recently he has written *Unequal City: London in the Global Arena* (London: Routledge, 2003). An examination of the economic and social changes that have occurred in London since the 1960s, the findings in this book reinforce the arguments advanced in this selection. Like other global cities, London's economic structure has been transformed from a city with a large manufacturing sector to one based on business and financial services. In contrast to Sassen's thesis, however, Hamnett argues that London has not become increasingly polarized but more professionalized as a result of a shrinking manual workforce and a growing new middle-class population associated with the rise of London's post-industrial economy.

[. . .]

THE GLOBAL CITIES THESIS

The last 10 years have seen a growing interest in the relationships between global economic restructuring, the social and spatial division of labour and changes in the urban hierarchy. Attention has been particularly focused on the global cities at the top of the urban hierarchy which play a key role in the control and coordination of international capitalism.

[. . .]

One of the leading proponents of the global cities thesis has been Saskia Sassen. Her earlier work on immigration and the informal sector in New York and Los Angeles and what she terms "the new labor demand in global cities" (1984, 1985, 1986, 1987), was followed by a major study of the mobility of labour and capital (1988), culminating in her book *The Global City* (1991). She argues that:

A combination of spatial dispersal and global integration has created a new strategic role for major cities. Beyond their long history as centers for international trade and banking, these cities now function as centers in four new ways: first, as highly concentrated command points in the organization of the world economy; second, as key locations for finance and specialized service firms, which have replaced manufacturing as the leading economic sectors; third, as sites of production, including the production of innovations, in these leading industries; and fourth, as markets for the products and innovations produced. (1991, 3)

Sassen (1991, p. 4) goes on to argue that:

These changes in the functioning of cities have had a massive impact upon both international economic activity and urban form. Cities concentrate control over vast resources, while finance and specialised service industries have restructured the urban social and economic order. Thus *a new type of city has appeared. It is the global city*. [Emphases added.]

SOCIAL POLARISATION IN GLOBAL CITIES: THE BASIS THESIS

[. . .]

The polarisation thesis has been developed by Sassen (1984, 1988, 1991) who argues that the changes in social structure are a direct result of the changes in economic base. She argues (1991, p. 9) that the evolving structure of economic activity in global cities, particularly the rapid growth of financial and business services and the sharp decline of manufacturing industry has "brought about changes in the organisation of work, reflected in a shift in the job supply and polarization in the income distribution and occupational distribution of workers".

[. . .]

Sassen (1991, p. 13) summarises her thesis as follows:

New conditions of growth have contributed to elements of *a new class alignment in global cities*.

The occupational structure of major growth industries characterised by the locational concentration of major growth sectors in global cities in combination with the polarised occupational structure of these sectors has created and contributed to growth of a high-income stratum and a low-income stratum of workers. It has done so directly through the organization of work and occupational structure of major growth sectors and it has done so indirectly through the jobs needed to service the new high-income workers, both at work and at home, as well as the needs of the expanded low wage work force. [Emphases added.]

[. . .]

SASSEN'S SOCIAL POLARISATION THESIS: A CRITICAL APPRAISAL

[. . .]

Central to her thesis is the assertion that global cities have been characterised by growing occupational and income polarisation which reflects the changes in industrial and employment structure. The thesis is internally consistent and well argued and it has become almost the conventional wisdom about contemporary social change in global cities, including the Randstad. The question, however, is not how attractive the thesis is, but whether it is theoretically and empirically valid. As I shall argue, there are a number of weaknesses in the thesis and its application which render it problematic. . . .

The ambiguity of social polarisation

Despite the centrality of the concept of polarisation in Sassen's thesis on employment and occupational change in global cities, it remains a most unclear and ill-defined concept which can mean very different things to different people. . . . Sassen is nowhere clear as to precisely what she means by polarisation. One of the best definitions is that of Marcuse (1989, p. 699):

The best image . . . is perhaps that of the egg and the hour glass: the population of the city is normally distributed like an egg, widest in the middle and tapering off at both ends; when it becomes polarized the middle is squeezed and the ends expand till it looks like an hour glass. The middle of the egg may be defined as intermediate social strata. . . . Or if the polarization is between rich and poor, the middle of the egg refers to the middle-income group. . . . The metaphor is not of structural dividing lines, but of a continuum along a single dimension, whose distribution is becoming increasingly bimodal.

As Marcuse rightly points out, polarisation is a process whereby "distribution is becoming increasingly bimodal". But Sassen herself tends to slide from arguments based on changes in occupational structure to arguments based on income distributions and she is vague whether polarisation is to be viewed in absolute or relative terms.

This may seem a trivial point about measurement procedure but it is crucial given the strong emphasis in her thesis that changes in the division of labour are producing *more* jobs at the top and at the bottom and *less* in the middle. There is a huge difference between asserting that changes are creating *large numbers* of low-skilled and or low-paid jobs and changes which lead to a *larger proportion* of low-skilled jobs but not larger numbers. Sassen suggests that a process of absolute polarisation is occurring. This may be true of New York and Los Angeles for reasons which are outlined below but in most Western capitalist countries the numbers of semi-skilled and unskilled jobs have been steadily shrinking for the last 20–30 years. While it may be possible to find relative occupational polarisation, this can occur as a result of differential shrinkage of the labour force. This is a very different process from the ones outlined by Sassen which are said to be creating larger numbers of less-skilled and low-paid workers.

Polarisation, professionalisation and proletarianisation

The social polarisation thesis can be criticised for failing to address the wider literature on the changing occupational structure of advanced capitalist societies. In particular, it seems to ignore the literature on the growth of professionalisation, and what has been termed the 'new middle class'. This thesis was first advanced by Bell (1973) in *The*

Coming of Post-Industrial Society. He argued that Western societies were characterised by the shift from industrialism to post-industrialism, from the production of goods to the production of services and by the development of a knowledge society organised around professional, managerial and technical skills. He argued that this was leading to a class structure characterised by an expanding professional and managerial workforce.

[. . .]

[T]here is evidence that a process of professionalisation is concentrated in a number of large cities with a strong financial/producer service base. As Ley (1992a, p. 10) noted in a Canadian metropolitan context:

> The positioning of the new middle class has a geographic as well as a sociological dimension. The astonishing pace of embourgeoisiement in the central cities gives a dramatic geographical form to the expansion of the new middle class jobs.

[. . .]

Nor is the evidence confined to North America. Ruth Glass (1973, p. 426) argued that:

> London is now being renewed at a rapid pace, but not on the model about which we are often warned. Inner London is not being Americanised: it is not on the way to becoming mainly a working class city, a polarised city, or a vast ghetto for a black proletariat. The real risk for inner London is that it might well be gentrified with a vengeance, and be almost exclusively reserved for a selected higher-class strata.

Subsequent work by Hamnett (1976, 1986, 1987) confirmed that London experienced an increase in the proportion of professional and managerial workers in 1961–1981, while the numbers and the shares of all other groups declined. There is no evidence for absolute social polarisation in London in the 1960s and 1970s, and the 1991 census is most unlikely to reveal a sudden reversal of fortune. To the extent that professionalisation is the dominant process of change in the occupational structure in international Western cities this raises the question of why Sassen is able to draw opposite conclusions from her analysis of New York and Los Angeles. . . .

Slaves of New York?

An examination of Sassen's previous work shows that the polarisation thesis is based on her work on migration and informalisation in New York and Los Angeles, both of which are cities with extremely high and continuing levels of immigration and a large low-wage immigrant labour supply. Sassen (1984, p. 140) argued that a process of "*peripheralisation at the core*" is leading to "the shaping of a new class alignment in advanced capitalist economies". She argues (1984, pp. 139–140) that:

It is growth trends, not decline trends, which are generating the polarization in the occupational structure, including a vast expansion in the supply of low-wage jobs and shrinking in the supply of middle-income jobs. . . . The large new immigration, directed mostly to a few major urban centers, can be shown to be primarily associated with this expansion of low-wage jobs.

In 1984, Sassen (p. 158) noted that:

The large influx of immigrants from low-wage countries over the last fifteen years, which reached massive levels in the second half of the 1970s, cannot be understood separately from this restructuring. The available evidence for New York City and Los Angeles . . . strongly suggests that a large share of immigrants in both cities represents an important supply of low-wage workers. New York and Los Angeles have the largest Hispanic populations of all SMSAs and the size of their Hispanic populations is significantly larger than that of the next series of cities.

[. . .]

The argument and the evidence is consistently repeated in Sassen's work, but she fails to make an obvious inference that the polarisation thesis, as it affects growth at the bottom end of the occupational and income distributions, may be contingent on the existence of large-scale ethnic immigration and a cheap labour supply. The absence of such high levels of in-migration and informalisation elsewhere may limit its generality. Indeed, it could be argued that it is the existence of such a large supply of cheap labour that enables a large-scale low-wage service sector and a "down-graded" manufacturing sector to exist. In countries lacking such a large supply of cheap immigrant labour, it may be necessary to invest in capital-intensive processes or to engage in what Gershuny (1978) has called the self-service economy. . . .

Sassen (1988, p. 146), however, draws the reverse inference, namely that:

The large influx of immigrants from low-wage countries over the last fifteen years which reached massive levels in the second half of the 1970s cannot be understood apart from this restructuring. It is a mistake to view this new immigration phase as a result mostly of the new legislation and as being absorbed primarily in declining sectors of the economy. The expansion in the supply of low-wage jobs generated by major growth sectors is one of the key factors in the *continuation* at ever higher levels of the current immigration.

This is not to argue, of course, that other global cities do not have a large migrant presence, or that these groups are not concentrated in low-skill and low-wage jobs. This would be wrong. The 1991 Census found that 20 per cent of Greater London's and 25 per cent of inner London's population were ethnic minorities, and aliens and ethnic minorities totaled 21 per cent of the four major Dutch cities and 24 per cent in Amsterdam (van Ammersfoort, 1992).

The argument rather is that the size of the recent ethnic immigration in other cities is much smaller than in New York or Los Angeles. The 1990 US Census found that of the 7.3m enumerated population of New York City, just 43 per cent were non-Hispanic whites (down from 63 per cent in 1970). The number of non-Puerto Rican Hispanics increased by 63 per cent in 1980–1990 as did the non-Hispanic Asian population. The total foreign-born population in 1990 was just over 2m or 28 per cent. The proportion of foreign-born migrants is similar in Los Angeles, and non-whites now make up over half the total population (Clarke, 1993).

These proportions are far higher than in other cities. To this extent, Sassen's social polarisation thesis may be a slave of New York which she has erroneously generalised to all global cities.

[. . .]

THE EVIDENCE FOR SOCIAL POLARISATION IN THE RANDSTAD

There have been frequent suggestions that the Randstad [comprising the cities of Amsterdam, Rotterdam, the Hague and Utrecht and their contiguous suburban areas] has experienced a process of polarisation in recent years ... [but] we will argue that the evidence for polarisation to date is weak. ...

The changing occupational structure of the Randstad

The occupational data for the Netherlands provided by the Labour Force Survey are very ambiguous and unsatisfactory for the purpose of occupational class analysis consisting as it does of a rather uneasy mixture of occupational and sectoral categories. There are nine occupational categories ... :

1. Qualified staff: scientific and educated specialists including doctors, economic and legal professionals and teachers and artists.
2. Managers.
3. Administrative occupations: book keepers, superintended transport and communications personnel, cashiers and other administrators.
4. Commercial occupations: managers of firms in wholesale and retail trade, purchasers, trade agents, salespersons, insurance agents, shop personnel.
5. Service occupations: managers and personnel in the hotel and catering industry, cleaners, fire brigade, police and security staff.
6. Others: farmers, fishermen and military.
7–9. Manual labour: production and construction workers, painters, drivers and other transport personnel, craftsmen.

It is clear from this list that there is no clear and unambiguous divide between managers and other occupations. Service occupations include managers as well as all those in the personal service sector. The commercial occupations include managers and administrative occupations cover a wide range of jobs. Even the qualified staff category is wide ranging, including as it does, artists and teachers along with lawyers and doctors. It should also be pointed out that the 1990 Labour Force Survey, unlike the previous survey, included information on persons working part-time less than 12 hours a week. Not surprisingly, this resulted in a very considerable increase in the number of employed workers, particularly women part-time workers.

[...]

Looking first at the four city regions as a whole, the total working labour force rose by 301,000 or 19.5 per cent from 1.54m to 1.84m [between 1981 and 1990]. ... The increase was comprised of 207,000 women and 94,000 men. The distribution of the changes 1981–90 ... shows that the great bulk of the increase – 231,000 (or 77 per cent) – was of managers and qualified staff. The numbers in this category grew by 61 per cent from 380,000 in 1981 to 611,000 in 1990. The other increases were all relatively small by comparison. The administrative category grew by 27,000 or 7.3 per cent from 1981 to 1990. The "commercial" category grew by 33,000 or 21 per cent (11 per cent of the total growth), and the service category grew by 51,000 (27 per cent) – 11 per cent of the total increase. Both the manual workers and the other category declined by just under 10 per cent. It is clear that, even given the unsatisfactory nature of the occupational categories and the inclusion of part-time work of under 12 hours a week, the overall trend was growing professionalisation. There is no indication in the occupational data for a trend towards growing polarisation in Randstad Holland, though it is possible that the content of different jobs has changed leading to *de facto* deskilling. ...

Nor is it the case that the overall professionalisation trend conceals a marked gender polarisation, with women becoming increasingly segregated in low-skill jobs. The number of women in administrative and commercial jobs, many of which are routine office jobs, rose faster than men and the proportionate increases were much greater – 24 per cent for administrative jobs (compared with 10 per cent for men) and 33 per cent for commercial occupations (compared with 13 per cent for men), and the increase (30 per cent) in the service category was also greater than for men (23 per cent). But the absolute increase in qualified staff was broadly equal: 124,000 men and 107,000 women. Given the smaller base of qualified women in 1980, the proportionate increase for qualified women (79 per cent) was much greater than for men (51 per cent).

Looked at overall, the proportion of women in qualified occupations rose from 25.6 per cent in 1980 to 32.9 per cent in 1990, while the proportion in all other occupational categories decreased. The proportion of men in qualified occupations rose by slightly more, from 24.1 per cent to 33.3 per cent: a total of 9.2 percentage points compared to 7.3 points for women, but women are by no means being concentrated in low-skilled jobs in the four city regions. . . .

This conclusion also holds when the four major cities are analysed separately from their surrounding urban regions. . . . [T]he number of managers and qualified staff in the four cities grew by 103,000 (62 per cent) in 1981–1990, compared to a total increase in the employed labour force of 58,000 and a decrease of 49,000 manual workers. The increases in the administrative, commercial and services occupational categories were very small, totaling 7,000 in all. The proportion of qualified staff in the labour force in the four cities rose sharply from 23 per cent to 34 per cent.

[. . .]

What can we conclude from this analysis of employment in the 1980s? Atzema and de Smidt (1992) argue that "The Randstad economy is developing a specialised employment structure (where) growth occurs in better qualified and high-wage jobs" (p. 284). They argue that there are major doubts whether the dual city concept can be applied to the Randstad. First they argue that the Randstad does not have a segmented labour market along American lines and they suggest that the process of economic restructuring has been slower in the Netherlands than elsewhere, partly because the economy of the Randstad has always been orientated more towards services and trade rather than the industrial sector. Finally, they point to the major role of the Dutch welfare state in ameliorating social inequality (Kloosterman and Lambooy, 1992).

The changing educational structure of the Randstad

The occupational classification is problematic as we have seen. Fortunately, the Labour Force Survey also provided data which enable us to assess the educational attainment of the labour force. This measure is a valuable complement to the occupational one. There are six categories: unknown, primary, lower, intermediate, higher and university.

Looking first at the changes in educational levels in the Randstad as a whole, the picture is very clear. Of a total increase in the labour force of 307,000 or 20 per cent, between 1981 and 1990, the 'unknown' and 'primary' categories fell by 36,000 (−77 per cent) and 77,000 (−26 per cent) respectively, and the 'lower' category fell by 5,000 (−1 per cent). Conversely, the other three categories all showed increases. The 'intermediate' category grew by 178,000 or 33 per cent, the higher category grew by 108,000 or 62 per cent and the university category grew by a remarkable 109,000 or 131 per cent. The proportion of graduates almost doubled from 5.4 per cent to 10.4 per cent, while the proportion of workers with higher education qualifications rose from 11.3 per cent to 15.3 per cent and the proportion with intermediate school rose from 35 per cent to 39 per cent. The other categories fell sharply. The evidence is clear. The educational composition of the Dutch labour force underwent an upward shift in the 1980s which is consistent with growing professionalisation. . . .

This upwards shift in educational attainment was not just a male prerogative. On the contrary, although the absolute increase in the number of males with university degrees (+71,000) was almost double the increase in the number of women (+38,000), the proportionate increase among women was far higher than men (211 per cent as against 109 per cent). And the absolute and proportionate increases in the numbers of women with intermediate and higher educational qualifications were also far higher than for men. The proportion of women with intermediate qualifications rose by 58 per cent (20 per cent for men) and the proportion of women with higher educational qualifications rose by 99 per cent compared to just 36 per cent for men. This suggests that Dutch women made very major progress in the educational system in the 1980s, and that they are rapidly narrowing the gap in the university sector. It is also very clear when we compare the cities with the city regions that, although the proportion of the labour force with intermediate, higher and university qualifications was higher in the city regions than in the cities in 1981, this situation has been reversed by 1990. The increase in both the number and the proportion of city residents with university degrees was much greater than in the regions, thus confirming Jobse and Musterd's (1992) findings that the more highly educated are concentrating in the cities. The educational data provide no evidence of polarisation.

The changing income structure of the Randstad

The evidence provided by income is far less sanguine than that from occupational data and, in some respects, it is quite contradictory. Kloosterman (1991) suggests that in the first half of the 1980s there was an increase in the number of low-wage jobs in the service sector in the Netherlands. His data suggest that the number of low-wage jobs grew very rapidly in the Netherlands as a whole between 1979 and 1988. They indicate that out of a growth of 633,000 jobs, 457,000 were low-paid, and 215,000 were middle-paid with a loss of 39,000 high-paid jobs. Numbers of low-wage jobs grew sharply in the hotel, catering and retail trades. But there was also an upwards shift in the income structure of manufacturing. . . . It should also be noted that the growth in the number of low-paid jobs and the fall in the number of high-paid jobs, particularly in the quaternary sector, partly reflects government policy in the early 1980s of civil service pay cuts (Kloosterman, 1991). As Kloosterman and Lambooy (1992, pp. 133–134) point out:

> The wage cuts in the public sector . . . which led to a sharp increase in the number of low-paid jobs has not affected the four cities much more than . . . the national economy. . . . In both the Greater Amsterdam and the Greater Rotterdam area, the wage cuts affected all pay scales. Besides inducing a rise in low-paid public sector jobs, these wage cuts also led to a substantial decrease in the number of high-paid jobs in the public sector. This decrease has more than offset the growth of high-salary positions in the Greater Rotterdam area.

These are national figures, but housing survey income data on the four major cities for 1981–1989 confirm the picture. . . . [T]he percentage of households in each of the lowest four income deciles rose over the period, while the proportion in each of the top six income deciles fell. The proportion of households with incomes in the bottom four deciles rose from 63.1 per cent in 1981 to 68.1 per cent in 1989: an increase of 5 points. When the proportion in each decile in the four cities is compared to the Netherlands average (10 per cent in each decile), it is clear that the major cities were strongly overrepresented in the lower income deciles and underrepresented in the higher income deciles in 1981, and that the concentration of low-income groups in the cities increased between 1981 and 1989. This partly reflects the suburbanisation of the high-income groups over the last 20 years, and data for the four city regions would show a less marked pattern. . .

[. . .]

Immigration in the Netherlands

Given my argument that Sassen's social polarisation thesis is, to a large extent, dependent on the existence of large numbers of low-skilled, low-paid immigrants in New York and Los Angeles, it is necessary here to say a few words about the scale of immigration in the Netherlands in general, and the Randstad in particular. Like most other Western European countries, the Netherlands has experienced a large influx of immigrants over the last 30 years (Roelandt and Veenman, 1992). Ammersfoort (1992, p. 442) states that "like all developed countries the Netherlands is an immigration country, though only on a modest scale". As of 1 January 1991, the Central Bureau of Statistics estimated that the number of persons who qualify as "ethnic minorities" totaled some 850,000 (5.7 per cent) of a total population of about 15m.

Most of the migrants have arrived over the last 30 years, and the major groups are those of Surinamese ethnic origin (244,200), Turks (207,000), Moroccans (167,000) and about 40,000 Moluccans (Ammersfoort, 1992). The Surinamese and the Moluccans are a product of the Netherlands colonial history. Not surprisingly, the immigrants are overwhelmingly concentrated in the four major cities. Ammersfoort (1992) states that the number of aliens and ethnic minorities in the four cities totals some 411,000 (21 per cent). This proportion excludes suburban jurisdictions, and the city boundaries are quite tightly defined. The proportions vary from 15 per cent in Utrecht, 20 per cent in the Hague and Rotterdam, reaching some 24 per cent in Amsterdam. Migrants are strongly concentrated in certain areas, particularly the older inner-city areas and some new estates (Dieleman, 1993), but Ammersfoort (1992, p. 450) states that "The Bijlmermeer is the only spot in Amsterdam where we can find in high-rise flats an absolute concentration resembling the segregation phenomena that are so familiar from American cities. In Amsterdam we see no concentration that according to our definition could be called a 'ghetto'."

But Ammersfoort and Dieleman both point to the fact that, because of the demographic structure of immigrant groups on the one hand and the Dutch population on the other, some neighbourhood schools have a very large ethnic composition. The fundamental problem facing many immigrants is that with the recessions of the 1980s there has been a sharp rise in unemployment, and ethnic minorities have much higher rates of unemployment than others. Roelandt and Veenman (1992, p. 139) conclude that while

> there is no ethnic underclass in the Netherlands ... an ethnically and culturally heterogeneous group of Turks, Moroccans and some Surinamese and Antilleans ... who are concentrated in the inner city neighbourhoods of the country's larger cities, are at risk of becoming an underclass.... There is little upward social mobility of any kind whatsoever and labour market prospects, especially for young Turks and Moroccans, are bleak.

It is not possible to be sanguine about the bleak prospects for immigrants in the Dutch labour market, but the situation is clearly very different in scale from that in the US.

CONCLUSIONS

It has been argued that Sassen's thesis of growing social polarisation in global cities is flawed on several counts. First, her concept of polarisation is vague and undefined, shifting between occupational and income polarisation and fails to distinguish between absolute and relative social polarisation. Secondly, it fails to engage adequately with existing work on social change, and the evidence of large-scale professionalisation in the occupational structure of Western societies and many global cities. Thirdly, it appears to be based primarily on evidence from New York and Los Angeles, both of which are distinguished by the presence of large-scale, continuing immigration and hence a large supply of low-wage workers. Its applicability in other contexts appears to be more limited and data on the Netherlands suggest that, although there is some evidence of income polarisation, occupational and education change in Randstad point strongly towards professionalisation. As Feinstein *et al.* (1992, p. 13) point out, "If the concept of the 'dual' or polarising city is of any real utility, it can serve only as a hypothesis, the prelude to empirical analysis, rather than a conclusion which takes the existence of confirmatory evidence for granted".

REFERENCES FROM THE READING

Ammersfoort, H. van (1992) "Ethnic residential patterns in a welfare state: Lessons from Amsterdam, 1970–1990," *New Community* 18: 439–456.

Atzema, O. and Smidt, M. de (1992) "Selection and duality in the employment structure of the Randstad," *Tijdschrift voor Economische en Sociale Geografie* 83, 4: 289–305.

Bell, D. (1973) *The Coming of Post-Industrial Society*, New York: Basic Books.

Clarke, W.A.V. (1993) "Urban restructuring from a demographic perspective," *Economic Geography* 63: 103–125.

Dieleman, F. (1993) "Multicultural Holland, myth or reality?" in R. King (ed.) *Mass Migrations in Europe*, London: Belhaven.

Feinstein, S., Gordan, I. and Harlow, M. (1992) *Divided Cities*, Oxford: Blackwell.

Gershuny, J.I. (1978) *After Industrial Society? Emerging Self-Service Economy*, London: Macmillan.

Glass, R. (1973) "The Mood of London," in D. Donnison and D. Eversley (eds) *London: Urban Patterns, Problems and Policies*, London: Heinemann.

Hamnett, C. (1976) "Social change and social segregation in inner London, 1961–71," *Urban Studies* 13: 261–291.

Hamnett, C. (1986) "The changing socio-economic structure of London and the South-East, 1961–81," *Regional Studies* 10, 5: 391–406.

Hamnett, C. (1987) "Social–tenurial polarisation in London and the South-East," *Environment and Planning A*: 537–556.

Jobse, R.B. and Musterd, S. (1992) "Changes in residential function of the big cities," in: F.M. Dieleman and S. Musterd (eds) *The Randstad: A Research and Policy Laboratory*, Dordrecht: Kluwer Academic Publishers.

Kloosterman, R.C. (1991) *A Capital City's Problem: The Rise of Unemployment in Amsterdam in the 1980s*, Amsterdam: Department of Economics, University of Amsterdam.

Kloosterman, R.C. and Lambooy, J.G. (1992) "The Randstad: A welfare region," in F. Dieleman and S. Musterd (eds) *The Randstad: A Research and Policy Laboratory*, Dordrecht: Kluwer Academic Publishers.

Ley, D. (1992a) "Gentrification in recession: Social change in Canadian inner cities, 1981–86," *Urban Geography* 13, 3: 230–256.

Ley, D. (1992b) *Gentrification and the Politics of the New Middle Class*, unpublished ms.

Marcuse, P. (1989) "Dual city: A muddy metaphor for a quartered city," *International Journal of Urban and Regional Studies* 13, 4: 697–708.

Roelandt, T. and Veenman, J. (1992) "An emerging ethnic underclass in the Netherlands? Some empirical evidence," *New Community* 19: 129–141.

Sassen, S. (1984) "The new labour demand in global cities," in M.P. Smith (ed.) *Cities in Transformation*, Beverley Hills, CA: Sage.

Sassen, S. (1985) "Capital mobility and labor migration: Their expression in core cities," in M. Tinkerlake (ed.) *Urbanization in the World Economy*, New York: Academic Press.

Sassen, S. (1986) "New York City: Economic restructuring and immigration," *Development and Change* 17: 85–119.

Sassen, S. (1987) "Growth and informalization at the core: A preliminary report on New York City," in J.R. Feagin and M.P. Smith (eds) *The Capitalist City*, Oxford: Blackwell.

Sassen, S. (1988) *The Mobility of Labour and Capital*, Cambridge: Cambridge University Press.

Sassen, S. (1990) "Finance and business services in New York City: International linkages and domestic effects," *International Social Science Journal* 125: 287–306.

Sassen, S. (1991) *The Global City: New York, London. Tokyo*, Princeton, NJ: Princeton University Press.

Van Kempen, R. and Van Weesep, J. (1989) "High incomes, low incomes. and the divided city: Social polarisation in the Netherlands." Paper presented at 7th Urban Change and Conflict Conference, Bristol.

Van Kempen, R. and Van Weesep, J. (1992) *Housing in the Dutch Welfare State*, Stetpro Working Papers.

Editors' references and suggestions for further reading

Fainstein, S. and Campbell, S. (1996) "Introduction: theories of urban development and their implications for policy and planning," in S. Fainstein and S. Campbell (eds) *Readings in Urban Theory*, Oxford: Blackwell, pp. 1–22.

Marcuse, P. and van Kempen, R. (eds) (2000) *Globalizing Cities: A New Spatial Order?*, Oxford: Blackwell.

Sassen, S. (2001) *The Global City: New York, London, Tokyo*, Princeton, NJ: Princeton University Press (first published 1991).

T
W
O

"From Colonial City to Globalizing City? The Far-from-complete Spatial Transformation of Calcutta"

from P. Marcuse and R. van Kempen (eds)
Globalizing Cities: A New Spatial Order? (2000)

Sanjoy Chakravorty

Editors' Introduction

For much of its history, urban geography has made a simple and largely taken-for-granted distinction between the "First World" cities of Western, capitalist nations and the "Third World" cities of developing areas such as Africa, Asia, and Latin and South America. This division of the world's cities clearly has some relevance given that there are a variety of distinctive issues that characterize Third World urban development, including the scale and scope of the informal sector, extensive shanty-style settlements, and rapid population increase but slow economic growth. However, the use of the terms "First" and "Third" World cities is also highly misleading. These terms imply a degree of *homogenization* in relation to the experiences and characteristics of urban development within both the "First" and "Third" Worlds which is simply unwarranted given the huge diversity that exists in terms of the historical legacies and contemporary socio-economic, political and cultural environments that exist within the developed and developing worlds. In addition, however, these terms also imply that there is a degree of separation between urban development in the First and Third Worlds when both past and present experiences of globalization indicate that there are myriad ways – economically, politically and culturally – in which cities in the developed and developing worlds are interconnected (see Robinson 2002). It is this interplay between urban development and the global economy which provides one of the central themes of Sanjoy Chakravorty's analysis of the changing spatial structure of Calcutta.

Although not a world or global city in the sense discussed in the readings by Beaverstock *et al.* and Hamnett in this section, Calcutta is one of the world's so-called mega-cities given its population of over eleven million people. Moreover, although Calcutta is "off the map" of the world city network (see Robinson 2002), Calcutta's historical and contemporary development has, as Chakravorty clearly shows, been influenced just as much by its connections with the global economy as by those cities that do form part of the world city hierarchy. During the colonial period and particularly from the eighteenth century onwards, the city rose to became a significant trading centre with the British Empire and by the end of the nineteenth century was the British capital in India. After independence in 1947, the city faced a period of what Chakravorty describes as "benign neglect" when Delhi replaced Calcutta as the political capital and Bombay/Mumbai became the centre of economic growth. From the mid-1980s onwards, however, Calcutta has benefited from political reforms that have made it easier for foreign capital to enter the country and the hope is that such political liberalization will dovetail with economic globalization to produce significant foreign investment. As Chakravorty shows, these changes in the nature of India's engagement with the world economy over time have been etched into Calcutta's' urban landscape. During the colonial period the city was deeply divided, reflected in the spatial segregation of colonizers and natives. In the immediate post-colonial period, these spatial divisions survived with the native upper class now occupying the spaces once reserved for the colonizers. The post-reform city that has developed over the past twenty years, however, displays a more complex socio-spatial geometry but includes important new elements, such as the building of New Calcutta, a new town on the fringes of the metropolitan area and targeted at a growing class of professional workers.

The connections which Chakravorty makes between the changing spatial structure of Calcutta and the city's links to the global economy nicely exemplify King's claim (1990) that "all cities are world cities." King's argument is that the "real development of London or Manchester" cannot be understood "without reference to India, Africa, and Latin America," just as "the development of Kingston (Jamaica) or Bombay [and we would add Calcutta] cannot be understood without reference to London or Manchester" (1990: 78). At the same time, however, Chakravorty also shows that it is important not just to privilege the role of global. Calcutta's colonial and post-colonial history reveals the complex and dynamic interplay between global influences and more "local" processes operating at national (India-wide) and regional (West Bengal) levels (see also Dick and Rimmer, 1998). Chakravorty's analysis also provides a necessary corrective to the "first world elitist bias to the globalization literature" (Short *et al.* 2000: 317). Chakravorty's frustration with the ethnocentrism that informs so many discussions of globalization is made clear in his introduction when he notes how the "default framework" for understanding contemporary urban change is the transition from a Fordist to a post-Fordist economy. This model is plainly inappropriate to India and many other developing countries given that "deindustrialization in the west actually implies its opposite in the Third World, i.e. increased industrialization, presumably in its urban centers."

Chakravorty's selection comes from *Globalizing Cities: A New Spatial Order?* edited by P. Marcuse and R. van Kempen (Oxford: Blackwell, 2000) which provides a series of case studies exploring the different ways in which globalization is reshaping the internal structure of cities in the developed and developing worlds. The traditional approach to studying cities in the developing world is exemplified in texts such as Josef Gugler's *The Urbanization of the Third World* (Oxford: Oxford University Press, 1988) and David Drakakis-Smith's *Third World Cities* (London: Routledge, 2000). A more recent attempt to place Third World cities in the context of world city approaches is contained in Josef Gugler's edited collection *World Cities Beyond the West: Globalization, Development and Inequality* (Cambridge: Cambridge University Press, 2004). In terms of post-colonialism, Jane Jacobs' *Edge of Empire: Postcolonialism and the City* (London: Routledge, 1996) provides a critical analysis of how the politics of post-colonial processes are etched into the landscapes of First World cities.

Sanjoy Chakravorty is Associate Professor in the Department of Geography and Urban Studies at Temple University in Philadelphia. After a degree in Civil Engineering at Jadavpur University in Calcutta he went on to complete a Ph.D. in Urban and Regional Planning at the University of Southern California. His main research interests focus on welfare and social change in less affluent societies.

■ ■ ■ ■ ■ ■

Like the proverbial Hindi deity, Calcutta has had many names: "city of palaces" (in the 19th century), "city of dreadful night" (Kipling's description at the turn of this century), "city of joy" (in the dreadful book and movie of recent years), "dying city" (by the late Rajiv Gandhi, Prime Minister of India 1984–89); its recent rulers have proclaimed that they would like the city to be known as the "gateway to the Asian tigers" (in media promoting the investment virtues of the city and the state).

In postmodern parlance these many names reflect the many histories and realities of the city – its colonial past, industrial decline, and hope for resurgence in the present and near future. Interestingly, these names also hint at the many geographies of the city – its palaces and hovels, wealth and poverty – and, analyzed chronologically, the names offer some insight into the spatial structure of the city.

[. . .]

[T]he default model [of a newly emerging spatial order within globalizing cities appears] to have been proposed, to a large extent, with the "western" city and economy in mind, especially its American variant. This default framework uses the new industrial divide, or the transition to a post-Fordist economy as the fundamental element defining the past and present structure of urban areas. The argument, very simply put, is that industrialization- and manufacturing-led economic growth created the "old" urban structure; deindustrialization- or service-sector-led global economic expansion is in the process of creating a "new" urban structure.

There are a number of reasons why this model cannot begin to apply to "Third World" contexts in general and its cities in particular. The so-called Third World encompasses a great diversity of development levels, political-economic structures and histories, and levels of integration into the global economy; their city (and country) sizes are diverse, their city functions are rarely comparable to developed nation city functions, their public sector is much more active in urban land markets, the CBD is more important as the locus of employment, rent gradients are more unilinear and steep moving away from the CBD – all factors leading to distinct monocentric cities as opposed to the clearly established polycentric cities of the west. Above all, deindustrialization in the west actually implies its opposite in the Third World, i.e., increased industrialization, presumably in its urban centers. In fact, a more appropriate argument is that there is no singular "Third World". . . . This case history is not the place to debate these arguments and questions; the focus is on the Calcutta story as a case history and should not to be used to draw generalizations about India or the Third World. . . .

It is clear that the two-stage Fordist/post-Fordist model cannot adequately describe the economic and urban development of India, particularly its colonial cities (and perhaps some other once-colonized Third World nations and cities). Rather, a three-stage model may be more appropriate, where the three stages are:

1. colonial economy during the first global period;
2. post-colonial (or command) economy during the nationalist period; and
3. post-command/reform economy, during the second global period.

The relationship between colonization and urban development in the colonized countries has been discussed quite exhaustively (see King, 1976). Many of these cities, usually ports, were created specifically for colonial extraction: i.e., to act as points of transhipment of commodities from the colonized region and processed goods to it, and as seats of administration. Their primary links were to the international economy rather than to the regional economy.

In the post-colonial or nationalist phase the idea of "development" as opposed to exploitation came to the fore; key ideas such as import-substituting industry, big push, infant industry protection, balanced growth, self sufficiency, etc., dictated the policies of the relatively inward-looking newly formed nation-states. And now, a combination of the failures of import-substituting industrialization in the south and the demand for new markets and production centers in the north, have forced many developing nations *back* to the global market. Post-Fordism does not describe this period as accurately as the term mixed-Fordism does, for large-scale capital-intensive production still has to take place somewhere – in the current global shift, in an ironic turn of events, it is the developing nations which are home to Fordist industry. I argue that these three stages are characterized by distinct modes and relations of production and investment, policy and goals, and consequently they also characterize distinct spatial forms. . . .

However, the Calcutta story would be poorly understood without mention of two factors that, though not unique to itself, are rather different from developed nation contexts. First, I must highlight the importance of the size and function of the "informal" sector in Calcutta's economy: by most estimates this collection of urban workers comprises 40 per cent or more of the Calcutta labor force, in occupations from garbage collection, material transport, home delivery of consumer products, to small crafts and manufacturing (leather products, printing, etc.). The notion of formal "flexible production and accumulation" . . . that some scholars argue is reshaping urban space in the developed world has long been an aspect of the conditions of production in Third World cities like Calcutta. One sub-sector of this informal economy is of particular interest in analyzing the spatial distribution of income and wealth: the domestic servants,

ubiquitous in upper- and upper-middle class residences, have lived and continue to live in close
proximity to their employers. I will argue that it is
essential to have an understanding of the spatial
distribution of the informal sector, particularly
the domestic service element, and its relationship
to capital and technology, to understand the geography of poverty and affluence in Calcutta.

[...]

THE SPATIAL STRUCTURE OF CALCUTTA

The colonial city

At inception, the colonizers (then merely traders)
sought to establish terms of trade favorable to
the home country; later, after gaining complete
territorial control, they used the colonized regions
as sources of raw material to be processed in
industrial England, and as captive markets for
the processed products. The colonial city was a
center of administration, a port, and a European
residential enclave. This city's structure (as shown
in Figure 1) was deeply divided – the important
spatial divide being that between the colonizers
(living in high amenity, well serviced areas) and
the natives (living in unplanned, congested, poorly
serviced areas). Describing a model of the South
Asian colonial city Dutt (1993, p. 361) writes:

> The European town ... had spacious bunga-
> lows, elegant apartment houses, planned streets,
> trees on both sides of the streets ... clubs for
> afternoon and evening get-togethers. ... The
> open space was reserved for ... Western recre-
> ational facilities, such as race and golf courses,
> soccer and cricket. When domestic water supply,
> electric connections, and sewage links were
> available or technically possible, the European
> town residents utilized them fully, whereas their
> use was quite restricted to the native town.

Calcutta city, then, started growing around an
empty core (the fort and Maidan), with the English
town growing south and south-west of Park Street
(see Figure 1), an area of Eurasian and mixed-
marriage residences (i.e., an Anglo-Indian enclave)
immediately to its north, and the area further north
and east being occupied by the natives (working

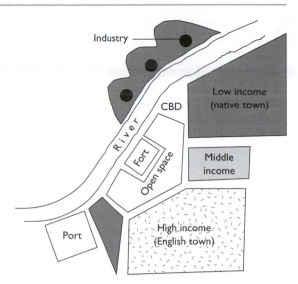

Figure 1 The spatial structure of colonial Calcutta.

as merchants and traders, and as clerks for the
British administrative system). However, it would
be wrong to presume that this spatial division by
race and class was strictly enforced. The first wave
of poor migrants to the city did not come to work
in factories (because none existed), but to service
the lavish lifestyles of the British. They settled in
small slums within the English town, "or how
else could the rich get servants, cooks, darwans,
chowkidars, cleaners, gardeners, dhobis and the
rest? Labour was abundant and cheap and it paid
to keep the slums within the city, in fact nearer the
mansions" (Munshi, 1975, p.111). Bardhan Roy
(1994) argues that the domestic services were
needed from early morning to late at night, and as
a result the dwellings of the poorest could be seen
within walking distance of most luxurious areas of
the city. I shall show later that this basic structure,
created in the 18th century, still dominates the
spatial pattern of work and home in the city.

The second wave of migrants came seeking
employment in the engineering units and jute and
textile mills that began west of the river from the
middle of the 19th century, and to work for large
transportation projects (Haora station in 1854,
Sealdah station in 1856, the Calcutta tramways, and
the Kidderpore docks). All of these activities (except
the tramways) took place on the fringes of, or outside the city boundaries of the time; the new slum
areas, as a result, began growing in these fringe
areas. Not surprisingly, the low-income occupations

were somewhat linguistically segregated: Bengalis in the clerical professions, Biharis as rickshaw pullers, porters, and factory labor, Oriyas in domestic service, and plumbing, gas, and electrical works. Their slums also tended to retain occupational and linguistic identities, as did Muslim slums (specializing in labor for soap and leather factories on the eastern and southern fringes). This structure within the city was replicated in miniature in the riverside industrial suburbs; these had small high-amenity areas (large estates as living space during the work-week for the British owners and managers), surrounded by a small middle-income area, and low-income areas where the factory labor lived.

The post-colonial city

With the achievement of independence in 1947, the spatial divisions of the colonial city (demarcated by class and race barriers) were largely retained, with the native upper class (capital and land owners, political leaders and top government officials) now occupying the privileged space once reserved for the colonizers. The refugee inflow from East Pakistan, however, introduced an unexpected spatial twist. As Goswami (1990, p. 92) writes:

> The influx of refugees really brought the city's elite face-to-face with the urban problems that were brewing for a long time. In the first place, unlike previous migrants, who were clearly subalterns, the typical displaced families were vocal and considered it a political right to be gainfully re-settled in the city. They belonged to the same culture background as the city's intelligentsia, and demanded to be heard. Second, they settled in areas that were perilously close to affluent South Calcutta neighborhoods: Behala and Chetla bordered Alipur, Kasba and Dhakuria were just next to Baligunj and Gariahat, Jadabpur was not too far from the mansions of Southern Avenue, and the brown and white sahibs could no longer go to play golf without seeing the slums in Tollygunj. The new urban poor could not be put out of sight in the unmentionable parts of north Calcutta.

The inherited (colonial) space was divided into quarters, or ghettoes: British, mixed-race, and native

town bordering the center (with slums interspersed in every quarter). The new (post-colonial) space retained much of this inheritance with the race divisions being replaced by class divisions. In addition, population pressure forced the city to grow outward, with the farthest areas being occupied by the low-income population. As depicted in Figure 2, the quartered structure of the colonial city was replaced by concentric half-circles, with income declining with distance from the center. However, as also suggested in Figure 2, the area north of the empty core is missing the middle-income ring. This area, adjacent to the CBD and Burrabazaar (a large wholesale market), was the native town during the colonial period; its infrastructure deficiencies have increased over time, and the absence of planning and investment here is evident in its congested lanes and by-lanes, open drainage, and generally miserable living conditions. As expected, this area (north Calcutta) does not house the elite or the upper-income population.

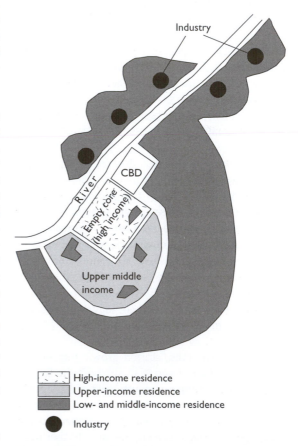

Figure 2 The spatial structure of post-colonial Calcutta.

The spatial distribution of the low-income population . . . [reveals] the proliferation of slum areas in the far south (Jadabpur). . . . This is not surprising, since as suggested earlier, north Calcutta has been completely congested for some time; any growth in poor areas has had to take place in the south (which is also a refugee stronghold, and a bastion of leftist politics). Two additional points should be noted: first, many slums have located on the least desirable public land (along railway tracks, and open sewerage and drainage lines); and second, the location of Calcutta's poorest, its pavement dwellers (or homeless population), is in many ways the opposite of that of the slum population. An extensive survey of the pavement dwellers was done in 1971 during the Census of India. This survey, and one carried out by the CMDA in 1987 found about 50–55,000 homeless persons in the city, and about four times that number in the metropolis. . . . The city homeless were (and still are) concentrated in the CBD, Burrabazaar, and Chowrangee, and are also found in the high-income areas of Park Street and Alipur. As expected, there are few pavement dwellers in the slums.

Some residential segregation by occupation, religion, caste, and ethnicity continued into the post-colonial period. Two points should be noted in this regard. First, the ghettoes are not large (i.e., unlike in the United States where the black population in cities like Detroit and Philadelphia is concentrated in large contiguous areas). For instance, one can find (in east Calcutta) a low-caste Hindu leather-worker bustee of say 15,000 people adjacent to an equally large Muslim leather-worker or tailor bustee. Second, this pattern of spatial separation is not confined to the poor: the business elite, which is generally non-Bengali, occupies the center; of special interest are the Marwaris (a group of very prosperous entrepreneurs from Rajasthan) who tend to live in enclaves in the Burrabazaar and Park Street areas. Professional South Indians tend to reside around the Lakes, and professional Bengalis live in south Calcutta.

The post-reform city

India's structural reform is a relatively recent event. Some spatial changes in the seven post-reform years are noticeable, but, at this moment, it is difficult to foresee the post-reform spatial structure with certainty. First, there is no guarantee that the reform process will continue in its present form, though it increasingly seems that a significant change has taken place in a society traditionally slow to change. The aftermath of the 1996 and 1998 elections suggests that regardless of the ideology of the group in power, the reforms will continue, and that it may be impossible to return to the centralized nationalist development ideology of the recent past. Second, our singular interest in economic outcomes in spatial terms may blind us to a perhaps more significant transformation in Indian society, where there is increasing (and more acceptable) social, cultural, and technological polarization. The reforms are significant (just as independence was) in more than economic terms. It has raised a number of unresolved intellectual and political questions about nationalism, regionalism, governance, decentralization, inequality, and secularism. The resolution of these questions may influence the spatial structure of urban society as deeply as do the economic actions of domestic and global actors. Therefore, given the absence of structural stability, and the lack of hard data for the post-reform period, I have to rely on declared intentions and plans, and my often idiosyncratic personal observations to formulate the following speculative analysis.

The most significant new spatial component of the reforms in Calcutta are its new town projects, particularly New Calcutta. . . . New Town projects in the Calcutta metropolis have a long history. In the late 1960s, following the recommendations of the Basic Development plan, one was created at Kalyani . . . at great expense and to resounding failure. Kalyani was a planned city where the state government would have relocated, but for the fact that the government employees refused to move. The city still has paved streets overgrown with weeds, and street lamps that were never lit – a perfect example of a planning disaster. Salt Lake, a new town closer to the city, was begun in the mid-1970s. This upper-income enclave is considered successful – it has a population of around 150,000 now, and is expected to grow to 250,000. Salt Lake has no slums; its residents' biggest and most persistent complaint concerns the difficulty of obtaining affordable and reliable domestic servants (the old bourgeois complaint that "good help is so hard to find"). Many state government offices have relocated to the Salt Lake township (the Chief Minister has moved his residence there from south Calcutta), and many of the region's electronic production units are also located there.

New Calcutta can be expected to be successful for the same reasons that Salt Lake has been successful: it is close enough to Calcutta city for a relatively easy commute (for employment or services), and as a planned development it will bypass the city's ills – poor infrastructure, slums, and poverty. This new town will have 100,000 dwelling units (spread over 8.4 sq. km.), 1.5 sq. km. for a new business district and commercial complexes, 2.2 sq. km. for "modern, pollution free industries", and 13.1 sq. km. of water bodies and green areas (including a golf course). A strong selling point of this new town is to be its proximity to Calcutta's recently expanded international airport – clearly the planners want this development to contain new (rather than relocated) industry, of the type that is high-tech and/or global in nature.

Can New Calcutta succeed without slums, or will its success depend on its ability to keep out slums? I believe that the answer to this question will partly lie in the degree of capital–labor substitution in the sphere of domestic production. Day (1992) discusses the capital–labor relationship in an analysis of housework in north America in the 20th century. She argues for a progressive model in which increasing industrialization led to higher wages, more women in the work force, and the availability of domestic appliances like the range, refrigerator, washer, dryer, and vacuum cleaner. As the supply of servants (usually recent immigrant women) fell, the wealthiest households continued to hire servants, but the middle-income groups did without, and substituted capital for labor. In India, the most visible signs of liberalization are colas, fast food, and domestic appliances – specifically the washer, dryer, and microwave oven (the refrigerator has been around for some time, and the vacuum cleaner is available, but generally considered unnecessary). New Calcutta is clearly designed for professional upper-income earners, the group most likely to adopt these household labor-saving devices. If that happens, New Calcutta will look like a "modern" city, what is sometimes called a postmodern city in the US context (Charlotte, NC, for example) – clean, spacious, and free of visible poverty.

A second major development is the growth of heavy industrial investment in Haldia (about 50 straight km. from Calcutta), and the so-far less successful Falta Export Processing Zone located between the two. As of July 1998 expected capital investment (largely in petrochemicals) in this port-city was about $4 billion (my calculations). If expectations are met, and it seems likely that they will be, Haldia will become a rather large industrial city. This city appears to be modeled after the Fordist growth centers like Durgapur and Bhilai, which, revealingly, are made up of "colonies" named after specific corporations; e.g., A VB colony, MAMC colony. That is, like most other planned developments in India, the city design will keep the informal sector and the poor spatially separated from the middle- and upper-income formal sector workers. As argued above, keeping the poor out may now be technologically feasible, but separating the dynamic and essential informal sector may be the seed of failure as a growth center, even if Haldia succeeds as an industrial enclave.

WHERE DOES THE CALCUTTA STORY FIT?

Calcutta's spatial structure cannot be separated from its political-economic history. This history has been influenced strongly by global and local events. On the one hand, Calcutta's genesis and early morphology was defined by the global force of colonialism: "chance selected, chance directed" the city grew as a center of colonial exchange and administration in inhospitable urban terrain – a silting river, salt marshes all around it, unstable soils unable to carry heavy loads, in a very poor rice-growing hinterland. It never acquired a strong industrial base; its predominant industrial commodity, jute, was made technologically obsolete not long into this century, and the city's global trading links withered with the downfall of its primary product. On the other hand, the more influential events of the 20th century, as far as the city is concerned, have been local or regional in character. While independence may be viewed as a global event (the end of the colonial system), the impact on Calcutta was in its lost hinterland, and the flood of refugees. In this Calcutta is different from even the other comparable Indian colonial cities,

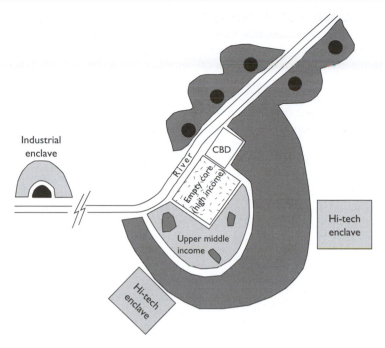

Figure 3 The spatial structure of post-reform Calcutta.

Mumbai and Chennai, not to speak of colonial cities worldwide. Thereafter local politics, policies and events (Freight Equalization, communist infighting, the refugee influx from the independence of Bangladesh) have held center-stage.

Now the city has been reintroduced, willy-nilly, to the global system of production and exchange. The question raised here is what will happen, or is happening, to the internal structure of the city as a result of its reintegration into the global economy. I believe that the answer will depend substantially on the degree of integration of the local economy in the global market. It appears unlikely that Calcutta will soon become a "world city" in Hall's (1966) terms, or that it is on the way to becoming a "global city" in Sassen's (1994) terms; that is, one cannot expect Calcutta to become either a global center of production (aircraft, ships, military hardware) or services (banking, insurance). The declared goals of the state's leaders are more modest – to become a center of large-scale production of petrochemicals, leather, pharmaceutical, metallurgical, and engineering items within the Southeast Asia region, and to compete globally in the electronics field (mainly computer software and hardware). The local state wants to evenly distribute the location of production facilities, but clearly expects such units to

converge around the infrastructure advantages in and around Calcutta (in Haldia, Falta EPZ, Kulpi, and the city's industrial suburbs).

There are several possible spatial outcomes. First, the city and state may utterly fail to integrate in the global economy ("a loser city"); this may imply spatial status quo and possibly increasingly miserable conditions for the city's poor. At the moment of writing this appears to be the outcome for the foreseeable future. Second, both goals (in the Fordist manufacturing and post-Fordist service arenas) may be successfully realized, leading to a spatial scenario as outlined in Figure 3: high-tech, high-income planned enclaves on the eastern edge of the city, and planned industrial enclaves further south, resulting in the creation of a considerably larger agglomerative field, or metro-region. Intermediate outcomes are also possible, whereby the city succeeds as a center of Fordist production, but fails in post-Fordist terms, or vice versa. (Truth be told, my feeling is that the city would rather be a Fordist success, if it could choose only one area of success.) And, given the city's history as a refugee haven, one should not rule out the possibility of events beyond the control of the local state (like war, famine, and natural disaster). Remember that Calcutta is the city of last resort for a largely rural population of over 300 million people (including Bangladesh). Through all these possibilities, except the disaster scenarios, the core city is unlikely to change much in terms of the spatial distribution of wealth and poverty.

Does the Calcutta story fit a model? I am not sure. The city is quite different from its colonial counterparts – the more segregated, hierarchical, monolingual Chennai, or the dynamic, polyglot, recently chauvinistic Mumbai. Calcutta's leftist political leaders do not demolish bustees in high-income areas (though, in a desperate move to clean up the city's image they have started conducting street sweeps to round up pavement hawkers – a fascinating story in itself). The city has the reputation of being more hospitable to refugees and migrants than any other in India. The Hindu–Sikh and Hindu–Muslim riots of the 1980s and early 1990s barely touched the city. Perhaps globalization will change all that, with increased hardening of spatial boundaries between income, language, caste, and religious groups. Certainly the bourgeois planning apparatus has worked and continues to work for the benefit of the upper classes. If the liberalization process continues, and is accompanied by a larger role for urban planning (as in planned developments in New Calcutta and Haldia), and the adoption of labor-saving household devices, one could see increased spatial separation between rich and poor in the new enclaves. Barring a dramatic economic turnaround, however, the city proper is likely to retain much of its present structure – perhaps not quite unique, but certainly one that cannot easily be fit into a model.

REFERENCES FROM THE READING

Bardhan Roy, M. (1994) *Calcutta Slums: Public Policy in Retrospect*, Calcutta: Minerva Publications.

Day, T. (1992) 'Capital–labor substitution in the home', *Technology and Culture* 33: 302–27.

Dutt, A. (1993) 'Cities of South Asia', in S.D. Brunn and J.F. Williams (eds) *Cities of the World: World Regional Urban Development*, New York: HarperCollins, pp. 351–87.

Goswami, O. (1990) 'Calcutta's economy 1918–1970: the fall from grace', in S. Chaudhuri (ed.) *Calcutta: The Living City*, Calcutta: Oxford University Press, pp. 88–96.

Hall, P. (1966) *The World Cities*, London: Heinemann.

King, A.D. (1976) *Colonial Urban Development: Culture, Social Power and Environment*, London: Routledge.

Munshi, S.K. (1975) *Calcutta Metropolitan Explosion: Its Nature and Roots*, New Delhi: People's Publishing House.

Sassen, S. (1994) *Cities in a World Economy*, Thousand Oaks, CA: Pine Forge Press.

Editors' references and suggestions for further reading

Dick, H.W. and Rimmer, P.J. (1998) "Beyond the Third World city: the new urban geography of South-east Asia," *Urban Studies* 35(12): 2303–21.

Jacobs, J. (1996) *Edge of Empire: Postcolonialism and the City* (London: Routledge).

King, A. (1990) *Urbanism, Colonialism and the World-economy* (London: Routledge).

Robinson, J. (2002) "Global and world cities: a view from off the map," *International Journal of Urban and Regional Research* 26(3): 531–54.

Short, J.R., Breitbach, C., Buckman, S., and Essex, J. (2000) "From world cities to gateway cities: extending the boundaries of globalization theory," *City* 4: 317–40.

Yeoh, B.S.A (1996) "Street-naming and nation-building: toponymic inscriptions and nationhood in Singapore," *Area* 28: 298–307.

Yeoh, B.S.A. (2001) "Postcolonial cities," *Progress in Human Geography* 25(3): 456–68.

"Cultural Globalization and the Identity of Place: The Reconstruction of Amsterdam"

from *Ecumene* (1999)

Jan Nijman

Editors' Introduction

Jan Nijman, a geographer at the University of Miami, returned to his home town of Amsterdam in 1996 to find a city that had noticeably changed during his decade-long absence. Amsterdam, ranked as a "secondary world city," functions as an important center in the world economy and consequently one would expect rapid growth and change. What intrigued Nijman, however, was the extent to which two spatially separate Amsterdams had evolved. These two Amsterdams are not the "dual cities" of social polarization examined in Hamnett's article (p. 74). Instead, one corresponds to the city as a basing point in the world economy while the other, the old center, operates as a defining place in the global cultural landscape by representing values distinctive to "local culture" (Nijman 1996; see also Claval 1993). Nijman explores this second Amsterdam in the following article by examining one aspect of cultural globalization, specifically considering the extent to which mass tourism has reconstituted local culture as it shapes the city's image.

Nijman focuses initially on the theoretical conceptualization of cultural globalization. As he notes, one fundamental challenge of the world city approach is the linking of the local to the global. Much of the literature, particularly that written by geographers, conflates cultural and economic globalization by emphasizing the materialist forces behind social change. Labelled "double-determinism," this perspective makes local culture primarily subject to global economics, implicitly de-emphasizing global cultural forces as dependent on economic determinants (see, e.g., Knox 1995). Thus, Nijman argues, the world city approach confuses the increased exposure of cities to both economic and cultural global processes with the special roles of particular cities as basing points in the wider economy. In other words, "the world city approach does not make a conceptual distinction between the city as object and subject in the process of globalization" (1996). Nijman urges that the processes which lead to the emergence of world cities as command centers in the world economy (the city as subject) be considered separately from the effects of cultural globalization on cities (the city as object). The geographical pattern that underlies the former is relatively simple, since some places are clearly more important than others in the geography of the world economy. In the latter case, however, the "geographical pattern of globalization is more complex, chaotic and diffuse" because cultural globalization involves a multitude of flows and counterflows. Thus, while the contribution of individual cities might be difficult to measure, the impact of cultural globalization on any particular city is potentially great.

The rapid pace of the globalization of ideas, values and lifestyles rests on both the relative ease of their distribution through information technologies and the increased flow of people, particularly migrants and tourists. Anthropologist Arjun Appadurai (1996) identifies the *relocation* of culture through flows of people and the *ethnoscapes* produced through this process, as the primary dimension of cultural globalization. Certainly, the impact of these transplanted and hybridized landscapes is more dramatic than cultural change among relatively isolated, sedentary people who are primarily dependent on access to mass information media. Nijman alters this concept of ethnoscape by suggesting that the flow of international tourists to Amsterdam influences the cultural landscape of the old center through the tourists' expectations of the city's tolerance for sex and drugs. While the cultural traits historically associated with Amsterdam are Calvinism, commercialism and tolerance of diversity, Nijman argues that commercialism exerts the most persistent influence that recasts and perhaps perverts the city center where historical tolerance is made to eclipse Calvinist values. Although international visitors assume they experience "authentic" aspects of Amsterdam's civic culture, Nijman suggests that increasingly tourists move about the streets primarily observing other tourists. Skillfully demonstrating the manner in which the old center has become a tourists' theme park that reflects the city's commodification as a tolerant place, he offers this Amsterdam as an illustration of the dynamic process of "glocalization." That is, the *local* entrepreneurial efforts capture the tourist trade by promoting *customary traits* that are then transformed in the context of *globalization*.

The city of Amsterdam is the focus of a number of essays that offer varying perspectives as they engage in a critical analysis of global restructuring and the world city hypothesis (Claval 1993; Fainstein 1996; O'Loughlin 1993; Soja 1993). To further explore the significance of tourism and efforts to marshal cities' cultural capital, see Kearns and Philo's *Selling Places* (1993), and Briavel Holcomb's work on marketing cities for tourism (1999).

Nijman continues his interest in world cities research with two major projects. One deals with Miami in the context of the urban order (2000a) as he challenges the traditions of the Chicago School (see Burgess, p. 19), and the possible postmodern alternative offered by the LA School (see Dear and Flusty, p. 138). The other reflects his concern with the Western bias in the globalization debate, as he and his research partner, Richard Grant, examine economic restructuring in cities in the less developed world (Grant and Nijman 2002). Nijman's particular contribution to this area of research focuses on Mumbai, India (i.e. 2000b) as he joins geographers such as Chakravorty (p. 84) in arguing that the globalization debate is not nearly as "global as it should be."

■ ■ ■ ■ ■ ■

In his opening speech at an urban studies conference a few years ago at the University of Amsterdam, the mayor of the city joked about Amsterdam's reputation abroad as the 'twin cities of Sodom and Gomorrah'. As he was addressing a group of academicians and urban planners, he did not dwell on the subject, though, and went on to elaborate on Amsterdam's more serious historic and modern achievements. The image of Amsterdam as a haven of moral permissiveness is indeed pervasive at home and abroad, and seems to overshadow whatever else this city is about. It is an image of relatively recent origin, even if it concerns a place with a much longer history.

Amsterdam provides a useful case study of the effects of cultural globalization on urban identities, in view of its centuries-long history as a city, and its reputation as a very international or global place. Moreover, the city has in past years been a popular destination for mass global tourism, and it appears to figure rather prominently (or at least recognizably) as a place in the global cultural landscape. In the *process*, the identity of Amsterdam seems to have been reconstituted to fit the description of the mayor's twin cities. This reidentification of place and reconstitution of local culture involves an intricate blend of authenticity and artificiality, to such an extent that it becomes difficult to know the 'real' Amsterdam.

The next part of the paper provides a theoretical discussion of cultural globalization as distinct from economic globalization. It uses the notion of 'glocalization' to underscore the global–local dynamics involved in the changing culture and identity of localities in the context of globalization. This section also relies on Appadurai's concept of 'ethnoscapes' to emphasize the importance of flows of people (particularly mass tourism) as vehicles of cultural globalization. The third part of the paper describes the historically grown urban identity of

Amsterdam since the sixteenth century and prior to the era of mass tourism of the last few decades. It concentrates on such widely observed cultural traits as Calvinism, commercialism and tolerance of diversity. This is followed by a discussion of Amsterdam's re-created identity in recent decades, mostly in relation to global tourism.

CULTURAL GLOBALIZATION AND THE IDENTITY OF PLACE

Popular and scholarly definitions of globalization tend to emphasize the economic dimension of the process. This certainly seems true for globalization as it relates to urban issues. In as far as the so-called 'world city' literature represents a corollary of the globalization debate, it too conforms to this penchant for economic determinism. Thus, the role of cities in the *process* of globalization is mainly viewed in economic terms. At the same time, *culture* tends to be viewed – often implicitly – as the dependent variable. The materialist orientation of much of the literature is accompanied by the notion that the social and cultural aspects of cities are often influenced by the world economy. . . . Such materialist perspectives tend to negate the importance and relative autonomy of global cultural forces that affect cities. When global cultural forces are considered, they are often viewed as dependent on economic globalization. . . . This bias often coincides with a lack of attention to the importance of localities in the globalization debate. The result may be dubbed 'double determinism': local culture is subordinate to global economics.

Alternatively, cultural globalization is viewed by some as a more autonomous process, and even as enabling some forms of economic globalization – although the relevance of cultural globalization to urban spaces has thus far received scant attention. Cultural globalization may be defined as an acceleration in the exchange of cultural symbols among people around the world, to such an extent that it leads to changes in local popular cultures and identities. According to a relatively small number of authors, globalization is indeed primarily about culture. For example, King states that globalization theory is concerned with 'questions of culture, identity, and meaning in-representations of the world as a single place'. The sociologist Malcolm

Waters points out that globalization has proceeded most rapidly 'in contexts in which relationships are mediated through symbols'. The exchange of symbols, in turn, depends in large part on available communication technologies, and has accelerated rapidly in recent decades. This also explains why, within the economic domain, the globalization of financial markets is proceeding much more rapidly than the globalization of material commodities markets. More pertinently, it also explains the impressive globalization of cultural values, ideas and lifestyles. . . .

An important component of this process is the consumption culture that originated in the United States and that has been mass-mediated to places around the world. This is not so much about similarities across the world in the consumption of particular *products* (e.g. Coca-Cola). It is rather about the similar social function of consumption in hitherto different societies. . . . Seen from this angle, one could argue that economic globalization, in the form of consumption of globally available commodities, is predicated and dependent on the globalization of cultural values and identities.

The acceleration of processes of cultural globalization in recent years can be explained through the increased availability of mass communication, transportation and information technologies (air travel, radio, newspapers, telephone, television, cinema, Internet). The availability of these media at the *mass level* is critical, if the process is to affect broad-based local urban or regional cultures. Increasingly, then, 'consumption' takes on a broader meaning: it is not just about the consumption of material commodities but also about the consumption of ideas, values and information.

Popular globalization debates put great emphasis on modern information technologies . . . as the key media of global cultural flows. An equally important development concerns the enormous expansion in the availability of transportation . . . that has facilitated a vast increase of *flows of people*. Appadurai . . . identifies the flows of people, particularly migrants and tourists, as the primary dimension of cultural globalization, what he terms 'ethnoscapes'. In the case of flows of people, the diffusion of cultural innovation (or the globalization of culture) takes in part the form of *relocation*: we are not talking now about people who remain in the same place and are subjected to global flows of information, but about flows of people themselves, in the process de-territorializing themselves. At the same time, it is people themselves (not computers or television) that serve as media of information, as the global transmitters of the images of localities.

The *geography* of cultural globalization and the role of cities remains something of a puzzle.... The emergence of world cities as *economic* command centres reflects the globalization of capital accumulation which is orchestrated and controlled from base points in the global economy. The geographical pattern that underlies this process is relatively simple, with control emanating from selected urban areas to the rest of the world's spaces. In the geography of 'control' of the world economy, a few places, such as New York, London or Tokyo, are much more important than most others. In contrast, the geographical pattern of *cultural* globalization is more complex, chaotic and diffuse. Places vary in their importance as cultural centres, but cultural globalization is not controlled from a small number of easily identifiable urban areas. Instead, cultural globalization concerns a complex multitude of flows and counter flows. It is increasingly difficult to determine the originality of a cultural flow, because it is likely to represent a bundling or modification of earlier cultural influences from other places.

Cities are important spatial entities in the process of cultural globalization because they are the places with concentrated mass populations and with the most advanced mass transportation and communication facilities. Cities constitute the main nodes in the global cultural network and are the main points of origin and destination of cultural flows. However, as argued earlier, the pattern of cultural globalization is highly diffuse and it is not as strongly controlled from a few selected urban centres. At the same time, every city that is more or less connected to global information networks will be more or less *affected* by cultural globalization: as a result, both its local culture and its image or identity are subject to change.

How can the effects of the globalization process on local (urban) cultures and place identities be understood? First, the increased access and exposure to mass communication media and international travel result in *individualization*: the assertion of people's individual identity as distinct from territory-based group identity. This process is closely related to changing lifestyles and consumption patterns. In the words of Knox:

> The lingua franca of this global metropolitanism is the patois of American television soap operas and comedy series; its dress code and world-view are taken from MTV and the sports pages; its politics is from the cyberpunk Mondo 2000; and its lifestyle is defined by promotional 'spots' for Budweiser, Coca-Cola, Levis, and Nike.

[...]

However, cultural globalization does not necessarily imply homogenization and the emergence of a single global culture. Instead, there is good reason to assume continued geographical variation of cultures. King argues that globalization does not simply eradicate local urban cultures. The process is rather more complex: The powerful forces of globalization ... on major world cities also result in the intensification of cultural nationalisms.... Clearly, the effects of globalizing forces are dependent on and vary with local context. This intricate relationship between globalism and localism has been dubbed 'glocalization', and while the term smacks of rhetorical overkill, it does capture the essence of the process under study. Globalization may simultaneously lead to dilution of local culture (globalization as the great 'equalizer') *and* to a deepening of particularity. These opposite effects concern different aspects of the local culture: some aspects of the local culture may disappear while others are strengthened.

The same is true for the identity of a place, particularly the ways in which it is known in other parts of the world. Some aspects of the 'personality' of a place may be reinforced in the process of globalization, while others may become underrepresented. Especially in localities with a long history and embedded identity, such as Amsterdam, globalization is likely to induce both continuity and change, each along different dimensions. The city, as a spatial entity, provides an interesting point of entry in debates on cultural globalization because it is the place where global cultural exchanges are concentrated, *and* at the same time the image of the city itself may constitute an item of cultural consumption in this global exchange....

AMSTERDAM'S HISTORIC IDENTITY

There can be little question that Amsterdam has since the sixteenth century been at the geographical core of Dutch society, whether it concerns economics, politics or culture. It is and always has been the largest city, and for a long time it was by far the most important port city. It was surpassed in the latter capacity only in the industrial era by Rotterdam, now considered to be the largest port in the world according to some measures. Amsterdam has been the gateway of a small and outward-looking country, heavily reliant on foreign trade. Importantly, the city of Amsterdam strongly dominated the country along almost any conceivable dimension at the time of the country's formation as a nation-state in the late sixteenth century, and during the Dutch 'Golden Age' in the seventeenth century. This is why Dutch culture in general and the urban culture of Amsterdam are so closely intertwined. The general characteristics of Dutch culture are often especially salient in Amsterdam and sometimes, indeed, 'made' in Amsterdam.

Amsterdam is part of the north-western region of Europe where, it is sometimes said, 'capitalism was born'. According to a well-known thesis of the sociologist Max Weber, this propensity for capital accumulation was closely related to the region's prevailing protestant work ethic, which encouraged saving and investment and frowned upon lavish spending and hedonistic life styles. This religion-based culture found its most succinct expression in Calvinism, a term that is still often used to describe one of the foundations of Dutch culture.

Calvinism inspires sobriety and simple living, and it has had profound effects on other aspects of Dutch culture, including the built environment. Amsterdam was the most powerful and richest city in the world economy in the early seventeenth century, but there is little in its urban design to remind us of that. If one considers the architectural grandeur of cities such as Venice, Paris, Rome, Madrid, or even London and Berlin, Amsterdam's landscape pales in comparison. The city's built environment is beautiful in many ways, but majestic it is not. Amsterdam lacks the spacious squares and wide avenues, the grand buildings, imposing statues or arches, and other public displays of wealth and artistic and cultural achievement. Even some of the most expensive mansions built during the Golden Age along the Herengracht . . . evoke the image of solidity and durability rather than glamour and splendour . . . a trait that pervades Dutch culture, and that is directly derived from Calvinism.

Moreover, Amsterdam was never dominated by royalty and was never the seat of national government. Especially during the Golden Age it was, in the words of the Dutch historian Jan Romein, 'a republic in a republic'. Amsterdam's relationship to The Hague as well as the cultural differences between the two cities are in some ways similar to the comparison between New York and Washington, DC. In past centuries, The Hague was the city of princes and kings, while Amsterdam was the city of the commercial patrician class – and there was not much love lost between them . . . Amsterdam was a hegemonic city, but it was never a great imperial city. Amsterdam has always been, in the words of historian Donald Olsen, 'a city entirely dedicated to making money'.

[. . .]

If Calvinism, commercialism and frugality are central ingredients of Amsterdam's historically grown culture, so is 'tolerance' . . . specifically tolerance of diversity in the form of religious beliefs and national backgrounds. Back in the sixteenth and seventeenth centuries, Amsterdam and the northern Netherlands successfully resisted Spanish rule and, with it, religious intolerance. The city's Jewish population grew rapidly at that time, especially after the fall to the Spanish of Lisbon in 1580 and of Antwerp in 1585. Through the years, these Jewish immigrants were joined by French Huguenots, Flemish, Lutheran Germans, Portuguese, Jews from east and central Europe, etc. In the final years of the sixteenth century, up to half of the city's population was foreign or of foreign extraction. The tolerant attitude of the city's population and leadership may at that time have been partially motivated by commercial considerations: in particular the Jews from Antwerp, many of whom had Portuguese origins, brought with them considerable wealth and entrepreneurial skills, and they are often said to have contributed significantly to Amsterdam's Golden Age. . . .

Finally, a fifth element of Amsterdam's urban culture that is often observed is 'egalitarianism'. Whether this is a quality of Amsterdam or pertains to Dutch political culture at large is not so clear. . . . The earliest roots of 'egalitarianism' date back to the late nineteenth century in the form of the emergence of democracy, the social agendas of political parties and the rise of the unions. In the twentieth century the Netherlands became a notable example of West European social democracy, with progressive laws on taxation, education, public housing and so forth. This form of egalitarianism reached its peak in the 1970s when the social welfare state – as it is now fashionable to argue – had expanded to unaffordable proportions. While all these trends found expression in the city of Amsterdam (e.g. at present about three-quarters of all housing in the city is public), they were all determined at the national scale and were not particular to Amsterdam.

What is special about Amsterdam, especially after the Second World War, is its role as an anti-establishment city (again, this is particularly clear in a comparison of Amsterdam with The Hague). In the late 1960s, it was by far the most important place for student demonstrations, political upheavals, the sexual revolution, grass-root activity, etc. In later years it was in Amsterdam that squatter movement first took hold and 'coffee shops' started to sell soft drugs. Part of this side of Amsterdam's urban culture was related to the large student population (it was the only city in the country with two large universities) – a student population, furthermore, that was in large part from a working-class background, made possible by the Dutch welfare state. During the 1960s, 1970s and 1980s, urban social movements such as the 'Provos', the 'Kabouters' and the squatters have played an important part in the changing image of the city, particularly inside the Netherlands, but their impact on the city's power structures is debatable. Indeed, one could argue naturally that from a historical point of view Amsterdam's egalitarian and revolutionary climate emerged quite late in its existence, but also that it may already be weakening. At the present, Amsterdam's grass-roots movements seem to be taking a back seat as the city's social climate is threatened by the dismantling of the Dutch welfare state while the urban and national economies are expanding rapidly. Amsterdam's economic growth machine and commercial establishment in the late 1990s seem more powerful than they have been in a long time.

GLOBALIZATION, MASS TOURISM AND THE REIDENTIFICATION OF AMSTERDAM

There can be little question that many inhabitants identify with Amsterdam as *their* city, a reflection of the importance of place-identity at the urban level. This is true even though many long-term Amsterdammers have in fact left the city to move to suburban towns like Almere. Indeed, it is the city itself that has through the ages acquired an identity, one that is acknowledged and adapted by many newcomers and temporary inhabitants from elsewhere in the Netherlands (e.g. the transient student population), and even by many foreign tourists.

Nowadays, the aspect of Amsterdam's urban culture and identity that receives by far the most attention is 'tolerance'. This is the case in the popular media, but also in the writings of urban scholars. One of the effects of cultural globalization for Amsterdam has been the commodification of its identity as a tolerant place. This is primarily related to tourism, one of the main fields of competition among European cities and a key source of income in Amsterdam. This process is a neat illustration of 'glocalization', in the sense that it constitutes 'a return to local culture' in the face of globalizatian. It is a local process that is at the same time reactionary to and exploitative of the constraints and opportunities that globalization offers, and this is where Amsterdam's traditional 'dedication to making money' comes into play. The end result is not a return to the culture of tolerance of the past, but the emergence of a globally conditioned modified version of that culture.

Foreign visitors no longer come to Amsterdam to seek refuge from persecution, in search of economic opportunity or to marvel at the city's achievements as a 'laboratory of modernity' as they did in the seventeenth century. Instead a growing number comes to 'let it all hang out' and enjoy the entertaining spectacle of 'tolerance'. Tolerance, perhaps Amsterdam's most prized commodity, is increasingly packaged and labelled to meet the demands of mass tourism and instant gratification. In the process it has become something of a perversion, in the sense that it turned into commercially motivated permissiveness that is in fact contrary to the city's Calvinistic roots. Abroad, Amsterdam is first and foremost associated with liberal attitudes towards sex and drugs, and this has become its niche in the competitive world of tourism. Local governments neither condemn nor openly advertise this image. Astonishment or criticism from abroad is usually greeted with a mixture of amusement and pity for the alleged narrow mindedness of the foreigners, and by a slightly more concealed sense of satisfaction with regard to the city's exceptionalism and marketability.

Global tourism has thus eroded some crucial parts of Amsterdam's place authenticity. Easily transmitted notions and images of Amsterdam as a city of tolerance may be deceptive. Fainstein remarks that in Amsterdam, 'tourist attractions mainly mix with the components of ordinary life rather than being segregated into a single zone'. She goes on to say that the various forms of entertainment 'give Amsterdam an edge that guards it from succumbing to the sanitized sameness of theme-park-style development'. I would argue, instead, that this mix of tourist attractions and ordinary life is not what it used to be, and that 'ordinary life' has been reshaped to form part of a decor of tolerance to accommodate the tourists. Along the way, a significant part of the city centre has evolved precisely into a sort of a 'theme park' – one with the appearance of authenticity that American master-developers could only dream about!

I should not take this point too far, but it is tempting to contrast the worn depictions of Amsterdam's 'authentic' character by foreigners (especially Americans), who are used to more blatantly 'artificial' forms of entertainment and who seem so easily charmed by the city's appearance, with a more sceptical view that considers the changed character of the city centre. At the risk of a little exaggeration, but only a little, the following is an account of how the central city presented itself to me as a former Amsterdammer upon my return in the autumn of 1996, after nine years' residence in the United States.

The theme of the 'park', which operates at maximum capacity in summertime, is not movies, water-world, high technology or Mickey Mouse, but drugs and sex under the pretence of normalcy. It is not a planned theme park, of course; instead it has emerged more or less spontaneously. It has no fence around it and there is no entry fee. Then again, this is not necessary, because payment is per consumption. Even if there is no gate, the main exit of the central station has the appearance of one. Large numbers of new visitors gather on the square in front of the station, exchanging information on things to do before they start their adventure with a walk down the main avenue into the 'park'.

The Damrak, where Amsterdam's traditional architecture is hidden behind a facade of advertisements, offers an overwhelming array of tourist services and, without wasting any time, shows off the city's famed tolerance by way of a sex 'museum'. Further into the 'park' the attractions include the Hash Museum, the Cannabis Connoisseurs' Club, innumerable coffee shops and houses of prostitution, sex shops, sex cinemas, condomeries, peep-shows, magic mushroom shops, etc. . . . The 'park' is largely located within Amsterdam's historic (seventeenth century) centre, and [is] about the size of planned theme parks in the United States. . . .

A key difference from American theme parks is the target audience: while the American variants are meant to provide family entertainment, Amsterdam is geared towards young adults. . . . It is in large part a self-selected crowd that 'fits' the city's image. As a consequence, these generally low-budget tourists tend to blend in and become themselves part of the entourage. Through their presence in the city, they play a role in the recreation of Amsterdam's image as a haven of 'tolerance', but one that bears little relation to the city itself, or at least to what it used to be.

Virtually all the labels and designations of the attractions are in the English language, but this is evidently too obvious a phenomenon to arouse suspicions about their authenticity. By now, a large majority of the 'window-prostitutes' are foreign and, like most of their customers, have no communal ties whatsoever with the area. Many sex shops sell, besides their specialized merchandise, foreign-language newspapers such as *USA Today* and *Das Bild*. On the Leidscheplein and other places, tourists look at tourists, not at Amsterdam. Or, perhaps more accurately, they are looking at what the centre of Amsterdam has become. This is quite a different 'centre of attraction' from the one that emerged from a study in the late 1960s. At that time, the centre was described as a place of and for Amsterdammers; foreign tourism never even entered the discussion. Of course, the important role of the inner city as a cultural centre for Amsterdammers and the Dutch population as a whole has continued, but this is in large part disconnected from the sex-and-drugs theme park. Few foreign tourists ever make it from the Bulldog coffee shop to the old Stadsschouwburg, a theatre some 50 yards across the Leidscheplein that draws a native crowd. The flow of people the other way around seems negligible as well.

. . . Geographically speaking, there has been a clear break in the historical development of the city. The Amsterdam that presently functions as a node or base point in the world economy is the Amsterdam of the periphery, increasingly on the south-eastern edge of the city. In Amsterdam, as elsewhere, there has been a rapid acceleration of the suburbanization of businesses and producer services, fuelled by problems of congestion in the centre and by the growing importance of up-to-date telecommunication infrastructure. In contrast to most American cities, however, burdened as they are with inner city decay, Amsterdam's historical centre is vital, and specializes increasingly in consumer services, particularly tourism, education and the entertainment industry.

The reconstruction of Amsterdam's culture of tolerance is not confined to the main tourist areas or to those who work in the tourism sector . . . [T]olerance cannot exist without cultural diversity. And even though most parts of the city are indeed quite diverse and Dutch culture at large still represents a notably tolerant and consensus oriented society, Amsterdam's city centre – of all places – presents a different picture.

The population residing in the centre of Amsterdam is not very diverse, particularly that part of the population that has the most visible role in Amsterdam local urban culture. First, it has increasingly become a demographically homogeneous population. The overwhelming majority are in their twenties and thirties, live alone and have no children. In addition, most are Dutch, with the immigrants (mostly Turks and Moroccans) residing primarily in the city's nineteenth century neighbourhoods or further out. Finally, there seems to be substantial homogeneity in terms of subcultural ways of life. For example, insofar as Amsterdam's urban civil culture has a distinct penchant for grass-roots activism, this has been most manifest in the inner city and among the inner-city population.

The centre of Amsterdam is not very tolerant or 'user-friendly' towards families with young children or the elderly. More pertinently, the city's famed tolerance seems increasingly geared towards tolerance of a very specific subculture and forms of behaviour. In the 1960s, Amsterdam shocked with its revolutionary elan and alternative social behaviours. Today, Amsterdam 'shocks' in a more predictable direction: the Gay Parades on the city's canals, the Cannabis Cup for best home-grown weed, or the Hash Bash, an outdoor marijuana fair, are examples. It is not new but rather more of the same, a more extreme version of the old.

[. . .]

The reconstruction of Amsterdam's identity is partially a matter of narratives that have become part of global information flows. Local advocates of the city emphasize what they think the foreign visitors want to see and hear. The mayor's joke about Sodom and Gomorrah is no exception. Tourist guides and brochures never cease to sell Amsterdam as a city where anything goes, where avant-gardism comes naturally, and where the status quo does not exist. Respectable local authors and journalists sometimes join the chorus. In the (English language!) *Penguin Guide to Amsterdam*, Martin Van Amerongen proclaims without hesitation that 'the Amsterdammer is an adherent of the doctrine of permanent revolution'. On its cover the book is advertised, rather ominously, as 'the ultimate guide to the real Amsterdam'. The foreign media repeat these clichés. Whenever the *New York Times* reports on Amsterdam, there is a good chance it has something to do with the red light district or with coffee shops.

There is no denying that Amsterdam today presents foreign visitors with a culture of tolerance and liberalism that is not found in most other places. Importantly, in Amsterdam these traits are said to [be] authentic. Tolerance is indeed a part of Amsterdam's historic identity, but that identity entailed much more. . . . It is *commercialism* that is the most persistent aspect of Amsterdam's continuing identity. In the seventeenth century this commercialism bred tolerance towards wealthy foreign migrants. In recent years it has been the driving force behind the transformation of the city centre into a sex-and-drugs theme park. Whatever sells. And sell it does, especially because foreign consumers have no doubt that they are getting the real thing.

CONCLUSIONS

Mass tourism and the globalization of information flows have in the past three decades or so contributed to the reconstruction of Amsterdam's identity. Globalization does not erase local cultures, but it does undermine their original form. In cities such as Amsterdam, with a long history and established identity, there is of course continuity in some ways. It is the weight of history that makes its presence felt, a history that is in part stored in the built environment. In the words of Anthony King, 'The identity of these localities and the built environment that helps to create it, are instrumental in explaining why such continuities persist.' But, as we have seen in the case of Amsterdam, there may be significant change as well. These changes are not always obvious; especially because so many cities (particularly in Europe) sell themselves to foreign tourists on the basis of their historical past and their authenticity.

Today's cultural representation of 'Amsterdam' is divorced from the city's commercial functions and character, something that was not the case in past centuries. This point is underscored by the present spatial separation of Amsterdam's economic and cultural roles in the world. . . . The urban functions of material production, command and control in the European and world economy, and producer services, are situated in the city's rapidly developing periphery, particularly in the south-eastern outskirts. In the seventeenth century these functions were concentrated in the city centre, an area that now specializes almost exclusively in consumer services and foreign tourism.

Amsterdam is not the only city to undergo these effects of cultural globalization in conjunction with mass tourism, even if it is an extreme case. Comparable examples might range from Key West to Kathmandu. What we can observe with a fair degree of certainty is the growing superficiality of urban identities. These identities are increasingly determined in the realm of globally transmitted sound-and-vision-bites, and this is in turn reflected in the local culture itself. It seems that mass tourism and the globalization of information flows result inevitably in an increasingly shallow understanding of local cultures and identities. In the process, the localities themselves turn into caricatures or mutant reflections of their past. While popular knowledge of the world's cultural landscape is increasingly global, understanding of individual places becomes increasingly shallow. The inevitable result, it seems, is the vulgarization of the world's cultural geography.

Editors' references and suggestions for further reading

Appadurai, A. (1996) *Modernity at Large: Cultural Dimensions of Modernity*, Minneapolis: University of Minnesota Press.

Claval, P. (1993) "The cultural dimension in restructuring metropolises: the Amsterdam example," in L. Deben, W. Heinemeijer, and D. van der Vaart (eds) *Understanding Amsterdam: Essays on Economic Vitality, City Life and Urban Form*, Amsterdam: Het Spinhuis, pp. 69–92.

Fainstein, S. (1996) *The Egalitarian City: Images of Amsterdam*, Amsterdam: Amsterdam Study Centre for the Metropolitan Environment.

Grant, R. and Nijman, J. (2002) "Globalization and the corporate geography of cities in the less-developed world," *Annals of the Association of American Geographers* 92: 320–40.

Holcomb, B. (1999) "Marketing cities for tourism," in S. Fainstein and D. Judd (eds) *The Tourist City*, New Haven, CT: Yale University Press.

Kearns, G. and Philo, C. (eds) (1993) *Selling Places: The City as Cultural Capital, Past and Present*, Oxford: Pergamon Press.

Knox, P. (1995) "World cities in a world system," in P. Knox and P. Taylor (eds) *World Cities in a World System*, Cambridge: Cambridge University Press, pp. 3–20.

Nijman, J. (1996) *The Global Moment in Urban Evolution: A Comparison of Amsterdam and Miami*, Amsterdam: Amsterdam Study Centre for the Metropolitan Environment.

Nijman, J. (2000a) "The paradigmatic city," *Annals of the Association of American Geographers* 89: 135–45.

Nijman, J. (2000b) "Mumbai's real estate market in the 1990s: de-regulation, global money, and casino capitalism," *Economic and Political Weekly* (India) 35, 7: 575–82.

O'Loughlin, J. (1993) "Between Sheffield and Stuttgart: Amsterdam in an integrated Europe and a competitive world-economy," in L. Deben, W. Heinemeijer, and D. van der Vaart (eds) *Understanding Amsterdam: Essays on Economic Vitality, City Life and Urban Form*, Amsterdam: Het Spinhuis, pp. 25–68.

Soja. E. (1993) "The stimulus of a little confusion: a contemporary comparison of Amsterdam and Los Angeles," in L. Deben, W. Heinemeijer, and D. van der Vaart (eds) *Understanding Amsterdam: Essays on Economic Vitality, City Life and Urban Form*, Amsterdam: Het Spinhuis, pp. 69–92.

PART THREE

Restructuring

Detroit's Renaissance Center with an old manufacturing site in the foreground. (Courtesy of Elvin Wyly)

INTRODUCTION TO PART THREE

The term "restructuring" tends to be applied mainly in the context of economic transformations and specifically in relation to the activities of capitalist firms and businesses as they seek to adjust to changing circumstances and maximize profit-making opportunities. As the previous section illustrated, such economic restructuring at a global level is having significant implications locally in terms of patterns of urban development in the developed and developing worlds. In this section, we develop this theme of the interplay between economic and urban restructuring further but narrow the spatial focus a little in order to emphasize the local dynamics of economic transformations and how these affect the configuration of urban space in advanced capitalist societies. Given the emphasis on capitalist economic processes, many urban geographers working on restructuring draw their theoretical inspiration directly from Marxist scholarship, and the first three readings in this section (by Harvey, Walker and Lewis, and Smith) exemplify this approach. Many of the more recent analyses of restructuring, however, have involved a critique of Marxism, and the fourth reading in this section (by Dear and Flusty) illustrates this in the form of a postmodernist perspective.

With Marxism providing such an important framework within which urban geographers have engaged with issues of restructuring, it is useful to consider briefly the historical context of Marxist analysis. London, where Marx was based for much of his life, was undergoing tremendous economic and social change in the nineteenth century and, witnessing these developments at first hand, Marx was in no doubt about the strong connections between economic and urban development. Not only does capitalism give rise to the industrial city but "the logic of capitalist development pulls into the metropolis the vast mass of wage labourers together with a 'reserve army' of the unemployed and underemployed on which the affluence of the bourgeoisie must depend" (Parker 2003: 103). This link between urban growth and economic development was given a sharper geographical focus by one of Marx's collaborators, Engels. In his 1844 study of Manchester, Engels (1971) highlighted how the city's residential structure, comprising a series of concentric zones radiating out from the city centre, was underpinned by the city's capitalist economic structure. The inner zone, nearest the city's central business district, contained "working peoples' quarters," the slums where the working class was crowded together. This was surrounded by the areas occupied by the middle bourgeoisie, while on the outskirts lived the upper bourgeoisie in "free wholesome country air, in fine, comfortable homes." Reflecting on this configuration of residential space, Engels observed wryly that the presence of "an almost unbroken series of shops" on the main routes into the centre of Manchester meant that "the members of the money aristocracy can take the shortest road through all the labouring districts without ever seeing they are in the midst of the grimy misery that lurks on either side to the right and left . . . the misery and grime which form the complement of their wealth."

This early attempt to incorporate space into the analysis of the capitalist city was largely overlooked within urban geography until the 1970s. Up to this point, as the Introduction indicates (pp. 1–2), urban geography was still strongly influenced by a mix of ecological, neoclassical and behavioral approaches (Bassett and Short 1989). However, by the mid-1970s, there were a growing number of critics of these perspectives, concerned at the privileging of individual choice and the neglect of wider constraints imposed by socio-economic structures. Of these critics, the most influential has been David Harvey, the author of the first reading in this section. Drawing on Marx's own analysis of capital accumulation, Harvey focuses on the pivotal role of the built environment of the city in the context of the circulation of capital and capitalist economic crises. In particular, Harvey shows how if there is a surplus of capital in the production or "primary circuit of capital," capital may then flow into the "secondary circuit" in the form of investment in land and the physical infrastructure of the city. Although this may aid capital accumulation and stave off an economic crisis at one point in time, such investment in the built environment may well become an obstacle to capital accumulation in the future. For Harvey, then, economic transformation and uneven urban development are viewed as crucially interlinked elements in the reproduction of capitalist society. Largely as a result of Harvey's Marxist-inspired theoretical and empirical work on the links between economic and urban restructuring under capitalism, "urban geography became a key entry point of radical ideas into the discipline" of human geography as a whole (Johnston and Sidaway 2004: 223), and the influence of Harvey's work may be clearly seen in Walker and Lewis' study of suburbanization (the second reading) and Smith's analysis of gentrification (the third reading).

With its origins in the early nineteenth century when the urban elite attempted to distance themselves from the crowding and pollution found in rapidly industrializing cities, suburbanization has long been understood from the perspective of the neoclassical and behavioral models discussed in the Introduction (p. 2). These approaches have emphasized the key role of innovations in transport technology and the residential preferences of affluent socio-economic groups in explaining suburban growth. By contrast, Walker and Lewis shift the focus from issues of culture and consumption to questions of economics and production in their analysis of suburban development in late nineteenth-century America. Industrialization, investment in real estate and the political influence of business leaders rather than the aspirations of middle-class groups are the key drivers of suburbanization for Walker and Lewis. Underlining the importance of economic transformation rather than cultural aspiration in explaining suburbanization, Walker (1981) has also investigated post-1945 suburban development in the USA in an account which also clearly shows the influence of Harvey's selection. According to Walker, post-war investment in suburban real estate reflected the way the economic boom at the end of World War II had created problems of over-accumulation (and thus declining profits) in the manufacturing sector, leading to a switch in investment into property development as a mechanism for, temporarily, resolving these problems and stimulating consumption of housing, cars and consumer durables.

Although in the third reading the spatial focus shifts from residential development on the urban periphery to the redevelopment of inner-city areas for affluent, middle-class populations as part of a process known as gentrification, the analytical focus remains firmly on the economic rather than cultural dimensions of urban development. Indeed, while at first glance gentrification might seem unrelated to suburbanization, viewed from Smith's neo-Marxist perspective both are firmly linked by underlying processes of economic transformation. Capital investment in suburban development in the post-1945 period in many developed countries starved the inner city of economic resources, contributing to the devalorization of the land market in central urban areas. Smith contends that by the 1970s this contributed to a significant "rent gap" (the gap between the rent currently being obtained and the potential rent under a higher use) which provided the opportunity for capital reinvestment in the inner city at the same time as an expanding middle-class population provided new demands for housing. The result was gentrification. Since this period, processes of gentrification have been identified in cities across the globe, ranging from the redevelopment of working-class areas in industrial cities such as Cleveland and Glasgow, to the conversion of factory spaces into domestic lofts in Manhattan (Zukin 1988; see also Atkinson and Bridge 2004). Indeed, Smith has recently described gentrification as a global urban strategy (2002). Nevertheless, as with suburbanization, while descriptions of the nature and scope of gentrification are broadly agreed, explanations of gentrification remain strongly contested. Consumption-based explanations, with their roots in neoclassical and behavioral approaches and which emphasize the cultural preferences of certain affluent social groups for inner-city living (see Ley 1996), continue to vie with production-based explanations, such as those of Smith, which stress the role of economic cycles of capital investment and disinvestment in creating opportunities for property developers in inner-city neighborhoods. More recent contributions by urban geographers, however, have emphasized that it would be misleading to view gentrification simply though the analytic lens of class relations. Knopp (1998), for example, has linked gentrification to expressions of gay sexual identity, while Bondi (see reading in Part 5, Difference) highlights the important contribution of women in professional and managerial occupations to the gentrification process.

These more pluralistic perspectives on gentrification are indicative of a broader shift in approaches to restructuring which emerged in urban geography in the 1990s. A growing number of geographers believe that Marxism, with its emphasis on the underlying logic of capital accumulation in explaining urban development, is increasingly out of touch with contemporary processes of urbanization. According to the geographer Edward Soja, the failure of Harvey and other Marxists to be "more open to alternative viewpoints" means that "the scope of their insights [into the contemporary capitalist city] has become increasingly narrowed as new streams of thought have entered into urban studies" (Soja 1995: 127). Soja's advocacy for a more pluralistic analytical perspective and his critique of Marxist "meta-theory" are hallmarks of postmodernism which, since the 1990s, has become an important theoretical framework within urban geography. Postmodernist approaches to restructuring eschew grand theory and the privileging of the economic in favor of drawing ideas and concepts from a variety of different analytical traditions. In addition to being a distinctive approach to studying the city, however, postmodernism also refers to a particular style that has influenced not only art and literature, but also, through its celebration of pluralism, contextualism and the human scale, the fields of architecture, planning and urban design. In Part 6, Form and Symbolism, Ley's reading provides a critical examination of these stylistic features of postmodern urban development. But postmodernism may also be thought of as an epoch, a major break with the past, and it is this understanding of postmodernism that is central to Dear and Flusty's argument in the fourth reading. Drawing on a range of research focused on the contemporary restructuring of Los Angeles, they contend that a new form of urbanism has emerged – the postmodern city – which is fundamentally different from the modern city in terms of the processes underpinning its development and the configuration of its spatial structure. The modern city, epitomized in the model of the Chicago School, and based around a central organizing core (see Burgess in Part 1, Foundations) has been displaced, according to Dear and Flusty, by the postmodern city, epitomized by Los Angeles, where "the urban periphery organizes the center within the context of globalizing capitalism." While the modern city was characterized by a series of neat concentric zones and a steady social gradient from inner-city areas of the poor and blue-collar workers out to the sprawling middle-class suburbs, the postmodern city is characterized by a seemingly much more random and complex urban structure. This includes gated communities and edge cities, carceral landscapes of fortified buildings and intensively surveilled public spaces, and fantasy worlds of shopping malls and theme parks. Yet not all urban geographers subscribe to the view that we are witnessing "a new trajectory of urbanization" (Soja 1995: 125) advanced by Dear, Flusty and other members of the so-called Los Angeles School of urban studies. Returning to the themes articulated in the first reading in this section, David Harvey maintains that far from being a radical break with modernism, the "condition of postmodernity" is part of the long-established process of restructuring by which capitalism attempts to stave off economic crises. Harvey accepts that there have been important changes in modes of economic organization in recent years, particularly with regard to a shift from mass production and mass consumption based around large industrial complexes to more flexible production systems, partly facilitated by technological changes. "But these changes," Harvey contends, "when set against the basic rules of capitalistic accumulation, appear more as shifts in surface appearance rather than as signs of the emergence of some entirely new postcapitalist or even postindustrial society" (1989: vii). In common with the other forms of restructuring discussed in this section, then, broad agreement over the contours of restructuring and the changing configuration of space in the capitalist city are often accompanied by deep divisions over the relative importance of economic, political and socio-cultural processes in explaining such restructuring.

References and suggestions for further reading

Atkinson, R. and Bridge, G. (2004) *Gentrification in a Global Perspective*, London: Routledge.

Bassett, K. and Short, J. (1989) "Development and diversity in urban geography", in D. Gregory and R. Walford (eds) *Horizons in Human Geography*, Harlow: Longman, pp. 175–93.

Engels, F. (1971) *The Condition of the Working Class in England*, Oxford: Blackwell.

Harvey, D. (1989) *The Condition of Postmodernity*, Oxford: Blackwell.

Johnston, R.J. and Sidaway, J.D. (2004) *Geography and Geographers: Anglo-American Human Geography since 1945*, London: Arnold.

Ley, D. (1996) *The New Middle Class and the Remaking of the Central City*, Oxford: Oxford University Press.

Knopp, L. (1998) "Sexuality and urban space: gay male identity politics in the United States, the United Kingdom and Australia", in R. Fincher and J. Jacobs (eds) *Cities of Difference*, New York: Guilford Press, pp. 149–76.

Parker, S. (2003) *Urban Theory and the Urban Experience: Encountering the City*, London: Routledge.

Soja, E. (1995) "Postmodern urbanization: the six restructuring of Los Angeles", in S. Watson and K. Gibson (eds) *Postmodern Cities and Spaces*, Oxford: Blackwell, pp. 125–37.

Smith, N. (2002) "New globalism, new urbanism: gentrification as global urban strategy", *Antipode* 34, 3: 428–50.

Walker, R.A. (1981) "A theory of suburbanization: capitalism and the construction of urban space in the United States", in M. Dear and A.J. Scott (eds) *Urbanisation and Urban Planning in Capitalist Society*, London: Methuen, pp. 383–430.

Zukin, S. (1988) *Loft Living: Culture and Capital in Urban Change*, London: Radius.

"The Urban Process under Capitalism: A Framework for Analysis"

from *International Journal of Urban and Regional Research* (1978)

David Harvey

Editors' Introduction

In the late 1960s and early 1970s, events in North American and West European cities had a profound impact on the discipline of human geography in general and urban geography in particular. Anti-war protests and student-worker uprisings made the traditional concerns of human geography seem increasingly out of touch with social and political realities, fuelling a frustration among some geographers at "the irrelevance of using quantitative methods to analyze spatial trivia like shopping patterns or telephone calls" (Peet 1998: 109). One of these geographers was David Harvey. Although *Explanation in Geography* (1969) had made him the guru of positivistic geography, Harvey's move to Johns Hopkins University in Baltimore in 1969 and exposure to the city's stark material inequalities reinforced his perception of an enormous gulf between the theoretical and methodological frameworks of spatial science and geographers' ability to say "anything meaningful about events as they unfold around us" (Harvey 1973: 128). As his essays in *Social Justice and City* (1973) vividly illustrate, such frustration led him initially to explore liberalism and then to embrace Marxism in an attempt to understand the interplay between urban spatial structures and social and economic processes. What specifically interested Harvey was how the expansionary character of capitalism encountered barriers and generated crises which had a fundamentally spatial dimension. In a series of publications in the 1970s and early 1980s, culminating in *Limits to Capital* (1982), Harvey used Marxist theory to develop an ambitious theoretical argument connecting urban restructuring to processes of capital accumulation and class struggle.

In "The Urban Process under Capitalism," Harvey's focus is the "double-edged nature of property for capital accumulation" (Savage *et al.* 2003: 49). Published in 1978, it examines what Harvey views as a fundamental problem of capitalist urban development, namely that investment in the physical infrastructure of the city may aid capital accumulation at one point in time, but, at a later stage, that infrastructure becomes an obstacle to further accumulation. Harvey's starting point is the contention that competition between individual capitalists creates a tendency towards over-accumulation in the "primary circuit" of capital (the manufacturing sector of the economy), that manifests itself in falling prices and rising unemployment. One way this problem can be resolved, Harvey suggests, is by switching investment into the secondary circuit of capital, which comprises "the built environment for production" (such as the transportion infrastructure, factories and offices), and "the built environment for consumption" (such as houses, shops, and streets). Switching investment between these circuits of capital, however, is fraught with problems. It may not be clear to individual capitalists which investments will prove productive, and investment in the built environment involves a long-term commitment to largely immobile assets. Most significantly, rather than resolving crises in capital accumulation, switching investment from the primary to the secondary circuit can cause further crises to develop. Investment in a city's transport infrastructure in one historical period, for example, may aid the efficient movement of commodities and so promote capital accumulation, but then may act as a barrier to accumulation later when new developments in transport technology cannot operate effectively in the existing built environment. More generally, as Harvey observes, "Under capitalism there is . . . a perpetual struggle in which capital builds a physical landscape appropriate to its own condition at a particular moment in time, only to have to destroy it . . . at a subsequent point in time." These contradictions and crises of accumulation are also closely linked to class struggle, because accumulation is the means whereby the capitalist class not only reproduces itself but also its domination over labour. Harvey's focus here is on working-class housing, arguing that policies of individual home ownership and suburbanization for more affluent workers may be interpreted as attempts by the bourgeoisie to encourage social stability ("the working class could be subjected to . . . the 'moral influence' of the suburbs") but also as strategies for using the housing sector as a vehicle for rapid accumulation through commodity production.

Harvey's argument advanced in this paper and developed further in *Limits to Capital* (1982) has proved highly influential within urban geography. His commitment to a structuralist and, more specifically, historical-materialist perspective identifying the underlying mechanisms and powerful interests producing particular spatial forms, clearly addresses some of the weaknesses of behavioral and humanistic geographical approaches to the city with their emphases on individual choice. Studies of suburbanization and gentrification (see readings by Walker and Lewis, and Smith in this section) in particular have drawn on Harvey's theoretical reasoning to illuminate how these processes of urban restructuring are inextricably linked to capitalist economic restructuring. Moreover, Harvey's ideas have been embraced well beyond geography's disciplinary borders by those who view his emphasis on the interplay between capital accumulation and the built environment as crucial to making sense of "dramatic episodes of contemporary urban change" (Savage *et al.* 2003: 51).

Yet critics point to important weaknesses in Harvey's arguments. These partly reflect the almost tautological character of his thesis, making it difficult to disentangle the evidence for the causes of urban change from evidence about urban change itself. As Savage *et al.* (2003: 52) explain,

> Decisions to invest in the built environment can be seen as resulting from a crisis in the primary circuit, which causes a shift of capital to the secondary circuit. How do we know there is a crisis in the primary circuit? Because capital is being switched into the secondary circuit.

Harvey's approach also seems to reify or abstract social processes in ways which gloss over the complexity and contingency of urban development. In particular, his argument assumes a high degree of rationality on the part of the bourgeoisie in the process of accumulation so that they appear as little more than puppets whose actions are determined by deeper structures. Finally, Harvey's argument is also guilty of a degree of reductionism and economic determinism that makes class the key focus of his analysis and ignores the significant role that other social divisions, such as age, gender and race, play in shaping urban landscapes.

In his more recent work Harvey continues to explore the characteristics of capitalist urbanization, and although his analysis remains grounded in an historical-geographical materialism, it is, in some respects, a more flexible form of Marxist analysis than that found in his earlier work. In *The Condition of Postmodernity* (1989), questions of culture occupy a more prominent position in his analysis than previously, and he acknowledges the need for sensitivity to issues of difference and "otherness." In *Spaces of Hope* (2000), he also pushes Marxist thought in new directions, examining the body as an "accumulation strategy," analyzing the consequences of globalization, and, in a case study of Baltimore (the city that had inspired many of his earlier contributions to urban geography), exploring the emergence of modern urban dystopias. Nevertheless, some critics insist that Harvey's work remains locked into the approach that characterized his earlier studies of urbanization in *Social Justice and the City* and *Limits to Capital*: a search for rationality and order in the urban landscape grounded in a universalist, historical-materialist discourse that fails to take seriously issues which cannot not be accommodated by the categories of "class" or "capital" (see Dear 1991). However, as his most recent book *Spaces of Capital* (2002) demonstrates, Harvey, who is currently Distinguished Professor in the Graduate Center of the City University of New York, remains a passionate defender of the insights offered by Marxist analysis and its continued and, he would argue, increasing relevance for understanding the socio-economic and spatial inequalities of capitalist urbanization in the twenty-first century.

My objective is to understand the urban process under capitalism. I confine myself to the capitalist forms of urbanization because I accept the idea that the 'urban' has a specific meaning under the capitalist mode of production which cannot be carried over without a radical transformation of meaning (and of reality) into other social contexts.

Within the framework of capitalism, I hang my interpretation of the urban process on the twin themes of *accumulation and class struggle*. The two themes are integral to each other and have to be regarded as different sides of the same coin – different windows from which to view the totality of capitalist activity. The class character of capitalist society means the domination of labour by capital. Put more concretely, a class of capitalists is in command of the work process and organizes that process for the purposes of producing profit. The labourer, on the other hand, has command only over his or her labour power which must be sold as a commodity on the market. The domination arises because the labourer must yield the capitalist a profit (surplus value) in return for a living wage. All of this is extremely simplistic, of course, and actual class relations (and relations between factions of classes) within an actual system of production (comprising production, services, necessary costs of circulation, distribution, exchange, etc.) are highly complex. The essential Marxian insight, however, is that profit arises out of the domination of labour by capital and that the capitalists as a class must, if they are to reproduce themselves, continuously expand the basis for profit. We thus arrive at a conception of a society founded on the principle of 'accumulation for accumulation's sake, production for production's sake'. The theory of accumulation which Marx constructs in *Capital* amounts to a careful enquiry into the dynamics of accumulation and an exploration of its contradictory character. This may sound rather 'economistic' as a framework for analysis, but we have to recall that accumulation is the means whereby the capitalist class reproduces both itself

and its domination over labour. Accumulation cannot, therefore, be isolated from class struggle.

[. . .]

THE LAWS OF ACCUMULATION

We will begin by sketching the structure of flows of capital within a system of production and realization of value. This I will do with the aid of a series of diagrams which will appear highly 'functionalist' and perhaps unduly simple in structure, but which nevertheless help us to understand the basic logic of the accumulation process.

[. . .]

The primary circuit of capital

In volume one of *Capital,* Marx presents an analysis of the capitalist production process. The drive to create surplus value rests either on an increase in the length of the working day (absolute surplus value) or on the gains to be made from continuous revolutions in the 'productive forces' through reorganizations of the work process which raise the productivity of labour power (relative surplus value). The capitalist captures relative surplus value from the organization of cooperation and division of labour within the work process or by the application of fixed capital (machinery). The motor for these continuous revolutions in the work process, for the rising productivity of labour, lies in capitalist competition as each capitalist seeks an excess profit by adopting a superior production technique to the social average.

[. . .]

The second volume of *Capital* closes with a 'model' of accumulation on an expanded scale. The problems of proportionality involved in the aggregative production of means of production and means of consumption are examined with all

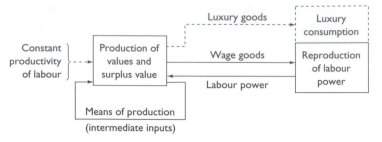

Figure 1 The relations considered for 'reproduction on an expanded scale'.

Source: *Capital*, Vol. 2.

other problems held constant (including techno-logical change, investment in fixed capital, etc.). The objective here is to show the potential for crises of disproportionality within the production process. But Marx has now broadened the structure of relation-ships put under the microscope (see Figure 1). We note, however, that in both cases Marx assumes, tacitly, that all commodities are produced and consumed within one time period. The structure of relations examined in Figure 1 can be characterized as the *primary circuit of capital*.

Much of the analysis of the falling rate of profit and its countervailing tendencies in volume three similarly presupposes production and consump-tion within one time period although there is some evidence that Marx intended to broaden the scope of this if he had lived to complete the work. But it is useful to consider the volume three analysis as a . . . cogent statement of the internal contradictions which exist within the primary circuit. Here we can clearly see the contradictions which arise out of the tendency for individual capitalists to act in a way which, when aggregated, runs counter to their own class interest. This contradiction produces a tendency towards *overaccumulation* – too much capital is produced in aggregate relative to the opportunities to employ that capital. This tendency is manifest in a variety of guises. We have:

1. Overproduction of commodities – a glut on the market.
2. Falling rates of profit (in pricing terms, to be dis-tinguished from the falling rate of profit in value terms which is a theoretical construct).
3. Surplus capital which can be manifest either as idle productive capacity or as money-capital lacking opportunities for profitable employment.

4. Surplus labour and/or rising rate of exploitation of labour power.

One or a combination of these manifestations may be present at the time. We have here a pre-liminary framework for the analysis of capitalist crises.

The secondary circuit of capital

We now drop the tacit assumption of production and consumption within one time period and con-sider the problems posed by production and use of commodities requiring different working periods, circulation periods and the like . . . I will confine myself to some remarks regarding the formation of fixed capital and the consumption fund. Fixed capital, Marx argues, requires special analysis because of certain peculiarities which attach to its mode of production and realization. These pecu-liarities arise because fixed capital items can be produced in the normal course of capitalist com-modity production but they are used as aids to the production process rather than as direct raw material inputs. They are also used over a relatively long time period. We can also usefully distinguish between fixed capital enclosed within the produc-tion process and fixed capital which functions as a physical framework for production. The latter I will call the *built environment for production*.

On the consumption side, we have a parallel structure. *A consumption fund* is formed out of com-modities which function as aids rather than as direct inputs to consumption. Some items are directly enclosed within the consumption process (consumer durables, such as cookers, washing machines, etc.)

while others act as a physical framework for consumption (houses, sidewalks, etc.) – the latter I will call the *built environment for consumption*.

We should note that some items in the built environment function jointly for both production and consumption – the transport network, for example – and that items can be transferred from one category to another by changes in use. Also, fixed capital in the built environment is immobile in space in the sense that the value incorporated in it cannot be moved without being destroyed. Investment in the built environment therefore entails the creation of a whole physical landscape for purposes of production, circulation, exchange and consumption.

We will call the capital flows into fixed asset and consumption fund formation the *secondary circuit of capital*. Consider, now, the manner in which such flows can occur. There must obviously be a 'surplus' of both capital and labour in relation to current production and consumption needs in order to facilitate the movement of capital into the formation of long-term assets, particularly those comprising the built environment. The tendency towards overaccumulation produces such conditions within the primary circuit on a periodic basis. One feasible if temporary solution to this overaccumulation problem would therefore be to switch capital flows into the secondary circuit.

Individual capitalists will often find it difficult to bring about such a switch in flows for a variety of reasons. The barriers to individual switching of capital are particularly acute with respect to the built environment where investments tend to be large-scale and long-lasting, often difficult to price in the ordinary way and in many cases open to collective use by all individual capitalists. Indeed, individual capitalists left to themselves will tend to under-supply their own collective needs for production precisely because of such barriers. Individual capitalists tend to overaccumulate in the primary circuit and to under-invest in the secondary circuit; they have considerable difficulty in organizing a balanced flow of capital between the primary and secondary circuits.

A general condition for the flow of capital into the secondary circuit is, therefore, the existence of a functioning capital market and, perhaps, a State willing to finance and guarantee long-term, large-scale projects with respect to the creation of the built environment. At times of overaccumulation, a switch of flows from the primary to the secondary circuit can be accomplished only if the various manifestations of overaccumulation can be transformed into money-capital which can move freely and unhindered into these forms of investment. This switch of resources cannot be accomplished without a money supply and credit system which creates 'fictional capital' *in advance* of actual production and consumption. This applies as much to the consumption fund (hence the Importance of consumer credit, housing mortgages, municipal debt) as it does to fixed capital. Since the production of money and credit are relatively autonomous processes, we have to conceive of the financial and state institutions controlling them as a kind of collective nerve centre governing and *mediating* the relations between the primary and secondary circuits of capital. The nature and form of these financial and state institutions and the policies they adopt can play an important role in checking or enhancing flows of capital into the secondary circuit of capital or into certain specific aspects of it (such as transportation, housing, public facilities, and so on). An alteration in these mediating structures can therefore affect both the volume and direction of the capital flows by constricting movement down some channels and opening up new conduits elsewhere.

The tertiary circuit of capital

In order to complete the picture of the circulation of capital in general, we have to conceive of a *tertiary circuit of capital* which comprises, first, investment in science and technology (the purpose of which is to harness science to production and thereby to contribute to the processes which continuously revolutionize the productive forces in society) and second, a wide range of social expenditures which relate primarily to the processes of reproduction of labour power. The latter can usefully be divided into investments directed towards the qualitative improvement of labour power from the standpoint of capital (investment in education and health by means of which the capacity of the labourers to engage in the work process will be enhanced) and investment in cooptation, integration and repression of the labour force by ideological, military and other means.

[. . .]

THE CIRCULATION OF CAPITAL AS A WHOLE AND ITS CONTRADICTIONS

Figure 2 portrays the overall structure of relations comprising the circulation of capital amongst the three circuits. The diagram looks very 'structuralist-functionalist' because of the method of presentation. I can conceive of no other way to communicate clearly the various dimensions of capital flow. We now have to consider the contradictions embodied within these relations. I shall do so initially as if there were no overt class struggle between capital and labour. In this way we will be able to see that the contradiction between the individual capitalist and capital in general is itself a source of major instability within the accumulation process.

We have already seen how the contradictions internal to the capitalist class generate a tendency towards overaccumulation within the primary circuit of capital. And we have argued that this tendency can be overcome temporarily at least by switching capital into the secondary or tertiary circuits. Capital has, therefore, a variety of investment options open to it – fixed capital or consumption fund formation, investment in science and technology, investment in 'human capital' or outright repression. At

particular historical conjunctures capitalists may not be capable of taking up all of these options with equal vigour, depending upon the degree of their own organization, the institutions which they have created and the objective possibilities dictated by the state of production and the state of class struggle. I shall assume away such problems for the moment in order to concentrate on how the tendency towards overaccumulation, which we have identified so far only with respect to the primary circuit, manifests itself within the overall structure of circulation of capital. To do this we first need to specify a concept of productivity of investment.

On the productivity of investments in the secondary and tertiary circuits

[. . .]

[C]apitalists as a class – often through the agency of the State – do invest in the production of conditions which they hope will be favourable to accumulation, to their own reproduction as a class and to their continuing domination over labour. This leads us immediately to a definition of a productive investment as one which directly or

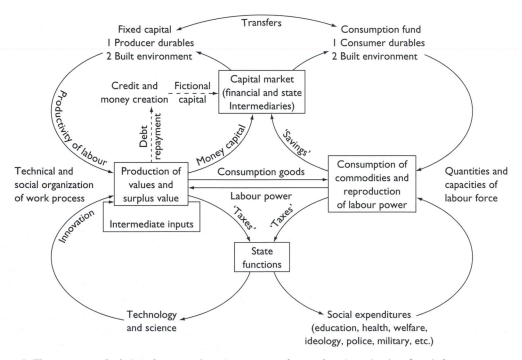

Figure 2 The structure of relations between the primary, secondary and tertiary circuits of capital.

indirectly expands the basis for the production of surplus value. Plainly, investments in the secondary and tertiary circuits have the potential under certain conditions to do this. The problem – which besets the capitalists as much as it confuses us – is to identify the conditions and means which will allow this potential to be realized.

Investment in new machinery is the easiest case to consider. The new machinery is directly productive if it expands the basis for producing surplus value and unproductive if these benefits fail to materialize. Similarly, investment in science and technology may or may not produce new forms of scientific knowledge which can be applied to expand accumulation. But what of investment in roads, housing, health care and education, police forces and the military, and so on? If workers are being recalcitrant in the work place, then judicious investment by the capitalist class in a police force to intimidate the workers and to break their collective power, may indeed be productive indirectly of surplus value for the capitalists. If, on the other hand, the police are employed to protect the bourgeoisie in the conspicuous consumption of their revenues in callous disregard of the poverty and misery which surrounds them, then the police are not acting to facilitate accumulation. The distinction may be fine but it demonstrates the dilemma. How can the capitalist class identify, with reasonable precision, the opportunities for indirectly and directly productive investment in the secondary and tertiary circuits of capital?

The main thrust of the modern commitment to planning (whether at the state or corporate level) rests on the idea that certain forms of investment in the secondary and tertiary circuits are potentially productive. The whole apparatus of cost-benefit analysis and of programming and budgeting, of analysis of social benefits, as well as notions regarding investment in human capital, express this commitment and testify to the complexity of the problem. And at the back of all of this is the difficulty of determining an appropriate basis for decision-making in the absence of clear and unequivocal profit signals. Yet the cost of bad investment decisions – investments which do not contribute directly or indirectly to accumulation of capital – must emerge somewhere. They must, as Marx would put it, come to the surface and thereby indicate the errors which lie beneath. We can begin to grapple with this question by considering the origins of crises within the capitalist mode of production.

On the forms of crisis under capitalism

[...]

We have already seen how the capitalists tend to generate states of overaccumulation within the primary circuit of capital and considered the various manifestations which result. As the pressure builds, either the accumulation process grinds to a halt or new investment opportunities are found as capital flows down various channels into the secondary and tertiary circuits. This movement may start as a trickle and become a flood as the potential for expanding the production of surplus value by such means becomes apparent. But the tendency towards overaccumulation is not eliminated. It is transformed, rather, into a pervasive tendency towards over-investment in the secondary and tertiary circuits. This over-investment, we should stress, is in relation solely to the needs of capital and has nothing to do with the real needs of people which inevitably remain unfulfilled. Manifestations of crisis thus appear in both the secondary and tertiary circuits of capital.

As regards fixed capital and the consumption fund, the crisis takes the form of a crisis in the valuation of assets. Chronic overproduction results in the devaluation of fixed capital and consumption fund items – a process which affects the built environment as well as producer and consumer durables. We can likewise point to crisis formation at other points within our diagram of capital flows – crises in social expenditures (health, education, military repression, and the like), in consumption-fund formation (housing) and in technology and science. In each case the crisis occurs because the potentiality for productive investment within each of these spheres is exhausted. Further flows of capital do not expand the basis for the production of surplus value. We should also note that a crisis of any magnitude in any of these spheres is automatically registered as a crisis within the financial and state structures while the latter, because of the relative autonomy which attaches to them, can be an independent source of crisis (we can thus speak of financial, credit and monetary crises, the fiscal crises of the State, and so on).

Crises are the 'irrational rationalizers' within the capitalist mode of production. They are indicators of imbalance and force a rationalization (which may be painful for certain sectors of the capitalist class as well as for labour) of the processes of production, exchange, distribution and consumption. They may also force a rationalization of institutional structures (financial and state institutions in particular). From the standpoint of the total structure of relationships we have portrayed, we can distinguish different kinds of crises:

a. *Partial crises* which affect a particular sector, geographical region or set of mediating institutions. These can arise for any number of reasons but are potentially capable of being resolved within that sector, region or set of institutions. We can witness autonomously-forming monetary crises, for example, which can be resolved by institutional reforms, crises in the formation of the built environment which can be resolved by reorganization of production for that sector, etc.
b. *Switching crises* which involve a major reorganization and restructuring of capital flows and/or a major restructuring of mediating institutions in order to open up new channels for productive investments. It is useful to distinguish between two kinds of switching crises:
   1. Sectoral switching crises which entail switching the allocation of capital from one sphere (e.g. fixed capital formation) to another (e.g. education);
   2. Geographical switching crises which involve switching the flows of capital from one place to another. We note here that this form of crisis is particularly important in relation to investment in the built environment because the latter is immobile in space and requires interregional or international flows of money-capital to facilitate its production.
c. *Global crises* which affect, to greater or lesser degree all sectors, spheres and regions within the capitalist production system. We will thus see devaluations of fixed capital and the consumption fund, a crisis in science and technology, a fiscal crisis in state expenditures, a crisis in the productivity of labour, all manifest at the same time across all or most regions within the capitalist system. I note, in passing, that there have been only two global crises within the totality of the capitalist system – the first during the 1930s and its Second World War aftermath; the second, that which became most evident after 1973 but which had been steadily building through the 1960s.

[. . .]

ACCUMULATION AND THE URBAN PROCESS

The understanding I have to offer of the urban process under capitalism comes from seeing it in relation to the theory of accumulation. We must first establish the general points of contact between what seem, at first sight two rather different ways of looking at the world.

Whatever else it may entail, the urban process implies the creation of a material physical Infrastructure for production, circulation, exchange and consumption. The first point of contact, then, is to consider the manner in which this built environment is produced and the way it serves as a resource system – a complex of use values – for the production of value and surplus value. We have, secondly, to consider the consumption aspect. Here we can usefully distinguish between the consumption of revenues by the bourgeoisie and the need to reproduce labour power. The former has a considerable impact upon the urban process, but I shall exclude it from the analysis because consideration of it would lead us into a lengthy discourse on the question of bourgeois culture and its complex significations without revealing very much directly about the specifically capitalist form of the urban process. Bourgeois consumption is, as it were, the icing on top of a cake which has as its prime ingredients capital and labour in dynamic relation to each other. The reproduction of labour power is essential and requires certain kinds of social expenditures and the creation of a consumption fund. The flows we have sketched, in so far as they portray capital movements into the built environment (for both production and consumption) and the laying out of social expenditures for the reproduction of labour power, provide us, then, with the structural links we need to understand the urban process under capitalism.

[. . .]

The contradictory character of investments in the built environment

We have so far treated the process of investment in the built environment as a mere reflection of the forces emanating from the primary circuit of capital. There are, however, a whole series of problems which arise because of the specific characteristics of the built environment itself. We will consider these briefly.

Marx's extensive analysis of fixed capital in relation to accumulation reveals a central contradiction. On the one hand, fixed capital enhances the productivity of labour and thereby contributes to the accumulation of capital. But, on the other hand, it functions as a use value and requires the conversion of exchange values into a physical asset which has certain attributes. The exchange value locked up in this physical use value can be re-couped only by keeping the use value fully employed over its lifetime, which for simplicity's sake we will call its 'amortization time'. As a use value the fixed capital cannot easily be altered and so it tends to freeze productivity at a certain level until the end of the amortization time. If new and more productive fixed capital comes into being before the old is amortized, then the exchange value still tied up in the old is devalued. Resistance to this devaluation checks the rise in productivity, and thus restricts accumulation. On the other hand the pursuit of new and more productive forms of fixed capital – dictated by the quest for relative surplus value – accelerates devaluations of the old.

We can identify exactly these same contradictory tendencies in relation to investment in the built environment, although they are even more exaggerated here because of the generally long amortization time involved, the fixity in space of the asset, and the composite nature of the commodity involved. We can demonstrate the argument most easily using the case of investment in transportation.

The cost, speed and capacity of the transport system relate directly to accumulation because of the impacts these have on the turnover time of capital. Investment and innovation in transport are therefore potentially productive for capital in general. Under capitalism, consequently, we see a tendency to 'drive beyond all spatial barriers' and to 'annihilate space with time' (to use Marx's own expressions). This process is, of course, characterized typically by 'long waves' of the sort which we have already identified, uneven development in space and periodic massive devaluations of capital.

We are here concerned, however, with the contradictions implicit in the process of transport development itself, Exchange values are committed to create 'efficient' and 'rational' configurations for spatial movement at a particular historical moment. There is, as it were, a certain striving towards spatial equilibrium, spatial harmony. On the other hand, accumulation for accumulation's sake spawns continuous revolutions in transportation technology as well as a perpetual striving to overcome spatial barriers – all of which is disruptive of any existing spatial configuration.

We thus arrive at a paradox. In order to overcome spatial barriers and to annihilate space with time, spatial structures are created which themselves act as barriers to further accumulation. These spatial structures are expressed in the form of immobile transport facilities and ancillary facilities implanted in the landscape. We can in fact extend this conception to encompass the formation of the built environment as a whole. Capital represents itself in the form of a physical landscape created in its own image, created as use values to enhance the progressive accumulation of capital. The geographical landscape which results is the crowning glory of past capitalist development. But at the same time It expresses the power of dead labour over living labour and as such it imprisons and inhibits the accumulation process within a set of specific physical constraints. And these can be removed only slowly unless there is a substantial devaluation of the exchange value locked up in the creation of these physical assets.

Capitalist development has therefore to negotiate a knife-edge path between preserving the exchange values of past capital investments in the built environment and destroying the value of these investments in order to open up fresh room for accumulation. Under capitalism there is, then, a perpetual struggle in which capital builds a physical landscape appropriate to its own condition at a particular moment in time, only to have to destroy it, usually in the course of a crisis, at a subsequent point in time.

[. . .]

CLASS STRUGGLE, ACCUMULATION AND THE URBAN PROCESS UNDER CAPITALISM

What, then, of overt class struggle – the resistance which the working class collectively offers to the violence which the capitalist form of accumulation inevitably inflicts upon it? This resistance, once it becomes more than merely nominal, must surely affect the urban process under capitalism in definite ways. We must, therefore, seek to incorporate some understanding of it into any analysis of the urban process under capitalism. By switching our window on the world – from the contradictory laws of accumulation to the overt class struggle of the working class against the effects of those laws – we can see rather different aspects of the same process with greater clarity. In the space that follows I will try to illustrate the complementarity of the two viewpoints.

In one sense, class struggle is very easy to write about because there is no theory of it, only concrete social practices in specific social settings. But this immediately places upon us the obligation to understand history if we are to understand how class struggle has entered into the urban process. Plainly I cannot write this history in a few pages. So I will confine myself to a consideration of the contextual conditions of class struggle and the nature of the bourgeois responses. The latter are governed by the laws of accumulation because accumulation always remains the means whereby the capitalist class reproduces itself as well as its domination over labour.

The central point of tension between capital and labour lies in the workplace and is expressed in struggles over the work process and the wage rate. These struggles take place in a context. The nature of the demands, the capacity of workers to organize and the resolution with which the struggles are waged, depend a great deal upon the contextual conditions. The law (property rights, contract, combination and association, etc.) together with the power of the capitalist class to enforce their will through the use of state power are obviously fundamental as any casual reading of labour history will abundantly illustrate. What specifically interests me here, however, is the process of reproduction of labour power in relation to class struggle in the workplace.

Consider, first, the quantitative aspects of labour power in relation to the needs of capitalist accumulation. The greater the labour surplus and the more rapid its rate of expansion, the easier it is for capital to control the struggle in the workplace. The principle of the industrial reserve army under capitalism is one of Marx's most telling insights. Migrations of labour and capital as well as the various mobilization processes by means of which 'unused' elements in the population are drawn into the workforce are manifestations of this basic need for a relative surplus population. But we also have to consider the costs of reproduction of labour power at a standard of living which reflects a whole host of cultural, historical, moral and environmental considerations. A change in these costs or in the definition of the standard of living has obvious implications for real-wage demands and for the total wage bill of the capitalist class. The size of the internal market formed by the purchasing power of the working class is not irrelevant to accumulation either. Consequently, the consumption habits of the workers are of considerable direct and indirect interest to the capitalist class.

But we should also consider a whole host of qualitative aspects to labour power encompassing not only skills and training, but attitudes of mind, levels of compliance, the pervasiveness of the work ethic and of possessive individualism, the variety of fragmentations within the labour force which derive from the division of labour and occupational roles, as well as from older fragmentations along racial, religious and ethnic lines. The ability and urge of workers to organize along class lines depends upon the creation and maintenance of a sense of class consciousness and class solidarity in spite of these fragmentations. The struggle to overcome these fragmentations in the face of divide-and-conquer tactics often adopted by the capitalists is fundamental to understanding the dynamics of class struggle in the workplace.

This leads us to the notion of displaced class struggle, by which I mean class struggle which has its origin in the work process but which ramifies and reverberates throughout all aspects of the system of relations which capitalism establishes. We can trace these reverberations to every corner of the social totality and certainly see them at work in the flows of capital between the different circuits. For example, if productivity fails to rise in the workplace, then perhaps judicious investment in 'human capital' (education), in cooptation (homeownership for the working class), in integration (industrial democracy), in persuasion (ideological indoctrination) or repression might yield better results in the long run. Consider, as an example, the struggles around public education. In *Hard Times*, Dickens constructs a brilliant satirical counterpoint between the factory system and the educational, philanthropic and religious institutions designed to cultivate habits of mind amongst the working class conducive to the workings of the factory system, while elsewhere he has that archetypal bourgeois Mr Dombey, remark that public education is a most excellent thing provided it teaches the common people their proper place in the world. Public education as a right has long been a basic working-class demand. The bourgeoisie at some point grasped that public education could be mobilized against the interests of the working class. The struggle over social services in general is not merely over their provision, but over the very nature of what is provided. A national health-care system which defines ill health as inability to go to work (to produce surplus value) is very different indeed from one dedicated to the total mental and physical well-being of the individual in a given physical and social context.

[. . .]

Some remarks on the housing question

The demand for adequate shelter is clearly high on the list of priorities from the standpoint of the working class. Capital is also interested in commodity production for the consumption fund provided this presents sufficient opportunities for accumulation. The broad lines of class struggle around the 'housing question' have had a major impact upon the urban process. We can trace some of the links back to the workplace directly. The agglomeration and concentration of production posed an immediate quantitative problem for housing workers in the right locations – a problem which the capitalist initially sought to resolve by the production of company housing but which thereafter was left to the market system. The cost of shelter is an important item in the cost of labour power. The more workers have the capacity to press

home wage demands, the more capital becomes concerned about the cost of shelter. But housing is more than just shelter. To begin with, the whole structure of consumption in general relates to the form which housing provision takes. The dilemmas of potential overaccumulation which faced the United States in 1945 were in part resolved by the creation of a whole new life style through the rapid proliferation of the suburbanization process. Furthermore, the social unrest of the 1930s in that country pushed the bourgeoisie to adopt a policy of individual homeownership for the more affluent workers as a means to ensure social stability. This solution had the added advantage of opening up the housing sector as a means for rapid accumulation through commodity production. So successful was this solution that the housing sector became a Keynesian 'contra-cyclical' regulator for the accumulation process as a whole, at least until the debacle of 1973.

[. . .]

The second example I shall take is even more complex. Consider in its broad outlines, the history of the bourgeois response to acute threats of civil strife which are often associated with marked concentrations of the working class and the unemployed in space. The revolutions of 1848 across Europe, the Paris Commune of 1871, the urban violence which accompanied the great railroad strikes of 1877 in the United States and the Haymarket incident in Chicago, clearly demonstrated the revolutionary dangers associated with the high concentration of the 'dangerous classes' in certain areas. The bourgeois response was in part characterized by a policy of dispersal so that the poor and the working class could be subjected to what nineteenth-century urban reformers on both sides of the Atlantic called the 'moral influence' of the suburbs. Cheap suburban land, housing and cheap transportation were all a part of this solution entailing, as a consequence, a certain form and volume of investment in the built environment on the part of the bourgeoisie. To the degree that this policy was necessary, it had an important impact upon the shape of both British and American cities. And what was the bourgeois response to the urban riots of the 1960s in the ghettos of the United States? Open up the suburbs, promote low-income and black homeownership, improve access via the transport system . . . the parallels are remarkable.

A CONCLUDING COMMENT

I shall end by venturing an apology which should properly have been set forth at the beginning. To broach the whole question of the urban process under capitalism in a short article appears a foolish endeavour. I have been forced to blur distinctions, make enormous assumptions, cut corners, jump from the theoretical to the historical in seemingly arbitrary fashion, and commit all manner of sins which will doubtless arouse ire and reproach as well as a good deal of opportunity for misunderstanding. This is, however, a distillation of a framework for thinking about the urban process under capitalism and it is a distillation out of a longer and much vaster work (which may see the light of day shortly). It is a framework which has emerged as the end-product of study and not one which has been arbitrarily imposed at the beginning. It is, therefore, a framework in which I have great confidence. My only major source of doubt, is whether I have been able to present it in a manner which is both accurate enough and simple enough to give the correct flavour of the potential feast of insights which lie within.

REFERENCES FROM THE READING

Marx, K. (1967) *Capital* (three volumes), New York: International Publishers.

Marx, K. (1967, 1968 and 1971) *Theories of Surplus Value*, Moscow: Progress Publishers.

Marx, K. (1973) *Grundrisse*, Harmondsworth: Penguin Books.

Editors' references and suggestions for further reading

Dear, M. (1991) "Review of *The Condition of Postmodernity*," *Annals of the Association of American Geographers*, 81, 3: 533–38.

Harvey, D. (1969) *Explanation in Geography*, London: Arnold.

Harvey, D. (1973) *Social Justice and the City*, Baltimore, MD: Johns Hopkins University Press.

Harvey, D. (1982) *Limits to Capital*, Chicago, IL: University of Chicago Press.

Harvey, D. (1989) *The Condition of Postmodernity: An Enquiry into the Origins of Cultural Change*, Oxford: Blackwell.

Harvey, D. (2000) *Spaces of Hope*, Edinburgh: Edinburgh University Press.

Harvey, D. (2002) *Spaces of Capital*, Edinburgh: Edinburgh University Press.

Peet, R. (1998) *Modern Geographical Thought*, Oxford: Blackwell.

Savage, M., Warde, A., and Ward, K. (2003) *Urban Sociology, Capitalism and Modernity*, London: Palgrave Macmillan.

"Beyond the Crabgrass Frontier: Industry and the Spread of North American Cities, 1850–1950"

from *Journal of Historical Geography* (2001)

Richard Walker and Robert D. Lewis

Editors' Introduction

Understanding the expansion and spread of cities, and in particular processes of suburbanization, has long occupied the attention of urban researchers. In Part 1, Foundations, for example, we saw how Burgess' ecological model of urban development suggested that the outward growth of cities was associated with the movement of more affluent groups from the urban core to the periphery. More recent accounts, such as Kenneth Jackson's *Crabgrass Frontier: The Suburbanization of the United States* (1985; see also Fishman 1987), have also tended to view this process in terms of the residential preferences of the middle classes for "homes in the park." The title of Walker and Lewis' paper, "Beyond the Crabgrass Frontier," clearly signals their belief that these accounts of urban growth are in need of revision. Specifically, Walker and Lewis argue that the "conventional story" of the spread of North American cities as a two-stage process of residential areas leading the way outward from a congested urban core and then eventually followed by industry is fundamentally misleading. Their contention is that North American suburbanization has involved "the simultaneous march of industry and cities outward" and has been underpinned by three key processes: urban geographical industrialization, investment in real estate, and political guidance by business and government leaders. Urban geographical industrialization refers to the way in which the expansion of cities is based on the ability of industrialization and capital accumulation to actually *create* places by, for example, building up territorial concentrations of related activities and attracting new labour forces. As cities develop, new industries emerge or existing industries restructure, leading to new industrial locales in outlying areas away from old centres.

On its own, however, the demand by industry for new spaces is necessary but not sufficient to explain the spread of North American cities. Also of crucial importance is investment in real estate. The profits to be made from investment in land on the suburban fringe make property developers key players in producing the "new constellations of employment, transport and residence" at the edge of existing cities. Indeed, Walker and Lewis observe that the property industry has been particularly inventive in creating "complete urban environments," such as industrial parks and regional malls, which fuel the outward extension of industrial space from the city centre.

A third set of factors identified by Walker and Lewis as crucial to the spread of North American cities concerns political guidance by business and government leaders. In the nineteenth century industrialists were successful in getting local, provincial, state and federal funding to build the infrastructure that allowed them to move production to the metropolitan outskirts. The development of planning regulations further helped to shape industrial suburban space by creating zones where manufacturers would be able to operate without interference "by complaining and litigious middle class neighbours."

The importance of the argument advanced in this paper lies partly in the way it situates the processes of suburbanization firmly within the context of the economic rather than within the social geography of the city. It therefore serves as a powerful critique of ecological and neoclassical models of urban development which tend to view the spread of cities as the "natural" outcome of factors such as the increasing affluence of consumers looking for more space. Moreover, by emphasizing the importance of strategies of capital accumulation to the spread of North American cities, this paper is clearly informed by the arguments advanced in Harvey's reading in this section, with suburbanization viewed as a means of staving off an accumulation crisis (see also Walker 1978) There are also important overlaps with Smith's paper on the gentrification of the inner city, also included in this section. This overlap is reflected partly in the imagery of the frontier which informs both these accounts of urban restructuring, but also, at a more theoretical level, both pieces clearly privilege production over consumption in the making of urban landscapes. Walker and Lewis are unequivocal in arguing that it is the search for increased profits among industrial interests and property developers which fuelled the suburbanization process between 1850 and 1950. As they explain, "Because profits increase with distance from the fringe, the search for profits in land speculation by property investors, developers and financial institutions exaggerate the demand for peripheral sites, tempting industry and residents to move to the suburbs and pulling the city outward into the space-extensive form characteristic of North American cities." A further important implication of their argument is that it calls into question claims made in the late 1980s and 1990s about the newness of so-called "edge cities." These urban complexes on the periphery of existing cities have become major employment centres with the dispersal of industry, offices, and retail malls, and therefore contain all the functions of the traditional city (Garreau 1991; Teaford 1997; and Dear and Flusty in this section). Far from being unprecedented developments, this phase of urban restructuring is, according to Walker and Lewis, "the latest episode in a long-running story of North American urbanization."

Richard Walker, a former student of David Harvey, is Professor of Geography at Berkeley, University of California, and co-author with Michael Storper of *The Capitalist Imperative: Territory, Technology and Industrial Growth* (Oxford: Blackwell, 1989) and, with Andrew Sayer, of *The New Social Economy: Reworking the Division of Labor* (Oxford: Blackwell, 1992). Robert Lewis is Associate Professor in the Geography Department, University of Toronto, and author of *Manufacturing Montreal: The Making of an Industrial Landscape 1850–1930* (Baltimore, MD: Johns Hopkins University Press, 2000).

■ ■ ■ ■ ■ ■

The conventional story of suburbanization in Canada and the United States portrays an outward movement of residences from the cities that only lately has been fuelled by the dispersal of employment to the urban fringe. In the classic studies, suburbia is conjured up as an image of "homes in a park," a middle landscape constituted as a way of life halfway between city and country. This conventional wisdom needs considerable revision. Residential areas have not singularly led the way outward from a previously concentrated city, but have always been joined at the hip by industry locating at the urban fringe. The outward spread of factories and manufacturing districts has been a decisive feature of North American urbanization since the middle of the nineteenth century. Suburban growth as a whole has been a mixture of industry and homes, the city sprawling ever outward from its initial point of establishment and repeatedly spilling over political, social and perceived boundaries. The result has been extensive, multi-nodal metropolitan regions. In this essay, we present a theoretical reinterpretation of industrial suburbanization. We argue that industrial decentralization has been repeatedly misinterpreted as new and unprecedented rather than an extension of past trends. In contrast to the prevailing interpretation, we claim that industrial suburbanization is the product of a combination of the economic logic of geographical industrialization, investment in real estate, and political guidance by business and government leaders.

[. . .]

The conventional logic of industrial dispersion

The conventional view begins with the assumption of overwhelming concentration of industry in the urban core in the nineteenth and early twentieth centuries. While the rate of decentralization has been debated, virtually all students of urban employment would agree with Allen Scott that centralized production was "characteristic of the large metropolis well into the twentieth century." For most urban and economic geographers, the suburbanization of industry did not occur until after World War II.

The principal factors behind industrial centralization in traditional intra-metropolitan location models are transportation costs and agglomeration economies. The movement of industry to the periphery, in this view, only came about with recent advances in transportation systems, industrial process technologies, and business organization that lowered the cost of locating away from urban centres, reduced the effects of agglomeration, and liberated factory and firm from the urban land nexus. The traditional emphasis on transport costs and agglomeration effects in urban land use models follows the theoretical lead of Alfred Weber. For most writers, the central manufacturing zones result from the minimization of transport costs to the urban market and to centrally placed shipping nodes such as ports and railway depots. This skeletal explanation is fleshed out with a theory of economies of proximity among many small firms concentrated in a limited area. Different versions of this account exist, but for Weber it is primarily transportation cost reduction among all firms that explains clustering, and secondarily access to a centrally located labour pool. The converse of this theory of concentration is the transport-driven model of industrial decentralization. In the classic version, cars and trucks lower costs of transport dramatically over rail and water and lessen dependence of urban manufacturers on ready access to central rail and harbor facilities. For example, "between 1915 and 1930, when the number of American trucks jumped from 158,000 to 3.5 million . . . industrial deconcentration began to alter the basic spatial pattern of metropolitan areas." Transportation becomes virtually universal in explaining suburbanization based on the argument that cars and trucks

provide unprecedented speed and flexibility ṇ ing workers and goods.

The product cycle model was grafted on Weberian location theory, adding the idea oi industrial "incubation." The central city, from this perspective, has a relative advantage as a source of innovation, thanks to maximum access to markets, new ideas, skilled labour, and finance. It serves to incubate new products and new firms that subsequently move to the suburbs (or backward regions). Agglomeration loses its grip as industry matures; the shift to mass production eliminates the reliance of small specialized producers on each other by standardizing input–output linkages and bringing a range of activities into large, integrated factories. The result is the dispersal of firms from the core to the city fringe.

The last addendum to the conventional theory of industrial decentralization allowed for the evolution of business organization from small, single-plant firms to the modern corporation. Theories of "corporate location" absorbed the product cycle into an overarching theory of the dispersal of branch plants to peripheral regions and countries from a corporate core. The causal mechanisms are standardized, large-volume flows of inputs and outputs, large-scale plants, and internalized transactions. The corporate umbrella severs the enterprise from external linkages (commodity trade, specialized labour skills, management inputs, etc.), breaking the collective logic of agglomeration and freeing corporate-owned factories to seek cheap land and labour far from the city.

Urban geographical industrialization

The conventional explanation of industrial location in city and suburb has serious problems. In the first place, transportation limits, but does not determine, the location of industry. Undeniably, transport costs influence the geography of urbanism: industry has always clustered near transportation nodes and corridors, whether harbours, rail lines or highways, in a way that leads to transport-tied corridors of industrial land use. Over time, improvements in transport have also allowed the city to spread out, but transportation access has been more widespread than conventional models allow. . . .

Weberian theory, including its account of agglomeration, suffers from undue emphasis on cost minimization with respect to input factors. Demand conditions are important, but even though capitalist firms try to keep down costs and weigh the relative prices of inputs, industrialization is not principally an optimization problem; it is a dynamic process in which new commodities and new ways of doing things continually displace the old, and today's prices based on further technical change displace yesterday's costs. . . .

Product cycle-incubator theory adds a needed element of innovation and productivity advance, but it does so in a highly stylized manner that assumes wrongly that the new is small and the old is large, industries mature in a systematic way, and they are well behaved in their locational choices. The evidence, thus, is mixed and quite often contradicts the model. . . .

In short, industrial development and location is not a monotonic, uniform process. What we see, instead, are successive eruptions of new industries, embodying new products and new technical bases, and a diverse array of production formats evolving and restructuring over time. Technical change has developed on a variety of material bases in different industries, moved along divergent industrial (and company) trajectories, and been altered radically by new discoveries from time to time. This has meant many patterns of initial location, agglomeration, and dispersal, giving North American cities quite distinctive industrial foundations, patterns of uneven spatial expansion of those sectors, and episodic additions of wholly new industries to the mix.

In fact, industry and the city have grown together as a unified process of geographic development. Industry does not locate in the city, it helps create the city. Urban expansion is based on the ability of industrialization and capital accumulation to create places at the same time as making commodities, building factories, raising up a labour force, and introducing new technologies. This process of "geographical industrialization" has the following principles. First, new industrial locales have the ability to break away from old centres and existing economies of agglomeration, thanks to both the rapid rates of accumulation and the experimental nature of their growth process. They are likely to avoid existing concentrations if they fear the effects of established labour practices, man-agement outlook, or worker militancy. Second, growing industries build up extensive territorial concentrations of related activities, such as specialized suppliers, merchants, financiers and educational institutions, and spin off new firms and even new industries in their process of expansion. Third, new industrial implantations attract and train new labour forces, steeped in the particular ways of working, technology and ethos of the industry. These fresh labour forces may have little in common with other segments of the labour market. Lastly, given the repeated and permanent nature of industrial revolution under capitalism, the space-economy has undergone many changes and upheavals.

Applied to the urban arena, this suggests that as cities develop new industrial sectors or their existing industries restructure and expand, successive nodes of growth erupt in outlying areas, growing in time to fill up the neighboring suburban territory. As cities have grown, layer upon layer of suburban development has been added to the built-up area, leaving former outlying districts well inside the metropolis and often erasing historic patterns of expansion by dispersion in the process. After many years, it is easy to mistake the older edge cities and secondary nodes for part of a single "central city." Modern metropolitan areas are so huge that even large and distant suburban edges of the past, such as Brooklyn, Oakland, or South Chicago, are now deeply embedded in the structure of the city. The study of North American urbanization thus requires a model that begins with the simultaneous march of industry and cities outward, rather than a two-stage process of building a dense concentration of activities in the core over the nineteenth century and then decanting them in the twentieth.

Industrial districts and the multi-nodal metropolis

The process of urban-industrial growth has another crucial dimension besides the outward flow and build-up of the city: the appearance of distinctive industrial districts within a multi-nodal metropolitan area. Classic agglomeration theory does not explain this phenomenon; the city is a single generic agglomeration with industry confined to the core. Conversely, traditional decentralization models allow only for the dispersal of large factories under the umbrella of the modern corporation. In both cases, too much is missing from the real fabric of urban industrialization; to recover it, we must consider the problem of industrial organization and the spatial division of labour.

[. . .]

The most sophisticated model of urban industrial clustering is that of Allen Scott, who tries to capture the dynamics of industrial agglomeration and decentralization in terms of "transactions costs." Oliver Williamson developed the theory of transaction costs to translate Alfred Chandler's insights into the rise of the modern corporation, based chiefly on the technical imperatives of scale, into the language of neo-classical economics. Scott realized that the same insight could be applied to geography, allowing for a reworking of Weberian agglomeration theory. He argued that urban concentration provided an alternative to the large firm. Complexes of vertically disintegrated producers within specific industrial sectors cluster to take advantage of mutual interaction. Complexes grow through the intensification of the division of labour, multiple linkages among firms, and flexibility in the face of changing markets. These generate economies of scope for individual specialists and collective economies for the entire industry.

Scott's work complemented that of European researchers who examined the vigorous industries of the Third Italy and rediscovered Alfred Marshall's idea of the "industrial district." While initially arguing for a small-firm model of clustering, Scott realized that both large and small factories and companies are embedded in industrial districts. Size would be decisive if external exchange were the only reason for agglomeration, as early transactions models implied. Yet, the benefits of interaction go to the heart of all extensive divisions of labour because they lower costs of interaction among dependent parts of production systems, reduce the risk of investment, lessen turnover time, and offer institutions of collective governance. Furthermore, they offer dynamic advantages by stimulating the collective process of learning and providing a milieu of problem-solving and innovation. Scott thus abandoned the simple model of central agglomeration and decentralization of large factories to the suburbs, in favour of one of multiple clusters throughout the metropolitan region as industrial districts can occur in any number of high-tech, large batch, or "new craft" sectors.

[. . .]

Building out the metropolis

Recent literature in industrial geography advances our understanding of spatial centration and dispersal significantly beyond the old models of urban centrality and suburbanization, but it has not made the further link to the build-out of the city. The way that urban areas expand through the mediation of property developers their sectoral logic of investment, production and profit must be examined because a principal dimension of urban industry or "production" is the construction of the city itself. Property capital's imprint on the suburban landscape can be discerned in various ways, including the shape of lot sizes, building placement, construction type, infrastructure, and improvements. This contributes to the array of urban forms that constitute the everyday vernacular landscape of the city, as well as to striking elements of homogenization across the North American urban system. The urban mosaic, or the mesogeography of urbanization, has four critical elements: property developers, building cycles, financial speculation, and uneven development. They combine to produce the repeated eruptions of new constellations of employment, transport and residence at the metropolitan fringe, and the great swaths of construction laid down in the form of peripheral belts, jutting wedges, industrial districts, satellite towns, and edge cities.

Cities have always grown at their edges, but it is erroneous to think that suburban industrial spaces, any more than residential areas, are built on demand without regard for the profits to be made from investment in land. The commodification of land, property investment, and speculative building have been hallmarks of urbanization and national expansion in the United States and Canada. Property investment at the suburban fringe creates the possibility of enormous gains through the maximization of the returns of capitalized rent. Because profits increase with distance from the fringe, the search for profits in land speculation by property investors, developers, and financial institutions exaggerate the demand for peripheral sites, tempting industry and residents to move to the suburbs and pulling the city outward into the space-extensive form characteristic of North American cities. This holds even for the industrial company acquiring land: suburban sites have offered not only low prices and easier assemblage of large plots for factories, but also the promise of speculative profit if surplus acreage is sold or developed. The property industry, moreover, has been particularly inventive in creating complete urban environments, from the housing tract to the regional mall to the industrial park. These condensed pieces of urbanity can be set down in the greenfields like seedlings, helping the city take root more quickly in fringe areas. To make sure investments are realized, promoters try to lever urban infrastructure, and other investors outward in order to "ripen" their investments. In this way, the extension of industrial space has been propelled outward from the city centre.

Waves of investment in property development that correspond to waves of capital investment, job creation and surging economic activity are another essential force in metropolitan expansion. Urban growth is neither incremental nor continuous in space and time, but occurs in bursts. The urban land market is notorious for boom-and-bust dynamics in subdivision, financing, and construction, with well-documented 12–25-year swings in activities such as aggregate building and transport expansion. This space-time rhythm appears as rings of building activity laid down around cities with each investment boom. The proximate mechanisms generating such property cycles are adjustments of supply to demand that overshoot because buyers and sellers of real estate compete fiercely and time-lags exist between initiation and completion of building projects. Technological and design changes in buildings, infrastructure and large-scale developments further modulate and accelerate the industrial land process at the urban fringe. The push of capital into real estate investment due to the build-up of surplus and fictitious capital in the financial system, however, is the most dramatic aspect of property booms, exaggerated in eras of financial frenzy in the economy at large such as the 1920s, or the 1980s.

[. . .]

Finally, city building through industrial suburban growth occurs within an economy that demonstrates persistent unevenness in rates of growth and capital accumulation among different industrial sectors and places. Capital flows triggered by unequal and fluctuating rates of profit and accumulation in the larger economic system give impetus to industrial shifts and property booms. As a result, in places where investment surges into new industrial suburban development, great swaths of cities can be laid down in short order before the hand of capital moves on. These temporal-spatial dynamics of capitalist growth have shown up clearly in the metropolitan record since at least the late eighteenth century in North America.

Politics and planning of industrial suburbanization

In addition to the economic logic of industrialization and property development, political intervention and conscious planning have also played a significant part in the intentional process of shaping and reshaping North American cities. Despite the apparent chaos of urban building, a prevailing vision of urban expansion and suburbanization has guided the plans of industrialists, developers and governments. The construction of cities is more than an exercise in economics; it is irreducibly about the search for geographic control, or the politics of space. Industrialists and other capitalists are acutely aware of the contradiction between the concentration of people and industry in the city, and they are keen to maintain their prerogatives in the arenas of investment, work and profitability, now termed the "local business climate." Location at the suburban fringe and outlying districts has offered the hope of combining the manifest benefits of access to the city and its agglomeration economies with a degree of freedom from the working class, city politics, and contending business interests. Because agglomeration effects can be created in outlying districts within reach of the urban centre and are operative at the metropolitan scale, this political elbow room can be created by means of industrial suburbanization and a space-extensive, multi-nodal city form. This was worked out during the nineteenth century and put decisively into place after the turn of the last century.

[. . .]

By the end of the last century, the level of labour militancy and political upheaval associated with reform movements rose. Capitalists became increasingly uneasy about their control of urban geography, and the politics of urban space became a subject of intense debate. Discourse both on the evils of "urban congestion," labour militancy, political corruption, and moral turpitude and on the virtues of the suburban solution to the dense city form became so heated that the viability of the labyrinthine spaces of big cities was thrown into question. The result was the rethinking by the capitalist class of its economic behavior and the growing desire for suburban escapism among the better off. Industrial dispersal could be seen, thereafter, as not only good for business, but as a social virtue and even a necessity to ward off revolution and degeneracy in the body politic. The attractions of decentralization increased correspondingly, and new outlying industrial sites began to multiply.

Planning was the handmaiden of politics in helping to create and shape suburban industrial space. The most limited form of planning for industrial sites at the urban periphery is the private assemblage of land for that purpose. The company town, such as Lowell, Pullman, and Homestead, was an early form of planning undertaken by a single company with a vision of housing provision and proper social life for "the hands," but they were expensive and usually found to be less conducive to labour peace than the distractions of urban life. The industrial park is another basic form of planning; land is carefully prepared, provisioned, and pre-planned by the developer, in concert with local authorities. At an even larger scale, entire industrial suburbs, such as the Chicago Stockyards or South San Francisco, were carefully planned as joint development efforts between industrialists and suburban governments.

As suburban jurisdictions proliferated after the turn of the century, many aimed to attract industry, most worked hand in glove with real estate promoters, and virtually all tried to provide the best business environment money could buy. Dozens of suburban governments around every big city became suppliers and boosters of industrial land away from the central city, often marketing themselves shamelessly and offering subsidies to capture new investors. In some cases, industry could wrap itself in the cloak of specialized city governments, such as West Allis, Gary, Vernon, Emeryville, and Maisonneuve, and turn its back on the exactions of civic politics and social demands for revenues and responsibility.

[. . .]

CONCLUSION

At the burgeoning edges of the metropolis are found a full panoply of workplaces, homes, infrastructure, and commerce that make up the economy and life of the city. These suburban nodes have ranged widely in size, character, and relative autonomy from the parent city, depending on circumstances of economic base, class base, political history, and the like. These extrusions of the growing city are not altogether random, but the complexity of metropolitan expansion requires the kind of non-determinate, non-uniformitarian theory now associated with interplanetary geophysics or hydrodynamics. There are no "normal" cities and suburbs, no uniform growth paths, no easy way out of the study of history; nonetheless, it is possible to capture the major forces at work behind diverse outcomes. We argue that the combination of geographical industrialization, land development, and metropolitan politics and planning is a theoretical framework that offers a means to advance beyond previous theories at the disposal of urban geographers and historians.

Editors' references and suggestions for further reading

Fishman, R. (1987) *Bourgeois Utopias: The Rise and Fall of Suburbs*, New York: Basic Books.

Garreau, J. (1991) *Edge City: Life on the New Frontier*, New York: Doubleday.

Jackson, K. (1985) *Crabgrass Frontier: The Suburbanization of the United States*, New York: Oxford University Press.

Teaford, J.C. (1997) *Post-suburbia: Government and Politics in the Edge Cities*, Baltimore, MD: Johns Hopkins University Press.

Walker, R. (1978) "The transformation of urban structure in the nineteenth century United States and the beginnings of suburbanization," in K. Cox (ed.) *Urbanization and Conflict in Market Societies*, Chicago, IL: Maaroufa Press, pp. 165–212.

"Gentrification, the Frontier, and the Restructuring of Urban Space"

from Neil Smith and Peter Williams (eds)
Gentrification of the City (1986)

Neil Smith

Editors' Introduction

In "Gentrification, the Frontier, and the Restructuring of Urban Space," Neil Smith, Distinguished Professor of Anthropology and Geography at the Graduate Center of the City University of New York, focuses on the processes which underpin the reinvestment of capital in inner urban areas in order to attract white-collar residents back to the city. Smith's starting point, however, is the imagery of the frontier. Although used originally in the nineteenth century to describe America's westward expansion, frontier imagery was used increasingly in the twentieth century in the context of the American city. As suburbanization expanded, the inner city came to be seen by the white middle class as an urban wilderness which only began to be tamed as urban pioneers, the new folk heroes of the urban frontier, engaged in processes of gentrification and urban renewal. Like all images, the use of "the frontier" to represent gentrification as the leading edge of an urban renaissance is not innocent or neutral. For Smith it falsely legitimizes a process of conquest in which the working-class inhabitants of the central city are displaced by more wealthy incomers.

There is a sense, however, in which the frontier image is an accurate one. According to Smith, the original frontier was, above all else, an economic phenomenon, providing the basis for the accumulation and expansion of capital. Gentrification, Smith contends, is about the redifferentiation of geographical space towards exactly the same end. It is driven by the needs of production and, in particular, the need to earn profit among builders, developers, landlords, mortgage lenders, government agencies and real estate agents. This, of course, begs the question: Why are some neighborhoods profitable to redevelop and others are not? For Smith, the crucial mechanism in determining where (and when) gentrification occurs is the so-called rent gap.

The rent gap refers to the difference between the actual ground rent obtained from the present land use of an area of the inner city and the potential rent that could be obtained from the "higher and better" use of the land given its central location. In the United States it has been the movement of capital into the construction of new suburban landscapes and the consequent creation of a rent gap that has created the economic *opportunity* for restructuring the central and inner city. Thus the rent gap provides the underlying economic driver for gentrification, although the determination of its specific form depends on other factors, including the growth in white-collar employment and the consumption patterns of these highly paid workers. Smith concludes his essay by noting that the kind of urban redevelopment associated with gentrification means that the Chicago model of urban structure (see reading by Burgess in Part 1) is now largely redundant. Instead of central areas of the city dominated by the relatively poor, the logical conclusion of current restructuring processes would be the creation of a "bourgeois playground" in the urban core.

Smith's argument in this paper is important for two main reasons. First, gentrification has been, and continues to be, a crucial process shaping the urban geography of cities around the world. During the 1970s and 1980s, gentrification represented one of the "leading edges" of metropolitan restructuring in cities in North America, Europe and Australia and, despite some concerns in the mid-1990s that the process of gentrification had run out of steam (Lees 2000), geographers are now referring to "post-recession gentrification" as central and inner-city neighborhoods in many cities in the advanced capitalist world continue to be a focus of redevelopment. Indeed, in policy terms, governments in the USA and the UK have increasingly represented gentrification as the basis for an "urban renaissance" (Lees 2000: 391). A second reason for Smith's importance is that gentrification represents "one of the key theoretical and ideological battle grounds in urban geography, and indeed in human geography as a whole" (Hamnett 1991: 174). On the one hand, there are the structural Marxists, like Smith, who stress the role of capital and class as part of a supply-side argument that focuses on the roles of builders, developers, landlords, mortgage lenders, government agencies, and real estate agents as they search for opportunities for profitable investment. On the other hand, liberal humanists stress the key role of choice, cultural consumption and consumer demand as the key forces behind the gentrification process. The work of David Ley has been particularly influential in developing these demand-side arguments, beginning with his paper on "Liberal ideology and the post-industrial city" (Ley 1980). In this and subsequent articles, Ley emphasizes the importance of a growing white-collar workforce whose cultural and consumption preferences provide the demand base for housing reinvestment in the inner city. Although Smith's work acknowledges that "demand-side" issues have a place in explaining gentrification, it is clear from the piece selected here that issues of culture and the role of gentrifiers are very much marginalized in his account.

Not surprisingly, perhaps, there have been several attempts to synthesize these so-called "demand" and "supply" explanations in the gentrification literature (see Hamnett 1991; Lees 1994). Hamnett argues, for example, that "a comprehensive and integrated explanation of gentrification must necessarily involve the explanation of where gentrifiers come from and why they gentrify, how the areas and properties to be gentrified are produced and how the two are linked" (p. 187). From this perspective it is therefore important to examine the role of institutional and collective social actors, including real estate agents, developers, and mortgage lenders, and the individual gentrifiers themselves. Similarly, Lees (1994) and Boyle (1995) have argued that it is important to integrate explanations of gentrification that focus on "economic" issues with those more concerned with "culture."

Interestingly, Smith (1996), too, in his more recent work on gentrification (*The New Urban Frontier: Gentrification and the Revanchist City*) has acknowledged that issues of supply and demand, production and consumption, are important to the rent gap theory. While still located firmly within a Marxist framework, Smith develops his earlier class-centred analyses to include discussions of gender and race in the gentrification process. As indicated by the title of his book, however, Smith also remains committed to understanding gentrification as a form of geographical revenge against the poor who "stole" the inner city from the respectable classes. By contrast, David Ley's (1996) more recent work, *The New Middle Class and the Remaking of the Central City*, continues to develop the argument that gentrifiers moved into the inner city because of its particular sense of place and that gentrification is a spatial expression of middle-class cultural values. Ley also emphasizes the importance of the "geography of gentrification" by which he means maintaining a sensitivity to the specific contexts and scales of gentrification rather than trying to reduce it to a single, generalizable process.

In addition to his specific research on gentrification, Neil Smith has a more general interest in the connections between space and social theory, reflected in his other books which include *Uneven Development: Nature, Capital and the Production of Space* (Oxford: Blackwell, 1991) and *American Empire: Roosevelt's Geographer and the Prelude to Globalization* (Berkeley, University of California Press, 2002).

In his seminal essay on "The significance of the frontier in American history," written in 1893, Frederick Jackson Turner wrote:

> American development has exhibited not merely advance along a single line, but a return to primitive conditions on a continually advancing frontier line, and a new development for that area. American social development has been continually beginning over again on the frontier.... In this advance the frontier is the outer edge of the wave – the meeting point between savagery and civilization.... The wilderness has been interpenetrated by lines of civilization growing ever more numerous.

For Turner, the expansion of the frontier and the rolling back of wilderness and savagery were an attempt to make livable space out of an unruly and uncooperative nature. This involved not simply a process of spatial expansion and the progressive taming of the physical world. The development of the frontier certainly accomplished these things, but for Turner it was also the central experience which defined the uniqueness of the American national character. With each expansion of the outer edge by robust pioneers, not only were new lands added to the American estate but new blood was added to the veins of the American democratic ideal. Each new wave westward, in the conquest of nature, sent shock waves back east in the democratization of human nature.

During the 20th century the imagery of wilderness and frontier has been applied less to the plains, mountains and forests of the West, and more to the cities of the whole country, but especially of the East. As part of the experience of suburbanization, the 20th-century American city came to be seen by the white middle class as an urban wilderness; it was, and for many still is, the habitat of disease and crime, danger and disorder. . . . Indeed these were the central fears expressed throughout the 1950s and 1960s by urban theorists who focused on urban "blight" and "decline," "social malaise" in the inner city, the "pathology" of urban life. . . .

Anti-urbanism has been a dominant theme in American culture. In a pattern analogous to the original experience of wilderness, the last 20 years have seen a shift from fear to romanticism and a progression of urban imagery from wilderness to frontier. Cotton Mather and the Puritans of 17th-century New England feared the forest as an impenetrable evil, a dangerous wilderness, but with the continual taming of the forest and its transformation at the hands of human labor, the softer imagery of Turner's frontier was an obvious successor to Mather's forest of evil. There is an optimism and an expectation of expansion associated with "frontier"; wilderness gives way to frontier when the conquest is well under way. Thus in the 20th-century American city, the imagery of urban wilderness has been replaced by the imagery of urban frontier. This transformation can be traced to the origins of urban renewal . . . but has become intensified in the last two decades, as the rehabilitation of single-family homes became fashionable in the wake of urban renewal. In the language of gentrification, the appeal to frontier imagery is exact: urban pioneers, urban homesteaders and urban cowboys are the new folk heroes of the urban frontier.

Just as Turner recognized the existence of Native Americans but included them as part of his savage wilderness, contemporary urban-frontier imagery implicitly treats the present inner-city population as a natural element of their physical surroundings. Thus the term "urban pioneer" is as arrogant as the original notion of the "pioneer" in that it conveys the impression of a city that is not yet socially inhabited; like the Native Americans, the contemporary urban working class is seen as less than social, simply a part of the physical environment. Turner was explicit about this when he called the frontier "the meeting point between savagery and civilization," and although today's frontier vocabulary of gentrification is rarely as explicit, it treats the inner-city population in much the same way. . . .

The parallels go further. For Turner, the westward geographical progress of the frontier line is associated with the forging of the national spirit. An equally spiritual hope is expressed in the boosterism which presents gentrification as the leading edge of an American urban renaissance; in the most extreme scenario, the new urban pioneers are expected to do for the national spirit what the old ones did: to lead us into a new world where the problems of the old world are left behind. . . . No one has yet seriously proposed that we view James

Rouse (the American developer responsible for many of the highly visible downtown malls, plazas, markets and tourist arcades) as the John Wayne of gentrification, but the proposal would be quite in keeping with much of the contemporary imagery. In the end, and this is the important conclusion, the imagery of frontier serves to rationalize and legitimate a process of conquest, whether in the 18th- and 19th-century West or in the 20th-century inner city. The imagery relies on several myths but also has a partial basis in reality. Some of the mythology has already been hinted at, but before proceeding to examine the realistic basis of the imagery, I want to discuss one aspect of the frontier mythology not yet touched upon: nationalism.

The process of gentrification with which we are concerned here is quintessentially international. It is taking place throughout North America and much of western Europe, as well as Australia and New Zealand, that is, in cities throughout most of the Western advanced capitalist world. Yet nowhere is the process less understood than in the United States, where the American nationalism of the frontier ideology has encouraged a provincial understanding of gentrification. The original pre-20th-century frontier experience was not limited to the United States, but rather exported throughout the world; likewise, although it is nowhere as rooted as in the United States, the frontier ideology does emerge elsewhere in connection with gentrification. The international influence of the earlier American frontier experience is repeated with the 20th-century urban scene; the American imagery of gentrification is simultaneously cosmopolitan and parochial, general and local. It is general in image if often contrary in detail. For these reasons, the critique of the frontier imagery does not condemn us to repeating Turner's nationalism, and should not be seen as a nationalistic basis for a discussion of gentrification. The Australian experience of frontier, for example, was certainly different from the American, but was also responsible (along with American cultural imports) for spawning a strong frontier ideology. And the American frontier itself was as intensely real for potential immigrants in Scandinavia or Ireland as it was for actual French or British immigrants in Baltimore or Boston.

However, as with every ideology, there is a real, if partial and distorted, basis for the treatment of gentrification as a new urban frontier. In this idea of frontier we see an evocative combination of economic and spatial dimensions of development. The potency of the frontier image depends on the subtlety of exactly this combination of the economic and the spatial. In the 19th century, the expansion of the geographic frontier in the US and elsewhere was simultaneously an economic expansion of capital. Yet the social individualism pinned onto and incorporated into the idea of frontier is in one important respect a myth; Turner's frontier line was extended westward less by individual pioneers and homesteaders, and more by banks, railways, the state and other speculators, and these in turn passed the land on (at profit) to businesses and families. . . . In this period, economic expansion was accomplished in part through absolute geographical expansion. That is, expansion of the economy involved the expansion of the geographical arena over which the economy operated. Today the link between economic and geographical development remains, giving the frontier imagery its present currency, but the form of the link is very different. As far as its spatial basis is concerned, economic expansion takes place today not through absolute geographical expansion but through the internal differentiation of geographical space. Today's production of space or geographical development is therefore a sharply uneven process. Gentrification, urban renewal, and the larger, more complex, processes of urban restructuring are all part of the differentiation of geographical space at the urban scale; although they had their basis in the period of economic expansion prior to the current world economic crisis, the function of these processes today is to lay one small part of the geographical basis for a future period of expansion. . . . And as with the original frontier, the mythology has it that gentrification is a process led by individual pioneers and homesteaders whose sweat equity, daring and vision are paving the way for those among us who are more timid. But even if we ignore urban renewal and the commercial, administrative and recreational redevelopment that is taking place, and focus purely on residential rehabilitation, it is apparent that where the "urban pioneers" venture, the banks, real-estate companies, the state or other collective economic actors have generally gone before. In this context it may be more appropriate to view the James Rouse Company not as the John Wayne but as the Wells Fargo of gentrification.

In the public media, gentrification has been presented as the pre-eminent symbol of the larger urban redevelopment that is taking place. Its symbolic importance far outweighs its real importance; it is a relatively small if highly visible part of a much larger process. The actual process of gentrification lends itself to such cultural abuse in the same way as the original frontier. Whatever the real economic, social and political forces that pave the way for gentrification, and no matter which banks and realtors, governments and contractors are behind the process, gentrification appears at first sight, and especially in the US, to be a marvellous testament to the values of individualism and the family, economic opportunity and the dignity of work (sweat equity). From appearances at least, gentrification can be played so as to strike some of the most resonant chords on our ideological keyboard. As early as 1961, Jean Gottmann not only caught the reality of changing urban patterns, but also spoke in a language amenable to the emerging ideology, when he said that the "frontier of the American economy is nowadays urban and suburban rather than peripheral to the civilized areas".... With two important provisos, which have become much more obvious in the last two decades, this insight is precise. First, the urban frontier is a frontier in the economic sense, before anything else. The social, political and cultural transformations in the central city are often dramatic and are certainly important as regards our immediate experience of everyday life, but they are associated with the development of an economic frontier. Second, the urban frontier is today only one of several frontiers, given that the internal differentiation of geographical space occurs at different scales. In the context of the present global economic crisis, it is clear that international capital and American capital alike confront a global "frontier" that incorporates the so-called urban frontier.... The circumspect observation of Gottmann and others has given way 20 years later to the unabashed adoption of the "urban frontier" as the keystone to a political and economic program of urban restructuring in the interests of capital.

The frontier line today has a quintessentially economic definition – it is the frontier of profitability – but it takes on a very acute geographical expression at different spatial scales. Ultimately, this is what the 20th-century frontier and the so-called urban frontier of today have in common. In reality, both are associated with the accumulation and expansion of capital. But where the 19th-century frontier represented the consummation of absolute geographical expansion as the primary spatial expression of capital accumulation, gentrification and urban redevelopment represent the most advanced example of the redifferentiation of geographical space toward precisely the same end. It is just possible that, in order to understand the present, what is needed today is the substitution of a true geography in place of a false history.

THE RESTRUCTURING OF URBAN SPACE

[...]

The most salient processes responsible for the origins and shaping of urban restructuring can perhaps be summarized under the following headings:

(a) suburbanization and the emergence of the rent gap;
(b) the deindustrialization of advanced capitalist economies and the growth of white-collar employment;
(c) the spatial centralization and simultaneous decentralization of capital;
(d) the falling rate of profit and the cyclical movement of capital;
(e) demographic changes and changes in consumption patterns.

In consort, these developments and processes can provide a first approximation toward an integrated explanation of the different facets of gentrification and urban restructuring.

Suburbanization and the emergence of the rent gap

[...]

The suburbanization process represents a simultaneous centralization and decentralization of capital and of human activity in geographical space. On the national scale, suburbanization is the outward expansion of centralized urban places, and this process should be understood in the most general way as a necessary product of the spatial central-ization of capital. It is the growth of towns into cities into metropolitan centers.

At the urban scale, however, from the perspect-ive of the urban center, suburbanization is a pro-cess of decentralization. It is a product not of a basic impulse toward centralization but of the impulse toward a high rate of profit. Profit rates are loca-tion specific, and at the urban scale as such, the economic indicator that differentiates one place from another is ground rent. Many other forces were involved in the suburbanization of capital, but pivotal in the entire process was the availability of cheap land on the periphery (low ground rent). There was no natural necessity for the expansion of economic activity to take the form of suburban development; there was no technical impediment preventing the movement of modern large-scale capital to the rural backwaters, or preventing its fundamental redevelopment of the industrial city it inherited, but instead the expansion of capital led to a process of suburbanization. In part this had to do with the impetus toward centralization (see below), but given the economics of centralization, it is the ground-rent structure that determined the suburban location of economic expansion. . . .

The outward movement of capital to develop suburban, industrial, residential, commercial, and recreational activity results in a reciprocal change in suburban and inner-city ground-rent levels. Where the price of suburban land rises with the spread of new construction, the relative price of inner-city land falls. Smaller and smaller quantities of capital are funneled into the maintenance and repair of the inner-city building stock. This results in what we have called a *rent* gap in the inner city between the actual ground rent capitalized from the present (depressed) land use and the potential rent that could be capitalized from the "highest and best" use (or at least a "higher and better" use), given the central location. This suburbanization occurs in consort with structural changes in advanced economies. Some of the other processes we shall examine are more limited in their occurrence; what is remarkable about the rent gap is its near univer-sality. Most cities in the advanced capitalist world have experienced this phenomenon, to a greater or lesser extent. Where it is allowed to run its course at the behest of the free market, it leads to the substantial abandonment of inner-city properties.

This devalorization of capital invested in the built environment affects property of all sorts, commer-cial and industrial as well as residential and retail. Different levels and kinds of state involvement give the process a very different form in different economies, and abandonment (the logical end-point of the process) is most marked in the US, where state involvement has been less consistent and more sporadic. At the most basic level, it is the movement of capital into the construction of new suburban landscapes and the consequent creation of a rent gap that create the economic *opportunity* for restructuring the central and inner cities. The devalorization of capital in the center creates the opportunity for the revalorization of this "underdeveloped" section of urban space. The actual realization of the process, and the determination of its specific form, involve the other trends listed earlier.

Deindustrialization and the growth of a white-collar economy

Associated with the devalorization of inner-city capital is the decline of certain economic sectors and land uses more than others. This is a product primarily of broader changes in the employment structure. In particular, the advanced capitalist economies (with the major exception of Japan) have experienced the onset of deindustrialization, whereas there has been a parallel if partial indus-trialization of certain Third World economies. . . . The corollary to this deindustrialization is increased employment in other sectors of the economy, especially those described loosely as white-collar or service occupations. Within these broad categor-ies, many very different types of employment are generally included, from clerical, communications and retail operatives to managerial, professional and research careers. Within this larger trend toward a growing white-collar labor force, there-fore, there are very different tendencies and these have a specific spatial expression, as we shall see in the next section. By themselves, the processes of deindustrialization and white-collar growth do not at all explain the restructuring of the urban centers. Rather, these processes help to explain, first, the kinds of building stock and land use most involved in the development of the rent gap, and, second, the kinds of new land uses which can be expected where the opportunity for redevelopment is taken. Thus, although the media emphasis is on recent gentrification and the rehabilitation of working-class residences, there has also been a considerable transformation of old industrial areas of the city. This did not simply begin with the conversion of old warehouses into chic loft apartments; much more significant was the early urban renewal activity which, although certainly a process of slum clearance, was also the clearance of "obsolete" (meaning also devalorized) industrial buildings (factories, warehouses, wharves, etc.) where many of the slum dwellers had once worked.

Although the devalorization of capital and the development of the rent gap explain the possibility of reinvestment in the urban core around which gentrifying areas are developing, and the transformation in economic and employment structures suggests the kinds of activity that are likely to predominate in this reinvestment, there remains the question as to why the burgeoning white-collar employment is, at least in part, being centralized in the urban core. The existence of the rent gap is only a partial explanation; there is, after all, cheap land available elsewhere, throughout the rural periphery.

Spatial centralization and decentralization of capital

[. . .]

It is a cliché today to suggest that the revolution in communications technology will lead to spatial decentralization of office functions. This annihilation of space by time, as Marx had it, has indeed led to a massive suburbanization of white-collar jobs following on the heels of industrial suburbanization. With the computerization of many office functions, this trend continues. But consistent with the ideology of classlessness which first sponsored the notion of white collar, this trend is generally treated as a suburbanization of any and all types of office work from senior executives to word-processor operatives. Yet the further the trend develops, the clearer it becomes that this is not so. Thus the simultaneous centralization and decentralization of office activities represents the spatial expression of a division of labor within the so-called white-collar economy. For the most

part, the office functions that are decentralized are the more routine clerical systems and operations associated with the administration, organization and management of governmental as well as corporate activities. These represent the "back offices," the "paper factories," or, more accurately, the "communication factories" for units of the broader system. . . .

Much less usual is the suburbanization of central decision making in the form of corporate or governmental headquarters. The office boom experienced by many cities in the advanced capitalist world during the past 15 years seems to have been of this sort; it has been a continued centralization of the highest decision-making centers, along with the myriad ancillary services required by such activities: legal services, advertising, hotels and conference centers, publishers, architects, banks, financial services, and many other business services. There are exceptions to the rule, and one of the most obvious is Stamford, Connecticut, which has attracted several new corporate headquarters. Yet Stamford is in no way typical. Rather it is unique, precisely in having attracted the decentralization of ancillary administrative and professional functions central to corporate headquarters, thus resulting less in a decentralization process than in a recentralization of executive functions in Stamford. . . .

The question we are left with, then, is why, with the decentralization of industrial and communications factories, there continues to be a centralization of headquarter and executive decision-making centers. Traditional explanations focus on the importance of face-to-face contact. However, although the face-to-face explanation begins to identify the relevant issues, it is too unspecific. It tends to evoke a certain sentimentality for personal contact, but we can be sure that no mere sentimentality is responsible for the overbuilt skyscraper zones of contemporary central business districts. Behind the sentimentality lies a more expedient reason for personal contact, and this involves the very different standards by which time is managed in different sectors of the overall production and circulation of capital. Briefly, in the industrial factory and in the communications factory, the system itself (either the machinery or the administrative schedule) determines the basic daily, weekly and monthly rhythms of the work

T
H
R
E
E

process. Serious change in this long-term stability comes either from external decisions or from only periodic internal disruptions such as strikes, mechanical faults, or systems failures. The temporal regularity of these production and administration systems, along with their dependence on readily available skills in the labor force and the ease of transportation and communication with ancillary activities, make suburbanization a rational decision. They have little to gain by a centralized location in the urban core, and with high ground rents they have a lot to lose.

But the temporal rhythm of the executive administration of the economy and of its different corporate units is not stable and regular in this fashion, much to the chagrin of managers and executives. At these higher levels of control, long-term strategic planning coexists with short-term response management. Changes in interest rates or stock prices, the packaging of financial deals, labor negotiations and bailouts, international transactions in the foreign exchange market or the gold market, trade agreements, the unpredictable behavior of competitors and of government bodies – all activities of this sort can demand a rapid response by corporate financial managers, and this in turn depends on having close and immediate contact with a battery of professional, administrative and other support systems, as well as with one's competitors. At this level, and in a multitude of ways, the clichéd expression that "time is money" finds its most intense realization.... Less commonly voiced is the corollary that space too is money; spatial proximity reduces decision times when the decision system is sufficiently irregular that it cannot be reduced to a computer routine. The anarchic time regime of financial decision making in a capitalist society necessitates a certain spatial centralization. It is not just that executives *feel* more secure when packed like sardines into a skyscraper can of friends and foes. In reality they *are* more secure when rapid decisions require direct contact, information flow, and negotiation. The more the economy is prone to crisis, and thus to short-term crisis management, the more one might expect corporate headquarters to seek spatial security. Together with the expansion of this sector *per se* and the cyclical movement of capital into the built environment, this spatial response to temporal and financial irregularity

helps to explain the recent office boom in urban centers. . . .

If, in the precapitalist city, it was the needs of *market exchange* which led to spatial centralization, and in the industrial capitalist city it was the agglomeration of production capital, in the advanced capitalist city it is the financial and administrative dictates which perpetuate the tendency toward centralization. This helps to explain why certain so-called white-collar activities are centralized and others are suburbanized, and why the restructuring of the urban core takes on the corporate/professional character that it does.

The falling rate of profit and the cyclical movement of capital

Given, then, the spatial character of the process, how are we to explain the timing of this urban restructuring? This question hinges on the historical timing of the rent gap and the spatial switch of capital back to the urban center. Far from accidental occurrences, these events are integral to the broader rhythm of capital accumulation. At the most abstract level, the rent gap results from the dialectic of spatial and temporal patterns of capital investment; more concretely it is the spatial product of the complementary processes of valorization and devalorization. The accumulation of capital does not take place in a linear fashion but is a cyclical process consisting of boom periods and crises. The rent gap develops over a long period of economic expansion, but expansion that takes place elsewhere. Thus the valorization of capital in the construction of postwar suburbs was matched by its devalorization in the central and inner cities. But the accumulation of capital during such a boom leads to a falling rate of profit, beginning in the industrial sectors, and ultimately toward crisis. . . . As a means of staving off crisis at least temporarily, capital is transferred out of the industrial sphere, and as Harvey [see pp. 112–13, this volume] has shown, there is a tendency for this capital to be switched into the built environment where profit rates remain higher and where it is possible through speculation to appropriate ground rent even though nothing is produced. Two things come together, then; toward the end of a period of expansion when the rent gap has emerged and has provided the opportunity for reinvestment, there is a simultaneous tendency for capital to seek outlets in the built environment.

The slum clearance and urban renewal schemes in many Western cities following World War II were initiated and managed by the state, and though not unconnected to the emergence of the rent gap cannot adequately be explained simply in these economic terms. However, the function of this urban renewal was to prepare the way for the future restructuring which would emerge in the 1960s and become very visible in the 1970s. In economic terms the state absorbed the early risks associated with gentrification, as in Philadelphia's Society Hill . . . which was itself an urban renewal project. It also demonstrated to private capital the possibility of large-scale restructuring of the urban core, paving the way for future capital investment.

The timing of this spatial restructuring, then, is closely related to the economic restructuring that takes place during economic crises such as those the world economy has experienced since the early 1970s. A restructured economy involves a restructured built environment. But there is no gradual transition to a restructured economy; the last economic crisis was resolved only after a massive destruction of capital in World War II, representing a cataclysmic devalorization of capital and a destruction prior to a restructuring of urban space. Today, 50 years later, we are again facing the same threat.

Demographic changes and consumption patterns

The maturation of the baby-boom generation, the increased number of women taking on careers, the proliferation of one- and two-person households and the popularity of the "urban singles" life-style are commonly invoked as the real factors behind gentrification. Consistent with the frontier ideology, the process is viewed here as the outcome of individual choices. But in reality too much is claimed. We are seeing a much larger urban restructuring than is encompassed by residential rehabilitation, and it is difficult to see how such explanations could at best be more than partial. Where such explanations might just be conceivable

for St. Katherine's Dock in London, they are irrelevant for understanding the London office boom and the redevelopment of the docklands. Yet these are all connected. The changes in demographic patterns and life-style preferences are not completely irrelevant, but it is vital that we understand what these developments can and cannot explain.

The importance of demographic and life-style issues seems to be chiefly in the determination of the surface form taken by much of the restructuring rather than explaining the fact of urban transformation. Given the movement of capital into the urban core, and the emphasis on executive, professional, administrative and managerial functions, as well as other support activities, the demographic and life-style changes can help to explain why we have proliferating quiche bars rather than Howard Johnsons, trendy clothes boutiques and gourmet food shops rather than corner stores, American Express signs rather than "cash only, no cheques."

[. . .]

DIRECTION AND LIMITS OF URBAN RESTRUCTURING

If the restructuring that has now begun continues in its current direction, then we can expect to see significant changes in urban structure. However accurate the Chicago model of urban structure may have been, there is general agreement that it is no longer appropriate. Urban development has overtaken the model. The logical conclusion of the current restructuring, which remains today in its infancy, would be an urban center dominated by high-level executive-professional, financial, and administrative functions, middle- and upper-middle-class residences, and the hotel, restaurant, moving, retail and cultural facilities providing recreational opportunities for this population. In short we should expect the creation of a bourgeois playground, the social Manhattanization of the urban core to match the architectural Manhattanization that heralded the changing employment structure. The corollary of this is likely to be a substantial displacement of the working class to the older suburbs and the urban periphery.

Editors' references and suggestions for further reading

Boyle, M. (1995) "Still top of our agenda? Neil Smith and the reconciliation of capital and consumer approaches to gentrification," *Scottish Geographical Magazine* 111: 95–107.

Hamnett, C. (1991) "The blind men and the elephant: the explanation of gentrification," *Transactions of the Institute of British Geographers* 16: 173–89.

Lees, L. (1994) "Rethinking gentrification: beyond the positions of economic or culture," *Progress in Human Geography* 18: 137–50.

Lees, L. (2000) "A reappraisal of gentrification: towards a 'geography of gentrification,'" *Progress in Human Geography* 24: 389–408.

Ley, D. (1980) "Liberal ideology and the post-industrial city," *Annals of the Association of American Geographers* 70: 238–58.

Ley, D. (1996) *The New Middle Class and the Remaking of the Central City*, Oxford: Oxford University Press.

Smith, N. (1996) *The New Urban Frontier: Gentrification and the Revanchist City*, London and New York: Routledge.

THREE

"Postmodern Urbanism"

from *Annals of the Association of American Geographers* (1998)

Michael Dear and Steven Flusty

Editors' Introduction

Postmodernism originated in philosophical discourses that questioned the grand claims and grand theory of the modern era, and it first captured the interest of geographers in the early 1980s, with Michael Dear among those who spearheaded discussions about its relevance for the discipline. In a paper on "Postmodernism and planning" (1986) he teased apart three key dimensions of postmodernism: style, method and epoch. Postmodernism as style began in literature and literary criticism but spread to other fields including architecture and urban design where it is associated with a sensitivity "to vernacular traditions, local histories, particular wants, needs, and fancies" (Harvey 1989: 66). As a method, postmodernism draws heavily on deconstruction, seeking to demonstrate the interplay between the multiple positions of authors and the writing and reading of their texts. As an epoch, postmodernism is taken to represent a fundamental discontinuity with modernism brought about by radical changes in the nature of global capitalism and, in particular, the transition from a Fordist to a post-Fordist phase of "flexible" capital accumulation.

While the impact of postmodernism on the discipline of geography has been highly uneven (something Dear anticipated in his 1988 paper "The postmodern challenge: reconstructing human geography"), it is in urban geography that its influence has been particularly significant. This owes much to important individual contributions, such as the studies by Ley, Knox and Goss discussed in Part 6, Form and Symbolism, but it also reflects the collective work of a group of researchers based in Southern California who have been labeled the Los Angeles School. Michael Dear and Steven Flusty are members of this School: Dear is Professor of Geography at the University of Southern California and Flusty lectures in geography at York University, Toronto. In their reprinted paper here they draw on a range of research regarding the contemporary restructuring of the Los Angeles metropolitan region, in order to argue for a "postmodern urbanism," a phrase signaling their contention that "we [have] arrived at a radical break in the way cities are developing." In contrast to the Chicago School's concept of the modern city "as an organic accretion around a central organizing core," Dear and Flusty argue that a postmodern urban process in which "the urban periphery organizes the center within the context of globalizing capitalism" now shapes cities. By synthesizing selected elements of Southern California's urban geographies, ranging from its edge cities to its cultures of heteropolis, Dear and Flusty develop the concept of a "proto-postmodern" urban process, driven by a global restructuring that is "permeated and balkanized by a series of interdictory networks." These processes yield a model of urban structure that is analogous, they contend, to a keno game card. On such a card there are numbered grids with some squares being marked in the course of a game while others are not, according to a random draw. Applied to the city, what characterizes postmodern urban structure is the apparently random development and redevelopment of urban land, the outcome of "exogenous investment processes" in which

"Capital touches down as if by chance on a parcel of land, ignoring the opportunities on intervening lots, thus sparking the development process."

By arguing that the distinctive features of postmodern urbanism mark a radical break from the modernist city and by seeking to move beyond existing conceptual frameworks in order to understand these new landscapes, Dear and Flusty's paper is an exciting and thought-provoking contribution to urban geography. With its decentered urban sprawl, gated communities and edge cities, Los Angeles is presented as the paradigmatic postmodern urban landscape, a landscape which stands in stark contrast to that associated with the modern city. Furthermore, by relying on a bundle of neologisms to capture the "newness" of postmodern urbanism – "commudities" (commodified communities), "cybergoisie" (an elite of chief executives and entrepreneurs), "protosurps" (marginalized surplus labor) – their argument also appears to mark a radical epistemological break with the established language used to understand urban processes.

In attempting to set such a path-breaking agenda, however, Dear and Flusty's ideas and arguments have attracted considerable comment and criticism. Some have quibbled about the status of Los Angeles as the quintessential postmodern city. Nijman (2000), for example, argues that Miami deserves this title because although, like LA, it contains an assemblage of postmodern features, such as privatopias and fortified enclaves, Miami is only one-fifth the size of LA, making the subjective experiences of postmodern urbanism much more intense. Other researchers have drawn attention to the intriguing parallels between the work of the Chicago School and that of the LA School despite Dear and Flusty's contention that important differences exist between them. Both schools, for example, focus exclusively on the urban experience of the United States, thus limiting the relevance of their models of urban processes and structures for other geographical contexts. Both schools also focus on issues of diversity, fragmentation, and dysfunction in the urban environment. For the Chicago School this was evident in their work on social disorganization and deviance in the city; in Dear and Flusty's analysis, it is manifest in their discussions of Southern California's "cultures of heteropolis" and the importance they attach to the "containment centers" (prisons), "gated communities," "interdictory spaces" and "street warfare" in their model of postmodern urban structure.

These points of overlap between the interests of the Chicago and LA schools feed into a bigger question which Dear and Flusty pose in the introduction to their paper: "Have we arrived at a radical break in the way cities are developing?" Their answer is unequivocally "Yes," but others are less convinced. Several observers of Los Angeles' urban landscape view many of its contemporary features firmly within a modernist rather than postmodernist framework. Ley (2001), for example, suggests that with its mix of freeways, high-rise buildings, and suburban sprawl LA's urban landscape is typical of the kind of urban society envisaged by high modernists such as Le Corbusier and Frank Lloyd Wright. Similarly, Caldeira (2000) argues that in LA "conventions of modernist city planning and technologies of security are being used to create new forms of urban space and social segregation" (p. 132) and that the fortified enclaves that Dear and Flusty use to help define postmodern urbanism are the outcome of modernist architectural design strategies. At a more general level there are those who are not convinced that the postmodern city constitutes a significantly new form of urbanization. Beauregard and Haila (1997) "see a more complex patterning of old and new, of continuing trends and new forces" in the new spatial elements of the contemporary city. "Edge cities" provide one example which, as Walker and Lewis suggest in their paper in this section, should be treated less as "something entirely new under the sun" and more "as the latest episode in a long-running story of North American urbanization." A slightly different and more equivocal assessment of whether the postmodern city should be viewed as a radical break with existing forms of urban development is put forward by Hannigan (1995). On the one hand, Hannigan argues that if postmodernism is viewed as a form of architecture and urban design of the built environment then the evidence for new urban landscapes is impressive. On the other hand, if postmodernism is viewed as involving fundamental changes in urban social relations then the fact that "the less fortunate as well as a sizeable chunk of the middle-class have opted out of or been excluded from the postmodern city" (p. 203) means the claim that it is a significant new element of urban life is less secure.

While there is continuing debate in some quarters over whether postmodern urbanism is a "reality" or a "fantasy" (see Cherot and Murray 2002), research on the postmodern city remains a flourishing area of urban studies with Michael Dear, Steven Flusty and other members of the LA School making significant contributions to this expanding field. Dear's (2000) *The Postmodern Urban Condition* elaborates on and extends the ideas and arguments contained in the article reprinted in this section, while two edited collections, *From Chicago to L.A.: Making Sense of Urban Theory* (Dear 2002) and *Spaces of Postmodernity* (Dear and Flusty 2001) provide more wide-ranging analyses of the relevance of postmodernism to the city and the discipline of human geography. However, it is not only contemporary processes of urbanization in the developed world that have come under the postmodern "spotlight." Taking up the invitation in the conclusion of Dear and Flusty's article to consider postmodern urbanism in other contexts, Quataert (2003) has examined postmodern cities in the Middle East, Myung-Rae (1999) has explored the postmodernity of the Korean capital, Seoul, and Edward Soja (2000), another prominent urban geographer and member of the LA School, has employed a postmodern perspective to revisit debates over the origins of urbanism in his *Postmetropolis: Critical Studies of Cities and Regions*.

In addition to his work on postmodern urbanism, Michael Dear has written widely on the city and political and social geography. His many books include *Urbanization and Urban Planning in Capitalist Societies* (London: Methuen, 1981, co-edited with A.J. Scott), *The State Apparatus: The Structure and Language of Legitimacy* (London: Allen & Unwin, 1984, co-authored with Gordon Clark), and, co-authored with Jennifer Wolch, *Landscapes of Despair: From Deinstitutionalization to Homelessness* (Princeton, NJ: Princeton University Press, 1987) and *Malign Neglect: Homelessness in an American City* (Los Angeles, CA: Jossey-Bass, 1993). Steven Flusty has research interests in globalization, surveillance and spatial justice, and is author of *De-Coca-Colonization: Making the Globe from Inside Out* (London: Routledge, 2003).

[...]

Have we arrived at a radical break in the way cities are developing? Is there something called a *postmodern urbanism*, which presumes that we can identify some form of template that defines its critical dimensions? This inquiry is based on a simple premise: that just as the central tenets of modernist thought have been undermined, its core evacuated and replaced by a rush of competing epistemologies, so too have the traditional logics of earlier urbanisms evaporated, and in the absence of a single new imperative, multiple urban (ir)rationalities are competing to fill the void. It is the concretization and localization of these effects, global in scope but generated and manifested locally, that are creating the geographies of postmodern society – a new time-space fabric. We begin this search by outlining the fundamental precepts of the Chicago School, a classical modernist vision of the industrial metropolis, and contrasting these with evidence of a nascent postmodern Los Angeles School. Next we examine a broad range of contemporary Southern California urbanisms, before going on to suggest a critical reinterpretation of this evidence that encompasses and defines the problematic of a postmodern urbanism. In conclusion, we offer comments intended to assist in formulating an agenda for comparative urban research.

FROM CHICAGO TO LOS ANGELES

[...]

The Chicago School

General theories of urban structure are a scarce commodity. One of the most persistent models of urban structure is associated with a group of sociologists who flourished in Chicago in the 1920s and 1930s. According to Morris Janowitz, the "Chicago School" was motivated to regard the city "as an object of detached sociological analysis," worthy of distinctive scientific attention:

> The city is not an artifact or a residual arrangement. On the contrary, the city embodies the real nature of human nature. It is an expression of mankind in general and specifically of the social relations generated by territoriality (Janowitz 1967: viii–ix).

The most enduring of the Chicago School models was the *zonal or concentric ring theory*, an account of the evolution of differentiated urban social areas by E.W. Burgess (1925) [and see Part 1, this volume]. Based on assumptions that included a uniform land surface, universal access to a single, centered city, free competition for space, and the notion that development would take place outward from a central core, Burgess concluded that the city would tend to form a series of concentric zones. . . . The main ecological metaphors invoked to describe this dynamic were invasion, succession, and *segregation*, by which populations gradually filtered outwards from the center as their status and level of assimilation progressed. The model was predicated on continuing high levels of inmigration to the city.

[. . .]

A "Los Angeles School"?

During the 1980s, a group of loosely associated scholars, professionals, and advocates based in Southern California began to examine the notion that what was happening in the Los Angeles region was somehow symptomatic of a broader socio-geographic transformation taking place within the U.S. as a whole. Their common but then unarticulated project was based on certain shared theoretical assumptions, and on the view that L.A. was emblematic of some more general urban dynamic. One of the earliest expressions of an emergent "L.A. School" was the appearance in 1986 of a special issue of the journal *Society and Space*, which was entirely devoted to understanding Los Angeles. In their prefatory remarks to that issue, Allen Scott and Edward Soja referred to Los Angeles as the "capital of the twentieth century," deliberately invoking Walter Benjamin's reference to Paris as the capital of the nineteenth. They predicted that the volume of scholarly work on Los Angeles would quickly overtake that on Chicago.

The burgeoning outlines of an L.A. School were given crude form by a series of meetings and publications that occurred during the late 1980s, and by 1990, in his penetrating critique of Southern California urbanism (*City of Quartz*), Mike Davis was able to make specific reference to the School's expanding consciousness. He commented that its practitioners were undecided whether to model themselves after the Chicago School (named principally for the city that was its object of inquiry), or the Frankfurt School (a philosophical alliance named only coincidentally after its place of operations). Then, in 1993, Marco Cenzatti published a short pamphlet that was the first publication to explicitly examine the focus and potential of an L.A. School. Responding to Davis, he underscored that the School's practitioners *combine* precepts of both the Chicago and Frankfurt Schools. Just as the Chicago School emerged at a time when that city was reaching new national prominence, Los Angeles has begun to make its impression on the minds of urbanists. Their theoretical inquiries focus not only on the specific city, but also on more general questions concerning urban processes. Cenzatti claims that one concern common to all adherents of the L.A. School is a focus on restructuring, which includes deindustrialization and reindustrialization, the birth of the information economy, the decline of nation-states, the emergence of new nationalisms, and the rise of the Pacific Rim. Such proliferating logics often involve multiple theoretical frameworks that overlap and coexist in their explanations of the burgeoning global/local order – a heterodoxy consistent with the project of postmodernism.

[. . .]

WAYS OF SEEING: SOUTHERN CALIFORNIAN URBANISMS

[. . .]

Taking Los Angeles seriously

[. . .]

One of the most prescient visions anticipating a postmodern cognitive mapping of the urban is Jonathan Raban's *Soft City* (1974), a reading of London's cityscapes. Raban divides the city into hard and soft elements. The former refers to the material fabric of the built environment – the streets and buildings that frame the lives of city dwellers. The latter, by contrast, is an individualized interpretation of the city, a perceptual orientation created in the mind of every urbanite. The relationship between the two is complex and even indeterminate. The newcomer to a city first confronts the hard city, but soon:

the city goes soft; it awaits the imprint of an identity. For better or worse, it invites you to remake it, to consolidate it into a shape you can live in. You, too. Decide who you are, and the city will again assume a fixed form around you. Decide what it is, and your own identity will be revealed (p. 11).

Raban makes no claims to a postmodern consciousness, yet his invocation of the relationship between the cognitive and the real leads to insights that are unmistakably postmodern in their sensitivities.

Ted Relph (1987) was one of the first geographers to catalogue the built forms that comprise the places of postmodernity. He describes postmodern urbanism as a self-conscious and selective revival of elements of older styles, though he cautions that postmodernism is not simply a style but also a frame of mind (p. 213). He observes how the confluence of many trends – gentrification, heritage conservation, architectural fashion, urban design, and participatory planning – caused the collapse of the modernist vision of a future city filled with skyscrapers and other austere icons of scientific rationalism. The new urbanism is principally distinguishable from the old by its *eclecticism*.

[. . .]

Raban's emphasis on the cognitive and Relph's on the concrete underscore the importance of both dimensions in understanding sociospatial urban process. The palette of urbanisms that arises from merging the two is thick and multidimensional. We turn now to the task of constructing that palette (what we earlier described as a template) by examining empirical evidence of recent urban developments in Southern California (Table 1). In this review, we take our lead from what exists, rather than what we consider to be a comprehensive urban research agenda. From this, we move quickly to a synthesis that is prefigurative of a protopostmodern urbanism, which we hope will serve as an invitation to a more broadly based comparative analysis.

Table 1 A taxonomy of Southern California urbanisms

Edge cities	Interdictory space
Privatopia	Historical geographies of restructuring
Cultures of heteropolis	Fordist/post-Fordist regimes of accumulation/regulation
City as theme park	Globalization
Fortified city	Politics of nature

Edge cities

Joel Garreau noted the central significance of Los Angeles in understanding contemporary metropolitan growth in the U.S. He asserts (1991: 3) that: "Every single American city that *is* growing, is growing in the fashion of Los Angeles," and refers to L.A. as the "great-granddaddy" of edge cities (he claims there are twenty-six of them within a five-county area in Southern California). For Garreau, edge cities represent the crucible of America's urban future. The classic location for contemporary edge cities is at the intersection of an urban beltway and a hub-and-spoke lateral road. The central conditions that have propelled such development are the dominance of the automobile and the associated need for parking, the communications revolution, and the entry of women in large numbers into the labor market.

[. . .]

Privatopia

Privatopia, perhaps the quintessential edge city residential form, is a private housing development based in common-interest developments (CIDs) and administered by homeowners' associations. There were fewer than 500 such associations in 1964; by 1992, there were 150,000 associations privately governing approximately 32 million Americans. In 1990, the 11.6 million CID units constituted more than 11 per cent of the nation's housing stock (McKenzie 1994: 11). Sustained by an expanding catalogue of covenants, conditions, and restrictions (or CC&Rs, the proscriptive constitutions formalizing CID behavioral and aesthetic norms), privatopia has been fueled by a large dose of privatization, and promoted by an ideology of "hostile privatism" (McKenzie 1994: 19). It has provoked a culture of nonparticipation.

[. . .]

Cultures of heteropolis

One of the most prominent sociocultural tendencies in contemporary Southern California is the rise of minority populations (Ong *et al.* 1994; Roseman *et al.* 1996; Waldinger and Bozorgmehr 1996).

Provoked to comprehend the causes and implications of the 1992 civil disturbances in Los Angeles, Charles Jencks (1993: 32) zeroes in on the city's *diversity* as the key to L.A.'s emergent urbanism: "Los Angeles is a combination of enclaves with high identity, and multienclaves with mixed identity, and, taken as a whole, it is perhaps the most heterogeneous city in the world." Such ethnic pluralism has given rise to what Jencks calls a *hetero-architecture*, which has demonstrated that: "there is a great virtue, and pleasure, to be had in mixing categories, transgressing boundaries, inverting customs and adopting the marginal usage" (1993: 123).

[. . .]

City as theme park

California in general, and Los Angeles in particular, have often been promoted as places where the American (suburban) Dream is most easily realized. Its oft-noted qualities of optimism and tolerance coupled with a balmy climate have given rise to an architecture and society fostered by a spirit of experimentation, risk taking, and hope. Architectural dreamscapes are readily convertible into marketable commodities, i.e., saleable prepackaged landscapes engineered to satisfy fantasies of suburban living. Many writers have used the "theme park" metaphor to describe the emergence of such variegated cityscapes. For instance, Michael Sorkin, in a collection of essays appropriately entitled *Variations on a Theme Park* (1992), describes theme parks as places of simulation without end, characterized by aspatiality plus technological and physical surveillance and control. . . . The phone and modem have rendered the street irrelevant, and the new city threatens an "unimagined sameness" characterized by the loosening of ties to any specific space, rising levels of surveillance, manipulation and segregation, and the city as a theme park.

[. . .]

Fortified city

The downside of the Southern Californian dream has, of course, been the subject of countless dystopian visions in histories, movies, and novels. In one powerful account, Mike Davis noted how Southern Californians' obsession with security has transformed the region into a fortress. This shift is accurately manifested in the physical form of the city, which is divided into fortified cells of affluence and places of terror where police battle the criminalized poor. These urban phenomena, according to Davis, have placed Los Angeles "on the hard edge of postmodernity" (Davis 1992a: 155). The dynamics of fortification involve the omnipresent application of high-tech policing methods to the "high-rent security of gated residential developments" and "panopticon malls." It extends to "space policing," including a proposed satellite observation capacity that would create an invisible Haussmannization of Los Angeles. In the consequent "carceral city," the working poor and destitute are spatially sequestered on the "mean streets," and excluded from the affluent "forbidden cities" through "security by design."

Interdictory space

Elaborating upon Davis's fortress urbanism, Steven Flusty observed how various types of fortification have extended a canopy of suppression and surveillance across the entire city. His taxonomy of interdictory spaces (1994: 16–17) identifies how spaces are designed to exclude by a combination of their function and cognitive sensibilities. Some spaces are passively aggressive: space concealed by intervening objects or grade changes is "stealthy"; space that may be reached only by means of interrupted or obfuscated approaches is "slippery." Other spatial configurations are more assertively confrontational: deliberately obstructed "crusty" space surrounded by walls and checkpoints; inhospitable "prickly" spaces featuring unsittable benches in areas devoid of shade; or "jittery" space ostentatiously saturated with surveillance devices.

[. . .]

Historical geographies of restructuring

[. . .]

In his history of Los Angeles between 1965 and 1992, Soja (1996) attempts to link the emergent patterns of urban form with underlying social processes. He identified six kinds of *restructuring*, which together define the region's contemporary urban process. In addition to *Exopolis* (noted above), Soja lists: *Flexcities*, associated with the transition to post-Fordism, especially deindustrialization and the rise of the information economy; and *Cosmopolis*, referring to the globalization of Los Angeles both in terms of its emergent world-city status and its internal multicultural diversification. According to Soja, peripheralization, post-Fordism, and globalization together define the experience of urban restructuring in Los Angeles. Three specific geographies are consequent upon these dynamics: *Splintered Labyrinth*, which describes the extreme forms of social, economic, and political polarization characteristic of the postmodern city; *Carceral City*, referring to the new "incendiary urban geography" brought about by the amalgam of violence and police surveillance; and *Simcities*, the term Soja uses to describe the new ways of seeing the city that are emerging from the study of Los Angeles – a kind of epistemological restructuring that foregrounds a postmodern perspective.

Fordist versus post-Fordist regimes of accumulation and regulation

Many observers agree that one of the most important underlying shifts in the contemporary political economy is from a Fordist to a post-Fordist industrial organization. In a series of important books, Allen Scott and Michael Storper have portrayed the burgeoning urbanism of Southern California as a consequence of this deep-seated structural change in the capitalist political economy (Scott 1988a, 1988b, 1993; Storper and Walker 1989). For instance, Scott's basic argument is that there have been two major phases of urbanization in the U.S. The first related to an era of Fordist mass production, during which the paradigmatic cities of industrial capitalism (Detroit, Chicago, Pittsburgh etc.) coalesced around industries that were themselves based upon ideas of mass production. The second phase is associated with the decline of the Fordist era and the rise of a post-Fordist "flexible production." This is a form of industrial activity based on small-size, small-batch units of (typically subcontracted) production that are nevertheless integrated into clusters of economic activity. Such clusters have been observed in two manifestations:

labor-intensive craft forms (in Los Angeles, typically garments and jewellery), and high technology (especially the defense and aerospace industries). According to Scott, these so-called "technopoles" until recently constituted the principal geographical loci of contemporary (sub)urbanization in Southern California (a development prefigured in Fishman's description of the "technoburb"; see Castells and Hall 1994; Fishman 1987).

[. . .]

Globalization

Needless to say, any consideration of the changing nature of industrial production sooner or later must encompass the globalization question (cf. Knox and Taylor 1995). In his reference to the global context of L.A.'s localisms, Mike Davis (1992b) claims that if L.A. is in any sense paradigmatic, it is because the city condenses the intended and unintended spatial consequences of post-Fordism. He insists that there is no simple master-logic of restructuring, focusing instead on two key localized macro-processes: the overaccumulation in Southern California of bank and real-estate capital, principally from the East Asian trade surplus, and the reflux of low-wage manufacturing and labor-intensive service industries, following upon immigration from Mexico and Central America. For instance, Davis notes how the City of Los Angeles used tax dollars gleaned from international capital investments to subsidize its downtown (Bunker Hill) urban renewal, a process he refers to as "municipalized land speculation" (1992b: 26). Through such connections, what happens today in Asia and Central America will tomorrow have an effect in Los Angeles. This global/local dialectic has already become an important (if somewhat imprecise) *leitmotif* of contemporary urban theory.

Politics of nature

The natural environment of Southern California has been under constant assault since the first colonial settlements. Human habitation on a metropolitan scale has only been possible through a widespread manipulation of nature, especially the control of water resources in the American West (Davis 1993;

Gottleib and FitzSimmons 1991; Reisner 1993). On one hand, Southern Californians tend to hold a grudging respect for nature, living as they do adjacent to one of the earth's major geological hazards and in a desert environment that is prone to flood, landslide, and fire (see, for instance, Darlington 1996; McPhee 1989). On the other hand, its inhabitants have been energetically, ceaselessly, and sometimes carelessly unrolling the carpet of urbanization over the natural landscape for more than a century. This uninhibited occupation has engendered its own range of environmental problems, most notoriously air pollution, but it also brings forth habitat loss and dangerous encounters between humans and other animals.

[. . .]

Synthesis: protopostmodern urbanism

If these observers of the Southern California scene could talk with each other to resolve their differences and reconcile their terminologies, how might they synthesize their visions? At the risk of misrepresenting their work, we suggest a schematic that is powerful, yet inevitably incomplete (Figure 1). It suggests a "protopostmodern" urban process, driven by a global restructuring that is permeated and balkanized by a series of interdictory networks; whose populations are socially and culturally heterogeneous, but politically and economically polarized; whose residents are educated and persuaded to the consumption of dreamscapes even as the poorest are consigned to carceral cities; whose built environment, reflective of these processes, consists of edge cities, privatopias, and the like; and whose natural environment, also reflective

of these processes, is being erased to the point of unlivability while, at the same time, providing a focus for political action.

POSTMODERN URBANISM

. . . Recognizing that we may have caused some offense by characterizing others' work in this way, let us move swiftly to reconstruct their evidence into a postmodern urban problematic (Table 2). We anchor this problematic in the straightforward need to account for the evolution of society over time and space. Such evolution occurs as a combination of deep-time (long-term) and present-time (short-term) processes, and it develops over several different scales of human activity (which we may represent summarily as micro- , meso- , and macroscales) (Dear 1988). The structuring of the time-space fabric is the result of the interaction among ecologically situated human agents in relations of production, consumption, and coercion. We do not intend any primacy in this ordering of categories, but instead emphasize their *interdependencies* – all are essential in explaining postmodern human geographies.

Our promiscuous use of neologisms in what follows is quite deliberate. . . . Neologisms have been

Table 2 Elements of a postmodern urbanism

GLOBAL LATIFUNDIA
HOLSTEINIZATION
PRAEDATORIANISM
FLEXISM
NEW WORLD BIPOLAR DISORDER
Cybergeoisie
Protosurps
MEMETIC CONTAGION
KENO CAPITALISM
CITISTÄT
Commudities
Cyburbia
Citidel
In-beyond
Cyberia
POLLYANNARCHY
DISINFORMATION SUPERHIGHWAY

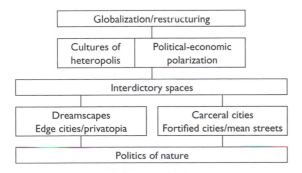

Figure 1 A concept of protopostmodern urbanism.

used here in circumstances when there were no existing terms to describe adequately the conditions we sought to identify, when neologisms served as metaphors to suggest new insights, when a single term more conveniently substituted for a complex phrase or string of ideas, and when neologistic novelty aided our avowed efforts to rehearse the break. The juxtaposing of postmodern and more traditional categories of modernist urbanism is also an essential piece of our analytical strategy. That there is an overlap between modernist and postmodern categories should surprise no one; we are, inevitably, building on existing urbanisms and epistemologies. The consequent neologistic pastiche may be properly regarded as a tactic of postmodern analysis; others could regard this strategy as analogous to hypothesis-generation, or as the practice of dialectics.

Urban pattern and process

We begin with the assumption that urbanism is made possible by the exercise of instrumental control over both human and nonhuman ecologies (Figure 2). The very occupation and utilization of space, as well as the production and distribution of commodities, depends upon an anthropocentric reconfiguration of natural processes and their products. As the scope and scale of, and dependency upon, globally integrated consumption increases, institutional action converts complex ecologies into

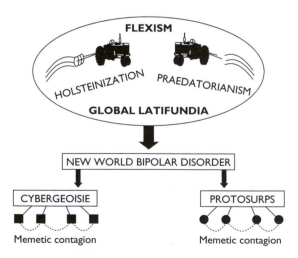

Figure 2 Elements of postmodern urbanism – 1.

monocultured factors of production by simplifying nature into a *global latifundia*. . . . Being part of nature, humanity is subjected to analogous dynamics. *Holsteinization* is the process of monoculturing people as consumers so as to facilitate the harvesting of desires, including the decomposition of communities into isolated family units and individuals in order to supplant social networks of mutual support with consumersheds of dependent customers. Resistance is discouraged by means of *praedatorianism*, i.e., the forceful interdiction by a praedatorian guard with varying degrees of legitimacy.

The global latifundia, holsteinization, and praedatorianism are, in one form or another, as old as the global political economy, but the overarching dynamic signaling a break with previous manifestations is *flexism*, a pattern of econo-cultural production and consumption characterized by near-instantaneous delivery and rapid redirectability of resource flows. Flexism's fluidity results from cheaper and faster systems of transportation and telecommunications, globalization of capital markets, and concomitant flexibly specialized, just-in-time production processes enabling short product- and production-cycles. . . . Globalization and rapidity permit capital to evade long-term commitment to place-based socioeconomies, thus enabling a crucial social dynamic of flexism: whereas, under Fordism, exploitation is exercised through the alienation of labor in the place of production, flexism may require little or no labor at all from a given locale. Simultaneously, local down-waging and capital concentration operate synergistically to supplant locally owned enterprises with national and supranational chains, thereby transferring consumer capital and inventory selection ever farther away from direct local control.

From these exchange asymmetries emerges a new world *bi-polar disorder*. This is a globally bifurcated social order, many times more complicated than conventional class structures, in which those overseeing the global latifundia enjoy concentrated power. Those who are dependent upon their command-and-control decisions find themselves in progressively weaker positions, pitted against each other globally, and forced to accept shrinking compensation for their efforts (assuming that compensation is offered in the first place). Of the two groups, the *cybergeoisie* reside in the "big house" of the global latifundia, providing indispensable, presently unautomatable command-and-control functions. They are predominantly stockholders, the core employees of thinned-down corporations, and write-your-own-ticket freelancers (e.g., CEOs, subcontract entrepreneurs, and celebrities).... Commanding, controlling, and prodigiously enjoying the fruits of a shared global exchange of goods and information, the cybergeoisie exercise global coordination functions that predispose them to a similar ideology and, thus, they are relatively heavily holsteinized.

Protosurps, on the other hand, are the share-croppers of the global latifundia. They are increasingly marginalized "surplus" labor providing just-in-time services when called upon by flexist production processes, but otherwise alienated from global systems of production (though not of consumption). Protosurps include temporary or day laborers, fire-at-will service workers, a burgeoning class of intra- and international itinerant laborers specializing in pursuing the migrations of fluid investment.... Subjected to high degrees of uncertainty by the omnipresent threat of instant unemployment, protosurps are prone to clustering into affinity groups for support in the face of adversity....

The sociocultural collisions and intermeshings of protosurp affinity groups, generated by flexist, induced immigration and severe social differentiation, serves to produce wild *memetic contagion*. This is a process by which cultural elements of one individual or group exert cross-over influences upon the culture of another, previously unexposed individual/group. Memetic contagion is evidenced in Los Angeles by such hybridized agents and intercultural conflicts as Mexican and Central American practitioners of Afro-Caribbean religion (McGuire and Scrymgeour 1998), blue-bandanna'd Thai Crips, or the adjustments prompted by poor African-Americans' offense at Korean merchants' disinclination to smile casually.... With the flexist imposition of global imperatives on local economies and cultures, the spatial logic of Fordism has given way to a new, more dissonant international geographical order. In the absence of conventional communication and transportation imperatives mandating propinquity, the once-standard Chicago School logic has given way to a seemingly haphazard juxtaposition of land uses scattered over the landscape.... The result is a landscape not unlike that

formed by a keno gamecard. The card itself appears as a numbered grid, with some squares being marked during the course of the game and others not, according to some random draw. The process governing this marking ultimately determines which player will achieve a jackpot-winning pattern; it is, however, determined by a rationalized set of procedures beyond the territory of the card itself. Similarly, the apparently random development and redevelopment of urban land may be regarded as the outcome of exogenous investment processes inherent to flexism, thus creating the landscapes of *keno capitalism*.

Keno capitalism's contingent mosaic of variegated monocultures renders discussion of "the city" increasingly reductionist. More holistically, the dispersed net of megalopoles may be viewed as a single integrated urban system, or *Citistät* (Figure 3). Citistät, the collective world city, has emerged from competing urban webs of colonial and postcolonial eras to become a geographically diffuse hub of an omnipresent periphery, drawing labor and materials from readily substitutable locations throughout that periphery....

Materially, Citistät consists of *commudities* (centers of command and control), and the in-beyond (internal peripheries simultaneously undergoing

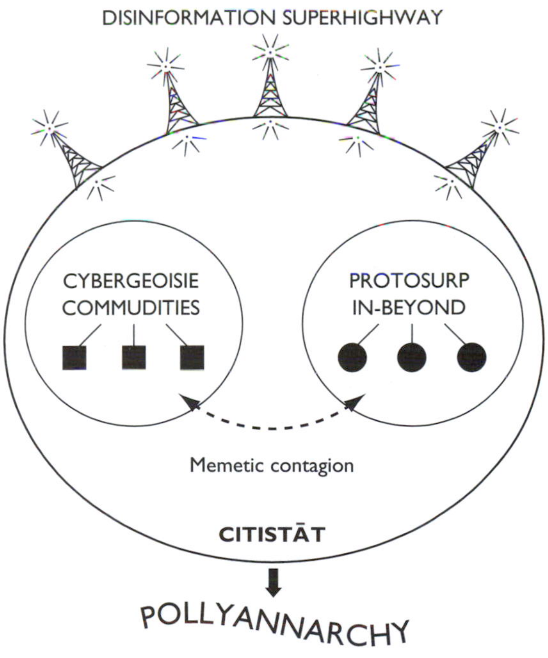

Figure 3 Elements of postmodern urbanism – 2.

but resisting instrumentalization in myriad ways). Virtually, Citistāt consists of *cyburbia*, the collection of state-of-the-art data-transmission, premium pay-per-use, and interactive services generally reliant upon costly and technologically complex interfaces; and *cyberia*, an electronic outland of rudimentary communications including basic phone service and telegraphy, interwoven with and preceptorally conditioned by the disinformation superhighway (DSH).

Commudities are commodified communities created expressly to satisfy (and profit from) the habitat preferences of the well-recompensed cybergeoisie. They commonly consist of carefully manicured residential and commercial ecologies managed through privatopian self-administration, and maintained against internal and external outlaws by a repertoire of interdictory prohibitions. Increasingly, these prepackaged environments jockey with one another for clientele on the basis of recreational, cultural, security, and educational amenities.... Citistāt's internal periphery and repository of cheap on-call labor lies at the *in-beyond*, comprised of a shifting matrix of protosurp affinity clusters. The in-beyond may be envisioned as a patchwork quilt of variously defined interest groups (with differing levels of economic, cultural, and street influence), none of which possesses the wherewithal to achieve hegemonic status or to secede. Secession may occur locally to some degree, as in the cases of the publicly subsidized reconfiguration of L.A.'s Little Tokyo, and the consolidation of Koreatown through the import, adjacent extraction, and community recirculation of capital....

Political relations in Citistāt tend toward polyanarchy, a politics of grudging tolerance of *difference* that emerges from interactions and accommodations within the in-beyond and between commudities, and less frequently, between in-beyond and commudity. Its more pervasive form is *pollyannarchy*, an exaggerated, manufactured optimism that promotes a self-congratulatory awareness and respect for difference and the asymmetries of power.... Pollyannarchy is evident in the continuing spectacle of electoral politics, or in the citywide unity campaign run by corporate sponsors following the 1992 uprising in Los Angeles.

Wired throughout the body of the Citistāt is the *disinformation superhighway* (or DSH), a mass

info-tain-mercial media owned by roughly two dozen cybergeoisie institutions. The DSH disseminates holsteinizing ideologies and incentives, creates wants and dreams, and inflates the symbolic value of commodities. At the same time, it serves as the highly filtered sensory organ through which commudities and the in-beyond perceive the world outside their unmediated daily experiences....

An alternative model of urban structure

[...]

[B]y now it is clear that the most influential of existing urban models is no longer tenable as a guide to contemporary urbanism. In this first sense, our investigation has uncovered an *epistemological radical break* with past practices, which in itself is sufficient justification for something called a Los Angeles School. The concentric ring structure of the Chicago School was essentially a concept of the city as an organic accretion around a central, organizing core. Instead, we have identified a postmodern urban process in which the urban periphery organizes the center within the context of a globalizing capitalism.

[...]

Keno capitalism is the synoptic term that we have adopted to describe the spatial manifestations of the postmodern urban condition (Figure 4). Urbanization is occurring on a quasi-random field of opportunities. Capital touches down as if by chance on a parcel of land, ignoring the opportunities on intervening lots, thus sparking the development process. The relationship between development of one parcel and nondevelopment of another is a disjointed, seemingly unrelated affair. While not truly a random process, it is evident that the traditional, center-driven agglomeration economies that have guided urban development in the past no longer apply. Conventional city form, Chicago-style, is sacrificed in favor of a noncontiguous collage of parcelized, consumption-oriented landscapes devoid of conventional centers yet wired into electronic propinquity and nominally unified by the mythologies of the disinformation superhighway. Los Angeles may be a mature form of this postmodern metropolis; Las Vegas comes to mind as a youthful example....

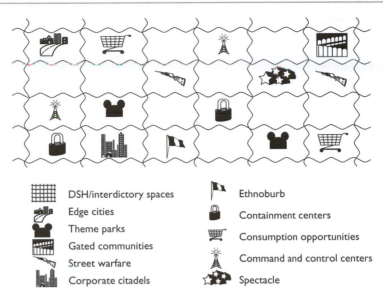

	DSH/interdictory spaces		Ethnoburb
	Edge cities		Containment centers
	Theme parks		Consumption opportunities
	Gated communities		Command and control centers
	Street warfare		Spectacle
	Corporate citadels		

Figure 4 Keno capitalism: a model of postmodern urban structure.

CONCLUSION: INVITATION TO A POSTMODERN URBANISM

[. . .]

We intend this essay as an invitation to examine the concept of a postmodern urbanism. We recognize that we have only begun to sketch its potential, that its validity will only be properly assessed if researchers elsewhere in the world are willing to examine its precepts. We urge others to share in this enterprise because, even though our vision is tentative, we are convinced that we have glimpsed a new way of understanding cities.

REFERENCES FROM THE READING

Boyd, I. (1997) *Am I Black Enough for You?*, Indianapolis: University of Indiana Press.

Burgess, E.W. (1925) "The Growth of the City." In R.E. Park, E.W. Burgess, and R.D. McKenzie (eds) *The City: Suggestions of Investigation of Human Behavior in the Urban Environment*, Chicago, IL: University of Chicago Press, pp. 47–62.

Castells, M. and Hall, P. (1994) *Technopoles of the World: The Making of the 21st Century Industrial Complexes*, New York: Routledge.

Cenzatti, M. (1993) *Los Angeles and the L.A. School: Postmodernism and Urban Studies*, Los Angeles, CA: Los Angeles Forum for Architecture and Urban Design.

Darlington, D. (1996) *The Mojave: Portrait of the Definitive American Desert*, New York: Henry Holt.

Davis, M.L. (1990) *City of Quartz: Excavating the Future in Los Angeles*, New York: Verso.

Davis, M.L. (1992a) "Fortress Los Angeles: The Militarization Of Urban Space," in M. Sorkin (ed.) *Variations on a Theme Park*, New York: Noonday Press, pp. 154–80.

Davis, M.L. (1992b) "Chinatown Revisited? The Internationalization of Downtown Los Angeles," in D. Reid (ed.) *Sex, Death and God in L.A.*, New York: Pantheon Books, pp. 54–71.

Davis, M.L. (1992c) "Think Green," in Aaron Betsky *Remaking L.A.*, *Los Angeles Times Magazine*, December 13.

Davis, M.L. (1993) "Rivers in the Desert," in *William Mulholland and the Inventing of Los Angeles*, New York: HarperCollins.

Dear, M. (1988) "The Postmodern Challenge: Reconstructing Human Geography," *Transactions of the Institute of British Geographers* 13: 262–74.

Dear, M., Schockman, H.E. and Hise, G. (eds) (1996) *Rethinking Los Angeles*, Thousand Oaks, CA: Sage Publications.

Fishman, R. (1987) *Bourgeois Utopias: The Rise and Fall of Suburbia,* New York: Basic Books.

Flusty, S. (1994) *Building Paranoia: The Proliferation of Interdictory Space and the Erosion of Spatial Justice*, West Hollywood, CA: Los Angeles Forum for Architecture and Urban Design.

Garreau, J. (1991) *Edge City: Life on the New Frontier*, New York: Doubleday.

Gottlieb, R. and FitzSimmons, M. (1991) *Thirst for Growth: Water Agencies and Hidden Government in California*, Tucson: University of Arizona Press.

Janowitz, M. (1967) "Introduction," in R.E. Park, E.W. Burgess and R.D. McKenzie (eds) *The City: Suggestions for Investigation of Human Behavior in the Urban Environment*, Chicago, IL: University of Chicago Press, pp. vii–x.

Jencks, C. (1993) *Heteropolis: Los Angeles, the Riots and the Strange Beauty of Hetero-Architecture*, London: Academy Editions; Berlin: Ernst and Sohn; New York: St Martin's Press.

Knox, E. and Taylor, P.J. (eds) (1995) *World Cities in a World System*, Cambridge: Cambridge University Press.

McGuire, B. and Scrymgeour, D. (1998) "Santeria and Curanderismo in Los Angeles," in P. Clarke (ed.) *New Trends and Developments in African Religion*, Westport, CI: Greenwood Publishing.

McKenzie, E. (1994) *Privatopia: Homeowner Associations and the Rise of Residential Private Government*, New Haven, CT: Yale University Press.

McPhee, J. (1989) *The Control of Nature*, New York: Noonday Press.

Molotch, H. (1996) "L.A. as Design Product: How Art Works in a Regional Economy," in A.J. Scott and E. Soja (eds) *The City: Los Angeles and Urban Theory at the End of the Twentieth Century*, Los Angeles: University of California Press, pp. 225–75.

Ong, E., Bonacich, E. and Cheng, L. (eds) (1994) *The New Asian Immigration in Los Angeles and Global Restructuring*, Philadelphia, PA: Temple University Press.

Park, E. (1996) "Our L.A.? Korean Americans in Los Angeles after the Civil Unrest," in M. Dear,

H.E. Schockman and G. Hise (eds) *Rethinking Los Angeles*, Thousand Oaks, CA: Sage Publications, pp. 153–68.

Raban, J. (1974) *Soft City*, New York: E. E. Dutton.

Reisner, M. (1993) *Cadillac Desert: The American West and Its Disappearing Water*, New York: Penguin Books.

Relph, E.C. (1987) *The Modern Urban Landscape*, Baltimore, MD: Johns Hopkins University Press.

Roseman, C., Laux, H.D. and Thieme, G. (eds) (1996) *EthniCity*, Lanham, MD: Rowman and Littlefield.

Scott, A.J. (1988a) *New Industrial Spaces: Flexible Production Organization and Regional Development in North America and Western Europe*, London: Pion.

Scott, A.J. (1988b) *Metropolis: From the Division of Labor to Urban Form*, Berkeley: University of California Press.

Scott, A.J. (1993) *Technopolis: High-Technology Industry and Regional Development in Southern California*, Berkeley: University of California Press.

Scott, A.J. and Soja, E. (eds) (1996) *The City: Los Angeles, and Urban Theory at the End of the Twentieth Century*, Los Angeles: University of California Press.

Soja, E. (1996) "Los Angeles 1965–1992: The Six Geographies of Urban Restructuring," in A.J. Scott and E. Soja (eds) *The City: Los Angeles and Urban Theory at the End of the Twentieth Century*, Los Angeles: University of California Press, pp. 426–62.

Sorkin, M. (ed.) (1992) *Variations on a Theme Park: The New American City and the End of Public Space*, New York: Hill and Wang.

Steinberg, J.B., Lyon, D.W. and Vaiana, M.E. (eds) (1992) *Urban America: Policy Choices for Los Angeles and the Nation*, Santa Monica, CA: RAND Corp.

Storper, M. and Walker, R. (1989) *The Capitalist Imperative*, Oxford: Blackwell.

Waldinger, R. and Bozorgmehr, M. (1996) *Ethnic Los Angeles*, New York: Russell Sage Foundation.

Editors' references and suggestions for further reading

Beauregard, R. and Haila, A. (1997) "The unavoidable incompleteness of the city," *The American Behavioral Scientist* 41: 327–41.

Caldeira, T. (2000) *City of Walls: Crime, Segregation and Citizenship in Sao Paolo*, Berkley: University of California Press.

Cherot, N. and Murray, M. (2002) "Postmodern urbanism: reality or fantasy?" *Urban Affairs Review* 37: 432–8.

Dear, M. (1986) "Postmodernism and planning," *Environment and Planning D: Society and Space* 4: 367–84.

Dear, M. (1988) "The postmodern challenge: reconstructing human geography," *Transactions of the Institute of British Geographers* 13: 262–74.

Dear, M. (2000) *The Postmodern Urban Condition*, Oxford: Blackwell.

Dear, M. (ed.) (2002) *From Chicago to L.A.: Making Sense of Urban Theory*, Thousand Oaks, CA: Sage.

Dear, M. and Flusty, S. (eds) (2001) *Spaces of Postmodernity: Readings in Human Geography*, Oxford: Blackwell.

Hannigan, J.A. (1995) "The postmodern city: a new urbanization?" *Current Sociology* 43: 152–214.

Harvey, D. (1989) *The Condition of Postmodernity*, Oxford: Blackwell.

Ley, D. (2001) "Book review of M. Dear *The Postmodern Urban Condition*," *Annals of the Association of American Geographers* 91: 577–9.

Myung-Rae, C. (1999) "Flexible sociality and the postmodernity of Seoul," *Korean Journal* 39.

Nijman, J. (2000) "The paradigmatic city," *Annals of the Association of American Geographers* 90: 135–45.

Quataert, D. (2003) "Historical and postmodern cities of the Middle East," *Journal of Urban History* 29: 347–53.

Soja, E. (2000) *Postmetropolis: Critical Studies of Cities and Regions*, Oxford: Blackwell.

THREE

Illustration XVIII.—Typical cottages of Polish laboring men. The children in the foreground were playing in the street until the camera proved an attraction. Several "little mothers" with their charges can be seen among them.

Plate 1 "Typical" Polish housing as illustrated in a 1906 housing study of Milwaukee, Wisconsin.

Plate 2 Piazza Tolomei in Siena, Italy. This medieval Tuscan town attracts international tourists interested in the beauty and charm of its twelfth-century public squares and buildings. (Courtesy of Thomas Harvey)

Plate 3 Looking through the window of a car showroom on to a street in Calcutta. With the liberalization of the Indian economy, car imports make inroads in the automobile market. (Courtesy of Scott Purl)

Plate 4 Reasserting local "traditions" in architecture, the law school of West Bengal National University, Kolkata (Calcutta) establishes its links with the past in Calcutta's suburban Salt Lake City. (Courtesy of Rina Ghose)

Plate 5 Loft housing along the Brooklyn shore of New York City as seen from the East River. (Courtesy of Harold Mayer Collection, American Geographical Society Library)

Plate 6 One of the towers of the Cabrini-Green public housing development stands in contrast to the skyline of downtown Chicago. Redevelopment of Cabrini-Green for "mixed-income housing" began with the demolition of public housing units in 1995. This housing tower no longer exists. (Courtesy of Jeffrey Zimmerman)

Plate 7 High-rise public housing in Harlem continues to be an important part of New York City's rental housing market with such developments providing shelter for 400,000 people. (Courtesy of Elvin Wyly)

Plate 8 A Prairie Crossing advertisement promises tradition and environmental conservation for Chicago area residents interested in living on the prairie in a new exurban development. (Courtesy of Jeffrey Zimmerman)

Plate 9 Neo-traditional architectural styles and New Urbanist design guidelines influence the development of Prairie Crossing in Chicago's exurbia. (Courtesy of Jeffrey Zimmerman)

Plate 10 The Arcade, a covered pedestrian alley built in 1922, is described as "the pride" of Letchworth Garden City's shopping. (Courtesy of Judith Kenny)

Plate 11 South Street Seaport in New York City offers shopping, restaurants, history and entertainment in a festival marketplace setting. (Courtesy of Harold Mayer Collection, American Geographical Society Library)

Plate 12 The Watergate complex in Washington D.C. was built in the late 1960s as one of the area's earliest and most prestigious mixed-use developments. (Courtesy of Harold Mayer Collection, American Geographical Society Library)

PART FOUR

Politics, governance, and inequality

Police and anti-war protestors meet on the streets of Cleveland. (Courtesy of Elvin Wyly)

INTRODUCTION TO PART FOUR

Parts 2 and 3 have shown how the dynamics of globalization and economic restructuring are of fundamental importance to understanding processes of urban development and the configuration of space in the city. In this section, we focus on the significance of politics and political power in shaping the urban landscape. In exploring these themes, the concerns of this section range from the global – how, for instance, has globalization affected city politics? – down to the local – what, for example, has been the impact of new institutions of urban governance on the inner city? To put these issues in historical context, however, it is useful to remember that the involvement of the state in managing the city is a relatively recent phenomenon. In their review of the changing nature of urban governance in Western societies, Knox and Pinch note that the period up to the first half of the nineteenth century was one of "virtual non-government" with urban affairs largely in the hands of merchants and patricians who had little interest in modifying the "organic growth of cities" (Knox and Pinch 2000: 133). Between 1850 and 1910, a much more interventionist approach developed, prompted largely by the problems of disease and disorder associated with the rapid growth of industrial cities. This period witnessed the beginnings of formal urban government which by the 1950s had developed into "large, vertically segregated bureaucracies of professional administrators geared to managing the city and its environment" (*ibid.*: 134). These city officials had a wide range of powers and responsibilities, ranging from planning policy and local taxation through to the provision of public housing and other social welfare services. Beginning in the 1970s, however, there have been some fundamental changes in the nature of urban governance. The rise of New Right political thinking (epitomized in the policies of President Reagan in the USA and Prime Minister Thatcher in the UK) challenged assumptions about the role of the state in welfare provision at both central and local levels and encouraged greater market provision wherever possible. One consequence has been a loss of power and autonomy by local government and the greater involvement of the private and not-for-profit sectors in the social and economic management of urban areas.

Against this background, the first reading by Clarke and Gaile provides a link back to Part 2 by considering the importance of globalization for city politics. Much of the discussion in this field has focused on "global homogenization" (Leo 1997), the contention that city politics is dominated by the pursuit of similar policies aimed at enhancing the attractiveness of local areas for footloose multinational capital. As Clarke and Gaile note, this is very much an extension of Peterson's (1981) *City Limits* thesis that "economic forces drive local policies," yet, as their survey of American cities reveals, such a conclusion is highly misleading. To be sure, many cities are involved in so-called "smoke-stack chasing" by driving down the local costs of land and labor, but it is also important not to underestimate "the politics of particularity" (Leo 1997). There are strong local ideas and policies concerning how cities should be governed based around goals such as sustainable development and community health rather than just short-term profits. Clarke and Gaile's analysis also draws attention to the political discourses that surround global–local power relations. Often the global is simply presented in terms of "strength, domination and action," while the local is viewed in terms of "weakness, acquiescence and passivity" (Gibson-Graham 2002). But, as Clarke and Gaile reveal, there is a spectrum of discourses (or "causal stories") about globalization which inform and frame the decisions of local city officials. Some of these "causal stories" are pessimistic and emphasize the loss of local autonomy in the face of globalization, but others are more sanguine, stressing the opportunities which emerge from the "new localism" that has accompanied global change.

It is clear from Clarke and Gaile's reading that globalization is in part responsible for ushering in a "new urban politics" (Cox and Mair 1988) characterized by a shift in emphasis for local authorities away from the provision of social welfare and towards a more pro-active commitment to local economic development. This shift from "urban managerialism" to "urban entrepreneurialism" (Harvey 1989) has prompted a period of considerable conceptual innovation as researchers grapple to understand the processes underpinning the emergence of the "entrepreneurial city" (Hall and Hubbard 1998). Two closely related theoretical frameworks – the growth machine thesis and urban regime theory – have proved particularly influential in urban geography and form an important part of the background to the second selection by Jessop, Peck and Tickell. The growth machine thesis (Molotch 1976) focuses on the activities of the business community in attempting to intensify urban development and boost the value of local property. To be successful, however, growth machines (or growth coalitions) also require sympathetic political leaders and need to be able to convince local populations that economic development will benefit the whole city rather than just a small elite. Focusing on these more informal aspects of coalition behavior, urban regime theory (Stone 1989) provides greater scrutiny of the lead role of city officials in establishing partnerships with the private sector. Crucially, regime theorists demonstrate that the power to shape cities is not given but has to be created or produced by different groups coming together and creating the capacity to govern. Moreover, unlike the growth machine approach, urban regime theory is not focused exclusively on economic regimes. "Progressive regimes," for instance, may seek to control development and limit environmental damage, while "maintenance regimes" are concerned to preserve the status quo rather than promote new development (Hall 1998). In the selection here, Jessop, Peck and Tickell focus their attention on the growth machine approach but subject it to critical scrutiny. The main thrust of their paper is a reformulation (the term they use is "retooling") of the theory in a way which addresses two important areas of weakness: the tendency towards "voluntarism" (the growth machine thesis typically emphasizes human agency at the expense of economic and political structures) and localism (the problems of underplaying the interconnections between local economic and political processes and structures and those that exist at broader geographical scales). A case study of the growth coalition that developed around Manchester's failed bid to host the 2000 Olympic Games provides an empirical application of this "retooled" growth machine.

While studies of growth machines and urban regimes capture significant aspects of the contemporary urban political geography of cities, it is also important to situate these in the context of longer term changes in the way cities are managed. This is the focus of the third reading by Goodwin and Painter. The challenge they confront is how to make sense of the transition that has occurred in the UK and elsewhere from a relatively uniform system of local government to a much more complex institutional framework of local governance characterized by the increasing involvement of private, voluntary (not-for-profit) and other non-elected organizations in policy making and service delivery in local areas. The conceptual framework that Goodwin and Painter draw upon to understand these changes is provided by regulation theory. An approach which has its roots in the work of French economists, regulation theory focuses on the social, political and economic conditions which help to stabilize the crisis-prone character of capitalism, highlighted by Harvey in Part 3. Focusing initially on the Fordist era (the period of sustained economic growth which in the UK lasted from the 1950s through to the 1970s), Goodwin and Painter show how urban government comprised a key element of the "mode of regulation," the political and sociocultural institutions and practices which contribute to the process of stabilizing capitalist development. To illustrate this, one needs only to think of the importance of public housing or the state provision of education and health care. These are vital to the private sector in terms of maintaining a skilled, healthy workforce and formed part of the "social wage," those "goods and services provided collectively to all or to those unable to afford them privately" (Painter 1995: 284). Following the economic crises that affected the UK and other industrialized nations in the 1970s, however, local government has faced growing demands on its welfare services as a consequence of rising unemployment. It has also had to confront national political strategies that have restructured urban policy to focus on enhancing economic competitiveness via a system of "centrally imposed and non-elected agencies" (Painter 1995: 287) that include Training and Enterprise Councils for improving the skills of the workforce and Urban Development Corporations to regenerate declining inner cities (the latter are discussed in Short's reading). From the perspective of regulation theory, the key question raised by these new institutions of local governance is whether they comprise elements of a new, post-Fordist "mode of regulation." Goodwin and Painter's response is that it is too early to say but also to emphasize that any answer to this question will need to be sensitive to existing evidence of strong geographical variations in the local governance of UK cities.

As indicated above, Urban Development Corporations (UDCs) provide one example of the new kind of institutions of local governance that during the 1980s and 1990s played an increasingly prominent role in the management of spaces within UK cities. Established mainly in places where the effects of deindustrialization had left large tracts of inner-city land derelict and underused, the aim of UDCs was urban regeneration. By suspending normal planning restrictions and using public money to undertake improvements in such areas as transport infrastructure and land reclamation, UDCs hoped to encourage private capital investment in these areas. In addition, the management of UDCs was not in the hands of locally elected officials but centrally appointed representatives, mainly from the business community. Some of the social and economic consequences of one of the first UDCs to be established, the London Docklands Development Corporation (LDDC), are examined in Short's reading. At one level, the LDDC was spectacularly successful in driving forward a program of regeneration for the area which has resulted in glittering new office blocks and new employment opportunities in financial and producer services and the media. But this economic success has been bought at a high price. As Savage *et al.* observe, "Disinherited local working-class communities lie alongside new gentrified enclaves; the congestion incurred by situating new plant and offices has required enormous unanticipated public investment in transport infrastructure; and considerations of social welfare . . . are completely neglected" (2003: 181). Indeed, as Short shows, the social landscape of Docklands is now highly polarized, resulting in fears of violent social conflict between the "winners" and "losers" in the regeneration of the area. Echoing some of the points made by Dear and Flusty in their account of postmodern urbanism (see Part 3, Restructuring), these social concerns are etched into the built environment in the form of gated communities and other residential security measures as the Docklands more affluent population of "yuppies" (young urban professionals) attempt to seal themselves off from their close neighbours of "yuffies" (young urban failures).

This juxtaposition of rich and poor in London Docklands is an important reminder that the local politics of growth cannot be separated from the local politics of welfare and social provision. Indeed, for all the emphasis on "growth machines" and the "entrepreneurial city," the expenditure of local authorities on economic development is still generally massively outweighed by expenditure on health, education and other welfare services. Thus to talk of a shift in local modes of governance from urban managerialism to urban entrepreneurialism is quite misleading: "most city governments adopt an amalgam of managerial (socially progressive) and entrepreneurial (growth-centred) policies" (Hubbard and Hall 1998). Against this background, a central theme of the final reading in this section by Wolch is the local politics of welfare. Focusing on homelessness in Los Angeles County in the 1980s, Wolch highlights some of the different local political responses to this problem. Most cities have been characterized by "civic silences," worried that doing anything might attract homeless people from other localities; other cities, with strong traditions of progressive politics and service delivery, have tried to tackle the problem, but typically "with inadequate resources and amid mounting pressures to back away from their commitments to the homeless." These findings clearly underline some of the themes of Clarke and Gaile's selection at the beginning of the section by illuminating the differentiated nature of "Local Politics in a Global Era." When this local dimension is combined with Wolch's analysis of the importance of global economic restructuring and changing national political priorities for understanding the causes of and responses to homelessness, this selection exemplifies the complex interconnections between globalization, restructuring, politics and governance which need to be unravelled in trying to make sense of social welfare in the city.

References and suggestions for further reading

Cox, K. and Mair, A. (1988) "Locality and community in the politics of local economic development," *Annals of the Association of American Geographers* 78: 307–25.

Gibson-Graham, J.K. (2002) "Beyond global vs local: economic politics outside the binary frame," in A. Herod and M. Wright (eds) *Geographies of Power: Placing Scale*, Oxford: Blackwell, pp. 25–60.

Hall, T. (1998) *Urban Geography*, London: Routledge.

Hall, T. and Hubbard, P. (1998) *The Entrepreneurial City: Geographies of Politics, Regime and Representation*, London: Routledge.

Harvey, D. (1989) "From managerialism to entrepreneurialism: the transformation of urban governance in late capitalism," *Geografiska Annaler* 71B: 3–17.

Hubbard, P. and Hall, T. (1998) "The entrepreneurial city and the 'new urban politics'," in T. Hall and P. Hubbard (eds) *The Entrepreneurial City: Geographies of Politics, Regime and Representation*, London: Routledge, pp. 1–23.

Knox, P. and Pinch S. (2000) *Urban Social Geography: An Introduction*, Harlow: Longman.

Leo, C. (1997) "City politics in an era of globalization," in M. Lauria (ed.) *Reconstructing Urban Regime Theory: Regulating Urban Politics in a Global Political Economy*, London: Sage, pp. 77–98.

Molotch, H. (1976) "The city as a growth machine: towards a political economy of place," *American Journal of Sociology* 82: 309–32.

Painter, J. (1995) "Regulation theory, post-Fordism and urban politics," in D. Judge, G. Stoker and H. Wolman (eds) *Theories of Urban Politics*, London: Sage, pp. 276–96.

Peterson, P. (1981) *City Limits*, Chicago, IL: University of Chicago Press.

Savage, M., Warde, A. and Ward, K. (2003) *Urban Sociology, Capitalism and Modernity* (2nd edn), London: Palgrave.

Stone, C. (1989) *Regime Politics: Governing Atlanta 1946–1988*, Kansas: Kansas University Press.

"Local Politics in a Global Era: Thinking Locally, Acting Globally"

from *Globalization and the Changing US City* (1997)

Susan E. Clarke and Gary L. Gaile

Editors' Introduction

What impact have the changes brought about by globalization (see Part 2) had on the politics of cities? Have city administrations reacted broadly in the same ways to the challenges of global economic change or is there evidence of a diversity of local political responses? These important questions have generated considerable controversy among urban geographers. There are those, such as David Harvey (1989), who argue that the new global regime of "flexible accumulation" leaves localities engaging in a reactionary, place-based politics with cities largely competing against each other to attract footloose multinational capital in order to secure a better position within the global urban hierarchy. Local politics in this context is heavily weighted towards entrepreneurial economic development issues while questions of social policy are marginalized. This interpretation of "local politics in a global era" has been criticized by those (e.g. Doreen Massey 1991) who view it as too reductionist because it assumes it is possible to "read off" local interests and institutional change from macroeconomic imperatives. Instead, Massey argues that globalization has created conditions for a diverse and progressive local politics of place, shaped by complex interrelationships between a range of different sociocultural, political and economic interests.

Against this background of a rather polarized debate, this reading, by political scientist Susan Clarke and geographer Gary Gaile, intriguingly suggests that there are elements of truth in both sets of arguments. Drawing on surveys of medium and large American cities, Clarke and Gaile show that in some places local officials continue to understand development problems in terms of an essentially economic model, comprising factor costs of land, labor and capital. Often referred to as "smoke-stack" chasing, these cities attempt to accommodate globalization by engaging in "locational strategies" that focus on reducing their factor costs relative to other places. Other cities, however, have explored alternative ways of promoting local development in a global era. "Asset-based human capital strategies," for example, focus on linking development with local job creation, education and training, while "sustainable development" approaches have replaced short-term economic growth goals with longer term concerns with urban ecology and community health.

Clarke and Gaile's paper, however, does more than simply disclose the diversity of local development strategies that American cities are adopting in a global era. In stark contrast to structuralist urban theorizing, Clarke and Gaile emphasize the socially constructed character of local responses to global economic change. While globalization has important material dimensions, its meaning and significance for local places is, Clarke and Gaile insist, something that is developed through the discourses (what they call the "causal stories") articulated by academics and policy makers that are then drawn upon by different groups and organizations to politicize issues linking the local and the global. Local politics is thus crucially concerned with the ideas and interpretations used to make sense of how and why globalization is defined as a "problem" for local communities. Some interpretations, or "causal stories," are quite sanguine about the implications of globalization. The "new localism" that has emerged as a result of global economic and technological changes has created opportunities for cities to upgrade their position within the global urban hierarchy by strategic improvements in their economic and social infrastructures. Other interpretations are more pessimistic. Drawing attention to the "dark side" of "new localism," these stories emphasize the growing socioeconomic inequality and erosion of local autonomy that has accompanied globalization. Whatever their relative strengths and weaknesses, however, the importance of these stories is that they frame issues for decision makers who use them to promote some solutions over others. "For local officials," Clarke and Gaile observe, "the dilemma is to sort out these causal stories and map the strategic consequences of these trends."

Echoing other contributors to this section (see Goodwin and Painter, and Jessop, Peck, and Tickell), another important theme in Clarke and Gaile's paper is the increasingly complex institutional environments that exist in cities associated with the shift from urban government to urban governance. Formal government at a local level now operates alongside a range of other actors from the private and non-profit sectors to create an institutional fabric that varies in thickness from city to city. Furthermore, as their survey of American cities reveals, different communities are clearly searching for different types of institutional arrangement to cope with changes in their economic and social situation. Although in some cities the traditional alliance between business and government interests remains unchallenged, evidence from both Chicago and San Francisco reveals an increasing local government emphasis on decentralizing power to community-based and non-profit organizations. However, as Wolch's reading in this section indicates, the capacity of these civil society groups to cope with a social problem such as homelessness is limited.

Clarke and Gaile's interests in the political dimensions of the local–global dialectic are shared by many other geographers working in the urban field. Kevin Cox has written extensively on this topic, and his edited volume *Spaces of Globalization: Reasserting the Power of the Local* (New York: Guilford Press, 1998) probes the complex relationships between the global and the local in the context of economic and political dynamics of advanced capitalist societies. More recently, Andrew Herod and M. Wright's *Geographies of Power: Placing Scale* (Oxford: Blackwell, 2002) also explores the different political and economic discourses that surround analyses of the global and local.

Susan Clarke is Professor of Political Science at the University of Colorado and, with Gary Gaile, co-authored *The Work of Cities* (Minnesota: University of Minnesota Press, 1998) and is co-editor (with Edward G. Goetz) of *The New Localism: Comparative Urban Politics in a Global Era* (Thousand Oaks, CA: Sage, 1993). Gary Gaile is Professor of Geography at the University of Colorado and, in addition to the volume with Susan Clarke, is co-editor (with Cort S. Willmott) of *Geography in America at the Dawn of the Twenty First Century* (Oxford: Oxford University Press, 2004).

Reconsidering local politics in terms of globalization processes is yielding fresh thinking that promises to reinvigorate the study of globalization as well as analyses of localities. In the continuing debate over the nature and extent of global economic integration and spatial reorganization of the economy, issues of scale are paramount. Bringing localities into analyses of globalization forces us to recognize "the concrete economic complexes situated in specific places" through which globalization exists. Taking localities seriously turns attention from the broad sweep of, among other things, economic shifts and technological changes constituting globalization processes to, as Sassen puts it, the array of local practices constituting and enabling globalization. This necessarily pushes the analytic focus beyond global cities to encompass the many types of places in which "the work of globalization gets done."

Similarly, taking globalization seriously challenges the modernist urban political economy framework. While still significant, these approaches hinder local political analyses to the extent that they emphasize economic interpretations of local global impacts. This limits our understanding of globalization and community: we overlook other systems of stratification and domination generating social complexity and tend to view "the global context of urban life" as "some objective structure existing 'out there'" rather than a socially constructed process. Given the contested meanings of the local global context, it is important to see local political processes as more than a matter of new interests or claims prompted by globalization or even new institutions such as public–private partnerships. The politics of ideas are fundamental: local politics center on creating, changing, and struggling over the ideas and interpretations that mediate our understanding of the local global context. This does not mean that globalization is not real; rather, it means our ideas about globalization, about how and why globalization is defined as a problem for local communities, are at the heart of local politics.

INTERPRETING GLOBALIZATION: THE EPISTEMIC COMMUNITY

Political change is as much about attention shifts and timing as it is about preferences and interests. In responding to their changing environment, local decision makers select certain dimensions of the situation, or context, for attention and direct problem-solving efforts to particular definitions of the problem of globalization. If we acknowledge that local officials seek to solve problems as well as satisfy preferences – their own and those of powerful interests – the interesting political questions center on which aspects of globalization practices are interpreted as causal stories about political problems rather than inadvertent conditions beyond intervention.

The politics of ideas involve the differing causal stories about globalization and its local impacts and the contested solutions to those problems. These causal stories emerge from a loose epistemic community of academics and policy makers with expertise and shared interests in understanding globalization processes; some are advocates of particular solutions almost independent of the problem (trade liberalization, job training), while others reflect more evolutionary policy thinking. The causal stories emerging from this epistemic community frame issues for local decision makers: for example, the stories attribute blame to certain aspects of the situation, they resonate with other values and beliefs important to citizens and decision makers, and they privilege some solutions over others.

Competing stories: the two faces of globalization

There are a multitude of stories to choose from, some more optimistic than others. Most present globalization as recasting distance and locational constraints on communities through economic and technological changes attendant to globalization. These changes alter economic space and raise the prospect of the "death of distance" as a significant factor in investment calculations. For many communities, this new space economy opens up economic opportunities. Furthermore, the most important value-added processes in the global economy depend on human capital and are inherently localized. As a consequence, there is a distinctive new geography of human capital in the United States and a new localism rejuvenating local politics. Understanding this new localism, however, depends on interpreting the causal stories behind the two faces of globalization.

The new localism story

As Margit Mayer sees it, the economic and spatial trends underlying globalization bring into question the ability of central governments to orchestrate the necessary local conditions of production required by global capital. To some, this is a "hollowing out" process akin to corporate restructuring: significant decision responsibilities move up to supranational and regional bodies and down to sub-national and local entities. Trade liberalization also reduces the prospects for national intervention in the flow of both capital and commerce and thus creates new economic spaces for cities. This new localism does not imply the withering away of the national state but it underscores the increased salience of other scales where globalization materializes.

These new localism causal stories resonate with cultural values of growth, wealth, power, opportunity, individualism, localism, and, increasingly, consumer sovereignty. From this perspective, the lesson for communities is to find a niche in the global urban hierarchy. City officials take on strategic broker roles, aiming to create settings for "the work of globalization." This new role centers on steering local governments toward improving the city's economic and social situation through negotiations with a wider set of actors. With its greater salience, local government acts as a catalyst of processes of innovation and cooperation. Underlying these causal stories is an assumption that global integration will bring a convergence in growth prospects and that cities can upgrade their position in a global urban hierarchy by their strategic interventions.

The dark side of the new localism story

A darker side of globalization makes the new localism more problematic. The hyper-mobility of capital, the international division of labor, and the "death of distance" due to information technologies appear to undermine local autonomy. In addition, international trade liberalization agreements such as the North American Free Trade Agreement reduce national authority, expose regions and cities to more global competition, and constrain local autonomy. Recent trade agreements allow pre-emptions of state and local government powers in economic development, environmental regulation, and other areas if they appear to privilege local firms or producers and thus would act as barriers to the "equal treatment" necessary for mobile trade and investment.

The concomitant rise in consumerism generates new local stratification patterns based on links to the global economy. An emergent world class of Cosmopolitans has weak community ties since their interests and resources transcend communities. In contrast, most citizens remain Locals, defined by particular places and limited opportunities; although many are weakly linked to a global web through their involvement in global practices, their well-being is dependent on decisions of world-class citizens. For Kanter, the smart question for individuals and cities is, What does it take to become world class? But to others, the rise of a world class signals an erosion in links between people, place, and identity; it contributes to an increased civic fragmentation

because there are fewer incentives to invest in a community's civic society.

This is obvious for Cosmopolitans, but Locals who see little gain from globalization may also find fewer incentives to contribute to civic society. The new economic changes appear to bring few benefits to central-city residents and newcomers: most new jobs go to suburban residents, and a two-tiered wage structure and segmented labor market offer little promise. Globalization practices at the local level often bring about social polarization, job displacement, wage compression, intensified property speculation, informal economies, immigration pressures, and the continued economic and social isolation of the poor in the central city. Despite gaining majority status in many large cities and achieving some political incorporation, the economic and social gains of African Americans and Hispanics relative to those of whites and even some recent immigrant groups remain low. The map of the local political terrain increasingly features greater social and economic polarization, intensified multi-ethnic competition for jobs, housing, and political resources, and eroding citizenship.

Yet shifting electoral demographics encourage the abandonment of cities by national policy makers. Through the 1980s and into the 1990s, national policies embraced a neo-liberal policy of facilitating capital mobility and cutting community programs that might distort private investment decisions. The Clinton administration's Empowerment Zone program includes more social funding than previous proposals, but stirs only modest local expectations. The remaining community programs are to be funded through "performance partnerships" tying local funds to achieving benchmarks rather than local needs. Place-based funding is increasingly replaced by transfer payments to individuals, particularly the non-urban middle class and elderly. In 1996 Clinton proposed that 61 per cent of federal aid to states and local governments go directly to individuals, compared to 35 per cent in 1960; 16 per cent in capital grants to state and local governments, compared to 47 per cent in 1960; and 22 per cent in other state and local grants, compared to 17 per cent in 1960.

These more pessimistic causal stories bring in cultural values of equity, obligation, rectitude, sustainability, and fairness. For some, local communities seem to have little choice but to cater to the demands of the global marketplace. To others, striving to become a world-class community is irrelevant; localities must address social citizenship issues. This darker side of globalization hints that growth prospects may become more uneven over time; cities initially positioned to benefit from globalization trends enhance their position at the expense of others, so the likelihood of upgrading a city's position in the global urban hierarchy is more remote.

Shifting institutional venues

Despite the conflicting assessments of the two faces of globalization, there is broader agreement within the epistemic community on the institutional changes promoted by globalization practices. These social and cultural aspects of economic change include more numerous and more diverse local actors seeking cooperation in the face of greater complexity, the emergence of a third sector of non-profit organizations, and the reconstruction of the local institutional infrastructure to accommodate different bargaining and negotiating processes.

The expansion of the local sphere and the increase in the different types of actors exacerbates the dilemma of generating "enough cooperation" to get things done. Bringing the necessary actors to the table and then moderating differences and negotiating cooperation is a new local government responsibility. This has prompted a variety of new bargaining systems and institutions. In each instance, the objective is to find new institutional and organizational arrangements with sufficient scope, responsiveness, and flexibility to accommodate these competing ideas and interests. Often, however, they operate outside formal government structures; the public sector may no longer be the center of negotiations and decision making about public resources.

Embedded in this expanded local sphere is the emergence of a third sector between state and market. To Drucker, this is the social transformation of our times. Less romantically, there is a clear increase in the use of non-profit organizations as service delivery mechanisms. As Mayer points out, local government becomes the enabler in negotiations with non-state actors rather than the provider of services or local field office of national

programs. In addition to their service delivery and catalytic role, non-profits, ironically, permit the enlargement of the sphere of local politics in the face of declining local public sector functions: local officials must play a more activist role in interactions with non-state sectors by moderating and managing areas of interest and providing resources on a conditional basis.

Some claim that the character of the local institutional infrastructure distinguishes between localities in terms of their effectiveness in coping with changing economic realities. Institutions are the legacy of past conjunctures of interests and ideas as well as the framework for current politics: they articulate working agreements on what can and cannot be done, how reforms will be implemented, and how costs and benefits are to be distributed. The thicker the institutional fabric, the more likely the local milieu will support features seen as necessary in a global era: institutional stability, commonly held knowledge, institutional flexibility, high innovative capacity, trust and reciprocity, and, potentially, a sense of inclusiveness. But it is not clear that institutional thickness is necessary or sufficient for local economic change. Indeed, it could support sclerosis and resistance to change as well as innovation.

It is evident that communities are searching for a new "institutional fix," for institutional arrangements compatible with a changing economic base and social structure. It appears easy to encourage more inter-institutional interactions but more difficult to generate collective representation or shared understandings. Not all institutions are of equal weight; non-economic groups may be plentiful and interactive but fragmented, underfunded, and unlikely to share a collective understanding of an agenda that does not support their constituencies.

IDEAS, INTERESTS, AND INSTITUTIONS: MORE COMPLEX LOCAL POLITICS

All these stories are true. For local officials, the dilemma is to sort out these causal stories and map the strategic consequences of these trends. In the process, attention shifts to new problem definitions and new solutions, prompting the local policy strategies that we will note in the following. Ideas do not bring about agendas and outcomes independently, of course, but they are constitutive of outcomes. Nor are ideas mere rationalizations of self-interested groups with fixed preferences; perceptions of interest can emerge from debates, and conflicts over ideas shape interests. As a consequence, local politics are shaped by the causal stories that different groups and organizations use to politicize issues linking the local and the global, to seek new institutional venues, and to promote some solutions over others. Our research indicates that the following local strategies are especially salient: classic locational approaches, the world-class community orientation, the entrepreneurial mercantilism strategy, asset-based human capital strategies, and the sustainable development orientation.

The city limits story

Analyses of local American politics are invariably marked by the long shadow of Paul Peterson's *City Limits*. Peterson's elegant deductive argument emphasizes the city limits, the constraints on local action built into the American decentralized federal system: cities are relatively free from central oversight but fiscally dependent on the health of the local economy for their tax revenues. In Peterson's model, economic forces drive local policies: they act as structural constraints and create a systemic bias toward business interests. In the context of these political and fiscal structures, businesses and residents seek places with the most favorable ratio of taxes paid to services received. Local developmental policies enhance that ratio; in Peterson's view, cities have a unitary interest in stimulating economic activity that employs residents, generates tax revenues, and contributes to attractive locational sites for capital and households.

Correspondingly, local redistributional policies distort that ratio by directing services and benefits to those paying fewer taxes. The logic of interjurisdictional competition and necessary sensitivity to cost–benefit ratios implies that redistribution of local funds is not in a city's interest, given the burden such transfers place on the productive members of the community. Such groups can threaten to exit if they perceive that the benefit–tax ratio is no longer favorable. Furthermore, in the absence of viable mechanisms such as coherent party structures

or stable social change coalitions, local demands for redistribution are not easily translated into redistributive policies.

To many, this notion of city limits is only exacerbated by globalization. The economic logic of federalism and globalization suggests that the menu of local choices will shrink in the face of greater global competition, with local communities putting increasing emphasis on entrepreneurial economic development and the subordination of social policies. It appears irrational for local communities to do otherwise. But this new local terrain is constituted by ideas, interests, and institutions unanticipated by the city limits model. These deductive models, for example, assume that policy choices stem from the self-interest of officials predominantly concerned with re-election chances and tax-based budgets. But we have seen that globalization has expanded the sphere of local political action, broadened local economic horizons, and escalated the competing understandings of localism and global competition. Increasingly, local politics are characterized by non-elected public, private, and non-profit actors as well as decision organizations and partnership arrangements that cannot be labeled as strictly public or private. Hewing to the city limits story ignores these contextual cues and inhibits the search for new solutions.

Locational strategies

Nevertheless, many cities balk at discarding the city limits story of a unitary interest in economic development. As Peterson's model predicts, local officials continue to diagnose local development problems in terms of an economic growth model emphasizing the importance of factor costs – basically, the costs of land, labor, and capital – in production processes. Cities adopt policies to reduce those costs relative to other cities, such as providing subsidies for firms, tax abatements, and below-market land costs. Although disparaged as "smoke-stack chasing" and "corporate-centered" strategies, these public policies are justified by the claim that the benefits spread to the community as a whole through jobs and tax revenues. Critics argue otherwise, pointing to the questionable cost-effectiveness of these deals and the skewed distributional benefits.

Given the prevalence of locational strategies, many cities clearly continue to cope with changes in their environment by applying decision rules and programmatic solutions that worked in the past. In that sense, city development responses may be path dependent: they lock in the protected interests of certain sectors of the business community as well as solution sets linked to factor costs. This limits the city's ability over time to adjust to changing constituencies and to address emergent problems unrelated to factor-cost issues. Local institutions continue to reflect this legacy of interests and economic growth models, particularly those articulated by past federal programs. They hamper new agreements on what can and cannot be done and new allocations of costs and benefits.

World-class community orientations

Although these classic locational strategies are pervasive, our research indicates that the worldview of many local officials is changing away from the focus on attracting firms encouraged by the city limits viewpoint, to a concern with facilitating economic growth processes. Cities still pursue development as predicted by the deductive models but in line with the new localism story: local economic well-being is attributed to gaining a niche in the global economy and becoming a world-class community. . . .

The world-class-community orientation singles out qualitatively different intervention styles featuring public entrepreneurship, public–private partnerships, the encouragement of research and development activities, new business starts, foreign investment, and small-business formation. This third wave demands entrepreneurial public roles and a broader range of non-governmental and market actors. From this perspective, local governments entertain more risk and seek new growth opportunities by marketing themselves in the global economy.

One element of this world-class local entrepreneur story involves developing information infrastructure relevant to the new global roles of cities. In our 1996 survey, over two-thirds of the communities recognized this as a new element on their economic development agenda. Nearly every community reported having or creating a home page on the World Wide Web; over 75 per

cent reported that some or all city hall offices were linked by E-mail; and most provided job and education information through publicly accessible computer terminals in libraries and at kiosks. This new path to global competitiveness is prompting the reconfiguration of local growth coalitions around "electronic public spaces" rather than traditional land-use issues.

Entrepreneurial mercantile orientations

Another twist on the new localism story entails solutions based on entrepreneurial mercantilism. Entrepreneurial mercantilism aims for more diversified growth built on local initiative and indigenous assets; selected intervention in the market encourages new and existing small businesses that would provide greater local benefits from economic development. In St. Paul, Mayor Latimer's Homegrown Economy Initiative in 1976 launched the city's mercantilist approach with substantial intergovernmental support and foundation funds. This early prototype sought to promote local ownership, job quality, economic diversification, and small businesses; in contrast to the world-class community path, the emphasis is on enhancing local self-reliance through emphasizing local benefits from investments, community and employer ownership, the net tax benefits of public subsidies, and attracting capital and funds from outside through export industries. In Minneapolis, officials report they are determined "to grow what we have, to expand what we have rather than pursue companies from outside the city and state." Similarly, Davenport, Iowa, recently refocused its incentives to what city officials describe as "primary business, those bringing wealth into our community."

Chicago's entrepreneurial mercantilism was neighborhood oriented. Responding to his electoral coalition of African Americans, white liberals, and Latinos, Mayor Harold Washington advocated increased job opportunities for local residents, promoted the balanced growth of the downtown and neighborhoods, assisted neighborhoods in participating in partnerships and coordinated investment, and encouraged more citizen participation. His "new populist" approach redirected funds toward smaller-scale neighborhood projects; it decentralized power to community-based and non-profit organizations to increase local autonomy and develop local assets. This agenda encapsulated the entrepreneurial mercantilist tenet that a local government can stimulate an alternative development strategy by providing seed money for community-based development and bringing neighborhoods and small businesses to the table on economic development deals.

Asset-based human capital strategies

In response to the dark side of globalization, there has been a resurgence of human capital initiatives and poverty-reduction efforts at the local level. These efforts are counter to the predictions of the city limits story. Indeed, these human capital strategies differ from past initiatives. While the federal Private Industry Council job training model remains important, many communities in our 1996 survey report local policy designs to target specific concerns. Minneapolis, for example, set up a task force to work on linking human capital and economic development; pushed by state legislation requiring reports on jobs created with public funds, the city is tying some economic development assistance to the creation of living-wage jobs. Other cities also report making development assistance contingent on specific job creation and training efforts. Portland, Oregon's efforts to link new and expanding businesses with local labor pools in the late 1980s evolved into a separate Workforce Development Department by the mid-1990s. With the incentive of state funding support, Oregon counties can form alliances to support workforce development and training programs. The sheer cost of human capital efforts limits what cities can do; cities are drawing on general funds, Community Development Block Grants, debt financing, state support, payroll taxes, and even lottery funds to support these programs.

Many local efforts are premised on explicitly reframing local poverty issues in terms of under-utilized assets rather than social needs and pathologies. Michael Porter faults past models for treating the central city as an island isolated from the rest of the city and focusing on public subsidies rather than on the creation of wealth via private investment. Porter advocates building on the competitive advantage of the central city by identifying the clusters of activities that would gain from a central-city location. He targets the development of critical masses of economic activities that would respond to local demand (such as specialty foods, financial services), link with other regional clusters, and export goods and services to broader markets. Porter argues that private sector models and leadership can build on the competitive advantages and underutilized assets of these distressed areas. While it could be argued that disinvestment by these private institutions created much of the difficulty facing these areas, Porter relegates government to a supportive role to business leadership.

In contrast, John L. McKnight and John P. Kretzmann's influential model of asset-based community organizing relies on strong public and civic leadership. They, too, reject past policy designs, characterizing them as "needs-driven" deficiency models creating client neighborhoods and ensuring only survival. In their alternative view, poor people are the underutilized assets. Kretzmann and McKnight argue explicitly for shifting attention away from deficiency cues and toward an internal focus on rediscovering local assets as the means to development solutions. This involves mapping and developing strategies to knit together the individual skills, associations, and institutions in the neighborhood. The literally hundreds of examples they provide include arrangements for street markets in Cleveland to accept food stamps to encourage local buying, church investments of pension funds in revolving loan funds for community-based housing projects in Massachusetts, and community development credit union loans to low-income residents for financing the purchase and repair of used cars in Chicago.

Sustainable development

Although often seeming more rhetoric than reality, issues of local sustainable development are increasingly visible. They are one of the areas where transnational organizations are influential in American local politics. Local sustainable development agendas stem from a definition of sustainability as current use of resources that does not threaten future generations; local decision making and governance are focused on sustainability values, including inter-generational and intragenerational equity concerns, rather than the city limits emphasis on short-term growth goals. In 1986, the World Health Organization inaugurated the Healthy Cities initiative, which now includes over 1,000 communities. These activities are tracked by organizations such as the Global Tomorrow Coalition in Washington, D.C., and the Community Sustainability Resource Institute in Maryland.

In cities as disparate as St. Paul, Jacksonville, San Diego, Chattanooga, Seattle, and Vancouver, sustainability is a key agenda issue. In Chattanooga, the sustainable development path was promoted as an insider strategy by local leaders seeing an urban revitalization strategy. In the 1970s, Chattanooga was one of the most polluted cities in the United States. With the help of local foundation support from a public–private task force, the city adopted an ecosystem approach to urban management that put sustainability as the primary agenda item. Reintegration of urban functions through clustering was seen as the key; in 1984 an initiative for urban ecology encouraged neighborhood networks and citizen participation to renovate the riverfront, attack air pollution, and close the production loop through recycling.

In cities where traditional business and government interests remain locked in, local groups have mobilized and pushed their objectives from the outside. In San Diego, citizens used "ballot box planning" to overcome the growth machine alliance; they continually threatened to put their growth management concerns on the ballot through the initiative process if elected officials refused to respond. Many such sustainable development initiatives are now supported through programs such as the Healthy Communities Initiative and the President's Council on Sustainable Development.

Sustainable development concerns also take regional forms, such as the Cascadia alliance in the Pacific Northwest. The commonality and interdependence of environmental interests in the Georgia Basin-Puget Sound bioregion ("one forest, one waterway, one air shed") led to the creation of new institutional structures to protect the quality of life in the area referred to as Cascadia in response to global growth pressures. With the advent of the North American Free Trade Agreement, the North–South trade links through Cascadia increased. This spurred the growth of bi-national alliances and multiparty groups and forums concerned with the impacts on the local environment. The Cascadia alliance, a regional alliance coordinating growth management and strategic planning efforts in Alaska, British Columbia, the Yukon, Alberta, Oregon, Washington, Montana, and Idaho, exemplifies a new institutional fix emerging in response to globalization and interdependence. The British Columbia Washington Environmental Cooperation Council was set up in 1993 to bring together regional government groups for cooperation on environmental and growth management strategies. To planners and citizen groups, these new institutions were essential: traditional structures of government appeared less effective in dealing with issues of interdependence and in responding to the demands of these new groups. The Cascadia alliance reflects the influence of transnational groups; these institutions also shape new transnational environmental and trade interests over time.

CONCLUSION

Redirecting our attention to the political dynamics of ideas, interests, and institutions highlights the ways in which communities are thinking locally and acting globally. Local officials make strategic choices between the causal stories generated by an epistemic community with expertise in globalization studies. They adapt these interpretations to fit the political and economic context in which they work. In doing so, they are reconstructing local policy agendas to take account of human capital needs, information technologies, and environmental issues. Many of these local responses to the spatial and political implications of economic change are unanticipated; local efforts to accommodate the work of globalization are contributing to political practices and institutional venues that challenge our current theoretical approaches.

Editors' references and suggestions for further reading

Harvey, D. (1989) *The Condition of Postmodernity: An Enquiry into the Origins of Cultural Change*, Oxford: Blackwell.

Massey, D. (1991) "A global sense of place," *Marxism Today*, June, pp. 24–9.

"Retooling the Machine: Economic Crisis, State Restructuring, and Urban Politics"

from Andrew Jonas and David Wilson (eds)
The Urban Growth Machine: Critical Perspectives,
Two Decades Later (1999)

Bob Jessop, Jamie Peck, and Adam Tickell

Editors' Introduction

With its glittering shopping malls, gentrified enclaves and gleaming offices, the post-industrial, postmodern city has been at the center of much urban geographical analysis since the late 1980s. Attention has focused partly on issues of spatial restructuring and urban design (see selections by Dear and Flusty (Part 3) and by Ley (Part 6)), and partly on concerns about increasing socio-economic polarization and intensified levels of surveillance and social control (see selections by Hamnett (Part 2), and Fyfe and Bannister (Part 7)). For many urban geographers, however, it is the changes in urban government and governance that raise some of the most intriguing questions about the contemporary city. Of particular interest is the way in which the restructuring of welfare provision and the increasingly global mobility of capital have led to the emergence of a "new urban politics" characterized by a relative shift in emphasis away from the local provision of welfare services and towards policies promoting local economic growth and development. Analysis of this transition from "managerialism" to "entrepreneurialism" (Harvey 1989) has been strongly influenced by two closely related conceptual frameworks: urban regime theory and the growth machine thesis. Originating in the work of urban political economists in the United States, both approaches focus on the processes of local coalition-building for the purposes of promoting local economic development. Urban regime theory (see Stone and Sanders 1987) looks at how urban local authorities build coalitions with other interests, mainly in the private sector, in order to facilitate local economic development. The growth machine approach (Logan and Molotch 1987) examines similar themes but from the other direction, focusing on the activism of entrepreneurs and the ways in which local business leaders define growth strategies and attempt to attract support from other interests, like urban government, in order to secure local economic prosperity. Both approaches, however, are united by their hostility towards "structuralist" accounts of urban change. Instead of viewing cities as being structured by some externally imposed logic, as is characteristic of the ecological approaches of the 1930s (see reading by Burgess in Part 1) and Marxist analysis in the 1970s (see reading by Harvey in Part 3), the machine and regime approaches focus on the intra-urban struggles and negotiations between different interest groups as they seek to enhance a city's economic fortunes.

On both sides of the Atlantic, the growth machine and urban regime literature have provided an attractive framework within which to make sense of the relationships between local authorities and business elites. In this reading the sociologist Bob Jessop and geographers Jamie Peck and Adam Tickell acknowledge the insights offered by the growth machine thesis but are also keen to address some strategic weaknesses that they argue characterize this approach; hence their title, "Retooling the Machine." At the core of their critique is a concern that the urban growth machine approach is characterized by both voluntarism and localism. It treats urban politics as existing in isolation from wider economic and political forces, and tends to focus on local actors and agency to the exclusion of economic and political structures. The result is what the authors describe as an "excessive localism" which suggests that variations in urban fortunes can almost be reduced to differences in the charisma of city leaders and their relative skills at urban networking. While such an approach acts as a "salutary corrective" to more structuralist accounts, of crucial importance to understanding the "new urban politics," Jessop *et al.* contend, is to focus on the interplay between the local level and the "supra-local" scale of global capitalism and the nation state. One way this can be achieved, they suggest, is through greater integration between regulationist approaches (of the type discussed in the reading by Goodwin and Painter) and the new urban politics literature. The regulationist emphasis on structures and macro-economic forces complements the more agency-oriented and meso-political focus of growth machine and urban regime analysis. For Jessop *et al.*, however, the weakness of the growth machine thesis can also be addressed by introducing insights from neo-Gramscian perspectives which emphasize the importance of wider sets of norms, values and visions in shaping local economic development strategies. This helps highlight the inter-relationships that exist between the "microdiversity" of urban coalitions and the "macronecessities" of economic and political pressures that exist beyond a city's boundaries. To illustrate what such an approach means in practice, the authors use the example of the city of Manchester's failed bid to host the Olympic Games in 2000. At one level, the bid involved a classic growth coalition as envisaged by the growth machine approach. Led by the private sector, and with strong representation from local construction companies and regional utilities, the bid had strong support from local governance institutions of the City Council and Urban Development Corporation. But Manchester's strategy was shaped not only by the internal dynamics of the local bid coalition, but also by economic and political decisions at a national level (by UK central government departments in London) and international level (by the International Olympic Committee based in Lausanne in Switzerland). In understanding the nature and dynamics of Manchester's bid, then, it is vital not to become trapped by the "excessive localism" of the growth machine thesis which would tend to overemphasize the role of local actors in shaping the local development strategies and underplay the importance of wider economic and political contexts within which such strategies are located.

Jessop *et al.*'s "retooling" of the growth machine thesis first appeared in A.E.G. Jonas and D. Wilson's (eds) *The Urban Growth Machine: Critical Perspectives, Two Decades Later* (New York: State University of New York Press, 1999) which assesses the contribution of the growth machine approach to studies of urban politics from a variety of disciplinary perspectives using a wide range of mainly US and UK case studies. For urban geographers, however, Jessop *et al.*'s reading also has a wider resonance in relation to understanding the dynamics of urban entrepreneurialism, the promotion of local economic development by local government in alliance with the private sector. Contributions to this important research agenda have been brought together in T. Hall and P. Hubbard's (eds) *The Entrepreneurial City: Geographies of Politics, Regime and Representation* (Chichester: John Wiley, 1998) which examines both the politics of pro-growth economic development and the parallel institutional shifts from urban government to urban governance in Western cities.

Bob Jessop is Professor of Sociology at Lancaster University and author of numerous books on capitalism and the state including *The Capitalist State: Marxist Theories and Methods* (Oxford: Blackwell, 1982), *State Theory: Putting the Capitalist State in its Place* (Cambridge: Polity Press, 1990), *The Future of the Capitalist State* (Cambridge: Polity Press, 2002), and co-editor with N. Brenner, M. Jones and G. MacLeod of *State/Space: A Reader* (Oxford: Blackwell, 2003). Jamie Peck is Professor of Geography at the University of Wisconsin, Maddison, and author of *Work-place: The Social Regulation of Labour Markets* (London: Guilford Press, 1996) and *Workfare states* (London: Guilford Press, 2001). With Adam Tickell, Trevor Barnes and Eric Sheppard, Peck is also co-editor of *Reading Economic Geography* (Oxford: Blackwell, 2003).

"You don't have to be a postmodernist," Harvey Molotch (1990: 175) argues, "to suspect efforts that cast all cities as uniform in their response to larger economic changes." Nor, of course, do you have to be an unreformed structuralist to discern intriguing parallels and telling similarities in the responses of contemporary cities to wider forces, such as neoliberalism and economic globalization. Nowadays most places, it seems, have their very own booster committees, complex networks of public–private "partnerships," and entrepreneurial urban strategies. Indeed, Harvey (1989) has even suggested that a generalized transition is under-way from an urban managerialism of the (Fordist) past to an urban entrepreneurialism of the (post-Fordist?) future. Harvey's concern is not simply with local features of this transition but also with the interurban context in which they are embedded. He writes that, as

> inter-urban competition becomes more potent it will almost certainly operate as an "external coercive power" over individual cities . . . [bring-ing] them closer into line with the discipline and the logic of capitalist development [while inducing] . . . repetitive and serial reproduction of certain patterns of development. (Harvey 1989: 10)

Such serial reproduction may extend beyond the seemingly ubiquitous world trade centers, postmod-ern shopping malls, and waterfront developments to embrace the very institutions and practices of urban governance. Structures and strategies of urban governance may be copied not so much because they demonstrably "work" but because their advocates have won out in the battle for ideas in response to shared problems. Yet as Molotch (1990: 176–77) notes:

> [E]ven if similar policies are pursued across place and time, those policies may be dictated not by underlying logics but by mundane polit-ics. Sometimes this repetition of the same polic-ies across places stems from pluralistic greed and similar class configurations, or more innoc-ently because of social contagion that spreads among decision-makers otherwise uncertain how to proceed. The failure of our grand theories to

explain these types of phenomena now leaves intact the task of explaining how differences in places are going to come about and how political initiatives, historical idiosyncrasies, physical constraints, or just plain luck affect outcomes.

In reaction to this alleged failure of "grand theory" to explain the macronecessities dictating new forms of urban regime, interest has burgeoned in the microlevel diversity of structures and strategies of urban governance (see Judge, Stoker, and Wolman 1995). This interest has diverse origins and is re-flected in various intellectual and methodological currents – for example, growth machine analysis and regime theory in the United States, and local governance research in the U.K. Notwithstanding differences in presentation and emphasis, there is widespread and overarching concern with the diverse "internal" architectures of urban coalitions and city elites. Indeed, regime theory deems it axiomatic "that policy is shaped by the composi-tion of the governing coalition and by the nature of the relationship between coalition members" (Stone 1991: 294), while growth machine approaches have tended to become more voluntarist over time as structuralist concerns (e.g., with value relations and the commodification of land) have receded into the background.

It is this slippage into voluntarism and localism that prompts our remarks in the current chapter. While accepting the value of much urban growth machine and urban regime analysis, we reject the tendency to treat urban politics as existing somehow in isolation from wider economic and political forces and processes. We also reject the tendency to focus on local actors and agency to the exclusion of local economic and political structures, and/or to regard the latter as highly malleable at the hands of local elites. In opposing this agency-centered localism, however, we are not advocating some form of global structuralism. Instead, we explore the dialectic among different spatial scales of economic and political organiza-tion and emphasize the mutual constitution of structure and agency across different levels. Our analysis draws upon several theoretical currents in regulation and neo-Gramscian state theory with a view to recontextualizing the analysis of urban politics.

GROWTH MACHINES AND AGENCY

The concern of the New Urban Politics with the internal architecture of the city has obvious advantages; it also involves obvious disadvantages. In particular, while these studies have produced deep and nuanced accounts of the structure and dynamics of urban coalitions, they reveal much less about the wider economic and political context within which urban strategies are embedded (cf. Stoker 1995; Ward 1995). The holding together of structure and agency was a key objective of the growth machine theorists, but equally they saw a need to distance themselves from some structuralist accounts of urban politics (see, in particular, Logan and Molotch 1987: 11). So while theoretically there was equal emphasis on urban actors as agents of change and as bearers of social relations, substantively the focus of the growth machine thesis has shifted to the former. Common cause is increasingly being made with urban regime approaches that also assign privileged roles to (local) politics and (elite) agency. Together, these approaches are beginning to define a new methodological orthodoxy in urban studies. As Molotch (1993: 31; cf. Judd and Parkinson 1990; Swanstrom 1993; Stone 1993) insists, "[t]here is plenty of human agency in this version of political economy. Where there is similarity across places, it derives from shared institutional contexts and parallel patterns of volition, rather than iron-like determinisms of hidden hands or exogenous constraints."

This emphasis on the autonomy of local elites and regimes is a salutary corrective to forms of structuralism that emphasize the autonomous logic of a global capitalism, the all-pervasive, homogenizing sovereign authority of the national state, or the hegemony of supra-local (if not always global) cultures and discourses. But this does not mean that the wider imperatives of interurban competition, state restructuring, and capital accumulation can be safely set aside while examining the internal dynamics of urban coalitions; nor, likewise, that local strategies can be examined as if they were somehow insulated from extralocal models, paradigms, or discourses. For there is always a two-way, if typically asymmetrical, flow between local economic, political, and ideological forces and those existing on supra-local scales (Harvey 1989; Preteceille 1990).

Yet the methodology of growth machine and urban regime analysis dictates that common structures and strategies of urban governance be interpreted not in terms of the impact of extra-local processes such as interurban competition, but instead in terms of basic similarities in local institutional contexts and norms of elite behavior. For example, DiGaetano and Klemanski (1993: 381) conclude their comparative study of the two "motor cities" of Birmingham, England, and Detroit, Michigan, by arguing that the two "regimes formed . . . in remarkably similar ways" (1993: 381). They also claim that:

> The difficulty in comparing urban politics across nations stems principally from the differences in intergovernmental systems. . . . However, regime theory approaches city governance from the bottom up. The local polity, not the nation state, is the object of analysis . . . [allowing the comparison of] governance in cities that operate in substantively different systems of central local relations and local government authority. (pp. 381–82)

Above all, the New Urban Politics appears to be "concerned with the possibility that local politics really does matter . . . [because] the economic imperative is pliable within limits" (Stone 1991: 290). Urban economic fortunes are always, of course, politically mediated at the local level. But this implies in turn that the economic and political are mutually, albeit indirectly, constitutive. Now, while analysts of urban politics are not unaware of this co-constitution, they often have a blinkered view of what it entails. Regime theorists, for example, proceed from the unquestioned assumption of "a liberal political economy [which combines] . . . a set of government institutions controlled to an important degree by popularly elected officials . . . [and an economy] guided mainly, but not exclusively, by privately controlled investment decisions" (Stone 1993: 2). In this context, Stone regards urban regimes as mediating "organisms," intervening between (given) imperatives of accumulation and (indeterminate) local political outcomes in such a way that "urban regimes are potentially an autonomous force" (p. 2). For many analysts of the New Urban Politics, therefore, the pattern of local political relations not only provides the point of methodological entry but also delivers the explanatory power.

The danger with this approach is that, instead of revealing the causal processes operating within growth machines or urban regimes more generally, it produces a series of inferences from case studies undertaken to verify, illustrate, or exemplify the value of the corresponding conceptual apparatus that informed the research. Harding (1995: 48), for example, maintains that the limited penetration of the growth machine concept in U.K. studies stems from its application "post hoc, to redescribe urban phenomena"; and Stoker (1995: 64) notes that regime analysis "suffers from the tendency of most of its main propositions to emerge inductively from observation of the urban scene." Rarely do case-study practitioners interrogate their conceptual apparati or try to relate their detailed findings to more abstract accounts of, say, capital accumulation or state restructuring (see Cox 1991). Instead their bottom-up methodology tends to assign causal power to local political networks and thereby suggest, unintentionally perhaps, that spatial variations in urban fortunes are merely a by-product of the geographies of charismatic city leadership or effective urban networking. Thus, coupled with the continued caricaturing (and often straightforward denial) of structural imperatives, studies of the New Urban Politics often fail to provide penetrating theoretical analyses with general import, succumbing instead to empiricism and "excessive localism" (see Harding 1995; Stoker 1995; cf. Cox 1991).

[. . .]

We would suggest that the growth machine model does, indeed, offer a concise, but powerful, explanation for urban growth dynamics in North America during the rise, golden age, and initial crisis of Fordism in local political systems in which supra-local (federal and state) governments had a relatively limited direct role to play in urban economic development. But while these very background conditions lend strength to the growth machine model they also limit its applicability beyond U.S. Fordism. . . .

Such arguments have not been lost on growth machine theorists. Indeed, Vicari and Molotch (1990) have recently emphasized that the approach should not be overextended, suggesting that U.S.-style urban growth machines are most likely to emerge where: (a) local governments have taxation powers, depend on the local tax revenue, and have primary responsibility of land-use control; (b) there is relatively weak integration between the tiers of government (so local government is relatively autonomous); (c) there is weak party organization; and (d) there is no political party that is "antigrowth." In these conditions, they argue, entrepreneurs are most able to manipulate spatial relations and, we would add, political systems. Further consideration of these four conditions helps us to "unpack" growth machine theory and see how far temporal and spatial shifts may affect the likely emergence of growth machines.

First, there is wide variation in the extent to which local governments have tax-raising powers as well as in the proportion of total income received from tax. In Britain, for example, local taxes of all kinds now typically account for less than one-quarter of all local government spending; and, in Germany, there has been a strong emphasis on progressive redistribution and the equalization of conditions across the *Länder*. Further, given the crisis of national Fordisms and the increased salience of structural competitiveness between space economies on various scales, control over land use may be less significant for growth coalitions even in the United States. Nowadays, the supply side – in its integral, social sense – matters far more to place-competitiveness than does the mere physical supply of land. Accordingly, a superior place product (and this now means more than a good building) starts to matter more than the capacity to determine the flow of people (as tenants, consumers, customers) past and/or to individual buildings, complexes, or built environments of similar (standardized, Fordist) design. Speculation in spatial structure takes new forms in which integral economic dimensions are highlighted and narrow, built environment issues are secondary (except, perhaps, at the most mobile, least skilled end of after-Fordist divisions of labor).

Second, the extent and form of integration among tiers of government clearly varies across space. For example, while France and Britain all have strongly centralized relationships, Germany privileges a complex central–regional balance of governmental power. More generally, political dynamics differ across national state systems. Even if British and American Fordism had been economically identical, the very different national political structures would ensure that urban political forms varied considerably. Furthermore, the temporal shift from Keynesian-welfarist state forms to more liberal, workfarist forms means that the relationship between the tiers of government – even in the United States – has become more complex and contested.

Third, party systems and political parties can make a difference. Parties often have distinct spatial programs and can develop hegemonic positions that define the terms of reference for business elites. Furthermore, a crisis in party systems at national or local level may create more opportunities for growth coalitions, but these opportunities may favor different forms to Fordist-style growth coalitions.

Fourth, antigrowth coalitions are more likely in economies and/or polities where growth is regarded as destructive of traditional ways of life and is restricted (whether for physical and/or regulatory reasons) to specific spatial areas. In addition, the end of Fordism has also seen the growth of an antimodernist, antigrowth (whether capitalist or socialized) politics, exemplified by the still disparate green movement.

REGULARIZING URBAN POLITICS?

Prompted perhaps by these theoretical and methodological shortcomings, there have been several recent attempts to deploy regulationist concepts in urban political analysis. Such attempts serve two different purposes. First, they provide a means of linking urban politics to broader economic tendencies (e.g., by referring urban regimes to Fordist or post-Fordist accumulation regimes, to national or regional variants thereof, or to different spatial logics). And second, they provide a means to link agency and structure in the study of urban governance via the concept of "mode of regulation" (with its implicit concern with the role of structural forms in providing a framework for "regularizing" practices and institutionalized compromises). . . .

These objectives cannot be achieved, however, just by noting some, intriguing complementarities between growth machine analysis and key regulationist concepts. Indeed, there are at least three obstacles to such a quick "conceptual fix." First, whereas the microeconomic foundations of the regulation approach are grounded in the labor process, the microeconomic foundations of growth machine theory are grounded in the transformation of the built environment. There is still considerable theoretical work to be done in providing clear linkages between these processes. Second, there are major tensions between the regulationist program with its residual structural biases and the still rather eclectic body of work on practices of urban governance (cf. Jessop 1995). And third, as the regulation approach is primarily relevant to the economic context (albeit in inclusive, integral terms) of growth machines, its use in studies of urban politics needs to be complemented by a more sophisticated analysis of the political domain. This latter task is the contribution of neo-Gramscian state theory (Jessop 1990).

[. . .]

Nonetheless, some distance remains between the respective approaches of regulationists and analysts of urban politics and governance. While not wanting to polarize the positions unnecessarily, we note that the regulation approach puts more emphasis on structure, macroeconomic forces and large-scale institutional mediations, while governance theories are more directly concerned with agency, meso-political forces and local-scale capacities. Although the regulationist approach is open, in principle, to issues of local agency, albeit embedded within a conception of uneven spatial development (see Painter and Goodwin 1995; Peck and Tickell 1995), residual antagonisms to structuralist accounts still pervade much contemporary research on urban governance. Much of this research continues to foster an exaggerated analytical opposition between "autonomous" local development (or "local leadership and politics matters") and structural determination of local fortunes (or "local leadership and politics is irrelevant"). Crudely, either cities are seen as "the helpless pawns of international finance, industry and commerce [or they] are in a position to mediate and direct their own destinies" (Parkinson and Judd 1988: 2). Problem cities are thus held to share the common characteristic of "leadership deficit," while it is seen as telling that cities more successful in capturing economic growth tend to possess a proactive elite group that can "speak for the city."

The neo-Gramscian approach may prove particularly useful in overcoming residual antagonisms to structural arguments. This approach is explicitly concerned with the role of political, intellectual, and moral leadership in actively conforming economic and political practices and institutions to the long-run demands of capital accumulation and political domination. In this sense, neo-Gramscian theorists could well ascribe a leading role to local "elites," but would also link their actions to broader structural constraints and social relations. In line with Gramscian analyses, we suggest that growth machines can be fruitfully analyzed in terms of strategically selective combinations of political society and civil society, government and governance, or "hegemony armoured by coercion" (Gramsci 1971: 271 and *passim*). This approach emphasizes the interdependence of ethico-political and economic-corporate forces (political society + civil society), allows more weight in the exercise of political power to non-state forms (government + governance), and stresses the importance of values, norms, vision, discourses, linguistic forms, popular beliefs, and so on, in shaping local accumulation strategies and their related modes of growth. In this spirit, we also argue that growth machines may be linked to a local hegemonic bloc (or, after Lipietz 1993, "regional armature"), and to a historical bloc (or that complex, discordant, contradictory unity of a mode of growth and its mode of economic regulation) (Jessop 1997a). "Bringing Gramsci back in" to analyses of urban politics provides a coherent theoretical framework that is both useful for exploring the various dimensions and modalities of urban regimes and consistent with the regulation approach (see Jessop 1992, 1997b).

Moreover, combining the regulation approach and neo-Gramscian state theory could prove useful in recoupling agency and structure in urban political analysis. Urban elites emphatically do possess agency. But it is essential to recognize that they also occupy positions within networks, structures, and meaning systems that significantly constrain their range of feasible actions and the longer-term consequences of the choices that they make. In this sense, local economic and political agents may well make their own history but they do not do so in circumstances of their own choosing nor, indeed, in circumstances that they can ever fully comprehend. Of crucial significance in this regard

are cities' positions within the spatial division of labor (itself a complex, overdetermined phenomenon) and the institutional forms and wider political forces that shape local politics (likewise complex and overdetermined). These structural contexts provide more than the scenic backdrop for local growth machines to act out petty dramas – they are *structuring* in complex yet significant ways.

We are arguing, then, for a thorough reconceptualization of structure and agency in urban politics. It is far from adequate just to summon up regulationist concepts – let alone simply to use regulationist language – as if this will provide all the answers. For the reasons noted above, this is unlikely to provide more than an abstract, structural context for studying urban governance. We need to move beyond this to consider the changing strategic contexts in which urban actors (individual and collective) shape urban fortunes.

While a range of case studies discloses that urban political forms are varied, thereby highlighting what we might refer to as "microdiversity" (see, e.g., Wagner, Joder, and Mumphrey 1995; DiGaetano and Klemanski 1993), there are still many features common to the political "will" of urban regimes within and across Western nations. Thus, there is a general emphasis nowadays on enhancing the role for business, a growing reliance on place marketing, rhetorics of urban entrepreneurialism, and so on. These shared features are indicative of common evolutionary trajectories (or what we might refer to as "macronecessity"). This coexistence within and across Western societies of both microdiversity and macronecessity has emerged because urban political systems are responding to a set of extra-urban transformations (economic, political, and so on) which provide structural and/or strategic constraints on local action. This indicates the need to explore the relationship between microdiversity and macronecessity. In pursuing this apparently self-contradictory theoretical goal, we suggest, first, that the real scope for microdiversity is constrained by significant macrolevel phenomena whose impact will emerge post hoc if it is not taken into account ex ante. And we suggest, second, that the "iron laws" of macronecessity are sufficiently contradictory in structural terms as well as dilemma-ridden in strategic terms that they typically operate through a discursive closure of opportunities, which still leaves some, albeit limited, scope for microlevel variations (cf. Jessop 1985).

There are four closely linked aspects that we wish to highlight in developing this approach to the paradox of microdiversity and macronecessity: (i) the structural constraints placed on actors, especially those that are beyond the scope of the actors' influence; (ii) the strategic context that is open to their influence within given spatial and temporal horizons of action; (iii) the strategic capacities of the various actors concerned, including learning capacities; and (iv) the actual strategies or tactics they pursue in specific conjunctures. In any complex situation, the strategic context depends upon acts of simplification, which attempt to isolate a subset of causal factors (and therefore ignore other causal factors) that may be susceptible to some influence. Here, the role of techno-economic paradigms, hegemonic modes of regulation, societal visions, and so on, can have a significant impact on local actors' interpretation of "what is to be done." Conversely, insofar as such discourses are widely shared and thereby constrain strategic actions, they will contribute to the cohesion and reproduction of the structures within which localities and local agents are embedded. Thus, strategic actions at the local level, whether in the pursuit of private growth (Logan and Molotch 1987) or in the pursuit of public grants (Cochrane, Peck, and Tickell 1996), may serve to reproduce the very structures from which they seek to escape. This is even more likely, as we shall see below in the case of the Manchester Olympic bid, when the key local actors and their strategic opportunities are heavily constrained by the exercise of extra-local authority and/or imperative allocation of extra-local resources essential to the success of local strategies.

[. . .]

RECONSIDERING BUSINESS ELITISM

[. . .]

Manchester's Olympic strategy . . . was not simply "Made in Manchester," despite claims to the contrary from those seeking to use the bid in the construction of a local hegemonic project (see Cochrane, Peck, and Tickell 1996). Rather, it was the outcome of a complex interplay between local political capacities and priorities on the one hand,

and wider (national and international) systems of rule setting – particularly those of the Department of the Environment in London and the International Olympic Committee in Lausanne – on the other. Relations of positionality (institutional, political, discursive) matter as well as relations of personality. This is not a plea for sterile analyses of office holding and formal political–bureaucratic relations, but for an appropriate contextualization of local political power. It is only to take a first step in this direction to observe that, "Office holding bestows authority, but the authority conferred is highly limited. . . . Energetic governance requires more than office holding can provide. The weakness of formal authority gives added importance to the personal leadership of prominent urban actors" (Stone 1995: 96). While Stone is right to highlight the existence of intellectual, moral, and political leadership, his account underemphasizes the fact that elites wield positional as well as personal power, and that dynamic personal leadership is insufficient to secure effective governance. An entrepreneurial urban agenda needs political as well as economic space for the operational autonomy of growth machines, as can be illustrated by the politics of Manchester's failed bid to host the Olympic Games.

At first sight, Manchester's Olympic bid process underlines claims that British urban politics are becoming more like those in the United States: pro-growth, dominated by business, departing from the British party-political norm, and despite the continued concentration of power at central government level, tending toward relative disintegration at local governance level. The bid was led by individuals from the private sector, although there was strong moral and material support from both local institutions of governance (the local authority, the local urban development corporations) and, later, national government. Support for the bid was not (party-)politicized and in a very short time acquired almost hegemonic status: it was regarded as self-evident that winning the Olympics would be good for Manchester, a truth proclaimed by local politicians and by companies supporting the bidding process. True to growth machine theory, there was heavy representation from local construction companies, regional utilities, the only bank with local headquarters, and the local media. Likewise, there is strong support from those institutions that Logan and Molotch (1987: 75) view as "auxiliary players": the universities, organized labor, and corporate capitalists. The Olympic bid also drew the City Council's senior officers and political leaders into a transformation of the modus operandi of urban economic development. The new approach would be based on elite networking, opportunism, and a more entrepreneurial approach on the part of officers and the whole organization.

Yet while the bid committee used the language and many of the tactics of the classic growth coalition, their actions also serve to demonstrate the limitations of local strategic capacity. This conclusion would still hold even if the Olympic bid had been successful: for success would have depended just as much on playing the game according to rules and on a playing field shaped by extra-local forces. Both sets of local partners – public and private – had only limited control over the bidding process. At its most benign, this might have resulted in a mutually beneficial local pooling of resources and influence; but just as often it ensured that both parties' goals were diluted and sometimes undermined. The internal machinations of the bid coalition played a part in shaping the Manchester strategy and perceptions of its "ownership," but they were also profoundly preconditioned by the city's weak position in the urban hierarchy and its dependence on (actual and potential) non-private funding streams from London or Lausanne. Obsessive concern with how power and control are being redistributed locally can mask the full extent to which the locus of power has already shifted away from the local scale altogether. Despite this, the new language of partnership requires that the process be represented as one of local negotiation and compromise:

> It's very difficult, in some of the partnerships that we have, to say that the control is absolutely here [with the City Council]. You can say, if it comes to some of the schemes, that the control is with the Council . . . you can say that control is with central government . . . you can say it's with the people who are actually running the bid . . . [In reality] they're partnerships where people have vetoes. (City councillor, Manchester City Council, summer 1993)

Few in the city were inclined (or, indeed, sufficiently knowledgeable) to question this informally regulated system of local checks and balances as long as there was still some prospect of winning the Games nomination. Eyes were being firmly fixed on the big picture – the possibility of winning the Games and the need to maintain a unified voice during the bidding period. It was deemed counterproductive and even treacherous to question the politics or finances of the bidding process during a time of maximum global exposure. This exerted a kind of local hegemonic discipline on urban political actors: it was taken as axiomatic in Manchester that the city's ability to "deliver" the Games depended on the continued strength and cohesion of its elite coalition.

Despite the emergence of what looked like a more vibrant urban politics, Manchester's Olympic project was heavily dependent on the decisions of national political actors. Manchester's much-trumpeted "new money" for the Olympic bid was, of course, anything but, since it was top-sliced from urban spending programs (*Financial Times*, June 23, 1993), while the city uniquely achieved its own expenditure line in the Department of the Environment's spending plans (DoE 1993). Thus, Manchester's Olympic bid must be seen as part of a wider reorganization in the funding and delivery of urban aid, as part of what Stewart (1994: 143) sees as a competitively (and centrally) orchestrated "new localism [based on the] decentralization of administration as opposed to the devolution of power and influence." In order to participate in this process, needless to say, it was crucial for Manchester to abide by a set of extra-local rules, which in this case were dictated as much by the Department of the Environment in London as by the International Olympic Committee in Lausanne. Certainly, there were wider and more complex processes at work here than the verve and charisma of the bid's private-sector leader (Sir) Bob Scott, or indeed the City Council's new-found entrepreneurialism. The emergence of both, in fact, was preconditioned by wider shifts in state strategies and business politics (Peck 1995; Peck and Tickell 1995). The detailed nature of Manchester's strategy was not predetermined by this strategic context, but it was forged within – and structured by – these strategic parameters. These may not quite be the "iron laws" of the market, but subtly coercive they are. Just because the rules of the game dictate that strategies must *appear* to be home grown and business-led, this does not mean that this is in fact their origin or their rationale.

In short, while there may be superficial similarities between urban machines in the United States and those emerging in the U.K., this convergence is due neither to the actions of autonomous political elites nor to the dominance of local property interests. Instead, they have arisen from changing structural constraints, the redefinition of strategic contexts, changed modes of political rationality on the part of the British central state and the European Commission, and the gradual adaptation of local states to these constraints as they seek to manage the repercussions of uneven development in ways that sustain local political legitimacy.

REFERENCES FROM THE READING

Cochrane, A., Peck, J. and Tickell, A. (1996) "Manchester plays games: Exploring the local politics of globalisation," *Urban Studies*, 33, 1317–34.

Cox, K.R. (1991) "Question of abstraction in studies in the New Urban Politics," *Journal of Urban Affairs*, 13, 267–80.

DiGaetano, A. and Klemanski, J. (1993) "Urban regime capacity: A comparison of Birmingham and Detroit," *Journal of Urban Affairs*, 16, 367–84.

DoE [Department of the Environment] (1993) *Annual Report 1993: The Government's Expenditure Plans 1993–94 to 1995–96*, Cm2207, HMSO, London.

Gramsci, A. (1971) *Selections from the Prison Notebooks*, Lawrence and Wishart, London.

Harding, A. (1995) "Elite theory and growth machines" in Judge, D., Stoker, G. and Wolman, H. (eds) *Theories of Urban Politics*, pp. 35–53, Sage, London.

Harvey, D. (1989) "From managerialism to entrepreneurialism: The transformation of urban governance in late capitalism," *Geografiska Annaler*, 71B, 3–17.

Jessop, B. (1985) *Nicos Poulantzas: Marxist Theory and Political Strategy*, Macmillan, Basingstoke.

Jessop, B. (1990) "Regulation theory in retrospect and prospect," *Economy and Society*, 19, 153–216.

Jessop, B. (1992) "Regulation und politik: Integrale Ökonomie und integraler Staat" in Demirovic, S., Krebs, H-P. and Sablowski, T. (eds) *Hegemonie*

und Staat, pp. 232–62, Westfälisches Dampfboot Verlag, Münster.

Jessop, B. (1995) "The regulation approach, governance, and post-Fordism: Alternative perspectives on economic and political change?" *Economy and Society*, 24, 307–33.

Jessop, B. (1997a) "A neo-Gramscian approach to the regulation of urban regimes: Accumulation strategies, hegemonic projects, and governance" in Lauria, M. (ed.) *Reconstructing Urban Regime Theory: Regulating Urban Politics in a Global Economy*, pp. 51–73, Sage, Thousand Oaks, Califorinia.

Jessop, B. (1997b) "The entrepreneurial city: Re-imaging localities, redesigning economic governance, or restructuring capital?" in Jewson, N. and MacGregor, S. (eds) *Transforming Cities: Contested Governance and New Spatial Divisions*, pp. 28–41, Routledge, London.

Judd, D. and Parkinson, M. (1990) "Urban leadership and regeneration" in Judd, D. and Parkinson, M. (eds) *Leadership and Urban Regeneration: Cities in North America and Europe*, pp. 13–30, Sage, Newbury Park, California.

Judge, D., Stoker, G. and Wolman, H. (eds) (1995) *Theories of Urban Politics*, Sage, London.

Lipietz, A. (1993) "The local and the global: regional individuality or interregionalism," *Transactions of the Institute of British Geographers*, 18, 8–18.

Logan, J.R. and Molotch, H.L. (1987) *Urban Fortunes: The Political Economy of Place*, University of California Press, Berkeley and Los Angeles.

Molotch, H.L. (1990) "Urban deals in comparative perspective" in Logan, J.R. and Swanstrom, T. (eds) *Beyond the City Limits: Urban Policy and Economic Restructuring in Comparative Perspective*, pp. 175–98, Temple University Press, Philadelphia.

Molotch, H.L. (1993) "The political economy of growth machines," *Journal of Urban Affairs*, 15, 29–53.

Painter, J. and Goodwin, M. (1995) "Local governance and concrete research: Investigating the uneven development of regulation," *Economy and Society*, 24, 334–56.

Parkinson, M. and Judd, D. (1988) "Urban revitalization in America and the U.K. – The politics of uneven development" in Parkinson, M., Foley, B. and Judd, D. (eds) *Regeneration of the Cities: The U.K. Crisis and the U.S. Experience*, pp. 1–8, Manchester University Press, Manchester.

Peck, J. (1995) "Moving and shaking: Business elites, state localism, and urban privatism," *Progress in Human Geography*, 19, 16–46.

Peck, J. and Tickell, A. (1995) "Business goes local: Dissecting the 'business agenda' in Manchester," *International Journal of Urban and Regional Research*, 19, 79–95.

Preteceille, E. (1990) "Political paradoxes of urban restructuring: Globalization of the economy and localization of politics?" in Logan, J.R. and Swanstrom, T. (eds) *Beyond the City Limits*, pp. 27–59, Temple University Press, Philadelphia.

Stewart, M. (1994) "Between Whitehall and town hall: The realignment of urban policy in England," *Policy and Politics*, 22, 133–45.

Stoker, G. (1995) "Regime theory and urban politics" in Judge, D., Stoker, G. and Wolman, H. (eds) *Theories of Urban Politics*, pp. 54–71, Sage, London.

Stone, C.N. (1991) "The hedgehog, the fox, and the New Urban Politics: A rejoinder to Kevin R. Cox," *Journal of Urban Affairs*, 13, 289–98.

Stone, C.N. (1993) "Urban regimes and the capacity to govern: A political economy approach," *Journal of Urban Affairs*, 15, 1–28.

Stone, C.N. (1995) "Political leadership in urban politics" in Judge, D., Stoker, G. and Wolman, H. (eds) *Theories of Urban Politics*, pp. 96–116, Sage, London.

Swanstrom, T. (1993) "Beyond economism: Urban political economy and the postmodern challenge," *Journal of Urban Affairs*, 15, 55–78.

Vicari, S. and Molotch, H.L. (1990) "Building Milan: Alternative machines of growth," *International Journal of Urban and Regional Research*, 14, 602–24.

Wagner, F.W., Joder, T.E. and Mumphrey, A.J. (eds) (1995) *Urban Revitalization: Policies and Programs*, Sage, London.

Ward, K. (1995) "Business elites and urban politics. A rough guide...to regime theory," *Working Paper 5, Business Elites and Urban Politics Research Programme*, The University of Manchester, Manchester.

Editors' references and suggestions for further reading

Harvey, D. (1989) "From managerialism to entrepreneurialism: the transformation in urban governance in late capitalism," *Geografiska Annaler*, 71B: 3–17.

Logan, J. and Molotch, H. (1987) *Urban Fortunes: The Political Economy of Place*, London: University of California Press.

Stone, C. and Sanders, H. (eds) (1987) *The Politics of Urban Development*, Lawrence: University Press of Kansas.

"Local Governance, the Crises of Fordism and the Changing Geographies of Regulation"

from *Transactions of the Institute of British Geographers* (1996)

Mark Goodwin and Joe Painter

Editors' Introduction

The late twentieth century was characterized by important changes in the management of cities in many areas of the developed world. An increasing emphasis on "proactive development strategies" emerged, focusing attention on the involvement of business interests in local politics as part of a strategy to encourage urban growth and investment. Related to this was an increasing trend towards the subordination of traditional social policies to those that supported economic development, described by David Harvey (1989) as a move from "managerialism" to "entrepreneurialism." This intersects with a third change, and the focus of this reading by Goodwin and Painter, concerning the way the management of cities shifted from fairly uniform systems of *local government* to much more complex arrangements of *local governance*. The term *local governance* refers to a broadening of those institutions around local government involved in policy-making and service delivery to include a range of private, voluntary (not-for-profit) and other non-elected organizations. In Britain, the development of local governance has been particularly marked. Training and Enterprise Councils, Urban Development Corporations, private firms, business elites, partnership organizations, community and voluntary groups are, along with local government, all part of a reconstituted local institutional framework involved in managing local economic and social development.

How can these changes from local government to local governance in Britain best be explained? In this reading, Mark Goodwin and Joe Painter argue that attention must focus on the broad economic and political contexts within which changes in local governance have occurred. To do this their analysis is informed by regulation theory. Originating in the work of a group of Parisian economists in the 1970s and 1980s, regulation theory focuses on the social and institutional conditions which attempt to mitigate the contradictions and crisis tendencies associated with capital accumulation. One of the key concepts for understanding how this process of maintaining stable capitalist development occurs is the "mode of regulation" (MOR). The MOR partly comprises the complex mix of social norms, conventions, customs and laws which help "normalize" the process of capital accumulation. As Goodwin and Painter show, however, the institutions and practices of local governance also play a strategically important role within the MOR. During the era of Fordist regulation (a period of sustained economic growth which in Britain lasted from the late 1950s to the early 1970s), local government contributed to the Fordist MOR in a variety of ways. It provided affordable housing for large numbers of low-paid workers, which, in turn, supported the mass consumption norms required to sustain Fordist growth. Local government also provided many of the welfare services which were unprofitable for the capitalist sector to produce. Following the economic crises that affected Britain and other industrialized nations in the mid-1970s, however, the Fordist MOR began to break down, and this, Goodwin and Painter contend, is associated with important changes in the nature of local governance. Of crucial importance, they argue, was the way that, during the 1970s and 1980s, local government came to be viewed as contributing to the crises of the Fordist MOR, and thus from being an *agent* of regulation, local government increasingly became an *object* of regulation. In other words, part of the central state's response to the crises of Fordism was to change the nature of local governance in Britain by reforming the roles and responsibilities of local government. From an organizational structure characterized by centralized service delivery, local governance increasingly involved a wide variety of public, private and not-for-profit providers; from a dominant role for elected local government, there was a shift to a multiplicity of agencies of local governance; and from a hierarchical, bureaucratic management structure, local governance became increasingly more devolved and performance driven.

While some regulation theorists like to characterize these changes in local governance as part of a shift from a Fordist MOR to a "post"- or "neo"-Fordist MOR, Goodwin and Painter are more cautious. They argue that simply describing changes in local governance is not sufficient to establish the existence of a new MOR. Rather, a regulationist perspective on the restructuring of local governance requires examining both the empirical extent and prevalence of the new developments as well as the nature of the relationship between these developments in order to assess whether they do in fact contribute to the functioning of a new MOR. Detailed empirical research is needed to address these issues, and one of the key strengths of Goodwin and Painter's article is that they provide an array of research questions to inform concrete research on this theme in local areas. Which organizations are now involved in producing and delivering local public services and what are the relationships between them? Have entitlements to services been changed? What are the principal economic and social goals of agencies of local governance? In what ways do these new developments interact with each other "in mutually reinforcing ways?" This last question is of crucial significance for assessing whether new forms of local governance are acting as a component of a new MOR, although Goodwin and Painter think this unlikely. This is largely because the much more geographically differentiated nature of local governance compared with old-style local government means that there is more geographical unevenness in the system of regulation, making it much more difficult to sustain stable economic development or social cohesion.

Several urban geographers in Britain have taken up Goodwin and Painter's call for concrete research on the restructuring of local governance, sparking off an intriguing debate about whether the shift from local government to local governance is as clear-cut as is often claimed. Focusing on the cities of Cardiff and Sheffield, Imrie and Raco (1999) argue that there are important continuities between "old-style" local government and new forms of local governance. Their case studies show, for example, that contemporary concerns about the limited democratic accountability of the institutions of local governance ignore the historical criticisms that were made of the lack of accountability in local government policy-making processes. Similarly, the entrepreneurial focus of local governance arrangements is, they argue, nothing new given historical evidence of entrepreneurial local government policies and the involvement of business elites in policy development. Other urban geographers have emphasized the significant discontinuities between the context in which local government operated before and after the 1980s. Ward (2000), for example, highlights how the reorganization of the state at national level in the 1980s meant that local government was required to incorporate a wider set of interests into the local political apparatus, leading to the emergence of new forms of local governance. By comparing Birmingham, Leeds and Manchester, Ward also shows (as suggested by Goodwin and Painter) that the precise form of the new local governance varies significantly between cities.

Mark Goodwin is Professor of Geography at the University of Exeter, and his books include *The Local State and Uneven Development: Behind the Local Government Crisis* (Cambridge: Polity Press, 1988) co-authored with Simon Duncan; and *Introducing Human Geographies* (London: Arnold, 1999) co-edited with Paul Cloke and Philip Crang. Joe Painter lectures in the Department of Geography at the University of Durham and is the author of *Politics, Geography and "Political Geography": A Critical Perspective* (London: Arnold, 1995).

INTRODUCTION

Local government in Britain has changed dramatically over the last fifteen years. There have been four key areas of change. These have involved increasing central control of local finance; the privatization and commodification of public services; the loss of local state autonomy over the remaining public services; and the expansion of non-elected sub-national agencies. The end result of well over 50 separate Acts of Parliament is a system of local administration and policy-making which has changed quite significantly in many of its key aspects. The former fairly uniform system of local government has been transformed into a more complex one of local governance, involving agencies drawn from the public, private and voluntary sectors. . . . But behind the alliterative phrase 'from government to governance' lies a series of crucial transformations in the social, political, economic and cultural relations which operate in and around the local state. There has been a restructuring of the institutions and mechanisms through which local governance operates, as well as changes in the content of political struggles, projects and alliances.

[. . .]

LOCAL GOVERNANCE, REGULATION AND CRISIS

Local governance and local government

To speak of local governance rather than local government implies a shift of focus which may be either analytic or substantive, or both. Analytically, the concept of 'governance' is broader than that of 'government'. It recognizes that it is not just the formal agencies of elected local political institutions which exert influence over the pattern of life and economic make-up of local areas. Within the political processes which affect the fortunes of any local area are a wide range of actors. These include the institutions of elected local government, to be sure, but also central government, a range of non-elected organizations of the state (at both central and local levels) as well as institutional and individual actors from outside the formal political arena, such as voluntary organizations, private

businesses and corporations, the mass media and, increasingly, supra-national institutions, such as the European Union (EU). The concept of governance focuses attention on the relations between these various actors. A substantive shift from government to governance implies not only that these other influences exist but also that the character and fortunes of local areas are increasingly affected by them.

These twin shifts (analytical and substantive) also problematize the 'local' in local governance. Local areas, however defined, are neither autonomous, nor isolated from wider state structures, political processes and economic links. This is particularly true in Britain, with its relatively centralized system of public administration. The institutions of elected local government are local in the sense that their territorial competence is limited but, even before the relative decline in its power, elected local government was subject to a national framework of regulations and standards. In addition, of course, central government was in any case fully involved in shaping local areas in a wide range of policy areas for which local authorities had only partial responsibility or none at all. All government activities and policies have implications for local areas, whether this is formally acknowledged by policy-makers or not. The governing of localities, in other words, is not only (nor necessarily mainly) a local matter.

The relative decline in the power of elected local government in Britain (at least in most policy areas) and the substantive shift towards a more diffuse 'local governance' have further complicated the relationship between the spatial scale of 'the local' and the processes and institutions which affect localities. Although this decline has included important elements of recentralization of political authority, notably in the area of public finance, it is not the case that recent changes have been solely a matter of a shift in power from elected local government to elected central government. The new local governance involves more power being exercised by a very varied range of institutions which operate at a range of spatial scales. At one extreme there are institutions which operate, at least in part, at a spatial scale lower than that of elected local government; examples include 'self-governing' schools and some voluntary organizations. By contrast, at the 'highest' spatial scale, global corporations may play a role in local governance by, for example, obtaining financial concessions associated with inward investment or negotiating a customized curriculum for workforce training with a local educational institution.

[. . .]

Regulation

Our approach to regulation theory is a methodological one. . . . Based on the writings of the Parisian 'regulation school', our starting point is a generic concept of regulation which does not in itself imply or presuppose any particular set of substantive claims (such as the existence of particular 'modes' of regulation). To summarize:

- Social systems are dynamic, complex and contradictory. In most social systems, therefore, there are tendencies towards crisis and rupture in the reproduction of their constitutive social relations through time and across space.
- The regulation of a social system refers to processes which mitigate contradictions, promote system reproduction and displace crises spatially or temporally.
- Regulation is not an inevitable, automatic or structurally necessary process but rather involves intentional social practices. However, the regulatory impact of such practices is frequently an unintended consequence of actions undertaken for other reasons.
- Regulation is an emergent property of social systems. That is, it is usually the result of the interaction of a number of separate elements, each of which is a necessary but insufficient condition for regulation. In the context of the effective regulation of capital accumulation (when this occurs), the regulatory mix is commonly referred to as the 'mode of regulation' (MOR).
- Since it is a social process, regulation is itself also dynamic and prone to contradictions and crises. Any given regulatory mix will be effective only for a period. Eventually, it too will be undermined.
- When regulation fails there are three possible outcomes: (i) the system being regulated is undermined and a crisis arises; (ii) a different

regulatory mix develops in its place; or (iii) the regulatory process becomes an object of regulation in its own right. Note that (iii) does not imply that such 're-regulation' will necessarily be successful: (iii) could coexist with (i). Logically, (iii) could also coexist with (ii).

It is important to distinguish the processes of regulation from the objects of regulation. Most writing on regulation theory to date refers to the regulation of the contradictory and crisis-prone process of capital accumulation. Although it is by no means the only element, central to these accounts is the role of the state in mitigating contradictions and managing crises. During the 1970s, political theorists began to point out that the state was itself contradictory and subject to crisis tendencies (some of which stemmed precisely from the regulatory role that the state had come to play). This led Claus Offe (1984) to write of a 'crisis of crisis management' in the capitalist state. At that time British local government moved from playing (among other things) a regulatory role to being bound up in the crisis tendencies of the Keynesian welfare state. We suggest that during the 1980s a particular political response to those tendencies, coupled to the desire of the right to defeat socialism and labourism, led to local government itself becoming increasingly an object of regulation. It was this which generated the substantive shift from government to other forms of governance. . . .

Crisis

. . . [T]he key aspects of the concept of crisis are as follows:

- A crisis is a rupture in the reproduction of a social system. Crises may be of two sorts: a crisis in the system which is a rupture in the reproduction of some part of the system, or a crisis of the system in which the system as a whole is under threat.
- All complex social systems exhibit crisis tendencies. The tendency to crisis, however, may be mitigated by regulation (see above).
- Crisis tendencies may lead to an actual crisis which, if resolved, produces a qualitative change

in the character of the system. In the absence of a resolution, it makes little sense to talk of a 'perpetual' crisis. It may be appropriate to refer instead to a failure of system reproduction (Hay and Jessop 1993). A crisis may thus lead either to a resolution or a failure, while a resolution may involve either changes in the system or a replacement of the system.

■ In a total breakdown of regulation, a crisis of the social system being regulated is, by definition, inevitable. However, the complete absence of regulation is very rare. On the other hand, the effectiveness of regulation can vary significantly.

■ The regulatory process itself is also crisis prone. A crisis of the regulatory mix may lead either to the failure of regulation or to the re-emergence of regulation on a different basis.

■ Crisis is not the only cause of regulatory failure. Failure may also occur through a long-term deterioration in the effectiveness of regulatory processes. Since both the processes and objects of regulation are constituted of dynamic social relations, it is possible for incremental change to lead to a qualitative shift in the success of regulation over time.

[. . .]

Crises and modes of regulation

. . . When regulation theory was first developed in the 1970s and early 1980s, it was frequently assumed implicitly that 'successful' regulation within a stable MOR was the norm. Each MOR was seen to have its own contradictions and crisis tendencies which would, in time, lead to its breakdown. But the assumption seemed to be that, after a relatively short period of crisis and restructuring, a new, qualitatively different MOR was likely to arise to take its place. Hence the widespread (but problematic) acceptance across many disciplines that the parameters of debate could legitimately be set by the twin terms of Fordism and post-Fordism.

In the 1990s, after twenty years of global economic upheaval and restructuring, it is less clear whether a new MOR will emerge to stabilize economic relations and promote sustained growth.

This suggests that successful regulation (and thus a MOR) is a relatively unusual phenomenon: a fortuitous and temporary socio-institutional pattern which, because of its partly 'accidental' character, is inherently rather unlikely to develop. While logically possible, modes of regulation might be rather rare. . . .

What are the implications of this for regulation theory? In our view, the absence of a new MOR would not invalidate the basic arguments of regulation theory but it would require some adaptation in the way they have commonly been applied. Specifically, it requires less stress to be placed on MORs as structured wholes and more concern with regulatory processes and mechanisms which may interact only rarely to secure sustained growth – hence our empirical concern with the changing structures of local governance. We believe the concept of regulation is worth retaining since socio-institutional supports are a necessary pre-condition for the reproduction of capitalist social relations to occur at all.

[. . .]

FORDISM AND LOCAL GOVERNANCE IN BRITAIN

Defining 'Fordism'

. . . According to Jessop (1992), as a MOR, Fordism involved: (i) a wage relation in which wages are indexed to productivity growth and inflation; (ii) a key role for the state in managing demand; and (iii) state policies which help to generalize mass-consumption norms. The state operated demand management policies in the fiscal sphere and underwrote a minimum level of working class consumption to complete the virtuous circle. We would also wish to include in our concept of the Fordist MOR, two key elements of what Jessop refers to as the mode of societalization. These are the trend to the mass consumption of key commodities, such as cars and televisions, and the trend towards bureaucratized state provision of collective public services. Both of these were simultaneously a social consequence of Fordism (and therefore a feature of 'societalization') and a condition of its continuation (and thus part of the MOR).

British local governance under the Fordist mode of regulation and its crisis

... Table 1 summarizes the main characteristics of local governance under British Fordism in relation to what we regard as the key institutional sites of regulation. We have organized these under two dimensions which Jessop (1992: 52) has labelled the 'structural' and 'strategic' moments of the mode of regulation: the former refer to the actual organ-ization[s], the latter to the strategic perspectives and discourses which are currently dominant. Thus any changes involved in a move away from Fordist local government will be both organizational/material and political/discursive.

The dynamics of the relationship between local governance and the Fordist mode of regulation in Britain have been outlined previously elsewhere (Painter 1991). To summarize some of its features briefly:

■ Elected local government under Fordism in Britain was a key element of the Keynesian welfare state.
■ As such, it provided important elements of the social wage, such as housing, which helped to

Table 1 Local governance in Britain under the Fordist mode of regulation

Key sites of regulation			British local governance in Fordism
Structural moments	Production relations	Financial regime	Keynesian
		Organizational structure of local governance	Centralized service delivery authorities
			Pre-eminence of formal, elected local government
		Management	Hierarchical
			Centralized
			Bureaucratic
		Local labour markets	Regulated
			Segmented by skill
		Labour process	Technologically undeveloped
			Labour-intensive
			Productivity increases difficult
		Labour relations	Collectivized
			National bargaining
			Regulated
	Consumption relations	Form of consumption	Universal
			Collective right
		Nature of services provided	To meet local needs
			Expandable
Strategic moments	Dominant politics	Ideology	Social democratic
		Key discourse	Technocratic/managerialist
		Political form	Corporatist
	Typical political goals	Economic	Promotion of full employment
			Economic modernization based on technical advance and public investment
		Social	Progressive redistribution/social justice

underwrite the mass consumption norm required to sustain Fordist growth.

- It also provided services for which there were political demands but which could not easily be mechanized and mass produced, and which were thus unprofitable for the capitalist sector to provide on a universal basis. Hence, the paradox that the Fordist state engaged in production using largely non-Fordist techniques.

- In other respects, however, local government in the Fordist mode of regulation paralleled Fordist organizational forms. It was, for example, frequently hierarchical, bureaucratic and corporatist.

- Local government also contributed to regulation through infrastructural provision and the planning system.

Although important parts of British local government contributed to the dynamic of the Fordist mode of regulation, the relationship was not a purely functional one. . . . [A]lthough it displaced certain crisis tendencies, for example, through its contribution to the social wage, it exacerbated others. Outside local government, a key element of Fordist regulation was the organization of labour relations in terms of collective bargaining. This was regulatory in the parts of the manufacturing sector where wage increases could be financed through productivity gains as a result of technical change. By contrast, when collective bargaining became widespread in the public-services sector, there was relatively little scope for such productivity growth and the organization of labour relations along these lines began to act dysfunctionally. . . .

BRITISH LOCAL GOVERNANCE AFTER FORDISM: ISSUES OF ANALYSIS AND QUESTIONS FOR RESEARCH

Modes of regulation can secure expanded reproduction only temporarily. In time, they are themselves subject to internal contradictions and eventually break down. However, the social and political institutions and practices of a mode of regulation can persist for many years after they have ceased to contribute effectively to sustained economic growth. Hence it is only now, twenty years after the emergence of economic crisis in Britain, that

we can be sure that we are seeing a sustained move away from the components of local governance under Fordism outlined in Table 1. This move has been produced in part by a shift in emphasis in the position of local government. Local government is simultaneously an agent and an object of regulation but the relative significance of these may vary over time. During the 1970s and, especially, the 1980s local government became increasingly an object of regulation in the attempts by the central state to develop strategies in response, in part, to the crisis of Fordism.

The shift may be seen in each of the areas identified in Table 1. Table 2 summarizes some of the new developments involved. . . . The elements in the right-hand column of Table 2 have been drawn from a range of academic and practitioner accounts of change in British local governance. They are assembled here as a heuristic device to provide a framework to guide research and do not represent research findings. The precise character of these new developments is a matter for empirical investigation. We focus in particular on those developments which contrast with the Fordist arrangements because we are interested in the impact of the breakdown of Fordism. In so doing, we recognize the constitutive role that social and political struggles are playing in determining the shape of the future. It is important to stress that we are not arguing that the emerging developments constitute a new MOR. First, as we have suggested, there is no necessity for one MOR to be followed rapidly, or ever, by another. Secondly, even if a new 'post-Fordist' MOR can be identified, it need not parallel Fordism in the central role it accords to local governance. These, too, are matters for further empirical investigation.

[. . .]

Some writers have indeed claimed that local governance in Britain is changing from a Fordist to a 'post-Fordist' or 'neo-Fordist' form . . . but this can amount to little more than cataloguing a series of changes which seem to fit with a supposed model of 'post-Fordist' industrial organization. An identification of local governance as 'post-Fordist' should involve more than this essentially descriptive activity. For us, the analytical power of regulation theory lies in its ability to analyse both the consequences of a particular MOR for the nature of local governance and the role that the structure

Table 2 New developments in British local governance

Site of regulation	British local governance in Fordism	New developments
Financial regime	Keynesian	Monetarist
Organizational structure of local governance	Centralized service delivery authorities Pre-eminence of formal, elected local government	Wide variety of service providers Multiplicity of agencies of local governance
Management	Hierarchical Centralized Bureaucratic	Developed 'Flat' hierarchies Performance driven
Local labour markets	Regulated Segmented by skill	Deregulated Dual labour market
Labour process	Technologically undeveloped Labour-intensive Productivity increases difficult	Technologically dynamic (information-based) Capital-intensive Productivity increases possible
Labour relations	Collectivized National bargaining Regulated	Individualized Local and individual bargaining 'Flexible'
Form of consumption	Universal Collective rights	Targeted Individualized 'contracts'
Nature of services provided	To meet local needs Expandable	To meet statutory obligations Constrained
Ideology	Social democratic	Neoliberal
Key discourse	Technocratic/managerialist	Entrepreneurial/enabling
Political form	Corporatist	Neocorporatist (labour excluded)
Economic goals	Promotion of full employment Economic modernization based on technical advance and public investment	Promotion of private profit Economic modernization based on low-wage, low-skill, 'flexible' economy
Social goals	Progressive redistribution/social justice	Privatized consumption/active citizenry

and practices of local governance play in processes of regulation. It is this second role that is often understated by existing accounts of 'post-Fordism' in local governance. For, while it is easy to chart a series of changes in the institutions and mechanisms of local government and label them 'post-Fordist', it is less clear what part these changed structures and practices might play in helping to stabilize the functioning of any new MOR.

In order to help answer this question, research on local governance needs to explore detailed case studies as well as build up a picture of extensive change. The key questions for such intensive work are centred on those identified in Table 3. What is especially significant is the extent to which these disparate (and sometimes initially unrelated) changes interact with each other in mutually reinforcing ways. This is something that can be revealed only on the basis of intensive qualitative and causal analysis in specific cases, since it involves examining the actual operation of potential regulatory processes in the ways that

Table 3 New developments in local governance and related research questions

Site of regulation	New developments	Research questions for each local area
Organizational structure of local governance	Wide variety of service providers Multiplicity of agencies of local governance	Which organizations are involved in the production and distribution of which local public services? (e.g. TECs, UDCs, voluntary bodies, private companies, local authorities)
Management	Developed 'Flat' hierarchies Performance driven	What new forms of management have been introduced? Have bureaucratic hierarchies been decentralized?
Local labour markets	Deregulated Dual labour market	How has regulation of the labour market changed? To what extent is the labour market divided into core and periphery?
Labour process	Technologically dynamic (information based) Capital-intensive Productivity increases possible	What forms of technical change and innovation have been introduced? What productivity increases have been made?
Labour relations	Individualized Local and individual bargaining 'Flexible'	Has national-level bargaining been replaced by local or individual negotiation? Have new pay structures and forms of remuneration been introduced?
Form of consumption	Targeted Individualized 'contracts'	Have entitlements to services been changed? Has 'customer service orientation' been introduced?
Nature of services provided	To meet statutory obligations Constrained	Have services been cut or their nature altered?
Ideology	Neoliberal	What are the political views and motives of local decision-makers?
Key discourse	Entrepreneurial/enabling	What are the key discourses informing local decision-making?
Political form	Neocorporatist (labour excluded)	What class and other alliances characterize local politics?
Economic goals	Promotion of private profit Economic modernization based on low-wage, low-skill, 'flexible' economy	What are the key strategic planning and local development objectives of agencies of local governance?
Social goals	Privatized consumption/active citizenry	What are the principal social goals of agencies of local governance?

they interact with others. We are thus arguing for a focus on the interaction and combination of a range of practices and mechanisms of regulation rather than measuring change against some ideal notion of what 'post-Fordist' local governance might look like.

[...]

CONCLUSION: LOCAL GOVERNANCE AND NEW GEOGRAPHIES OF REGULATION

The relatively stable regulation of economic activity provided by Fordism was premised on the possibility of national modes of growth. This implied that a degree of geographical coherence in regulation was possible at the scale of the nation-state. Such coherence was in turn bolstered by a commitment to regional balance within the nation-state. The dissolution of Fordism has been closely connected with the twin processes of globalization and localization which have systematically undermined that possibility. In some cases, national governments have been far from innocent victims of such changes. The British government, for example, led the field in promoting the deregulation of international financial markets. Whatever the causes, however, the consequence has been to place much greater emphasis on competition between places for resources. . . .

Under Fordism, welfare levels were politically defined as a right to certain minimum standards. Increasingly, they are now seen as subordinate to, and defined by, economic performance and development, both nationally and locally. The new institutions and agencies of governance allow different welfare strategies to be developed more easily in different places, according to an area's social and economic structure.

The rise of the new local governance is not, therefore, merely a consequence of the breakdown of Fordism but is also a causal factor in the dissolution of the possibility of pursuing Fordist strategies. Moreover, its geographical unevenness is not just a by-product of change but is part of what makes the changes possible.

Any combination of MOR and accumulation system under capitalism will be unevenly developed and, while some regional economies, for

example, will be favoured by national accumulation strategies . . . others will not. (Peck and Tickell 1992: 10).

Moreover, the differentiated spaces of regulation within a nation arise not only as a result of localized conditions of production and consumption, and local constellations of social forces and cultural practices but also because local agencies are often the very medium through which regulatory practices are interpreted and ultimately delivered. In other words, as we suggested in our earlier discussion of 'the local', mechanisms and components of regulation operate locally and regionally . . . as well as nationally and internationally.

Under Fordism, these components were often used to carry out a national strategy based on a commitment to regional balance and even growth. New towns, overspill towns, explicit regional policy, office and industrial development permits, urban redevelopment schemes and infrastructural provision were all used in an attempt to mitigate the worst effects of uneven development. Indeed, the 'one nation' policies of the post-war social democratic settlement could be said to refer to a spatial as well as a political consensus – one nation geographically as well as socially. Dunford and Perrons (1994) presented evidence that explicit interregional resource transfers were made during the Fordist era in an effort to reduce spatial disparities. They concluded that Fordism was a period of convergence as well as of growth. This consensus and convergence no longer holds and throughout the 1980s we witnessed increasing social and spatial inequalities, measured through a variety of socio-economic indicators. . . . While this would have been constructed as a serious social and economic problem within the dominant discourses of Fordism, the rhetorics of local governance understand inequality either as irrelevant or as an active element in the promotion of economic dynamism.

Places are now expected to compete for scarce central resources. Local–central government relations have been reconstituted so that the centre no longer underwrites local expansion as it once did through Keynesian growth packages. It forces places instead to fling themselves into the competitive process of attracting jobs and investment by bargaining away living standards and regulatory controls (Peck and Tickell 1994: 280–1).

Instead of local authorities implementing universally agreed levels of service provision, a plethora of different agencies now struggle against one another. At its most extreme, this emphasis on competitiveness has forced some places to redefine their welfare strategies to fit in with different local economies and social formations (see Cochrane 1993). This process has been facilitated by the restructuring of the welfare state which has opened up the possibility for markedly different welfare strategies to be pursued in different places.

At first sight, therefore, there seems to be a good fit between the emerging geographically differentiated system of local governance and the global–local economic processes of the world after Fordism. A dynamic relationship between the two in which local flexibility provided by the new local governance promoted strong economic growth, would fulfil at least the minimal criteria for the identification of successful regulation. However we believe no such conclusion can be drawn. There are clear objections on empirical grounds and we have already indicated that there are few signs of economic stabilization in the medium term.

Theoretically, there are also good reasons why the emergent forms of local governance will not help to secure a temporary coherence in a nascent MOR. The geographical differentiation of local governance is as much a hindrance as a help to regulation. The contradictions thrown up by capitalist uneven development, which continue to lead to problems of political legitimation and economic performance, seem less amenable to constraint through a system of governance than they did through a system of government. If anything, these contradictions are exacerbated by the emerging system of local governance which actively promotes uneven development. Initially, the widening of inequality in Britain caused deepening economic and political problems in the inner cities and the older industrial regions but, more recently, the overheating of the southeast economy in the late 1980s produced growing problems in the area's housing and labour markets. Indeed, Peck and Tickell (1992: 20, original emphasis) argue that 'it was the *absence of effective regulation in the South* which brought home the contradictions of uneven development and Thatcherism' and they go on to claim that this lack of effective regulation in the

south caused the failure of the whole Thatcherite experiment by helping to undermine its social, economic and geographic base.

What we are witnessing with the demise of Fordism is the emergence of much greater geographical unevenness in the system of regulation. The abandonment of national redistributive strategies and the emerging global mosaic of regional economies have led to the development of a parallel mosaic of differentiated spaces of regulation. These are constituted from a mixture of regulatory and anti-regulatory processes operating at different spatial scales. While the new local governance is part of that mixture, there is little evidence so far that it is capable of helping to sustain economic development or social cohesion in the medium term.

REFERENCES FROM THE READING

Cochrane, A. (1993) *Whatever Happened to Local Government?*, Buckingham: Open University Press.

Dunford, M. and Perrons, D. (1994) 'Regional inequality, regimes of accumulation and economic development in contemporary Europe', *Transactions of the Institute of British Geographers* 19, 2: 163–82.

Hay, C. and Jessop, B. (1993) 'The post-Fordist local state: some sceptical remarks', Paper presented at the ninth Urban Change and Conflict Conference, University of Sheffield, September.

Jessop, B. (1992) 'Fordism and post-Fordism: a critical reformulation', in M. Storper and A.J. Scott (eds) *Pathways to Industrialization and Regional Development*, London: Routledge, pp. 46–69.

Offe, C. (1984) ' "Crises of crisis management": elements of a political crisis theory', in C. Offe (ed.) *Contradictions of the Welfare State*, London: Hutchinson.

Painter, J. (1991) 'Local government and regulation theory', *Local Government Studies* 17, 6: 23–44.

Peck, J. and Tickell, A. (1992) 'Local modes of social regulation? Regulation theory, Thatcherism and uneven development', SPA Working Paper 14, School of Geography, University of Manchester.

Peck, J. and Tickell, A. (1994) 'Searching for a new institutional fix: the after-Fordist crisis and global–local disorder', in A. Amin (ed.) *Post-Fordism: A Reader*, Oxford: Blackwell, pp. 280–315.

Editors' references and suggestions for further reading

Harvey, D. (1989) "From managerialism to entrepreneurialism: the transformation of urban governance in late capitalism," *Geografiska Annaler* 71B: 3–17.

Imrie, R. and Raco, M. (1999) "How new is the new local governance? Lessons from the United Kingdom," *Transactions of the Institute of British Geographers* 24: 45–64.

Ward, K. (2000) "A critique in search of a corpus: re-visiting governance and re-interpreting urban politics," *Transactions of the Institute of British Geographers* 25: 169–85.

"Yuppies, Yuffies and the New Urban Order"

from *Transactions of the Institute of British Geographers* (1989)

John R. Short

Editors' Introduction

Lying just to the east of London's "Square Mile" with its dense cluster of international financial services is London Docklands. Once the world's busiest port in terms of the value and volume of trade, by the 1970s it had become a vast area of abandoned docks, vacant land and derelict warehouse buildings. Beginning in the 1980s, however, Docklands underwent a process of regeneration that provides a spectacular example of corporatist-style politics in urban development (Hall 2002). Drawing on policy ideas originating in the USA, the UK government designated Docklands an Enterprise Zone under the control of an appointed Urban Development Corporation, replacing the locally elected councils as the planning authorities in the area. With powers to acquire land and the resources to invest in infrastructure provision, the Docklands Development Corporation had, by the late 1980s, engineered a dramatic transformation in the economic, social and built environment of the area. It is these changes that provide the inspiration for this reading by John Rennie Short, Professor of Geography at the University of Maryland, Baltimore. Set within the wider context of the economic restructuring and resulting social polarization that many have claimed are the hallmarks of world cities (although see Hamnett's reading in Part 2), Short's analysis focuses on the growing inequalities and conflicts within Docklands between the "new middle classes" or "yuppies" (young upwardly mobile people or young urban professionals) and an "underclass" of "yuffies" (young urban failures). Most of the new jobs in the glittering office blocks of Docklands are in the highly paid fields of FIRE (finance, insurance and real estate), producer services and the media, bringing in people from outside the local area. This, in turn, has created a demand for new housing resulting in what Short terms the "yuppification" of large areas of Docklands. At the same time, employment opportunities for the existing population of Docklands are much more limited, contributing to a growing number of yuffies. Young, alienated and unemployed, yuffies have been likened to the "dangerous classes" of nineteenth-century cities and are seen as a potent threat by the new middle classes who have moved into Docklands. This juxtaposition of extreme affluence and poverty in the same social space is, Short contends, illustrative of a "new urban order" that characterizes the world cities of "disorganized capitalism."

While there are important parallels with Hamnett's reading on "Social Polarisation in Global Cities," Short's analysis highlights the important *political* dimensions that underpin the emergence of, and resistance to, a new urban order. Changes by the UK government to the income tax and benefits regime in the 1980s reinforced the social inequalities brought about by processes of economic restructuring by reducing the levels of taxation on those earning the highest incomes and cutting the benefits available for those who were unemployed. In addition, government policy was crucial to the remaking of the landscape of London Docklands. By designating the area as an Enterprise Zone, central government displaced the planning role of locally elected councils, creating a local political environment in which the concerns of local communities could be marginalized and private sector interests were expected to flourish. This shift in the balance of political control in Docklands clearly intersects with wider debates about the relationships between "urban government" and "urban governance" considered in this section (see Goodwin and Painter reading). While Short's paper highlights some of the negative social consequences for Docklands of this political strategy, Fainstein (1994) has assessed its economic impacts and is similarly pessimistic. "The whole Docklands experience," she contends, "exposes the fatal weaknesses of relying heavily on property development to stimulate regeneration – government-supplied incentives to the development industry inevitably beget over-supply if not accompanied by other measures to restrict production" (p. 213). Other assessments of the Docklands development are in Peter Hall's (1998) *Cities and Civilization* (London: Weidenfeld and Nicolson) and Janet Foster's (1999) *Docklands: Cultures in Conflict, Worlds in Collision* (London, UCL Press).

Over the past twenty years, John Rennie Short has contributed many widely used texts in urban geography. Among his books in this field are *An Introduction to Urban Geography* (London: Routledge, 1984), *The Urban Arena: Capital, State and Community in Contemporary Britain* (London: Palgrave Macmillan, 1984), *The Humane City* (Oxford: Blackwell, 1989), *The Urban Order* (Oxford: Blackwell, 1996) and, with Yeong Kim, *Globalization and the City* (London: Longman, 1999).

■ ■ ■ ■ ■ ■

[. . .]

CHANGING SOCIAL RELATIONS

World cities are losing their status as manufacturing centres and are becoming centres for the tertiary and quaternary sectors of the economy. Since the mid-1960s a new international division of labour has emerged in which world cities have lost much of their manufacturing employment through closure, mechanization, suburbanization of industry, more efficient work practices and what Peter Dicken (1986) refers to as the global shift of industrial employment from the core to the semi-periphery and peripheral countries of the world economy. New employment growth in world cities is dominated by service employment especially in the producer services category (Marshall 1988). The biggest and fastest growth has occurred in financial services. The measurement, monitoring, moving and management of money is now a major growth industry. The changes have not been uniform across space. . . . Britain because of its poor competitiveness, lack of investment and a financial structure which easily allowed foreign investment, showed some of the biggest losses of industrial employment. British capital found it easy and more profitable to invest in manufacturing abroad. . . .

The decline in Britain was greatest in the big cities. Between 1971 and 1981 1.8 million manu-facturing jobs, over one-third of all manufacturing employment were lost in the conurbations of Britain; in London alone 200,000 manufacturing jobs have been lost every five years since 1961. . . . The net result was a form of deindustrialization with specific implications for gender and class relations. An examination of the figures in *Employment Gazette* reveals that between 1971 and 1981, the number of jobs done by men fell by 1.7 million: in contrast, female employment has increased; over the period between 1961 and 1981 the number of women in paid jobs increased from 7 million to 8.5 million, from 35 per cent of the total workforce to 45 per cent. The increase in part-time employment by one million jobs in the past 15 years has been almost entirely female labour. While cultural images of work and popular representatives of employment may still make a distinction between male workers and female house-wives the new realities are very different. A majority of adult women have either part-time or full-time employment in addition to their domestic chores. While the male working class saw the closure of traditional avenues of employ-ment, female members of the working class saw increased employment opportunities albeit in rout-ine jobs. Labour organizations have traditionally been forged from the experience of male working class. Deindustrialization has meant a decline in the social and political power of the male working class.

Table 1 Employment in the UK 1971 to 1986

	Millions	
	1971	1986
Manufacturing	8.06	5.23
Services	11.62	14.49
banking, finance, insurance and business services	1.33	2.20
All industries and services	22.13	21.59
Unemployed	0.75	3.28

Source: Social Trends (1988, Tables 4.9 and 4.19)

Table 1 shows the general pattern of a decline in manufacturing employment over the period 1971 to 1986 and the absolute and relative increase in service employment. The table also reveals that over 30 per cent of this increase came from the financial services category. In this sector employment is bifurcated between a high paid managerial section, predominantly although not completely male, and mostly female workers in the routine office work. Both manufacturing loss and financial services gain have been urban-based. The predominant centre is London accounting for 60 per cent of all financial and producer service employment. . . .

In summary, there has been a loss of manufacturing employment and an increase in service employment all against a background of rising unemployment. The social effects have been a reduction in the power of the traditional male working class, an increase in female employment and the emergence of a new middle class. These trends have been given popular recognition in the terms *yuppie* and *yuffie*, themselves part of a plethora of new words coined in the 1980s including *buppies*, *swells* and (my favourite) *lombards*. A yuppie is a young upwardly mobile person though the 'u' can also denote urban. Yuffies are young urban failures. If the yuppies are the successful new middle class, yuffies are the stranded and blocked working class. The other terms? Buppie is the yuppie's black equivalent, swell is single women earning lots in London, a term which summarizes the rise of the female executive and perhaps the beginnings of the end for the monopoly of the male domination

of senior and responsible positions. Lombard is lots of money but a right dickhead, a term of abuse whose real quality is only recognized if you know that one of the main streets in the City of London is Lombard Street.

[. . .]

As myths yuppies have become a powerful model, a peg for advertising campaigns and dedicated followers of fashion. Some developers are also building for the yuppies, parts of cities are being yuppified. This has both an empirical and a symbolic element. Empirical in the sense that changes can be seen in the form of new housing stock, leisure facilities etc. and symbolic in the sense that the meaning of particular places is being transformed. As fact yuppies are an emerging social group with particular forms of employment and consumption. Their existence is due to the rise of non-manual and especially managerial and professional categories of employment. Yuppies are the higher paid members in the technical and management levels of the control centres of international corporations, the expanding financial services sector, producer services and the media industry. They are particularly found in London where these sectors are concentrated. The term yuppie is a loose one, it is suggestive of a new social group, not so much a class more a constellation of groups whose emergence has been noted but not fully identified. The term is useful as a shorthand, a generalization which stands in contrast to that other social grouping of contemporary society, the yuffie.

Yuffies are young urban failures. The 'failure' is their inability to get a job. It does not imply personal failing or irresponsibility. The main problem of the yuffie is that they were born at the wrong time. Life chances of any social group vary through time, we know this from our experience as academics. Born in 1935 you managed to avoid serving in the war and if you followed an academic career you were at your most marketable when the higher education sector experienced its biggest expansion. Born in 1965 your chances of an academic career are severely limited by the financial restrictions on universities and colleges. Same occupation, contrasting opportunities for different cohorts. The same with broad social categories. A semi-skilled worker in Britain had more opportunities when looking for a job in 1968 than in 1988. The yuffies are the unlucky cohort.

As a slump hit the world economy in the 1970s firms responded by either going out of business or by shedding labour. Labour was either sacked, or 'lost' through natural wastage. By not employing new labour, young people coming into the job market for the first time bore the brunt of the recession. Their problems were exacerbated by the response of the British State from the mid-1970s to redirect public expenditure. The public sector was thus not able to soak up the unemployment produced by deindustrialization. The net effect was an increase in unemployment in the UK. In 1988 over 3 million were out of work, almost one million of whom were under 25. . . . In London a third of the population are aged under 26 and of these one in eight are unemployed.

The young unemployed do not necessarily become yuffies. Inability to gain formal employment does not necessarily mean a complete reliance on government income support. Opportunities are available in the twilight world of the informal economy. Research in this area is difficult but the results we do have suggest that the amount of informal work done by the unemployed is very slight. The work of Ray Pahl (1984) suggests that either those in employment or newly made redundant make greater use of the black economy because they have easier access to potential customers, materials and networks of social communication. Young people with no, or limited, work experience are thus in a more difficult position than most. There is a distinction between individuals and households. A young person may be unemployed but not necessarily poor if they are in a household with a worker in paid employment (see Pahl 1988). Yuffies are not simply the young unemployed, they are the alienated young unemployed. They are the people who rioted in 1980 and 1981. These disturbances were not so much race riots as youth riots. . . .

Yuffies pose a problem for the State. Governments in liberal democracies have the problem of providing an income support which is not too low to cause social unrest but not too high so that recipients are discouraged from looking for work. Yuffies also pose a threat in the mythologies which circulate in town and cities. They are perceived as the main source of crime, they threaten people and property, they are the muggers and burglars, the leading roles in the law and order script which reads as follows: there was once a golden age where cities had less crime, now it is not safe to walk the streets as these young hooligans threaten life and property. The nostalgic reference to the past and the calling for more policing is not new. Geoffrey Pearson (1983) has shown it to be a recurring theme of urban Britain at least from the Victorian period, an enduring fear which always fastens on to the young male 'lower orders'. The latest folk devil is the young black unemployed male now a popular symbol of criminal intent, social disorder and moral disruption. Yuffies are the id of the urban imagination.

[. . .]

There is a basis of social experience for these cultural expressions. In Britain there is a growing bifurcation of life chances. For those in employment, income tax has been reduced. Since the Conservatives came to power in 1979 the basic rate of income tax has been reduced from 33 per cent to 25 per cent while the maximum rate has been slashed from 83 per cent to 40 per cent. The wealthiest have benefited most from these changes. The corollary is that those in low income employment face high marginal rates of taxation while the unemployed are facing reductions on benefits. From April 1988 the Social Security Act of 1986 came into effect in which young people were to receive no benefits if they refused a place or a youth training scheme. Under this new poor law arrangement a young single unemployed person aged under 25 would receive £23.55 per week compared to £31.35 under the previous system. By contrast, the average weekly take home pay of an accountant in the City of London in April of the same year was £350. The net result is a growing gap between rich and poor (see Table 2). The richest 10 per cent of the population now have more income than the bottom 50 per cent. This is a reversal of a 50 year trend towards greater equality of income. Not only is there greater inequality but it is becoming more visible. The term young in yuppie is not only a function of age but a position in the great divide in attitudes to credit. The deferred gratification of the old middle class reared on ideas of sacrifice, saving and waiting has given way to the conspicuous consumption of the new middle class emerging in an era of credit cards, buy-now-pay-later slogans and a banking system which is encouraging personal indebtedness.

[. . .]

Table 2 Share of total income after tax in the UK 1975 to 1976, 1984 to 1985

	1975 to 1976		1984 to 1985	
	% share	Average income (£)	% share	Average income (£)
Top 1%	3.9	9,010	4.9	31,060
Next 9%	19.2	4,910	21.6	15,240
Next 40%	50.3	2,900	48.6	7,640
Lowest 50%	26.6	1,230	24.9	3,160

Source: Social Trends (1988, Table 5.14)

RESTRUCTURING SPATIAL RELATIONS

A major cause of the spatial restructuring in world cities is the growth of the financial service industries which is causing a demand for office space. The mid to late 1980s has seen a commercial office boom in the world cities similar to the early 1980s (Short 1988). Where the pressure cannot be met by intensification of existing spaces there is pressure for the extension of commercial space. The tight clustering of such industries means that the extensions cannot be too far away; firms renting space too distant lose credibility and vital contacts. In London the pressure has built up around the edges of the City.

Pressure has also come from new housing demands. The new middle class are more than end points of structural changes. As E.P. Thompson (1968) reminds us in the title and the text of his most famous work, classes make themselves. As part fact, part myth yuppies make themselves in their lifestyles especially in attitudes to time and space. The Filofax is the yuppie icon. It indicates the problem of time yet its successful management. It suggests a life full of work, commitments, movement and meetings. It represents a full life. The problem with the unemployed, in contrast, is how to fill time. The cruel paradox of modern life is those with more resources have less time while those with most time have least resources. The attitude to time is matched by an assessment of space. Yuppies are inner city-dwellers. Their jobs are in the central areas. Not for them the trek to the suburbs made by their parents and dreamed of by their grandparents. A central location saves time in journey to work, entertainment and contact with friends and influences. It is also symbolic of wider attitudes.

The suburbs are, in essence, places for children, they indicate the willingness of people to lead their lives 'for the children'. Suburbs are places of sacrifice, sites for the reproduction of the family. The garden, the lower density and the search for better schools are the essential ingredients of the suburban choice. Yuppie households, if they have children, are concerned as much with the wants of the adults as the perceived needs of the children. The emergence of the yuppies is signalled by new forms of housing consumption. On the one hand there is gentrification as middle-class groups move into low-income neighbourhoods close to the city centre. . . . On the other hand there are the new-build schemes. Aware of the market possibilities developers are now consciously meeting the demand for centrally located dwellings for young middle- and upper-income households with a mix of dwelling size designed to interest non-child and single-person households. . . .

This spatial restructuring involving the extension of commercial space and new forms of residential space is not just a demand-led phenomenon. It is also a result of the growing power and influence of financial institutions seeking long-term investment. It is as much investment as demand-led. Much of urban research in the last two decades has been concerned with the relationship between the mode of production and urban structure. . . . Too little attention has been paid to the importance of the mode of investment. Finance capital has had an enormous impact on the landscape and life of cities. Financial institutions, including insurance companies and pension funds, have been the recipients of the growth of personal savings. Money has flooded as incomes have increased, the number of people in occupational pension schemes has quadrupled in 30 years and more recently in Britain there has been the move towards privatization of old age pensions whereby individuals have been encouraged to invest in a financial institution rather than the State pension scheme. The result has been a huge increase in investing institutions (Plender 1982). Urban property is an attractive investment for these institutions:

(1) it takes up big chunks of money, a handy characteristic for hard-pressed investment managers seeking to make as few decisions as possible;
(2) it is a long-term investment which can be used to balance shorter term investments in gilts, stocks and shares;
(3) its scarcity is assured by the nature of absolute space; in the urban context this is often reinforced by planning controls.

Favoured investments have been commercial properties. . . . There is nothing an investment-fund manager likes more than a tall office block on a prestige site in a favoured location. Increasing asset value is assured. In the world cities property investment has been attracted to the commercial core and surrounding areas, particularly along the fault lines along the edge of CBD. Here the conversion of residential use to commercial users and low-income housing to high-income housing provides the greatest returns. But while individual speculative ventures may give high returns, as in the case of one company obtaining permission for an office block in an adjacent residential area, such deals are risky and time-consuming. Pension fund managers are risk averters rather than risk takers. They seek high returns but on a very secure basis as in the more orderly, organized extension of commercial space and residential renewal such as the construction of Battery Park city on the southern tip of Manhattan, the commercialization of the Rocks area of Sydney and the biggest of them all, the transformation of London Docklands.

London Docklands

Located just east of the capital's financial centre, the 16 square miles of Docklands was the commercial water frontage of London. . . . It was also the home of working-class communities, almost 40,000 initially based on dock-working. By the 1960s the docks were being closed because they were unable to cope with the bigger container ships. The port functions moved east to Tilbury and, in Docklands, registered dock employment fell from 25,000 to 4,100 between 1960 and 1981. . . .

London Docklands was well placed for spatial restructuring. Recent years have seen employment growth in London's City. . . . Close to the City it gave opportunities to developers for the recommodification of derelict land into offices and residences. There was an alignment of investment-rich institutions, a demand from a buoyant City for office property and new housing requirements of the growing new middle class which all led to the recommodification and yuppification of the area. The successful prosecution of these aims required three things:

(1) incentives to private capital;
(2) political power;
(3) central organization.

(1) In the 1980 Budget the Chancellor of the Exchequer announced the creation of enterprise zones to promote private redevelopment of inner city areas. Under this scheme incentives were provided over the period 1981 to 1991 which included exemption from rates and land taxes on site disposal, tax allowance for building construction and relaxation of planning controls. Eleven zones were designated, one of them, the Isle of Dogs, in Docklands. . . .

(2) The commercial transformation of large areas will favour lucky landowners and astute developers but it will not directly benefit the local people. Any truly democratic local representation will thus tend to resist such changes. For the developments to take place, power must be taken out of local hands. This is the rationale behind the creation of London Docklands Development Corporation (LDDC). The LDDC was established by a Conservative government in 1981. It replaced the Docklands Joint Committee established in 1974 and made up of representatives of five dockland boroughs. That Committee was concerned with the needs of local residents. The non-elected, government appointed LDDC has no need to court local political support. Its aim has been to 'develop' Docklands for the private sector.

(3) Individual companies are unwilling and unable to undertake such large and speculative ventures. The LDDC has acted as central organizer of the project – assembling land, making environmental improvements and providing the vital, initial infrastructure investment. The LDDC spent £130 million between 1981 and 1985, almost £200 million from 1985 to 1987 including £35 million on a light railway system which links the area to the City, London's financial centre, and has asked for £531 million for the period 1988 to 1993. The area is now an attractive location for office users, it is now 'closer' to the City and all of central London, yet rents are only quarter of what they are in the City.

[. . .]

In their own terms the LDDC has been successful (Church 1988). Almost £2,200 million of private investment has been attracted. The whole area has been transformed and major developments are planned or in the process of construction. . . . Almost 500,000 square metres of office development is completed or under construction. By the year 2000, it is estimated that 2 million square metres will be constructed. At Canary Wharf is planned the biggest single development, a £3,000 million complex of office and shopping space which eventually will employ 72,000 people. In London Docklands there are the equivalent of fifteen Canary Wharf size developments under construction. Between now and the end of the century office development in Docklands will increase London's commercial office space by 50 per cent.

Housing has also been built. Thirteen thousand dwellings were completed or under construction by 1987. . . . The LDDC plan is to complete 25,000 dwellings by the end of the century. Selling points have been the water frontages and the relative cheapness considering the easy access to the City. Developments on this scale do more than just meet a demand. Giddens (1973) has suggested that an important influence on class formation is residential segregation. The residential element of Docklands development will create the environmental context for the new middle class being as much a class 'for itself' as 'in itself'.

In effect there has been a transformation of the landscape of Docklands. . . . The industrial buildings of the past are being recycled both in terms of use and meaning. The Docklands as Victorian economic resource is giving way to Docklands as postmodern landscape of offices, from old working class to new middle class. Docklands has become a spectacle for the display of post-industrial employment and the presentation of housing forms for the new middle class. The transformation of the Docklands is not only a change in use but a change of meaning.

[. . .]

A NEW URBAN ORDER?

A new urban order is emerging from the contestation for political power and social meaning in world cities. Commercial pressure for central locations, the growing power of the mode of investment and the emergence of new social groupings are taking place at the same time as deindustrialization, pressure on the poor and the emergence of an underclass marginalized by economic change. Returning to Docklands again, Church and Ainley (1987) show how local unemployment especially amongst the young continues to persist despite the boom of Docklands. I use my words precisely. *Taking place* is exactly what is happening as the two social forces are meeting in the same social spaces. Yuppification involves the destruction of an existing community and its replacement by a new one with consequent changes in the meaning and use of space. There is local resistance. The press releases of the Isle of Dogs Neighbourhood Committee, for example, provide an antidote to the publicity machine of the LDDC. They point out that few of the jobs created have gone to local residents. When the average local income was £8,500 per household the average price of a two-bedroomed property in the area was £185,000. More radical has been the attitude of *Class War*. The East End has reached a point of no return; according to them, it's either resist or die. In their newspaper *Class War* and bill posters they urge local people to mug a yuppie, scratch BMW cars and make life as unpleasant as possible for the affluent incomers. . . .

Class War is unusual, perhaps the best (and only?) example of an urban social movement condensing the yuppie/yuffie bifurcation: more common are the unorganized random acts of resistance/vandalism. As a correspondent of the *East London Advertiser* wrote:

> I was delighted the other day when sitting with my younger sister on the Isle of Dogs and saw some youngsters ripping up newly planted trees and using them to attack yuppie homes. Hopefully some young people locally will still have some fight in them and will repel these new Eastenders by making life unbearable for them. (Kane 1987)

As the letter suggests, young people constitute a point of resistance. They have energy, anger and have not yet learnt to accept their fate. In Docklands, however, it looks as if this resistance will ultimately fail. It is the death throes of a community undergoing marginalization and eventual disintegration. The organized yuffies of *Class War* constitute a nuisance and a threat but not a permanent block to the changes. The power of finance capital in alliance with a central government committed to private enterprise and big business is too big an opponent for a small working-class community with few political friends and limited resources.

But the yuffies still have power. Their very existence in the collective urban imagination has produced effects. First, there is the fear of crime; 'colonization' of space involves the invasion of someone else's place. In the imperial past overseas colonization was underwritten by the British Army and Navy. Now it is the police who defend the urban colonizers. It is not that crime is any more prevalent in gentrified areas although the contrast between rich and poor does provide greater opportunities. It is more a case of the new middle classes having the right language and the necessary confidence to demand better policing. Demands for more effective policing are greatest in areas undergoing gentrification. Secondly, the fear of the yuffies is apparent in the new built forms. There is a contemporary urban enclosure movement which is blocking off and minimizing public open space. . . . Riverside frontages are being alienated, walls are being constructed and barriers being created all in order to keep out the urban folk devils. . . . The security arrangements of residential blocks are a major selling point while commercial properties are so designed that their frontages ward off rather

than invite. The attraction of water frontages is only partly the scenic views, for on one side, at least, they can be easily defended against the yuffies. This bunker architecture is concerned more with security than display, personal safety more than show and the exclusion of indigenous communities rather than their incorporation.

The fear of the underclass has always been a major element in the life of London as in all world cities. In the past this has been managed by segregation, people knowing and keeping (in) their place. When different groups are in the same places the emphasis switches to the architectural design of the buildings, the location of buildings and the construction of defensible spaces. In London Docklands and selected areas of other world cities economic restructuring is causing a change of use, a change of meaning and a contestation for the social control of urban spaces. The new urban order of disorganized capitalism will arise from this struggle; its eventual shape a function of conflict and compromise, its final form a mark of victory. And of defeat.

REFERENCES FROM THE READING

Church, A. (1988) 'Demand-led planning, the inner city crises and the labour market: London Docklands evaluated', in Hoyle, B. and Husain, S. (eds) *Revitalising the Waterfront: International Dimensions of Waterfront Redevelopment*, London: Belhaven Press.

Church, A. and Ainley, P. (1987) 'Inner city decline and regeneration: young people and the labour market in London's Docklands', in Brown, P. and Ashton, D.N. (eds) *Education, Unemployment and Labour Markets*, London: Falmer Press.

Dicken, P. (1986) *Global Shift*, London: Harper and Row.

Giddens, A. (1973) *The Class Structure of the Advanced Societies*, London: Hutchinson.

Harvey, D. (1985) *The Urbanization of Capital*, Oxford: Blackwell.

Kane, F. (1987) 'The new eastenders', *The Independent*, 26 September.

Marshall, J.W. (ed.) (1988) *Uneven Development in the Service Economy: Understanding the Location and Role of Producer Services*, Oxford: Oxford University Press.

Pahl, R.E. (1984) *Divisions of Labour*, Oxford: Blackwell.

Pahl, R.E. (1988) 'Some remarks on informal work, social polarization and the social structure', *International Journal of Urban and Regional Research* 12: 247–67.

Pearson, G. (1983) *Hooligan: A History of Respectable Fears*, London: Macmillan.

Plender, J. (1982) *That's the Way the Money Goes: Financial Institutions and the Nation's Savings*, London: Andre Deutsch.

Short, J.R. (1988) 'Construction workers and the city: analysis', *Environment and Planning A* 20: 719–32.

Thompson, E.P. (1968) *The Making of the English Working Class*, Harmondsworth: Penguin, 2nd edn.

Editors' references and suggestions for further reading

Fainstein, S. (1994) *The City Builders: Property, Politics and Planning in London and New York*, Oxford: Blackwell.

Hall, P. (2002) *Cities of Tomorrow* (3rd edn), Oxford: Blackwell.

FOUR

"From Global to Local: The Rise of Homelessness in Los Angeles during the 1980s"

from Allen J. Scott and Edward W. Soja (eds)
The City: Los Angeles and Urban Theory at the End of the Twentieth Century (1996)

Jennifer Wolch

Editors' Introduction

Intense competition between "entrepreneurial" cities for jobs and capital investment has meant that many local political and business elites seem increasingly fearful that the economic vitality of downtown regeneration attempts will be compromised by the visible presence of marginalized and stigmatized social groups in public spaces. In some cities this has resulted in a "war" being waged against the homeless. Smith (1996) has highlighted the police tactics deployed to "take back" Tompkins Square Park in New York, stolen from gentrifiers by the homeless; Mitchell (1997) has drawn attention to the use of anti-homeless laws to "cleanse" public spaces of homeless people; and Macleod (2002) has focused on a range of "interdictory architectures and technologies" which are displacing the homeless from Glasgow's downtown areas. For many urban geographers, then, interest in the homeless reflects a wider concern about the rise of increasingly punitive or revanchist (a term coined by Smith and deriving from the French word *revanche* meaning revenge) urban policies. However, as this reading by Wolch illustrates, the issue of homelessness also needs to be set within wider economic, political and geographical contexts. Not only do local "problems" of homelessness reflect the complex interplay of national and international processes but local policy responses are also highly differentiated, ranging from revanchist to much more progressive approaches.

Wolch's account begins with the "powerful systemic forces," operating at spatial scales that range from the global to the local, that have contributed to Los Angeles becoming the homeless capital of the USA in the 1980s. Processes of economic restructuring brought about a significant growth in the number of working people in poverty as a result of low pay, unemployment and unstable working conditions. These provided crucial preconditions for growing homelessness that were exacerbated by efforts at the federal, state and local political levels to restructure the welfare state. The result has been not simply a growing homeless population but one whose spatial and social contours have become increasingly complex. Historically, the homeless were concentrated in the Skid Row district of LA and, although this remains the largest single concentration of homeless people in the region, comprising mainly African-American males under age 35, other urban areas now have significant homeless populations often with very different demographic profiles. In addition to the 1980s witnessing the emergence of more widely distributed and socially heterogeneous homeless populations, there was also a growing diversity of responses across the Los Angeles region to the homelessness crisis. Some cities chose not to respond in any significant way either by claiming they had no problem or minimizing the extent of the problem. Others adopted a more "revanchist" approach, enlisting the police to enforce laws against creating a public nuisance or trespassing in an effort to deal with panhandling. Some cities, however, responded constructively to the problems of homelessness. These tended to be places with a tradition of progressive politics and service delivery by both government and non-profit organizations, high levels of technical expertise in fields such as housing, social services and community development, and a history of public concern about homelessness which served to mobilize local government action.

Wolch's analysis of homelessness provides an excellent demonstration of how local social issues are intertwined with complex economic and political processes operating at much larger geographical and temporal scales. Moreover, by highlighting the differentiated character of local responses to homelessness, her study also shows how important place is to understanding variations in urban policy. As Wolch indicates in her introductory comments, however, her analysis is largely focused on the structural level of economic and political processes that have given rise and shaped responses to homelessness. Missing are those "homeless voices" describing personal experiences of homelessness and providing insights into how individuals cope with life on the street. This more agency-centered as opposed to structural perspective on the problem of homelessness does, however, feature strongly in other work by Wolch. For example, with Rowe she has examined the social networks of homeless women in Skid Row in Los Angeles, showing how individual women attempt to address their vulnerability to assault by establishing relationships and informal communities with others on the street (Rowe and Wolch, 1990). Other significant contributions to the study of homelessness in urban areas by geographers include work by Veness (1994) and Brinegar (2003) on homeless shelters, Ruddick's (1996) study of homeless youth, and Daly's (1996) comparative study of homeless in the UK, USA and Canada.

Jennifer Wolch is Professor of Geography at the University of Southern California, and much of her research has focused on problems of service-dependent and homeless people in American cities, social policy and human service delivery, and the role of the voluntary non-profit sector in the American welfare state. These interests are reflected in a range of publications. With Michael Dear she has co-authored both *Landscapes of Despair: From Deinstitutionalization to Homelessness* (Cambridge: Polity Press, 1987) and *Malign Neglect: Homelessness in an American City* (Jossey Bass, 1993), and co-edited *The Power of Geography: How Territory Shapes Social Life* (London: Unwin Hyman, 1989). She is also the author of *The Shadow State: Government and Voluntary Sector in Transition* (Foundation Center, 1990). A further area of research is in animal geography, and Wolch has examined issues relating to cultural diversity and attitudes towards animals and the impacts of urbanization and urban design on animal life. With Jody Emel she has co-edited *Animal Geographies: Places, Politics and Identity in the Nature–Culture Borderlands* (London: Verso, 1998).

■ ■ ■ ■ ■ ■

Los Angeles became the homeless capital of the United States in the 1980s. In alarming numbers, Angelenos were cast away from traditional anchors of family, job, and community as waves of economic and social polarization resulted in spreading homelessness. In 1990 to 1991 an estimated 125,600 to 204,000 people were homeless in Los Angeles County at some point during the year, and between 38,420 and 68,670 people were homeless on any given night. Many thousands more were precariously housed, living in fear of eviction or foreclosure, doubled up with family or friends, or constantly on the move as livelihoods and life-sustaining relationships eroded and personal vulnerabilities came to outweigh strengths.

There are many pathways to homelessness in Los Angeles, and as many poignant and disturbing variations on those pathways as there are homeless people. Without detracting from the authority of those homeless voices, it is clear that Angelenos became homeless in record numbers because of powerful systemic forces that shaped their lives in profound ways. These forces, operating at spatial scales ranging from global to local, led to a restructuring of the regional economy, loss of critical welfare state supports, and a shrinking supply of low-cost housing. Combined, they created a swelling population of economically marginalized and precariously housed people. Some of these people became homeless, outcasts from the city's riches and entitlements.

[. . .]

ECONOMIC MARGINALIZATION AND THE SPREAD OF POVERTY

Economic expansion characterized much of the post-World War II period in Los Angeles, transforming the Southland into the largest manufacturing region in the country, with one of the most important centers of international financial and business services. The Los Angeles "job machine" was not immune from broader structural changes in the economy, however. Periods of economic decline during the 1970s and 1980s, linked to world oil and banking crises and rising global competition, presaged the region's shift from a manufacturing center characterized by a mix of traditional Fordist industries, craft production, and aerospace sectors to a post-Fordist service and manufacturing economy. For many workers, the result was poverty, unemployment, and insecure working conditions. Some of the most severely marginalized became homeless.

[. . .]

The successive waves of economic restructuring (deindustrialization, reindustrialization, public sector retrenchment, and service sector expansion) left the regional economy in a vulnerable condition. By the early 1990s, California and Los Angeles were in the throes of a severe downturn, more precipitous than that experienced by the rest of the nation. Unemployment rates soared, with record numbers of claims filed for jobless benefits. Some key sectors were hit the hardest. The cold war ended and the national defense budget was downsized, leading to the loss of tens of thousands of defense-related jobs in Southern California. Multiplier effects of this dramatic decline hurt other sectors of the economy. Key nondurable goods sectors that had helped sustain manufacturing growth over the previous decade began to lose employment. Apparel, which had long been the largest job gainer among nondurables, became the largest job loser in 1990 (at least as far as reported

employed was concerned). With both business and consumer spending down, other major sectors such as retail and service employment also lagged. And, symbolizing the end of an era in Los Angeles, the local automobile manufacturing industry became extinct when the General Motors plant in Van Nuys closed in the summer of 1992.

[. . .]

THE RISE OF A REGRESSIVE WELFARE STATE

The surge in poverty linked to economic marginalization was not remedied by the nation's welfare state, despite the rhetoric of a "social safety net" and a "kinder, gentler nation." Rather, mirroring a trend throughout advanced capitalist countries, public resources were shifted from social needs to investment capitalists in the hope of improving the U.S. position in the international economy. Funding reductions and regressive administrative changes were enacted in many welfare programs during the 1980s. This remaking of welfare occurred at federal, state, and local levels. In Los Angeles, a particularly reactionary county government dealt with the swelling ranks of needy people by acting to restrict the level and availability of poor relief and other key social services. More and more people were impoverished, fueling the city's homelessness crisis.

Federal and state welfare restructuring

California's tax revolt movement of the late 1970s led to sharp reductions in locally generated funds. Then, in 1981, federal cutbacks in a wide range of social spending and intergovernmental transfer programs further reduced California's fiscal resources and directly affected public assistance recipients. For instance, major changes in the Aid to Families with Dependent Children (AFDC) program in 1981 altered eligibility rules and benefit rates and eliminated work and child care allowances, thus denying benefits to large numbers of recipients who also lost Medicaid and Food Stamp benefits. Federal assistance to state and local governments, mainly in housing, health, job training, human services, community development, and income support, also dropped dramatically throughout the early 1980s. Various federal social service and health programs were converted to block grants and their funding levels slashed, and funding for Food Stamps, child nutrition, and unemployment insurance benefits was cut deeply.

With the recession of 1981 to 1982, California's fiscal woes deepened. Despite the recovery that followed, overall budget growth was slow, and welfare recipients were favored targets for spending reductions. In the face of rising joblessness and skyrocketing demands on health and welfare services (the state's welfare population grew three times faster than the general population), California's basic response was to slash safety net expenditures. Health services were privatized and scaled back and responsibility for funding shifted to the counties; by the mid-1980s, California's per capita spending on Medi-Cal patients was the lowest among the ten largest states in the nation, and by 1990, the state, once a leader in health services to the poor, was ranked forty-seventh out of the fifty states in terms of per capita spending on indigent patients. Mental health services were deeply cut, and by 1986 California was near the bottom (42nd) in terms of its care of the mentally disabled.

[. . .]

FROM HOMED TO HOMELESS

[. . .]

Local geographies of the homeless

In Los Angeles, homeless people historically congregated in the Skid Row district, cheek by jowl with the downtown's glittering "trophy" office towers housing centers of international finance, trade, and business services. Although many of its single room occupancy (SRO) hotels have been demolished in recent decades, the Skid Row district remains relatively intact, compared to other cities in which skid row areas were eradicated through urban renewal. Thus not surprisingly, Skid Row today houses the largest single concentration of homeless people in the Los Angeles region. However, over the 1980s, homeless people became widely distributed across the urban area,

as Angelenos living in various parts of the polycentric city found themselves on the streets. Their residence in outlying neighborhoods prior to the onset of homelessness led them to stay within the broad confines of this home "turf" or locale, where they had social ties and more knowledge of community resources. Residence in a neighborhood other than Skid Row was also facilitated by a City of Los Angeles zoning ordinance that permitted by-right homeless shelters of thirty beds or fewer in a variety of commercial, industrial, and high-density residential zones, and by the increasingly restricted opportunities for new shelter and service development in Skid Row itself. Last, the homeless population was no longer composed of older white alcoholic males who historically gravitated toward Skid Row. Rather, the growing diversity of the homeless population, especially the emergence of homeless women and children as a fast-growing component, reinforced spatial decentralization trends.

Systematic surveys of the Los Angeles homeless population have been restricted to Skid Row, while other efforts that threw a wider net, such as the S-night count of the 1990 census, are widely discredited. However, 1991 GR [General Relief] program data on the number of homeless applicants for GR, by regional DPSS [Department of Public Social Services] GR office, and information on the number of homeless families applying for special homeless assistance funds at DPSS AFDC offices reveal the increasingly decentralized and complex local geographies of Los Angeles County's homeless population. Local wisdom suggests that approximately half the total is located in Skid Row; that is, there are between 10,000 and 15,000 on the streets, in hotels and shelters, or living in other transient circumstances (in cars, on rooftops, etc.). However, it is clear that this is an exaggeration; there are large numbers of homeless people outside of downtown, especially the South Central, South Bay, and west side areas.

Considering homeless families applying for AFDC assistance, the largest proportions of this population applied in offices located in the South Bay, central, and south-western areas of the county; offices near downtown served lower proportions of these families. However, the sizes of DPSS district office service areas vary widely, as does population density. Thus, on a per capita basis, the highest

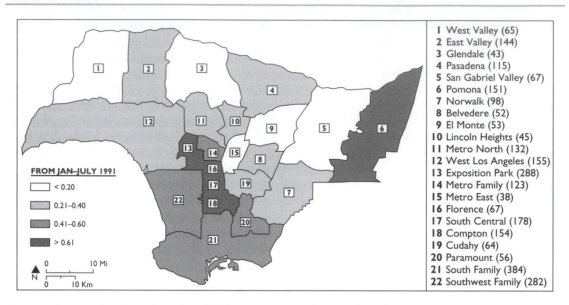

Figure 1 Los Angeles County homeless AFDC applicants per 1,000 population by district.

rates of homeless family applicants were in South Central and southwest of downtown (see Figure 1). A striking finding is that rates were also very high in the Pomona Valley, the easternmost district service area of the county. Many communities in the San Gabriel Valley to the west of Pomona offer little in the way of shelter and emergency services and refer homeless families to shelters in the Pomona area. Thus the high concentration of such families applying at the Pomona DPSS office may reflect a process in which families in the greater San Gabriel Valley become homeless, are referred to shelters in the Pomona area, and social workers in those shelters refer the families to the local DPSS office for AFDC assistance.

The distribution of homeless single people applying for GR was also decentralized, but there were clear differences from the distribution of applicants for AFDC homeless family assistance. The largest proportion of the county's GR homeless applicant pool was in the greater South Bay area, followed by the west side and downtown. The dominance of the South Bay primarily results from district office closures that greatly enlarged this service area. On a per capita basis, the South Bay receded in importance, while the territory extending from downtown to the west side had the highest homeless GR applicant rates (Figure 2; it should be noted that the westernmost portion of district area 10, served by the Rancho Park office,

is comprised of Malibu and the Santa Monica Mountains, and so the map over-emphasizes its importance). DPSS district offices at the Civic Center, which serves Skid Row; Metro Special, just south of downtown, serving South Central; and Rancho Park, serving West Los Angeles, Santa Monica, Culver City, and Venice, are the primary sites where single homeless people applied for homeless relief. The Pomona area, so dominant in terms of homeless families, had much lower per capita homeless GR applicant rates.

Homeless people applying for assistance in different parts of Los Angeles County differed from each other in terms of basic demographic and socio-economic characteristics. A 1987 survey of homeless GR applicants in six major sub-areas of the county (the Civic Center, which encompasses Skid Row; the inner city; South Central; Long Beach; the west side; and the San Fernando/San Gabriel Valley) suggests these basic variations by place. Although they are in line with the overall regional distribution of homeless GR applicants, the survey's sample sizes are small, thus allowing only a few comparisons. Homeless GR applicants from the Civic Center catchment area, which serves Skid Row, were predominantly single African-American males, over half of which were under thirty-five years old. In contrast, although most homeless applicants in the west side and South Central districts were also males, both districts

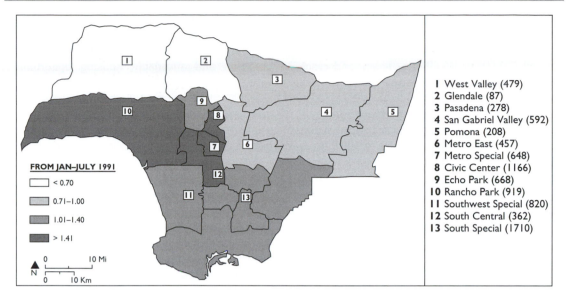

Figure 2 Los Angeles County homeless GR applicants per 1,000 population by district.

had higher shares of women. In Long Beach and the west side, homeless respondents were more likely to be white; and in the valley, Latino. The west side had by far the highest proportion under twenty-five years old (75 per cent), the valley the lowest. Respondents from the valley were most apt to be married; west side applicants had the highest rate of college attendance, the Civic Center the highest share never graduating from high school, and Long Beach the highest rate of high school graduates. These demographic differences reflect the general populations of the county's sub-areas and their poor populations, suggesting that homeless people often stay within their traditional geographic zones in the urban region.

THE LOCAL RESPONSE TO HOMELESSNESS

Most cities in Los Angeles County strenuously avoided confronting the homelessness crisis. Most had neither developed a specific policy statement on homelessness nor taken specific actions regarding the issue as they entered the 1990s. And while the largest and most diverse jurisdictions (such as the county and the cities of Los Angeles and Long Beach) had some of the most extensive sets of policies and programs directed toward the homeless, the likelihood of response bore little

relation to city size. Several smaller localities had also articulated complex responses to homelessness and had targeted relatively substantial resources to shelter and service provision. But some of the region's larger cities had done nothing to respond to the homelessness crisis, either because they were determined to ignore the problem or because they feared attracting homeless people from other localities. As one local official put it, "Our city would like to do something to help the homeless . . . but frankly we don't want to end up being a magnet for homeless people."

Civic silences

The vast majority of cities maintained an official silence on the issue of homelessness. They had no policy toward homeless people, homelessness prevention, or homeless programs. Only four jurisdictions in the county had issued a specific policy statement on homelessness: the county and the cities of Los Angeles, Long Beach, and Santa Monica.

The county's 1985 policy called for more lobbying of state and federal governments, greater local coordination and resource creation through public/private partnerships, a variety of specific instructions to county departments to increase access of homeless people to existing resources, and a commission to oversee county efforts on behalf of the homeless. Later, the county established a homeless coordination unit, took on administration of the federally funded Cold/Wet Weather program, and created the Los Angeles Coordinating Council on the Homeless (in 1991). However, the policy did not recommend additional county expenditures.

The City of Los Angeles adopted a formal policy statement with recommendations in 1987. By that time, the city had committed over $1.2 million to homeless-related services, including funding 695 shelter beds and providing food, clothing, and emergency referral services to over 18,000 individuals. Filled with exhortations to "urge," "encourage," or "support" services to be provided by other levels of government, the policy outlined a fifteen-point "Homeless Action Plan." This plan advocated the continuation of ongoing activities, directed consideration of "regional service centers for the homeless," supported the expansion of job training for the homeless, urged federal, state, and county governments to develop "comprehensive" homeless policies, and suggested that the county increase its commitment to improving General Relief. (The city later sued the county in an attempt to compel such an improvement in GR.) Some concrete proposals emerged, including the establishment of a homeless coordinating unit, a homeless services steering committee, an emergency shelter contingency plan, a mobile ombudsman program, and a "move-in" loan program. Many of these actions were taken.

In the following year, Long Beach followed suit and adopted a formal policy statement on homelessness. This statement consisted of a very general resolution by the city council recognizing the extent and origins of the homelessness crisis and adopting "a policy on homelessness that has as its goal adequate housing, food, and medical services for every resident of the City." To demonstrate its commitment, Long Beach called on the county to fulfill its welfare responsibilities to the homeless and indigent, supported state and federal lobbying efforts of the county, and directed city administrators to apply for nonlocal grant funds, establish an advisory committee, and fund a homeless coordinator. The latter two units were in place by 1991.

Santa Monica's policy statement emerged much later, in the context of an emotional debate over the presence of homeless encampments in the city. The policy called for a delicately balanced set of proposals combining greatly enlarged service provision and affordable housing opportunities with rigorous pursuit of public safety measures, for example, prosecution of illegal activities (such as drug dealing), adoption of an anticamping ordinance to prevent people from sleeping in City parks and beaches, and a ban on outdoor meal programs.

Although none of the other cities in the county had a formal policy statement, as required by state law, many made mention of homelessness in the Housing Elements of their General Plans. Typically, these statements were brief and general. For instance, Arcadia's Housing Element claims that "most of the 'homelessness' in Arcadia is generally temporary displacement-job layoffs, eviction or family disputes," the implication being that the city has no real homeless people; their policy is to refer "displaced" persons to agencies located in other nearby cities. In other cities, the Housing Element detailed the support provided to a nonprofit organization to pick up and deliver homeless people to a shelter in a nearby city.

In two cases, though, General Plan statements were more extensive. Pasadena's draft Housing Element, for example, provided an extensive discussion of homelessness, proposed a new zoning ordinance for shelters and SRO hotels, and promoted a range of new and existing city activities to provide shelter and services to the homeless. West Hollywood's General Plan had not only a Housing Element but a Human Services Element as well. As a result, the city created a nonprofit organization to operate a shelter and adopted a strong anticamping ordinance that they began enforcing when city shelter facilities were opened. Much less elaborate but nonetheless surprising were the cases of Huntington Park and La Verne, both small cities with Housing Elements with proactive policies toward the homeless. The new Housing Element in the small, low-income, largely Latino city of Huntington Park provided a detailed discussion of the extent of homelessness there and adopted policies to facilitate transitional housing and SRO hotel development. La Verne, a middle-income Anglo city of 30,000 in the eastern San Gabriel Valley, had been able to find few homeless people but nonetheless had adopted a Housing Element that called for an annual city grant to a nonprofit shelter provider, zoning changes to allow shelters, counselling centers and other services for the homeless to operate in the city with a conditional use permit, and the development of an information/ referral brochure for homeless people.

Exclusion through zoning and policing

Most cities (about three-fourths) had no specific provisions for shelters, service facilities (such as soup kitchens or drop-in centers), or single room occupancy housing developments. In these instances, such developments were simply prohibited, or else a prospective service provider or developer would have to run the conditional use permit (CUP) gauntlet. Of the jurisdictions with existing or new draft zoning ordinances, ten had SRO ordinances, eight had shelter ordinances, and four had policies directed toward other types of homeless service and housing facilities (including transitional housing). These included the county and its largest cities (Long Beach and Los Angeles) and a scattering of other cities ranging in size and complexity. In most instances, the zoning ordinances were altered to include shelters, SROs, or other service facilities as specifically allowed uses, typically with a CUP; some involved performance standards to be met through the conditional use permitting process, to minimize the impact of public opposition. Interestingly, several cities had recently adopted or were considering transitional housing ordinances, to allow such housing to locate by-right in multifamily residential areas; others, like Huntington Park, were interested in developing SRO units in their downtown centers.

In addition to zoning codes making the siting of homeless facilities difficult, many cities had municipal codes restricting access to public parks during the night, banning loitering and soliciting, and prohibiting trespassing on private property. Laws prohibiting panhandling were also common. But a court decision struck down the state's Penal Code section that had outlawed panhandling, on the grounds that it violated rights to freedom of speech. In any event, these codes predated the homelessness crisis and were rarely used proactively to control local homeless presence or social behavior. Probably the dominant response of cities was complaint-driven enforcement policy. Most cities

had no complaints, and so little need for enforcement. On rare occasions, they prosecuted homeless people, usually for loitering, trespassing, or some other minor offense, when their behavior generated a complaint from a resident or local business. More often than not, formal citations were not issued; rather, homeless people would simply be asked to "move along," or would be referred to shelters or services by city police or county sheriffs.

Some cities, however, experienced mounting pressure to actively control the location and behavior of homeless people. To circumvent their inability to prosecute people for panhandling, some cities reported using laws against creating a public nuisance or trespassing to deal with panhandlers. One city, West Hollywood, distributed flyers informing residents that they were entitled to make a citizen's arrest of panhandlers, on a variety of legal grounds; the catch was that the individual making the arrest was responsible for detaining the panhandler until law enforcement officials could arrive and take the panhandler into custody – an unappealing prospect. Concerns about people sleeping in public places or erecting encampments were also voiced by some local jurisdictions, which adopted a variety of responses. California's general vagrancy law was ruled unconstitutional in the early 1980s, leading some cities concerned with the issue to adopt ordinances that prohibited camping or erecting permanent structures in public places, especially parks. By far the most common rationale for such ordinances was that encampments created health, safety, zoning, and fire code infractions. Actual removals were carried out by a variety of public sector jurisdictions. For example, the City of Los Angeles shut down "Justiceville," a 60-person, organized encampment located on a former playground in Skid Row, removed a group of homeless living on the steps of City Hall, and bulldozed a camp of 125 people living behind the Union Rescue Mission.

In addition to such sporadic removals, a small number of cities routinely conducted law enforcement "sweeps" to remove homeless people from public sidewalks and other spaces. For example, the City of Los Angeles in 1987 began routine sweeps of Skid Row streets to eliminate encampments for health and safety reasons. These sweeps were by far the most highly publicized and bitterly contested destructions of homeless encampments in the county. Eventually, the city adjusted its policy of street sweeps, publishing guidelines and posting police and sanitation sweep schedules involving three sweeps per week in Skid Row. Later, these practices were stepped up; all fifty blocks of Skid Row became off-limits to allow for street cleaning Monday through Friday. In addition, the police attempted to confine homeless individuals to those sidewalks in front of the district's six large missions, in the process ousting people from doorways, breaking up sidewalk gatherings, and waking up people sleeping on side streets and taking them to mission area sidewalks.

Less publicized sweeps also occurred in Long Beach, West Hollywood, and Santa Monica. Long Beach adopted an anti-encampment policy based on health and safety rationales and conducted sweeps regularly. Like Los Angeles, Long Beach posted sweep days and hours and before each sweep, sent social service personnel to provide verbal notification of sweeps and referrals to shelter and service resources. In West Hollywood and Santa Monica, relatively lenient municipal ordinances and law enforcement practices themselves became targets for community backlash against the homeless and resulted in anti-encampment policies and associated sweeps.

Paucity of funding for homeless programs

Los Angeles County received the majority of non-local funds for homeless programs and spent the largest number of locally generated dollars as well. A study of funding for homeless programs in the county reported that in 1988, the county received $10 million from the federal government and $14 million from the state of California. Federal funds originated with the McKinney Act; state funds were allocated through a variety of departmental programs, the largest of which was from the Mental Health Department, followed by the Emergency Shelter Program, AFDC Homeless Assistance Program, and smaller programs targeted to homeless youth, the homeless mentally disabled, and victims of domestic violence. In addition, the county administered Community Development and Community Services block grant funds, much of which it distributed to cities to use

for various purposes, including homeless programs. Not surprisingly, given its statutory responsibilities for population welfare, the county provided the largest amount of nonlocal funding for shelter and services for homeless people. Hamilton, Rabinovitz and Alschuler, Inc. estimated in 1988 that the county allocated $80.9 million, the largest amounts being channelled through the General Relief Program and other Department of Public and Social Services programs ($65.7 million), followed by health services ($12.2 million). The following year, the county spent an additional $730,000 administering the AFDC Homeless Assistance Program. Apart from these statutory, unavoidable responsibilities, however, the county spent little of its own General Fund revenues on the homeless and fought legal actions attempting to compel them to increase entitlement benefits for GR recipients to reduce their risk of homelessness.

More than half of all municipalities failed to spend any local or nonlocal funds on programs targeted to homeless people. Of those cities that did spend funds to provide shelter or services, about one-third spent local funds, one-third spent state and federal funds, and the remainder combined local and nonlocal funding. Virtually all funds were allocated to service-providing nonprofit organizations. Of the fewer than forty cities reporting some expenditures in the homeless area, only a dozen indicated spending more than $50,000.... Not surprisingly, given its enormous population and widespread homelessness, the largest municipal spender was the City of Los Angeles. By the late 1980s, the City of Los Angeles had been allocating over $1 million in General Fund monies each year to homeless programs. But the city eliminated most of its local funding for the homeless in the early 1990s as additional nonlocal revenues became available and its fiscal problems intensified. By 1991, almost $4 million was being spent on homeless programs, virtually all from nonlocal sources. However, the city's separately financed Community Redevelopment Agency, mandated by state law to spend 20 per cent of revenues on low-income housing, appropriated 1.5 million to purchase 102 mobile homes for transitional occupancy by homeless families in 1987; spent $39 million on homeless-targeted capital and service programs in Skid Row between 1977 and 1986; and in fiscal year 1990 to 1991, budgeted $8.56 million on Skid Row for shelters, transitional housing, hotel rehabilitation, and service provision.

[. . .]

How can we summarize this picture of local response? First, the majority of cities did not respond in any significant way to the homelessness crisis. Not surprisingly, some of these cities were affluent, home-owner enclaves (such as Palos Verdes Estates) that reported having no homeless people living within their jurisdiction and therefore felt little inclination to take action. Other cities minimized the extent of their homeless populations, sometimes claiming that homeless people were just "passing through" on their way to shelters and services in downtown Los Angeles or the South Bay. Homeless people requesting assistance or encountered by law enforcement officers were typically sent to local nonprofit agencies or (more commonly) to services located in nearby jurisdictions. Second, those cities that did respond to the crisis varied dramatically with respect to population size and urban complexity. The cities with the largest-scale homeless coordinating staff and service and shelter grant programs and/or policies (either stand-alone or General Plan Housing Elements) were Los Angeles, Long Beach, Santa Monica, Pasadena, Pomona, and West Hollywood – cities of vastly different scale and scope, ranging from 36,000 residents to over 3 million.

Rather than ecological features, three commonalities appear to have conditioned the extent and nature of the most responsive cities: traditions of progressive politics and service delivery; level of professional and technical staff expertise; and history of public concern about the homeless. Cities such as Santa Monica, West Hollywood, and Pasadena became dominated by progressive forces in the 1980s, emphasizing redistributive service provision. Los Angeles had a long tradition of federal antipoverty programs and nonprofit service delivery (especially in Skid Row) and at least a nominal commitment to serving the poor. Long Beach, much more conservative, nonetheless was a port city and thus home to the typical panoply of rescue missions and soup kitchens. These cities also appear to have had employees with substantive technical expertise in housing, planning, social services, and community development, familiarity with the range of political issues surrounding homelessness, and knowledge about

funding programs and working with nonprofit organizations. Last, in each of these cities, public controversy served to mobilize local government action. This prompted local politicians to embark on homeless programming and planning.

Among these proactive places, Santa Monica and West Hollywood stand out as the most responsive. Simply in terms of per capita spending on homeless programs, these two cities allocated between $13 and $15 – eight to nine times the city of Long Beach, and perhaps as much as four times the city of Los Angeles. Both cities boasted an ideologically based commitment to generous community service delivery. During the late 1980s, they found themselves forced to cope with rapidly increasing demands for homeless services. This was partly because of their service policies, but also because of their location (near the beach), social climate (high tolerance for diversity), and, crucially, the practices of other cities who failed to respond to the homelessness crisis, except to "dump" their homeless people onto those jurisdictions offering support facilities and resources. Both Santa Monica and West Hollywood tried to respond to growing demands, but despite their reputations as progressive cities, local officials faced a vitriolic public backlash against the homeless that prompted increasingly stringent social control measures.

CONCLUSION

During the 1980s, Los Angeles was enmeshed in the powerful dynamic of economic restructuring driven by complex changes at both global and local scales. The shift to post-Fordism meant the elimination of thousands of jobs from traditional manufacturing industries. Additional jobs were lost as the public sector retrenched, and defense downsizing led to massive loss of aerospace jobs in the region. Those manufacturing industries that survived tended to reorganize production (e.g., by using flexible production techniques) in order to enable quick adjustments in workforce and production processes as market conditions required. A simultaneous reindustrialization witnessed a spectacular growth in service-related and high-technology industries plus low-technology industries such as garment manufacturing. The latter were often low-skill and low-waged, the former high-skill and well-paid. This development contributed to an increasing bimodality in the region's income distribution. . . .

If poverty, unemployment, and unstable working conditions associated with post-Fordist Los Angeles were crucial preconditions for mounting homelessness, so too were efforts at federal, state, and local levels to remake and partially dismantle the welfare state that had historically protected people from the ravages of the labor market. Over the 1980s, the poor and homeless faced ever-dwindling federal and state health and welfare supports and increasingly hostile and penurious local welfare systems. . . .

Given spreading economic marginalization and welfare state dismantling, it is hardly surprising that decent housing was increasingly beyond the reach of more and more households. To make matters worse, the stock of affordable housing diminished over the course of the 1980s. Housing and rental costs surged ahead of the nation at an alarming rate. The rate of housing production could not match the rates of immigration and new household formation in the region. But other important factors conspired to cause a significant deterioration in the position of low-income renters. Demolitions of affordable housing accelerated, the lost units being replaced by upmarket rentals; local regulations and price inflation inhibited the conversion of units through the shadow market. Housing simply became beyond the means of large segments of the urban population. Available affordable units were often overcrowded and of declining quality. Vacancy rates were negligible at the low end of the housing market. The working poor and those on welfare were obliged to compete for the diminishing number of affordable units. People were forced to invent unconventional methods of putting a roof over their heads, including everything from converted garages to cardboard boxes.

With notable exceptions, local government officials remained spectators at the unfolding homelessness crisis. Their (in)actions ranged from suing one another over perceived dereliction of duties to simply transporting homeless people to neighboring cities. Only a small handful of cities tried to tackle the problem, with inadequate resources and amid mounting pressure to back away from their commitments to the homeless. Most cities spent little or nothing on the homeless and did little in the way of adjusting their plans and policies to encourage the delivery of homeless services or to prevent people from becoming homeless in the first place. In a seemingly endless shell game, localities strove to avoid taking responsibility and shifted burdens of providing support for homeless people to other jurisdictions in the region.

Thus have hundreds of thousands of people been impoverished by the complex interaction of global, national, and local economic forces; become marginally housed in crowded, unaffordable dwellings; and found the social welfare safety net swept out from beneath them without much local assistance in sight. They now teeter on the edge of homelessness with little except chance to hold them back. For those that lose this high-stakes game of chance, a new mendicancy awaits on street corners throughout Los Angeles, where homeless people cluster to beg coins from passersby.

Editors' references and suggestions for further reading

Brinegar, S.J. (2003) "The social construction of homeless shelters in the Phoenix Area," *Urban Geography* 24, 1: 61–4.

Daly, G. (1996) *Homeless: Policies, Strategies and Lives on the Street*, London: Routledge.

Macleod, G. (2002) "From urban entrepreneurialism to a 'revanchist city?' On the spatial injustices of Glasgow's renaissance," *Antipode* 34: 602–24.

Mitchell, D. (1997) "The annihilation of space by law: the roots and implications of antihomeless laws in the United States," *Antipode* 29: 303–35.

Rowe, S. and Wolch, J. (1990) "Social networks in time and space: homeless women in Skid Row, Los Angeles," *Annals of the Association of American Geographers* 80: 184–204.

Ruddick, S. (1996) *Young and Homeless in Hollywood*, London: Routledge.

Smith, N. (1996) *The New Urban Frontier: Gentrification and the Revanchist City*, London: Routledge.

Veness, A. (1994) "Designer shelters as models and makers of home: new responses to homelessness in urban America," *Urban Geography* 15: 150–67.

PART FIVE

Difference

Anti-racist public message. (Courtesy of Elvin Wyly)

INTRODUCTION TO PART FIVE

Residents of cities, as sociologist Richard Sennett notes, are always "people in the presence of otherness" (1990: 123). Both consciously and unconsciously, an individual's experience of the city is influenced by categories of difference, such as race, ethnicity, class, gender, sexuality, disability and age. Since the development of the modern industrial city, this degree of diversity at an intimate scale, coupled with the intensity of urban life, has drawn scholarly interest focusing on how social groups experience the city in different ways. Burgess' model of the early twentieth-century industrial city (p. 19) provides one early examination of "otherness" as it reflects both period conditions and the contemporary moral geographies of the city. The research agenda of the Chicago School of Sociology, which echoed the concerns of society in general, celebrated the vitality and opportunity associated with urban life, while it expressed concern about the breakdown of community, shared values, and the resulting loss of social control.

In Western cities of the early twenty-first century, the impacts of economic restructuring, globalization, including an unprecedented level of migration, and a new sensitivity to the politics of difference provide the current context for our scholarly agendas. A recent review of urban geographic research concluded that geographers' interest in the spatial construction of social life, particularly research that highlights difference, is the most enduring hallmark of the subdiscipline in the 1990s (Aitken *et al.* 2003). Inspired by the political struggles of women, minorities and other oppressed people, as well as by contemporary social theory, geographers have moved beyond an earlier preoccupation with patterns of segregation to challenge categories of difference and the role of space/place in critical theory. Since the 1980s, in particular, the contributions of feminist theory and post-colonial studies, combined with postmodernism's celebration of different identities and voices from the margins, have drawn attention to power relations and the politics of identity in urban geography.

The important consequences of this shift from generalizing theories associated with patterns of segregation and the city to experiences of difference and multiple contexts are suggested in geographers Ruth Fincher and Jane Jacobs' provocative introduction to *Cities of Difference*. They ask, "What happens to studies of housing, suburbia, the inner city, ghettos, gentrification, social polarization, and urban social movements when framed not by a theory of 'the city' but by theories of difference?" (1998: 2). Fincher and Jacobs advocate a new framework for the examination of difference, the formation of urban life and the structuring of urban space – cultural political economy. This framework recasts the traditional dichotomy of cultural and political economy perspectives by focusing on the intertwined nature of "representations of groups and urban places in public discourses . . . with the production and reproduction of the material imperatives of city lives – through urban labor markets and the restructuring of entitlements through the state" (1998: 3).

Although not identified explicitly by their authors as examples of cultural political economy, each of the following readings offers an analysis of difference informed by cultural politics and material circumstances. The quality and degree of productivity inspired by this recent interest in geographies of difference is quite impressive and thus makes it difficult to select only five entries for this section. Despite the array of work from which we choose these entries, we are confident that the following selections demonstrate progress in research related to three categories of difference – race and ethnicity, gender, and sexuality – while they also explore several research themes that characterize recent advances in the study of difference and the shaping of cities' social, cultural and economic geographies.

The first two articles examine race/racism as a dynamic force in place-making, contributing to the literature of post-colonial studies in the first instance and the literature of environmental racism in the second. In "The Idea of Chinatown," Kay Anderson confronts a place previously viewed unproblematically as an ethnic enclave reflecting a cultural heritage. Her study of Vancouver, BC's late nineteenth- and early twentieth-century Chinatown demonstrates that it developed instead as a *Western* landscape type influenced by dominant beliefs that defined Chinese immigrants as biologically different and inferior and actively shaped by civic authority and local regulation as a consequence. In "An Archaeology of Environmental Racism in Los Angeles," Laura Pulido, Steve Sidawi, and Robert Vos examine the historical evolution of discriminatory pollution patterns in two LA area communities composed of different Latino populations. Underscoring the need for qualitative analysis to conceptualize race and class as social relations, they detail in their case studies the multiple and distinct forms of racism that intertwined with a racialized division of labor to develop and maintain these industrial landscapes. Their analysis does a particularly good job in demonstrating the extent to which class difference is complexly intertwined with race and ethnicity.

Both articles represent a break with earlier studies of race and ethnicity in urban geography via their rejection of categories of racial difference as given and fixed. Instead, race is viewed as a social construction. Race is frequently enclosed by quotation marks to emphasize its problematic character. This anti-essentialist perspective requires that "race" itself be deconstructed as a naturalized hierarchy of distinctive human groups. While the biological sciences dismissed essentialist racial categories as meaningless decades ago, race continues in common usage with real material and ideological consequences that vary over space and time. It is the very construction of race and the processes of racialization, a "representational process whereby social significance is attached to certain biological and/or cultural characteristics" (Walter 1999), that become the focus of analysis.

With this progressive agenda, it is perhaps not surprising that Fincher and Jacobs express weariness at the "overexposure" of Burgess' concentric zone model and its unexamined categories of ethnic and racial difference (1998: 5). Although not carrying the same weight of *difference* as race, we should note that ethnicity is also a problematic term. Ethnicity was first used as a noun in the early 1940s when researchers sought a replacement for "race," which had become associated with the Nazis' policies of genocidal cleansing (Hiebert 2000). Although race and ethnicity tend to be distinguished by whether the categories are assigned based on physical or cultural characteristics or assumed by the group itself in the case of ethnic identity, the persistent violence of "ethnic cleansing" provides the extreme example of the links between these two concepts of difference and the implicit power relations contained in their common usage. At any rate, the Chicago School's relatively optimistic assumptions of assimilation for various ethnic groups certainly gave way over time and, by the 1970s, persistent segregation became the focus of a range of studies influenced by human ecology within social geography. Describing instances of voluntary and enforced segregation based on assumptions of either choice or constraint, scholars refined their vocabulary to distinguish among: (1) colonies, a temporary stage in the assimilation of groups into the wider socio-spatial fabric of a city; (2) enclaves, more permanent ethnic settlements based on the internal cohesion of their members; and (3) ghettos, settlements that form in response to the negative conditions and external forces facing a particular ethnic or racial group (Boal 1976). Recent empirical studies which build upon this concern for the settlement patterns of racial and ethnic minority populations include *Ethnicity in Contemporary America: A Geographical Appraisal* (McKee 2000) and *EthniCity: Geographic Perspectives on Ethnic Change in Modern Cities* (Roseman et al. 1996). *EthniCity* explores these issues for major cities in the world system.

The growing attention recently given to race in geography has been described as a renaissance (Peake and Schein 2000), while another review of progress in human geography notes the contributions of those who are producing "anti-racist geographies" (Nash 2003). Owen Dwyer's bibliographic essay on the attention given to race in geographic research during the twentieth century provides an important historiographic evaluation as well (1997). In addition to providing useful progress reports, these essays suggest areas for further research, including the call for studies exploring "whiteness" as a racial category explicitly stating the need to address the racialized geographies beyond the inner city. By moving beyond the conventional geographies of race and racism, as Laura Pulido argues, we might explore the acceptance of white privilege interwoven with the historical processes of suburbanization (Pulido 2000).

Given the historical gender association of the suburbs and women, this reference to the suburbs offers a segue into the second category of difference dealt with in this section. Gendered forms of exclusion and marginality in the city have been the focus of feminist scholarship in geography since the mid-1970s when geographers took their inspiration from the women's movement of the 1960s. Introducing concepts of gender and feminist scholarship in *Gender, Identity and Place: Understanding Feminist Geographies*, Linda McDowell defines feminism as both a political movement and a theoretical field of analysis (1999: 9). Early feminist work in geography emphasized the material inequalities between men and women and insisted that women's experience of the city should not be subsumed into the universalizing category of "man" that dominated modernist social science. Women's roles were to be counted. The meaning of the term *gender*, however, evolved through the 1970s and 1980s, shifting from theories of gender roles to an examination of the social construction of gender relations in all areas of life (see Pratt 2000). By the 1990s, influenced by the "cultural turn" in the discipline as well as advances in feminist theory, feminist scholarship questioned gender categories by asking how masculinity and femininity were constructed at different times and in different places.

The significance of waged work and gender has been a key area of scholarship over several decades and, as Linda McDowell evaluates the literature of feminist geography, it may very well be the most significant focus of research to date (1999: 123). One of the most influential research projects on waged work conducted to date is addressed in the third reading in this section. Geraldine Pratt and Susan Hanson's study of the home/work dynamic of dual-headed households in Worcester, Massachusetts reflects the critical concerns of feminist geographers and the increasing complexity attached to the interrogation of gender relations. They challenge not only the assumptions common in the many geographic studies that leave gender divisions unexamined, but also the feminist theorizing that emphasizes the malleability of gender identities without sufficient attention to empirical research. In their case study, Pratt and Hanson observe a "stickiness to identity grounded in the fact that many women's lives are lived locally" with the result that their commitment to home responsibilities creates a limited job market producing "spatial containment" in the local labor market. Although her study does not appear in this collection, Kim England also examines the extent to which women are enmeshed in a complex web of localized relations that produce "pink collar ghettos." Or, as she refers to it – do they experience "spatial entrapment"? (England 1993).

In the context of globalization and economic restructuring, the attention given to the gendering of work, the social and cultural dimensions of workplace dynamics, and the role of gender in the organization of firms affords significant insight into the rising participation of women in waged work. For further reading, the work of Doreen Massey (1994) and Linda McDowell (1997) are particularly worthy of attention.

Whereas Pratt and Hanson's study of Worcester, Massachusetts uncovered the extent to which gender roles and residential location might restrict the size of the labor market for women, Liz Bondi's examination of gentrification in the article that follows suggests the significance of workplace and life course in women's residential choices. As suggested by the number of readings in this volume which focus on gentrification (see also Smith, Part 3; Ley, Part 6), this process represents an important aspect of contemporary urban spatial restructuring. Social change is also implicated in what are actually multiple processes that contribute to gentrification. In 1991, Bondi reviewed an extensive literature to evaluate the degree to which gender is constitutive of gentrification processes. Although these studies highlighted the significance of financially independent middle-class women with lifestyles and outlooks similar to professional middle-class men, she noted that additional empirical work was required particularly in places other than global cities. By studying Edinburgh neighborhoods, Bondi's research contributes to our understanding of the different ways in which gender, class and life course are interwoven in the gentrification of particular cities and neighborhoods.

Sexuality is another category of difference that has inspired research on the processes of gentrification. Although significant contributions to our understanding of the sexuality of urban space are relatively recent, Larry Knopp's research on gentrification influenced by gay men exemplifies the influential work that seeks to map the geographies of gay men and lesbians (Knopp 1992). More recently, research influenced by queer theory has begun to explore queer geographies, geographies that challenge the uncritical conceptualization of gay and lesbian identity in societies that posit heterosexuality as the norm (Bell *et al.* 1994; Bell and Valentine 1995).

The complex relationships among gender, sexuality, and space are explored in the final reading in this section with specific attention being given to the social meaning of public space and the politics of identity. Explaining that the compulsory heterosexuality of the "everyday street" constructs space by defining proper or normal conditions, Gill Valentine notes that "lesbians . . . are only allowed to be gay in specific spaces and places" (p. 263). As mentioned in the volume introduction, to assume the existence of public space assumes a unified public realm without differences of acceptability. In her exploration of both the subtle and "in your face" strategies for renegotiating this normative space of the street, Valentine explores the means of resisting the heterosexual norm. Quoting Don Mitchell, she states, "by claiming space in public, by creating public spaces, social groups themselves become public" (Mitchell 1995, in Valentine 1996: 152). By identifiably performing their sexual identities in public, however, the "subversive acts" of individual lesbians become overtly political.

References and suggestions for further reading

Aitken, S., Mitchell, D. and Staeheli, L. (2003) "Urban geography," in G. Gaile and C. Wilmont (eds) *Geography in America at the Dawn of the 21st Century*. Oxford: Oxford University Press.

Bell, D. and Valentine, G. (eds) (1995) *Mapping Desire: Geographies of Sexualities*. London and New York: Routledge.

Bell, D., Binnie, J., Cream, J. and Valentine, G. (1994) "All hyped up and no place to go," *Gender, Place and Culture* 1: 31–48.

Boal, F. (1976) *Urban Ethnic Conflict: A Comparative Perspective*. Chapel Hill: University of North Carolina Press.

Bondi, L. (1991) "Gender divisions and gentrification: a critique," *Transactions of the Institute of British Geographers* 16: 190–8.

Duncan, N. (ed.) (1996) *BodySpace: Destabilizing Geographies of Gender and Sexuality*. London and New York: Routledge.

Dwyer, O. (1997) "Geographical research about African Americans: a survey of journals, 1911–1995," *The Professional Geographer* 49(4): 441–51.

England, K. (1993) "Suburban pink collar ghettos: the spatial entrapment of women," *Annals of the Association of American Geographers* 83: 225–42.

Fincher, R. and Jacobs, J. (1998) *Cities of Difference*. New York: Guilford Press.

Hiebert, D. (2000) "Ethnicity," in R. Johnson *et al.* (eds) *Dictionary of Human Geography*. Oxford: Blackwell.

Knopp, L. (1992) "Sexuality and the spatial dynamics of capitalism," *Environment and Planning D: Society and Space* 10: 651–69.

McDowell, L. (1997) *Capital Culture: Gender at Work in the City*. Oxford: Blackwell.

McDowell, L. (1999) *Gender, Identity and Place: Understanding Feminist Geographies*. Minneapolis: University of Minnesota Press.

McKee, J. (ed.) (2000) *Ethnicity in Contemporary America: A Geographical Appraisal*. Lanham, MD: Rowman and Littlefield.

Massey, D. (1994) *Space, Place, and Gender*. Minneapolis: University of Minnesota Press.

Mitchell, D. (1995) "The end of public space? People's Park, definitions of the public and democracy," *Annals of the Association of American Geographers* 85(1): 108–33.

Nash, C. (2003) "Cultural geography: anti-racist geographies," *Progress in Human Geography* 27(5): 637–48.

Peake, L. and Schein, R. (2000) "Racing geography into the new millennium: studies of 'race' and North American geographies," *Social and Cultural Geography* 1(2): 133–42.

Pratt, G. (2000) "Gender and geography" and "Feminist Geography," in R. Johnson *et al.* (eds) *Dictionary of Human Geography*. Oxford: Blackwell.

Pulido, L. (2000) "Rethinking environmental racism: white privilege and urban development in Southern California," *Annals of the Association of American Geographers* 90: 12–40.

Roseman, C., Laux, H. and Thieme, G. (eds) (1996) *EthniCity: Geographic Perspectives on Ethnic Change in Modern Cities*. Lanham, MD: Rowman and Littlefield.

Ruddick, S. (1997) "Constructing difference in public space: race, class, and gender as interlocking systems," *Urban Geography* 17: 131–51.

Sennett, R. (1990) *The Conscience of the Eye: The Design and Social Life of Cities*. New York: Knopf.

Valentine, G. (1990) "Women's fear and the design of public space," *Built Environment* 16: 288–303.

Walter, B. (1999) "Race," in L. McDowell and J. Sharpe (eds) *A Feminist Glossary of Geography*. London: Arnold.

F
I
V
E

"The Idea of Chinatown: The Power of Place and Institutional Practice in the Making of a Racial Category"

from *Annals of the Association of American Geographers* (1987)

Kay J. Anderson

Editors' Introduction

Using a nomenclature of colonies, enclaves and ghettos to distinguish among instances of choice and constraint, geographers have shown a long-standing concern for the causes and consequences of urban ethnic segregation. Cultural geographer Kay Anderson's seminal work on Vancouver's Chinatown offers a challenge to this taxonomy of territorial difference by confronting the unexamined concept of race itself. Chinatown, she argues, is not simply an enclave of the Orient in the West, not simply an ethnic landscape of streets and buildings that provide a natural connection between its residents and the traditions of their cultural homeland. Instead, Anderson interprets Chinatown as a racial category, the material embodiment of a historically specific racial ideology. Indeed, she concludes that Chinatown became a *Western* landscape type through the exercise of power by civic authorities influenced by dominant discourses. Discourses may be defined as social frameworks that enable and limit ways of thinking and acting. In *Vancouver's Chinatown: Racial Discourse in Canada, 1875–1980* (1991), Anderson explores the changing discursive practices that shaped the definition and regulation of that district over time. Negative racial stereotyping from the 1880s through the 1930s helped create a vice district. Conflicting images during the 1930s and 1940s overlay the vice district reputation with an emerging tourist market then associated with Vancouver's "Little Corner of the Far East." Post-World War II modernization nearly destroyed the neighborhood during the 1950s and 1960s as the promise of urban renewal threatened to remove the "slums" of Chinatown while, beginning in the 1970s, the same area was celebrated as both an ethnic heritage district and a major tourist destination. David Ley's (p. 304) discussion of the development goals and politics in Vancouver addresses adjacent neighborhoods during this same period.

In "The Idea of Chinatown," Anderson focuses on the early Chinese settlement in Vancouver, and effectively demonstrates the links among racial ideology, institutional practice, and Chinatown's vice district reputation. During the late nineteenth and early twentieth centuries, Chinatown's image as a vice district became solidified as civic authorities selectively enforced civil codes deemed appropriate to the moral codes of two distinct populations – the Chinese and the Europeans. Drawing upon post-colonial theory, with specific reference to Edward Said's *Orientalism* (1979), Anderson examines how Europeans' "imaginative geographies" during the imperial age helped intensify their own cultural identity by heightening their difference from the other cultures. She illustrates this by citing reports from nineteenth-century Christian missionaries in China who noted both their frustration with the few conversions and the existence of "rampant idolatry, infanticide, slavery in women, polygamy, opium obsession, noonday orgies, treachery, and endemic gambling" (1987: 591). Assuming such standards of behavior as being "Chinese," Vancouver officials patrolled for evidence of these vices most vigorously in Chinatown.

Beyond a critique of cultural difference, during this late nineteenth-century "Age of Race," the discourse of race linked culture, biology and environment to underscore irreconcilable difference. Furthermore, the influence of Social Darwinism suggested that "higher races" could be vulnerable to contamination by contact with immigrants' deficiencies. Efforts to control the contagious effect of Chinese populations extended beyond Vancouver, particularly in cities located on the west coast of North America. The roots of contemporary zoning laws, in fact, may be found in such discrimination. Perceiving Chinese laundries as havens for undesirables, local legislation in the San Francisco area sought to destroy such businesses that had spread beyond Chinatown. In 1886, local officials attempted to close down more than 300 of them, arguing that they represented nuisances and fire hazards. Federal courts (Yick Wo vs. Hopkins, Sheriff) struck down the statute, ruling that it gave arbitrary powers of racial discrimination.

Anderson's analysis of Chinatown as a racial and spatial category problematizes the concepts of race and ethnicity, challenging explanations which assume essentialism – the idea that such differences are intrinsic rather than social constructs. Instead, she argues, the concepts are products of specific historical and geographical forces rather than biologically given ideas with meanings dictated by nature. This is not to suggest that ideas about race and ethnicity are not powerful sources of identity for an individual. Such an understanding requires, however, that racism practiced by dominant society be evaluated as the primary explanatory variable in segregation.

Anderson holds a research position at the University of Sydney where she continues to center her work on spatialities of race. Her body of work focuses not only on Chinatowns in Western cities, but also analyzes the narratives surrounding aboriginal Redfern, an urban space in Sydney (Anderson 1993). Anderson has also developed several self-critiques to extend her earlier work. First, by examining the raced and gendered discourses involved in the spatial category of Chinatown, she critiques work that disengages race identities from other categories of historically situated oppressions such as class, gender, and sexuality (Anderson 1996). She extends the critique further to integrate more thoroughly the economic stakes in racialized place meanings by contextualizing the concepts of class and capital in contemporary Chinatowns in New York and Melbourne and in aboriginal Redfern (Anderson 1998). Anderson's most recent research project (2002) explores the history of nineteenth-century Western science to examine the racist dimensions of the construction of humanity and animality, and culture and nature in colonial biology.

■ ■ ■ ■ ■ ■

They come from southern China ... with customs, habits and modes of life fixed and unalterable, resulting from an ancient and effete civilization. They form, on their arrival, a community within a community, separate and apart, a foreign substance within but not of our body politic, with no love for our laws or institutions; a people that cannot assimilate and become an integral part of our race and nation. With their habits of overcrowding, and an utter disregard for all sanitary laws, they are a continual menace to health. From a moral and social point of view, living as they do without home life, schools or churches, and so nearly approaching a servile class, their effect upon the rest of the community is bad.... Upon this point there was entire unanimity. (Canada 1902: 278)

It would be easy to interpret the words of Royal Commissioners Clute, Munn, and Foley in 1902 as further evidence, if more were needed, of the weight of racial discrimination in British Columbia during the stern years of the late nineteenth and early twentieth centuries. Like many other official utterances at the turn of the century, their words strengthen the claim that the Chinese, because of their distinctiveness, were subjected to many forms of victimization at the hands of a vigorously

Figure 1 The "Unanswerable Argument" 1907.

nativistic white community. Largely in response to that prejudice, overseas Chinese formed Chinatowns, or so a tradition of liberal discrimination studies has held. . . .

It is possible, however, to adopt a different point of departure to the study of Chinatown, one that does not rely upon a discrete "Chineseness" as an implicit explanatory principle. "Chinatown" is not "Chinatown" only because the "Chinese," whether by choice or constraint live there. Rather, one might argue that Chinatown is a social construction with a cultural history and a tradition of imagery and institutional practice that has given it a cognitive and material reality in and for the West. It is, as Ley (1977) describes the elements of human apprehension, an object for a subject. For if we do not assume that the term "Chinese" expresses an unproblematic relationship to biological or cultural constants but is in one sense a classification, it becomes apparent that the study of the Chinese and their turf is also a study of our categories, our practices, and our interests. Only secondarily is the study about host society attitudes; primarily it concerns the ideology that shaped the attitudes contained in the opening quotation. This step beyond "white" attitudes is critical because it is not prejudice that has explanatory value but the racial ideology that informs it. Such an argument is not unimportant for the conceptualization of Chinatown. Indeed it requires a more fundamental epistemological critique of the twin ideas of "Chinese" and "Chinatown," of race and place.

[M]y aim here is to argue the case for a new conceptualization of Chinatown as a white European idea with reference to one context, that of Vancouver, British Columbia. There, one of the largest Chinatowns in North America stands to this day, in part as an expression of the cultural abstractions of those who have been in command of "the power of definition," to use Western's (1981) valuable phrase. But the thrust of the paper is not limited to the study of ideas. Indeed the significance of "Chinatown" is not simply that it has been a representation perceived in certain ways, but that it has been, like race, an idea with remarkable social force and material effect – one that for more than a century has shaped and justified the practices of powerful institutions toward it and toward people of Chinese origin. . . .

A SKETCH OF THE SETTLEMENT AT DUPONT STREET, 1886–1900

By the time the City of Vancouver was incorporated in 1886, Chinese settlement in the city was severely proscribed. The senior levels of state had already intervened in the "Chinese question" and ensured that by the mid-1880s, there would be limits on the participation of Chinese-origin people in political life, their access to Crown land, and their employment on public works. In 1885, after the completion of the trans-Canada railway, the federal government in Ottawa took a decisive step by imposing a head tax on Chinese entrants, and in 1903 Wilfrid Laurier's administration raised it to an almost prohibitive level of $500. Thus by the time of Vancouver's first municipal election in May 1886, when 60 Chinese-origin men were chased from the polls and denied the vote (Morley 1961: 73), a culture of race was fully respected in separate statutory provisions for "Chinese" by the provincial and federal administrations.

From the late 1850s, when gold was first discovered in British Columbia, people from China lived and worked on Burrard Inlet. Most were men employed in unskilled jobs at the Hastings and Moodyville sawmills, but a minority opened stores to service the mill employees. By 1884, the population of Chinese on the inlet was 114 (Morton 1977: 144) and a number of settlements had been established. One of these was built in the vicinity of Dupont Street where woods and a rocky outcrop afforded protection (Yip 1936: 11)

and where nearby industries on False Creek offered employment opportunities. . . .

One such camp on the Brighouse Estate in the West End was particularly provocative to Vancouver's early European residents. Resentment was intense against the laborers who cleared land there at low cost, and on the night of February 24, 1887 some 300 rioters decided to escalate their intimidation strategies. Unimpeded by local police, they raided and destroyed the camps of Chinese laborers; they then attacked the washhouses, stores, shacks, and other structures in the vicinity of Dupont Street (*VN* [*Vancouver News*]). The day after, some 90 Chinese from that area were moved to New Westminster (*VN*). So lax were the local authorities in controlling the violence that the Smithe administration in Victoria, hardly known for its sympathy to Chinese, annulled Vancouver's judicial powers and dispatched special constables to take charge of what the attorney general described as Vancouver's decline into "mob rule" (Roy 1976; *VN*).

In the context of such hostility the Chinese returned to Vancouver and re-established a highly concentrated pattern of residence. Most of those who had fled returned directly to the original Dupont Street settlement. . . . It was a swampy district, with an adverse physical quality that paralleled the peripheral legal, political, social, and economic status of the pioneers it housed. Some lived more comfortably than others, however. Laborers mostly resided in wooden shacks, in conditions a Chinese statesman found "distressed and cramped" on his visit in 1903 (Ma 1983: 34), but merchants usually lived in elevated brick structures on the north and south sides of Dupont Street.

By the turn of the century, the total population of the settlement was 2,053 men (of whom 143 were merchants and the rest workers), 27 women (16 of whom were wives of merchants), and 26 children (Canada 1902: 13). Family life was the preserve of a small economic and political elite, some members of which established a property base in the area from the 1890s (see Yee 1984). As W.A. Cumyow, a British Columbia-born court interpreter testified in 1901, "a large proportion of them would bring their families here were it not for the unfriendly reception . . . which creates an unsettled feeling" (Canada 1902: 236). Such was the marginal turf from which the Chinese launched their contested claim to Canadian life in the twentieth century.

CHINATOWN AS A WESTERN LANDSCAPE TYPE

How was it that the streets of Dupont, Carrall, and Columbia in Vancouver became apprehended as "Chinatown"? Whose term, indeed in one sense whose place was this? No corresponding term "Anglo town" existed in local parlance, nor were the residents of the likes of Vancouver's West End known as "Occidentals." Why then was the home of the pioneers known and intelligible as "Chinatown"? Consistent with the prevailing conceptualization of Chinatown as an "ethnic neighborhood," we might anticipate the response that Chinese people – a racially visible and culturally distinct minority – settled and made their lives there through some combination of push and pull forces. One view, then, might be that the East lives on in the West and Chinatown expresses the values and experiences of its residents.

That people of Chinese origin, like other pioneers to North America, brought with them particular traditions that shaped their activities and choices in the new setting can hardly be disputed. Indeed an important tradition of scholarship has outlined the significance of such traditions for North American Chinatowns as overseas Chinese colonies. Needless to say, Chinese residents were active agents in their own "place making" as were the British-origin residents in Vancouver's Shaughnessy. My decision not to give primary attention to the residents' sense of place then is not to deny them an active role in building their neighborhood nor any consciousness they may have had as Chinese. Some merchants from China might have even been eager to limit contact with non-Chinese, just as China had obviated contact with Western "barbarians" over the centuries. Others, given a choice, might have quickly assimilated.

. . . Regardless of how each of the residents of such settlements defined themselves and each other – whether by class, occupation, ethnicity, region or origin in China, surname, generation, gender, or place of birth – the settlement was apprehended and targeted by European society through that society's cognitive categories. Without needing the acknowledgment or acceptance of the residents, Chinatown's representers constructed in their own minds a boundary between "their" territory and "our" territory.

In his important discussion of "imaginative geographies" such as Europe's "Orient," Edward Said (1978: 55) argued that this distinction is one that "helps the mind to intensify its own sense of itself by dramatizing the distance and difference between what is close and what is far away." This process suggests the argument that although North America's Chinese settlements have often been deliberately isolated, "Chinatown" has been an arbitrary classification of space, a regionalization that has belonged to European society. Like race, Chinatown has been a historically specific idea, a social space that has been rooted in the language and those of its representers and conferred upon the likes of Vancouver's Dupont Street settlement.

The word "arbitrary" is not unimportant here. "Chinese" have been residentially segregated and socially apprehended in North America on capricious grounds. Such a claim rests on the view that any classification of the world's populations into so-called "races" is arbitrary and imperfect. . . . [M]ost contemporary biologists agree that genetic variability between the populations of Asia, Europe, and Africa is considerably less than that within those populations. Apart from the visible characteristics of skin, hair, and bone by which we have been socialized to "see" what is popularly called a difference of "race," there are, as Appiah (1985: 22) notes, "few genetic characteristics to be found in the population of England that are not found in similar proportions in Zaire or China." The important point is that because genetic variation is continuous, "racial" difference cannot be conceptualized as absolute. "Racial categories form a continuum of gradual change, not a set of sharply demarcated types" (Marger 1985: 12), a point that leads biologists to argue: "Any use of 'racial' categories must take its justification from some other source than biology" (Lewontin *et al.* 1984: 127).

[. . .]

Almost no attention has been given to the process by which racial categories are themselves constructed, institutionalized, and transmitted over time and space. Banton (1977: 19) suggests as much in his statement: "Though much has been said about the evils associated with racial classification, there has been little systematic study of the process."

Chinatown has not been incidental to the structuring of this process in the example of the classification "Chinese" or "Oriental." Indeed by situating one such place "in process, in time" (Abrams 1982), it is possible to demonstrate that as a Western idea and a concrete form Chinatown has been a critical nexus through which a system of racial classification has been continuously constructed. Racial ideology has been materially embedded in space . . . and it is through "place" that it has been given a local referent, become a social fact, and aided its own reproduction.

In itself, the idea of Chinatown would not be so important or enduring but for the fact it has been legitimized by government agents who make cognitive categories stand as the official definition of a people and place. In the Vancouver case, "Chinatown" accrued a certain field of meaning that became the justification for recurring rounds of government practice in the ongoing construction of both the place and the racial category. Indeed the state has played a particularly pivotal role in the making of a symbolic (and material) order around the idiom of race in Western societies. By sanctioning the arbitrary boundaries of insider and outsider and the idea of mainstream society as "white," the levels of the state have both "enforced" and "propagated" a white European hegemony. . . .

THE "CELESTIAL CESSPOOL": SANITARY DIMENSIONS OF THE CHINATOWN IDEA, 1886–1920

Shortly before the anti-Chinese riot of 1887, a reporter for the *Vancouver News* wrote: "The China Town where the Celestials congregate is an eyesore to Civilization" and if the City could be "aroused to the necessity of checking the abuse of sanitary laws which is invariably a concomitant of the Chinese, [it] will help materially in preventing the Mongolian settlement from becoming permanent." Four months later, a row of "hateful haunts" on Carrall Street was specifically singled out for the attention of Council. There, warned the *News*, "in the nucleus of the pest-producing Chinese quarter . . . strict surveillance by the City will be necessary to prevent the spread of this curse."

[I]t was the "ordinary Chinese washhouse scattered over the city" that was an early target of civic concern. For a "race" so dirty, there was certainly plenty of work in the business of cleanliness, and by 1889 as many as 10 of the 13 laundries owned by merchants from China were located outside Dupont Street (*Henderson's* 1889: 426). One medical health officer found the spread so fearful as to condemn the washhouse "an unmixed evil, an unmitigated nuisance" and from the late nineteenth century, Council sought means of keeping the "Chinese" laundry in its proper place.

Important judicial limits hampered the City of Vancouver, however. For one, Vancouver's municipal charter (and ultimately the British North America Act of 1864) did not grant legal competence to Council to deny business licenses to "particular nationalities or individuals." The city's challenge was to circumvent such legal restrictions on its political will, and in the case of the "Chinese" laundry, numerous indirect strategies were devised. One alderman, for example, arrived at an artful solution. According to his 1893 bylaw, no washhouse or laundry in Vancouver could be erected outside specified spatial limits, "that is to say beyond Dupont Street and 120 feet on Columbia Avenue and Carrall Street, southerly from Hastings."

During the late nineteenth century, an equally vigorous assault was launched in the name of sanitary reform on the wooden shacks of the Dupont Street settlement. In 1890, fear of cholera gripped the city and the local press demanded the city take action against "the people of Dupont Street" given that "in Chinese style . . . they will not fall into line for the purpose of maintaining cleanliness" (*VDW* [*Vancouver Daily World*]). Fear of contamination from "the degraded humanity from the Orient" (McDonald 1893) was widespread in Vancouver society, and it was customary for letters to the editor to argue that although the "white" race was superior, "Oriental" afflictions would eventually subvert it.

The city fully shared this twist of Darwinist logic and in the mid-1890s – in a significant act of neighborhood definition – Council formally designated "Chinatown" an official entity in the medical health officer rounds and health committee reports. Along with water, sewerage, scavenging, infectious disease, slaughter houses, and pig ranches, Chinatown was listed as a separate category and appointed "a special officer to supervise [it] under the bylaws."

[. . .]

Identity and place were inextricably conflated, and the process of racial classification was corroborated with every official expedition . . . [Yet] the "Chinese" disease-bearing capacity was never borne out by actual disease or epidemic outbreaks recorded in the health inspector's reports or in the local press.

At the same time, a number of Chinese-origin merchants made known their willingness to establish an amenable environment for business and residence. At odds with the typifications projected on the area, some merchants complained to Council about the poor condition of Dupont Street and its sidewalks; in 1899, 24 firms requested Dupont Street be sprinkled twice daily in the summer and back lanes be repaired; and in 1905, a group of businessmen asked the Board of Works to pave Shanghai Alley. Far from passive victims steeped in some fixed standard of living, or for that matter, hapless victims of "white" prejudice, the entrepreneurial sector of Chinatown effectively used its understanding of civic politics to try to elevate the physical condition and social profile of the neighborhood. The Lim Dat Company was so dissatisfied with the City's refuse collection in the area that in 1906 it applied for a license to conduct its own street cleaning operation.

The local unit of knowledge called "Chinatown" was carried forward in government practice and rhetoric well into the new century. . . . Whether or not the image of Chinatown as unsanitary was accurate, the perceptions of image makers intent on characterizing the area as alien were the ones that continued to have consequences. . . . By 1910 . . . a circle of city officials . . . sought to achieve "full control of conditions in Chinatown." They hoped to "reform" the area with wider powers of bylaw enforcement that would stifle "Chinamen [who] manage to fight bylaws by successful applications for injunctions" (*Province*). Fortunately for the residents, the provincial government was not inclined to concede such powers to the city.

[. . .]

[T]he idea of "Chinatown" was being inherited by successive rounds of officials who adopted the conceptual schemes of their predecessors. The health committee of Council described the area as a "propagating ground for disease" in 1919, and, true to old remedies, an inspection team was set up to monitor the area despite the fact that still no concrete evidence confirmed that Chinatown was a threat to public health. Within ten months, the owners of more than 20 lodgings were threatened with orders to condemn their buildings, including the Chinese Hospital at 106 Pender Street East (*Chinese Times*). Indeed, well into the 1920s the city operated assertively in the idiom of race, indiscriminately raiding Chinatown and harassing residents about bylaw compliance (e.g., *Chinese Times*, 1921).

VICE-TOWN: MORAL DIMENSIONS OF THE CHINATOWN IDEA, 1886–1920

Matters of hygiene were only part of the vocabulary out of which this [Chinatown] idea was being constructed. Equally significant and perhaps more effective were moral associations. Because the "Chinese" were inveterate gamblers, "Chinatown" was lawless; as opium addicted, Chinatown was a pestilential den; as evil and inscrutable, Chinatown was a prostitution base where white women were lured as slaves. "Is there harm in the Chinaman?" Reverend Fraser asked a meeting of the Asiatic Exclusion League in 1907. "In this city," he said, "that would be answered with one word, 'Chinatown,' with its wickedness unmentionable" (*Province*).

[. . .]

The plight of the fallen woman disappearing into the clutches of procurers in segregated "Oriental" vice districts was, from the turn of the century, a pressing concern of Vancouver's moral reform groups (Roy 1980: 82). Not surprisingly, therefore, anxiety was heightened in Vancouver by the location of the "restricted area" (where prostitution was tolerated by the police) right next to the Chinese quarter from the time the city was incorporated. But the reform groups' worst fears were realized in 1906 when prostitutes moved en masse to Shanghai Alley following a Council request for their eviction from the prior location. . . . Later, in the face of much local protest about the unhappy combination of prostitutes and "Chinamen" in the one location, the restricted area was moved to other areas in the East End (Nilsen 1976, ch. 3).

Of these various niches where prostitution enjoyed a blind eye in Vancouver, an especially evil construction was cast upon the practice only in "Chinatown." . . . Most often, petitions to Council concerning prostitution dwelt on the risk to property values; however in Chinatown, the voice of the nineteenth-century Protestant missionary reverberated. One resident contended: "It is a disgrace to our city to have *that* evil in *that* location" (cited in Nilsen 1976: 37) . . . Chinatown was intelligible only in terms of a few (unflattering) criteria. . . . Indeed Council simply ignored a 1906 petition from the Chinese Board of Trade, which, in protesting the unimpeded movement of prostitutes into Shanghai Alley and Canton Street, reflected concerns not far removed from the most traditional of Christian mission ministers in Chinatown. "We the undersigned (30) merchants and others," the Board wrote,

> beg leave to respectfully call your attention to the fact that several of the women of ill repute . . . are moving into Shanghai Alley and Canton Streets. This we consider most undesirable. It is our desire to have our children grow up learning what is best in Western civilization and not to have them forced into daily contact with its worst phases. (Shum Moon 1908)

Some Chinese merchants mounted a campaign in Vancouver against another perceived vice out of which the non-Chinese concept of place was constructed. In 1908, the merchants' anti-opium league sent a petition to Ottawa asking the federal government to "decisively exercise its authority and powers to prohibit the importation, manufacture and sale of opium into Canada so that the social, physical, and moral condition of both Chinese and Europeans may be vastly improved." But try as some merchants did . . . the drug that Britain had introduced to China in the 1840s was now a powerful metaphor for neighborhood definition. And like the construction put upon "white" participation in Chinatown's bawdy houses, the large extent of non-Chinese use of opium . . . uncovered in [a] 1908 investigation only confirmed the belief that Chinatown was a menace to civilized life. White drug use did not prejudice, but rather validated, the more comforting racial and spatial category.

Once the image of "Chinatown" as an opium den was consolidated, no amount of counter-evidence could acquit it and all manner of accusations could be adduced, especially by politicians, to support the neighborhood image. By the 1920s, when the race idea was being feverishly exploited in British Columbia, the old opium image fed and was assimilated into an image of Chinatown as a narcotics base and "Chinese" as dangerous drug distributors. In March 1920, for example, an editorial warned . . . "if the only way to save our children is to abolish Chinatown, then Chinatown must and will go, and quickly."

In the context of rising anti-Chinese sentiment in the House of Commons, Consul General Yip and the Chinese Benevolent Association of Vancouver formed a Self Improvement Committee to try to elevate the reputation of their neighborhood. [The] president of the Chinese Benevolent Association . . . spoke out against the *Sun*'s vendetta, calling attention to "the suggestion . . . that Chinese vendors are merely conveniently used and that the traffic is controlled by persons other than Chinese" (Cumyow 1920). But the irrepressibly anti-Chinese Member of Parliament for Vancouver Centre, H. Stevens (also secretary of the Vancouver Moral Reform Association) was not to be deterred, and in a series of speeches in the House of Commons in the early 1920s, he transmitted these most recent charges against Vancouver's Chinatown to the senior level of government. "The basis of the pernicious drug habit on the Pacific Coast is Asiatic," said Stevens in 1921. "We have seen in Vancouver almost innumerable cases of clean, decent, respectable young women from some of the best homes dragged down by the dope traffic and very, very largely through the medium of the opium dens in the Chinese quarter."

[. . .]

Known to police for "inveterate" gambling, the "heathen Chinese" was actively pursued by officers of the Vancouver police force from the 1890s to the late 1940s. By that time, the extent of the harassment had become embarrassingly transparent even to the city. . . .

Just as the opium den raids vindicated widespread assumptions about the moral laxity of the Chinese, the formidable scrutiny that Chinatown experienced from the City for gambling sprang from and confirmed popular assumptions about a generically addicted "Chinaman." And one vice bred another, as Alderman McIntosh observed in 1915. Gambling and opium in Chinatown required constant civic vigilance he claimed, because they were associated with tuberculosis and slavery in women.

Yet gambling was not restricted to the Pender Street area. One letter to the editor in 1900, appealing for greater control of gambling in the city, said: "Everyone knows that gambling goes on promiscuously all over Vancouver, in clubs, in hotels, in saloons, in rooms connected with saloons and in private houses." But only in "Chinatown," was a neighborhood image built around both its practice and the attempts of police – confronted by "ingenious Oriental systems of spring doors and getaway rat tunnels" – to suppress it. So perturbed were Chinese merchants by this harassment that in 1905, the Chinese Board of Trade protested:

> The members of our board are law abiding citizens. Many of them have been residents of this country for a number of years and are large holders of real estate, payers of taxes and other civic assessments. The members . . . have been constantly annoyed by what we believe to be an unjustifiable intrusion of certain members of the Vancouver Police Force . . . in the habit of going into our stores and rooms where our families live, showing no warrant whatsoever, nor do they claim any business with us . . . We are subjected to indignities and discriminating treatment to which no other class would submit and to which *your* laws, we are advised, we are not required to submit. (Young 1976: 65)

CONCLUSION

I have argued that "Chinatown" was a social construct that belonged to Vancouver's "white" European society, who, like their contemporaries throughout North America, perceived the district of Chinese settlement according to an influential culture of race. From the vantage point of the European, Chinatown signified all those features that seemed to set the Chinese irrevocably apart – their appearance, lack of Christian faith, opium and gambling addiction, their strange eating habits, and odd graveyard practices. That is, it embodied the white Europeans' sense of difference between immigrants from China and themselves, between the East and the West. This is not to argue that

Chinatown was a fiction of the European imagination; nor can there be any denying that gambling, opium use and unsanitary conditions were present in the district where Chinese settled. The point is that "Chinatown" was a shared characterization – one constructed and distributed by and for Europeans, who, in arbitrarily conferring outsider status on these pioneers to British Columbia, were affirming their own identity and privilege. That they directed that purpose in large part through the medium of Chinatown attests to the importance of place in the making of a system of racial classification.

Studies of the social meaning of place in human geography have too rarely taken measure of the role of powerful agents, such as the state, in defining place. Yet those with the "power of definition" can, in a sense, create "place" by arbitrarily regionalizing the external world. In the example here, Chinatown further became the isolated territory and insensitive representation its beholders understood in part through the legitimizing activities of government. Perhaps not all places are as heavily laden with a cultural and political baggage as "Chinatown." But Chinatown is important in pointing up once again the more general principle that a negotiated social and historical process lies behind the apparently neutral-looking taxonomic systems of census districts. More importantly perhaps, the manipulation of racial ideology by institutions is additional testimony to the fact that a set of power relations may underpin and keep alive our social and spatial categories.

The importance of these "imaginative geographies" cannot be underestimated because, as we have seen, they organize social action and political practices. Indeed the idea and influence of "Chinatown" is further evidence of the growing consensus in human geography that our landscape concepts, as symbolic resources, have a critical structuring role in the making of wider social processes. In the course of its evolution, Chinatown reflected the race definition process, but it also informed and institutionalized it, providing a context and justification for its reproduction. Pender Street has been the home of the overseas Chinese to be sure, but "Chinatown" is a story, which, in disclosing the categories and consequences of white European cultural hegemony, reveals more the insider than it does the outsider.

REFERENCES FROM THE READING

Abrams, P. (1982) *Historical Sociology*, Bath: Pitman Press.

Appiah, A. (1985) 'The uncompleted argument: DuBois and the illusion of race', *Critical Inquiry* 12, 1: 21–37.

Banton, M. (1977) *The Idea of Race*, London: Tavistock.

Canada (1902) *Report of the Royal Commission on Chinese and Japanese Immigration*, printed by S. E. Dawson.

Cumyow, W.A. (1920) 'In defense of Chinatown', *Sun*, March 24: 6.

Henderson's BC Directory (1889) Victoria: L. G. Henderson.

Lewontin, R., Rose, S. and Kamin, L. (1984) *Not in our Genes: Biology, Ideology and Human Nature*, New York: Pantheon Books.

Ley, D. (1977) 'Social geography and the taken-for-granted-world', *Transactions of the Institute of British Geographers* 2: 498–512.

Ma, L. (1983) 'A Chinese statesman in Canada, 1903: Translated from the travel journal of Liang Ch'ich'ao', *BC Studies* 59: 28–43.

Marger, M. (1985) *Race and Ethnic Relations*, Belmont, Calif.: Wadsworth.

McDonald, W. (1893) 'The degraded Oriental', *Vancouver Daily World*, March 23: 4.

Morley, A. (1961) *Vancouver: From Milltown to Metropolis*, Vancouver: Mitchell Press.

Morton, J. (1977) *In the Sea of Sterile Mountains*, Vancouver: J. J. Douglas.

Nilsen, D. (1976) 'The "social evil": Prostitution in Vancouver, 1900–20', BA Honours essay, Department of History, University of British Columbia, Vancouver.

Roy, P. (1976) 'The preservation of peace in Vancouver: The aftermath of the anti-Chinese riot of 1887', *BC Studies* 31: 44–59.

Roy, P. (1980) *Vancouver: An Illustrated History*, Ontario: James Lorimer.

Said, E. (1978) *Orientalism*, New York: Vintage Books.

Shum Moon (1908) 'Chinese deny responsibility', *Province*, February 3: 16.

Western, J. (1981) *Outcast Capetown*, Minneapolis: University of Minnesota Press.

Yee, P. (1984) 'Business devices from two worlds: The Chinese in early Vancouver', *BC Studies* 62: 44–67.

Yip, Q. (1936) *Vancouver's Chinatown*, Vancouver: Pacific Printers.

Young, D. (1976) 'The Vancouver police force, 1886–1914', BA Honors essay, Department of History, University of British Columbia, Vancouver.

Editors' references and suggestions for further reading

Anderson, K. (1991) *Vancouver's Chinatown: Racial Discourse in Canada, 1875–1980*, Montreal: McGill University Press.

Anderson, K. (1993) "Constructing geographies: 'race,' place and the making of Sydney's Aboriginal Redfern," in P. Jackson and J. Penrose (eds) *Constructions of Race, Place and Nation*, Minneapolis: University of Minnesota Press.

Anderson, K. (1996) "Engendering race research," in N. Duncan (ed.) *Body Space*, London: Routledge.

Anderson, K. (1998) "Sites of difference: beyond a cultural politics of race polarity," in R. Fincher and J. Jacobs (eds) *Cities of Difference*, New York Guilford Press.

Anderson, K. (2002) "The racialization of difference: enlarging the story field," *The Professional Geographer* 54, 1: 25–31.

Boal, F. (1976) *Urban Ethnic Conflict: A Comparative Perspective*, Chapel Hill: University of North Carolina Press.

Said, E. (1979) *Orientalism*, New York: Vintage Press.

"An Archaeology of Environmental Racism in Los Angeles"

from *Urban Geography* (1996)

Laura Pulido, Steve Sidawi, and Robert O. Vos

Editors' Introduction

Despite the long tradition that geography as a discipline has of focusing on the interaction of humans and the environment, little attention has focused on urban environments until recently. The question as to why that should be warrants attention, as a growing number of geographers, including those involved in research related to issues of environmental justice, have acknowledged. Although it provides only a partial explanation of the discipline's oversight of urban environmental research, we might note that the concept of environmental justice itself was articulated only relatively recently. The focus on environmental justice grew from an initial focus on environmental racism, which also has a relatively short history as an operating research concept. Benjamin Chavis, then head of the United Church of Christ's Commission on Racial Justice, coined the term in 1982 to describe the discriminatory targeting of communities of color in locating hazardous waste sites. Intense public pressure in the USA from environmental and civil rights activists during the 1980s helped raise awareness of the unequal exposure of various communities to environmental degradation; and studies such as that conducted by the United Church of Christ's Commission for Racial Justice (1987) provided much needed empirical evidence to support claims of environmental racism. In 1992, the United States' Environmental Protection Agency (EPA) produced a report that revealed a similarly strong correlation between the location of commercial hazardous-waste facilities and the percentage of minority residents in those same communities. Increasingly the debate shifts from whether environmental racism exists to how different populations become vulnerable.

The following article by Laura Pulido (geographer at the University of Southern California) working with University of Southern California graduate students Steve Sidawi (geography) and Robert O. Vos (political science) not only contextualizes that question in the complex patterns of ethnic and racial communities of Los Angeles, it also advocates a changed methodological approach. As their title's reference to archaeology suggests, Pulido and her co-authors focus on the historical evolution of discriminatory patterns in Los Angeles, explicitly examining racism as a dynamic force in place-making. To explore this topic effectively, they argue the need for a new methodological approach to get beyond environmental justice researchers' stock question in defining the causal mechanism of inequity that is "which came first, the people or the hazard." Urging researchers to move beyond the race versus class debate which dealt with these concepts as mutually exclusive and static categories, they conceptualize race and class as intertwined social relations. Furthermore, they argued that these historical processes cannot be understood without employing qualitative methods to break apart the unitary concept of racism into its many forms, and to acknowledge it as both an ideological and material force. Thus they present an alternative to the quantitative methodology (based on the correlation of toxins and race and/or class) that had dominated research on environmental racism, as they recommend an integration of concepts from contemporary work in the area of critical human geography and the human/environment research tradition.

In launching their study, Pulido, Sidawi and Vos conducted a spatial analysis of air toxins in urban Los Angeles County and then chose two of the most polluted communities for historical examination – East Los Angeles/Vernon and the City of Torrance. The edited paper provided in this volume begins with the historical analysis of the two communities. In its entirety, "An Archaeology of Environmental Racism in Los Angeles" presents an informative comparison of insight gained by quantitative and qualitative methodologies.

The three authors of this paper continue to work on the practical and theoretical aspects of urban geography. After completing their doctoral degrees, Steve Sidawi began a career in labor organizing and Robert Vos joined the research staff at the University of Southern California's Center for Sustainable Cities. Professor Pulido contributes to research on issues of environmental racism, integrating race, ethnic studies, and political activism in her work as well. Her book *Environmentalism and Economic Justice: Two Chicano Struggles in the Southwest* (1996) critiques mainstream environmentalism as it documents ongoing debates over the definition of environmental issues. Her critique supports a growing challenge to traditional understandings of nature and the environment as well as the concerns of marginalized people. As one African-American Angelino put it, "the environment is not just forests and wetlands; the environment is where you live, so the housing crisis should also be considered an environmental issue" (Pulido 1996: xiv).

This attention to the reframing of nature and environmentalism is shared by urban political ecologists, as reflected in Kaika and Swyngedouw's article in this volume (p. 343). The critical human geography perspective of urban political ecology links the specific analysis of urban environmental problems to larger socio-ecological theory. The aim of this emerging field is to expose the processes that bring about highly uneven urban environments. For further reading in this rich body of literature, see the special issue of *Antipode* (2003) edited by Swyngedouw and Heynen which sets out the contours of urban political ecology.

Michael Dear and Steven Flusty, who examine the postmodern landscape of Los Angeles (p. 138), suggest the growing presence of a Los Angeles School of urbanism. Certainly those following the development of urban political ecology cannot help but note the influence of several geographers in the Los Angeles area (Keil 2003). Professor Pulido and her colleagues at the University of Southern California, in particular Stephanie Pincetl and Jennifer Wolch, are making significant contributions to urban political ecology. Concerned with the "history of growth and associated denaturing of Southern California," they set out an agenda for urban ecological work in "Urban Nature and the Nature of Urbanism" (Wolch *et al.* 2002).

To advance this environmental agenda, Pulido urges an examination of the various forms of racism and the manner in which they contribute to environmental racism (Pulido 2000). She argues, for example, that the malicious intent associated with locating facilities in communities of color is not the only form of environmental racism. The consequences of a less conscious and hegemonic form of racism – the acceptance of white privilege – may be seen in the historical processes of suburbanization and decentralization. Elsewhere she has argued that the study of environmental racism came late to the discipline of geography. She outlines how this opportunity to lead in an area that traditionally belonged to geography (human interaction with the environment) went instead to cognate disciplines such as sociology (see Pulido 2002). While this brings us back to that question of why geographers have lagged in their attention to urban environments, Pulido urges an effort to remedy this failing.

INTRODUCTION

A key methodological problem posed by the environmental racism literature involves the question

of which came first, the people or the hazard. This question is considered necessary to ascertain whether empirically measurable racism created patterns of disproportionate exposure, or whether

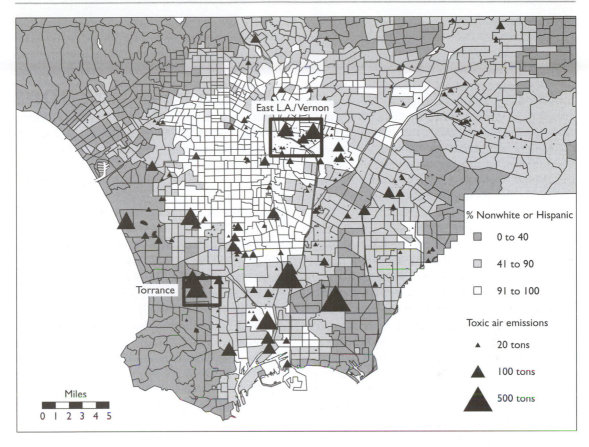

Figure 1 Non-white or Hispanic population by census tract (1990); and toxic release inventory emissions (1992), Southern LA County.

such patterns are simply a function of larger social and economic structures and forces. Cutter (1995) presents the problem in the following way:

> Were the LULUs [locally undesirable land uses] or sources of environmental threats sited in communities because they were poor, contained people of color, and/or politically weak? Or, were the LULUs originally placed in communities with little reference to race or economic status, and over time, the racial composition of the area changed as a result of white flight, depressed housing prices, and a host of other social ills? In other words, did the residents come to the nuisance or was the nuisance imposed on them (voluntarily or involuntarily)?

These questions are significant not only because they underscore the need for historical analysis, but also because they intimate other problems in current approaches to environmental racism research. In addition to a general absence of solid historiography, there is a tendency to treat the two primary categories of analysis within the environmental racism literature, "race" and class, as static categories, instead of as social relations; although this allows researchers to operationalize questions concerning "which came first" and to compare the significance of race and class in creating inequitable patterns, it does so only by ignoring the historical and spatial contingency of racial categories and the dynamic complexity of urban sociospatial patterns. By contrast, recent work in critical human geography has insisted upon conceptualizing urban space, economic relations, and racial categories as active social processes. In the case of environmental racism, this means, first focusing on the *simultaneous evolution* of racism (versus race), class formation, and the development of industrial landscapes, instead of assuming linear causality (as in the question of which came first). In other words, we seek to uncover how peoples and places become racialized (the process of grafting a racial meaning onto various phenomena within the context of unequal social relations), how they are integrated into local economic structures, and the spatiality of those dynamics. Given this framework, places are not simply sites, but are relations themselves that partially constitute the landscape. Thus, instead of seeking the definitive answer as to whether environmental racism actually exists, an equally important research agenda is to study such processes.

This paper offers a methodological alternative to the current literature by emphasizing social relations and processes in the history of environmental racism in Los Angeles. Through historical analysis we reveal the processes by which [two] communities became disproportionately affected. We contest normative conceptualizations of race and class by approaching both as historically and geographically specific social relations that are spatially constituted. Thus, instead of defining class strictly in terms of income categories, we also consider the importance of the division of labor and one's relation to the means of production. Race is approached in a similar way. Instead of accepting racial groups a priori, we investigate the process of racialization.

[. . .]

By paying attention to the spatiality of racial groups, including how they are constructed and "classed," we illuminate how race operates in the creation of place, and ultimately in the formation of inequality. Although we present only two case studies, we hope these empirical analyses will not only suggest the diversity of ways in which racism operates, but also stimulate further alternatives in environmental racism research.

AHISTORICISM AND THE OPERATIONALIZATION OF RACE AND CLASS

Recent reviews of the environmental racism literature point to the many methodological uncertainties that currently characterize this body of work. As it is a relatively recent field of inquiry, it is inevitable that both conceptual and technical

issues, such as scale of analysis, subpopulations, or equity criteria, will be debated in order to reach consensus. Yet, even as quantitative approaches to environmental racism research become increasingly sophisticated, two issues stand out as being particularly intractable without a move to more qualitative forms of research: understanding environmental racism in historical context and understanding race and class as social relations. We will discuss each in turn.

The value of historical research

There is a clear need for historical investigation to supplement studies of contemporary socioeconomic and pollution patterns. It has been argued that troublesome issues of intent and motive can be discerned only through history. For instance, was a facility purposely located in a non-White community, or did the community subsequently become non-White through residential migration? While there are a growing number of historical studies, only Sidawi (1995) has problematized race. He does so by uncovering the racialized (and racist) nature of a planned community's spatial and economic organization. Focusing on the local planning discourse, he traces how certain groups were cast as non-White, and were then excluded and exploited.

Been's (1994) analysis of Houston landfills indicates some of the challenges posed by historical investigation. Most environmental equity research compares pollution and demographic (usually census) data, yet neither census tracts nor "racial"/ethnic categories have remained consistent over time. Accordingly, an in-depth historical analysis must rely upon an array of sources. Despite these data issues, several researchers have contributed important insights or methodological interventions by focusing on the history of places. For example, Hurley's (1995) study of air pollution in Gary, Indiana is significant in that he located his study within the context of the larger forces of urban development and industrialization. Krieg's (1995) study of Boston suggests that, at times, racism may account for environmental inequities, while at other times, class does, depending upon how a place develops. Both studies are important in shifting the debate from whether environmental racism exists to uncovering *how* different populations become vulnerable to environmental hazards. We believe that the emphasis on process, rather than dichotomous outcomes, enables us to reconceptualize race and class as social relations. This reconceptualization, in turn, allows us to move beyond the limiting race-versus-class debate, which is based on static categories.

Dominant conceptualizations of race and class

Research on environmental racism is similar to other forms of outcome or pattern research, such as studies of housing discrimination, in that it seeks to document spatial distributions – in particular, inequitable access/exposure to both positive and negative attributes. As mentioned above, while only a limited number of studies have examined how such patterns evolved, even fewer have sought to unpack those structures, forces, and categories themselves. There are two particular problems associated with the dominant conceptualization of race. First, racial categories are treated as fixed, and second, racism is conceived in a unitary fashion.

Because most positivist research on environmental racism relies upon the census and its categories, "race" is simply agreed to refer to those groups designated as minorities. While there is an obvious truth to this, such a practice also denies the fluid nature of racial categories (Menchaca 1993; Rodriguez 1994; Sacks 1994).... This process of racialization is inherently spatial in that racism is a dynamic force in place-making, and places (at least in the U.S.) are inherently racialized. It is precisely these assumptions that need to be recognized as political processes and uncovered.

A second problem is posed by employing a unitary concept of racism, that is, assuming a single racism exists instead of multiple and distinct forms. Because positivist environmental racism research often compares race and class as two distinct categories and pays only limited attention to process, racism is conceded only when race emerges as a statistically significant variable.... [T]here are many forms of racism, and they vary in terms of their content, nature, severity, ideology,

and scale of operation. Allowing racist distinctions not only moves us beyond reproducing racism as a dualism (i.e., it either does or does not exist), but also allows us to understand how racism articulates with other relations to create unique places. Seen in this way, there is only *racism* to be studied. Racism (including racial categories), however, can be fully grasped only by studying how it works in conjunction with other forces and processes to create landscapes of inequality....

By focusing on process and relations, it becomes fairly meaningless to speak of race or class, as we must now confront the racialized nature of the economy. In other words, how are various racial groups inserted into the local economy? To date, most research operationalizes class as income, which serves to reduce a complex relation to a static category. For instance, although Anderton *et al.* (1994) employed a variety of economic variables, including types of work, these were not individually analyzed, but aggregated into a general socioeconomic standing.... Income, however, along with other Weberian indicators such as educational attainment or housing value, does not convey the full story.

Conceiving of class as a social relation necessitates a different type of analysis. The essence of a relation is that it can be understood only within the context of other groups or structures. Although the term *class* is frequently used to refer to a variety of phenomena, in this context we will use it to refer to workers who exist in opposition to capital. But class is not the only antagonistic economic relation that exists. Sayer and Walker (1992) have pointed out that the division of labor is a distinct but often ignored analytical category that offers a different set of insights. Although a large group may be designated as workers, there will be tremendous variation within that population according to their place in the division of labor. Different positions carry with them varying degrees of power, money, status, and security. Typically, the division of labor is structured along "natural" fractures, such as ethnicity and gender (Kobayashi and Peake 1994).

Using such an approach, an alternative project is to investigate the racialized nature of the economy, its spatiality, how it changes over time, and the implications for vulnerability to environmental hazards.

THE EVOLUTION OF ENVIRONMENTAL RACISM IN TWO LOS ANGELES COMMUNITIES

. . . [W]e selected two of the most polluted areas for further analysis and investigation. These communities were selected primarily because of total TRI emissions. The area we have defined as East Los Angeles/Vernon accounted for 12 per cent of the total county emissions in 1992 (2,242,976 pounds) and is composed primarily of small firms. In contrast, the northern portion of the City of Torrance contains large firms, including the third- and fourth-largest emitters in the county, the Mobil Oil Refinery and Reynolds Metals, and is responsible for 10 per cent of total county emissions (1,856,777 pounds). In addition to having high levels of emissions, each area also represents a different historical trajectory: one is a planned industrial community, the other an amalgam of political entities with a more complex history of industrialization.

[. . .]

Each site brings into focus some of the complexity that surrounds the creation and interpretation of environmental racism. First, both sites suggest that standard spatial units of analysis can at times hide as much as they reveal, problems that can be resolved only through qualitative investigation. Moreover, each case reveals the different processes that may contribute to environmental inequities, suggesting that distinct forms of racism are at work. These range from the intentional siting of polluting facilities in non-White communities to the deliberate development of industrial areas in racialized minority urban spaces. Also included, however, are racialized structural processes, such as class formation, that are pivotal to understanding environmental inequity. Each case also illustrates the role of racial ideology. By paying attention to racial meanings and how they intersect with material structures and processes, we can better grasp the diversity and significance of the ideologies informing these larger patterns. . . .

Torrance

Torrance is significant because at first glance, it would appear that non-Whites are not disproportionately exposed to the city's air toxics. Historically,

Torrance has been a White city, and strict zoning since its inception in 1912 has kept residential land uses highly segregated from heavy industry (Phelps 1995). However, the industrial district's toxic pollution historically has had a disproportionate impact on the city's Latino community, a pattern that persists today. The *Mexicano* residential area, historically known as "the Pueblo Lands," is inserted into the industrial district and is situated immediately downwind of Mobil Oil and Dow Chemical facilities. Because the community is much smaller than a census tract, it is easily overlooked in standard analyses, yet its existence as an environmentally marginalized community has become apparent during recent years. In a 1979 refinery explosion and fire it was the only community to be evacuated (Robak 1979). By examining the spatial and socio-historical development of the Latino community, we can begin to understand how this pattern is a function of racism and the spatiality of that racism. The Torrance case specifically demonstrates how a planned industrial suburb created a racialized division of labor that was expressed both in a racist planning discourse and in extreme residential segregation.

Originally conceived in 1911 by a consortium of investors hoping to create an open-shop alternative to Los Angeles's existing industrial district Torrance was designed by its developers to attract both industry and White workers. Anchored by a subsidiary of Union Oil Company, the development attracted several other large industrial firms, including the Pacific Electric Shops and the Llewellyn Iron Works. City-sponsored advertising campaigns aimed at attracting White residents stressed that Torrance would be "America's first great industrial garden city" where "flowers and sunshine and a little bungalow seem possible for everyone" (*Torrance Herald*, 1914). The city's salubrious and orderly environment was also emphasized. An advertisement by the Torrance Realty Company in 1914 promised a "perfect sewer system . . . wide, beautifully paved streets . . . [and a] big, clean organized development." By 1922, the city had attracted over 1,000 residents, 95 per cent of whom were White.

Even though Torrance was conceived and marketed as a White workers' paradise, a small percentage of factory positions was reserved for non-White workers, primarily of Mexican origin.

That *Mexicano* workers were referred to as "Cholos and other low-grade laborers," "needed in the rough work of the factories" suggests a racialized division of labor (Olmsted 1912; Bartlett 1913). As was typical in Los Angeles industry at the time, positions associated with high skill and wage levels were reserved for White workers, while many positions associated with low skill and wage levels were reserved for Latinos.

Closely linked to the racialized division of labor was the creation and enforcement of racially segregated residential areas. Evidence from planning documents reveals a highly deliberate confinement of Latino residents to a district otherwise zoned for heavy industry. Planners' racial stereotypes, such as "Mexicans like to . . . be out of sight of 'Gringos'" and "they do not care for grass or garden plots," provided a discursive basis for the creation of segregated environments. The spatial separation between racial groups, with industrial land use serving as a buffer zone, was legitimized "to guard against possible race antagonism" (Willard 1913, p. 302). The segregation set forth by the planners was subsequently enforced by racially restrictive covenants on real estate and a general atmosphere of intimidation, including a police department with close ties to the Ku Klux Klan (Shanahan and Elliott 1984). A racialized division of labor within a city generally hostile to non-White residents underscores the historical interconnectedness between one's economic position, the creation and reproduction of racial categories, their spatiality, and how all three are mutually constituted.

During the 1940s and 1950s, Torrance partook of Southern California's larger industrial and population expansion, as it developed into one of Los Angeles's largest bedroom suburbs. Nevertheless, it remained an overwhelmingly White city. Of 22,000 residents in 1950, 2 per cent were Japanese, 5 per cent were Latino, and, despite proximity to the rapidly growing African American communities near South Central Los Angeles, only 0.15 per cent were Black. Moreover, the general socio-spatial patterns imposed by the early developers held, with Latinos concentrated in the industrial northeast of the city and Whites in the housing tracts expanding to the south and west. By the 1950s and 1960s, Latinos had made tremendous gains in unionized factory employment, especially in the local steel and aluminum

plants, as the earlier divisions of labor underwent some dissolution. Yet discriminatory practices in housing markets maintained the exclusion of Latinos and African Americans from most residential areas (Shanahan and Elliott 1984). Thus, the particular mix of forces that served to expose Latinos disproportionately changed over time.

In the 1970s and 1980s, metals-related employment dropped markedly as regional employment shifted to aerospace and service industries (United States Bureau of the Census, 1970, 1980, 1990). This shift has refocused attention on the hazardousness of the remaining industrial facilities, most of which are petroleum and chemical related. The Mobil Refinery, for example, has become the center of a local political debate over environmental issues after a series of industrial accidents (Rae-Dupree and Schoch 1991). Eventually the city sued Mobil to ensure stricter safeguards against toxic releases (Schoch 1995). The Oil, Chemical, and Atomic Workers Union has blamed the increase in accidents at the refinery on the increased use of nonunion contract workers (Rabin 1988), who are disproportionately Latino (Lee 1990; Kochan *et al.* 1992). Although Torrance has become more racially diverse (Whites now constitute only 66.7 per cent of the population), Latinos are still relatively concentrated in the industrial northeast and thus are more vulnerable, due to both their residential location and their unequal position in the factories.

The Torrance case study demonstrates how city planning and the racialized division of labor combined to create environmental inequalities early in the city's history. It also demonstrates how a multiplicity of forces including housing discrimination, regional economic changes, planning ideology, and economic structures complicates any effort to attribute environmental racism to a single cause or form of racism. Instead, the development of Torrance shows how social relations based on race intertwine with larger political economic processes in complex ways.

East Los Angeles/Vernon

In the mid-1980s, the industrial city of Vernon was selected as a site for a major hazardous-waste incinerator. Charging that the area's residents were already burdened disproportionately by environmental hazards, the grassroots group Mothers of East Los Angeles (ME LA) led a successful political struggle to block the incinerator's construction (Gutierrez 1994). While the siting is often considered a clear case of environmental racism, the focus on a single siting decision itself belies the complexity underlying environmental inequalities. Once again, reliance on standardized spatial units is a problem. While it is true that the area generally suffers from severe air pollution, much is made of the fact that the most "toxic zip code" is sandwiched between Black and Latino communities. Such catchy phrases mask the fact, however, that the actual zip code is home only to 350 residents (92 per cent of whom are Latino). In contrast, what *is* significant about Vernon is its proximity to the heavily populated *barrios* of East Los Angeles and Boyle Heights. Considering the magnitude of Vernon emissions, prevailing wind patterns, and proximity to freeways (City of Los Angeles, 1988; Pansing *et al.* 1989), the area we have defined as East Los Angeles/Vernon is one of the most polluted spots in Southern California. Here, public housing projects and apartment buildings are adjacent to highly hazardous and toxic metal-plating shops (Sahagun 1989). Thus our task is to uncover how and why such a polluted area developed in close proximity to the region's largest *barrio*.

To understand the simultaneous evolution of industry, pollution, and racism in East Los Angeles/Vernon, we start with the original *pueblo* of Los Angeles, which has served as the entry port for Mexican immigrants since the 1850s. The *pueblo* was itself a highly segregated *barrio* and was the first site of industrial activity in the city (Griswold del Castillo, 1979). As *el pueblo* became more urban and industrialized, *Mexicanos* were displaced eastward, marking the beginning of the East Los Angeles *barrio* (Romo 1983). Historically, the residential patterns of the Chicano/*Mexicano* population have been closely linked to particular industries, a fact that reflects their role as workers in a limited number of sectors. Sanchez (1993) describes the role of industrial development on both old and new *barrios* and the problems it posed for local *Mexicanos*.

> Mexicans could hardly settle down permanently in a community when control of their neighborhoods was firmly entrenched in the hands of

Anglo American industrial and commercial interests. The residential practices of immigrants were always tempered by the zoning practices and labor needs of the city's establishment.

Indeed, as the Eastside *barrio* took shape, transportation and zoning planners were laying the groundwork for intensive industrial development in and around the new neighborhood (Olmsted 1933; Sanchez 1993).

Mexicanos were initially drawn to the Eastside by employment opportunities and a lack of strictly enforced building codes, making home ownership a possibility for some. The development of railroad yards in and around Boyle Heights, the early growth of light industries in Vernon, and the emergence of the pottery, brick, and clay industries adjacent to the Los Angeles River allied to a boom in the construction materials industry. As a result, the early Boyle Heights *barrio* was surrounded by railroad yards, brickyards, and light manufacturing, all industries with limited occupational mobility but open to Chicanos/*Mexicanos*.

It is important to point out that the Eastside has not always been limited to Latinos. At different times it has also been home to vibrant Jewish, Armenian, and Russian communities, to name but a few (Sanchez 1993). But beginning around 1940, Chicano/*Mexicano* in-migration to the Boyle Heights area increased rapidly. Chicanos/*Mexicanos* were drawn both by employment opportunities and by an older, relatively affordable housing stock, and whole blocks are reported to have changed from Jewish to Mexican in one week's time (Gustafson 1940). Elsewhere in the expanding *barrio*, Mexican immigrants, described as living in an isolated cluster in "small dilapidated shacks on low flat land," had located in the Belvedere area of unincorporated East Los Angeles. Known to residents as "La Barria," this area was marked as one of lower social status, and school boundaries were manipulated to segregate children.

. . . A 1940 thesis depicted Boyle Heights in the following way:

> Originally endowed with a charming topography and a climate with little fog, admittedly the finest in the city, the section could have been one of idyllic beauty, but it has been butchered

by the hand of man. Yellow, eroded walls of earth line the streets, with shanties perched precariously on the eminences above, and huddled in the hollows. (Gustafson 1940, p. 10)

Such representations not only had repercussions for the future development of the area, but also served to limit Latinos' efforts to achieve some level of control and autonomy. For instance, County supervisors squelched an incorporation drive in 1931 because local commercial interests opposed being part of a "Mexican area". Police Chief Parker engaged in a less thinly veiled racist discourse in his explanation of Boyle Heights crime in 1960: "Some of the people have been here since before we were, but some are not so far removed from the wild tribes of Mexico" (Acuna 1984, p. 89). Even a 1990 planning document notes, "The negative 'East LA.' image deters new and existing employment" (L.A. Design Action Planning Team). In short, the combination of a subordinated racial group and a dirty industrialized landscape served to characterize East Los Angeles as a highly undesirable place. Moreover, it should be clear that this landscape – its industrial nature in particular – developed partly because of the role of Chicanos/*Mexicanos* as low-wage manufacturing workers.

Although the northern part of the study area has been overwhelmingly Latino for several decades, the southern portion has a somewhat different history. The southern municipalities of Maywood, Huntington Park, and Bell were founded between 1906 and 1924 exclusively to house the White working class employed in the Fordist industries of Vernon (Shevky and Williams 1949). Today, however, these cities are almost entirely Latino, as the White working class began leaving in the early 1970s. Whites left in response both to the decline of Fordist industries and to growing minority activism, including the Watts riot. Consequently, the racialized and aversive attitudes directed towards East Los Angeles/Vernon were extended to these areas as they became heavily Latino. This "image problem" has influenced not only the type of reindustrialization that has recently shaped East Los Angeles/Vernon, but also its legacy of industrial pollution. One analysis of economic development pointed out: "Cleaner industries are dissuaded from locating in the area because of the

toxic contamination" (L.A. Design Action Planning Team, 1990, p. 12).

In effect, the image of East Los Angeles/Vernon as dangerous, polluted, and home to undesirable Latinos seems to have limited the area's economic options in the wake of deindustrialization. Other places in Los Angeles have had more varied redevelopment opportunities as housing or commercial areas after losing Fordist production (e.g., Fontana or parts of Downtown) (Davis 1992). However, the popular perception of East Los Angeles/Vernon as a negatively racialized and polluted area was one factor in precluding similar redevelopment schemes. . . . [T]he Los Angeles Community Redevelopment Agency (1993) believes that further development of "light" industry is the only logical and practical alternative for the Boyle Heights community. In its most recent industrial role, Vernon has become a platform for post-Fordist craft specialty production. The auto, tire, and steel-related mass production industries that fled in the 1960s and 1970s were largely replaced in the 1980s by low-wage apparel, food, and furniture industries (Valle and Torres 1994).

In addition to this type of reindustrialization, Vernon also has become more closely associated with specific hazardous and polluting activities. The 1995 Vernon Industrial Directory lists at least nine firms as "hazardous-waste processors." Moreover, many of the city's toxic air polluters are of relatively recent origin. A comparison of the 1992 TRI data and the 1961 Industrial Directory reveals that two-thirds of present reporting facilities have located in Vernon within the last 30 years.

It is within the historical context of the past 100 years that we must understand the highly publicized efforts of Vernon politicians to woo a large hazardous-waste incinerator and a hazardous-waste treatment facility in the late 1980s (Russell 1989; Gutierrez 1994). Vernon politicians were so eager for these facilities that they attempted to undermine public participation by requesting ad hoc exemptions to legal requirements for environmental impact assessments. Thus, what was widely cited as clear evidence of environmental racism *is* racism, but it cannot be reduced to a simple siting decision. As can be seen, decades of racist political economic history were embodied in that single event.

CONCLUSION

. . . Both of the case studies we have presented show some of the limitations of dominant approaches to the study of environmental racism. First, it is difficult to make inferences about environmental racism as a *process* without detailed historical investigation. This means moving beyond not only studies of contemporary outcomes or patterns, but also studies that focus on events like hazardous facility sitings without acknowledging that racial and environmental meanings accumulate in places over much longer periods of time. Moreover, these meanings can and do change, as in periods of economic restructuring.

Second, both case studies suggest some of the problems inherent in employing standard units of spatial analysis. In the Torrance case study, the predominantly Latino community is isolated in a much larger industrial district, which itself is part of a larger census tract. In the East Los Angeles/Vernon case study, the most polluted census tract represents the industrial City of Vernon, while most affected residents live in neighboring tracts. The historical inconsistency of both standard spatial units and the definitions of racial categories presents additional difficulties for quantitative analysis. In any case, the inescapable tendency of quantitative studies to accord explanatory power to racial categories themselves, instead of to racism, is itself a reason to turn to more qualitative forms of analysis.

Most importantly, the case studies suggest that varying forms of racialization and racism operate over time and space, a fact that complicates the reduction of environmental racism to discrete and measurable acts of discrimination. The early development of Torrance was characterized by a highly deliberate and conscious set of racist practices and ideologies. In this case, Latinos were disproportionately affected because of a complex pattern of employment and residential discrimination that became codified via the city planning process. Suburban industrial development involved the exploitation of a racially segmented labor force, which led to the residential confinement of Latino workers. In East Los Angeles/Vernon, the Chicano/*Mexicano* population arose in conjunction with certain industries dependent on their labor. The early industrial tone of the

barrio continued as more Latinos came to the Los Angeles area and provided a ready pool of cheap, desirable labor. The racial meaning embedded in the *barrio* was very real, as East Los Angeles became a stigmatized place, which in turn limited its redevelopment options. Environmental racism thus cannot be understood except as deeply embedded in racialized processes of urban and industrial development.

Certainly we do not wish to negate the need for continued quantitative analyses of environmental racism issues, but we believe that such work needs to be complemented by more qualitative analyses that will not only increase our understanding of the processes underway, but also contribute to the larger field of urban geography.

REFERENCES FROM THE READING

Acuna, R. (1984) *A Community Under Seige: A Chronicle of Chicanos East of the Los Angeles River*, Los Angeles, CA: Chicano Studies Research Center, University of California, Los Angeles.

Anderton, D., Anderson, A., Oakes, J., and Fraser, M. (1994) "Environmental equity: The demographics of dumping," *Demography* 31: 229–48.

Bartlett, D. (1913) "Torrance: An industrial garden city," *The American City* 9: 310–14.

Been, V. (1994) "Locally undesirable land uses in minority neighborhoods: Disproportionate siting or market dynamics?" *Yale Law Journal* 103: 1383–422.

City of Los Angeles (1988) *Boyle Heights Community Plan: Draft Supplemental Environmental Impact Report*, Planning Department.

Cutter, S. (1995) "Race, class and environmental justice," *Progress in Human Geography* 19: 107–18.

Davis, M. (1992) *The City of Quartz*, London, England: Verso.

Griswold del Castillo, R. (1979) *The Los Angeles Barrio, 1850–1890*, Berkeley, CA: University of California Press.

Gustafson, C. (1940) *An Ecological Analysis of the Hollenbeck Area of Los Angeles*, Unpublished Master's thesis, Department of Sociology, University of Southern California.

Gutierrez, G. (1994) "Mothers of East Los Angeles strike back," in R. Bullard (ed.) *Unequal Protection*, San Francisco, CA: Sierra Club Books, pp. 220–33.

Hurley, A. (1995) *Environmental Inequalities*, Chapel Hill, NC: University of North Carolina.

Kobayashi, A. and Peake, L. (1994) "Unnatural discourse: 'Race' and gender in geography," *Gender, Place and Culture* 1: 225–44.

Kochan, T.A., Wells, J.C., and Smith, M. (1992) "Consequences of a failed IR system: Contract workers and the petrochemical industry," *Sloan Management Review* (summer): 79–89.

Krieg, E. (1995) "A socio-historical interpretation of toxic waste sites: The case of Greater Boston," *The American Journal of Economics and Sociology* 54: 1–14.

L.A. Design Action Planning Team (1990) *Boyle Heights, Los Angeles* (Planning Document, Box C 1518), Los Angeles, CA: Los Angeles City Archive, Department of City Planning.

Lee, P. (1990) "Are refineries unsafe?," *Los Angeles Times* (April 15): D1, D19.

Menchaca, M. (1993) "Chicano Indianism: A historical account of racial repression in the United States," *American Ethnologist* 20: 583–603.

Olmsted, F.L., Jr. (1912) Letter to Thomas Fellows. Library of Congress, Manuscript Division, Olmsted Associates – Series B, File 5354, January 12.

Pansing, C., Rederer, H., and Yale, D. (1989) *A Community at Risk: The Environmental Quality of Life in East Los Angeles*, Client Project for the Mothers of East Los Angeles. Department of Urban Planning, University of California, Los Angeles.

Phelps, R. (1995) "The search for a modern industrial city: Urban planning, the open shop, and the founding of Torrance, California," *Pacific Historical Review* 64: 503–35.

Rabin, J. (1988) "Second Mobil letter apologizes again for refinery woes," *Los Angeles Times* (December 10): A8.

Rae-Dupree, J. and Schoch, D. (1991) "Torrance and the Mobil refinery," *Los Angeles Times* (September 12): B3.

Robak, W. (1979) "Tired evacuees not afraid to go home," *South Bay Daily Breeze* (December 5): A3.

Rodriguez, C. (1994) "Challenging racial hegemony: Puerto Ricans in the United States," in S. Gregory and R. Sanjeck (eds) *Race*, New Brunswick, NJ: Rutgers University Press, pp. 131–45.

Romo, R. (1983) *History of a Barrio: East Los Angeles*, Austin, TX: University of Texas Press.

Russell, D. (1989) "Environmental racism," *The Amicus Journal* (spring): 22–32.

Sacks, K. (1994) "How did Jews become White folks?" in S. Gregory and R. Sanjeck (eds) *Race*, New Brunswick, NJ: Rutgers University Press, pp. 78–102.

Sahagun, L. (1989) "Toxic neighbors," *Los Angeles Times* (August 28): B1.

Sanchez, G. (1993) *Becoming Mexican-American: Ethnicity, Culture and Identity in Chicano Los Angeles, 1900–1945*, New York: Oxford University Press.

Sayer, A. and Walker, R. (1992) *The New Social Economy: Reworking the Division of Labor*, Cambridge, England: Blackwell.

Schoch, D. (1995) "Torrance asks for tougher Mobil refinery safety plan," *Los Angeles Times* (February 16): B3.

Shanahan, D. and Elliott, C. (1984) *Historic Torrance: A Pictorial History of Torrance, California*, Redondo Beach, CA: Legends.

Shevky, E. and Williams, M. (1949) *The Social Areas of Los Angeles: Analysis and Typology*, Los Angeles, CA: University of California Press.

Sidawi, S. (1995) *The Historical Roots of Environmental Inequalities in a Planned Industrial Suburb*, Paper presented at the Annual Meeting of the Association of American Geographers, Chicago, IL, March.

Torrance Herald (1914) "Torrance ideal site of industrial plants" (January 1): 1.

United States Bureau of the Census (1970) "Census of Population and Housing: Census Tracts Los Angeles-Long Beach, CA," Standard Metropolitan Statistical Area, Par 1. Washington, DC: Author.

United States Bureau of the Census (1980) "Census of the Population and Housing: Summary Tape File 3," Washington, DC: Author.

United States Bureau of the Census (1990) "Census of Population and Housing: Summary Tape File 3," Washington, DC: Author.

Valle, V. and Torres, R. (1994) "Latinos in a 'post-industrial' disorder," *Socialist Review* 23, 4: 1–28.

Willard, W. (1913) "Moving the factory back to the land," *Sunset* 30: 299–304.

Editors' references and suggestions for further reading

Cutter, Susan (1995) "Race, class and environmental justice," *Progress in Human Geography* 19: 107–18.

Keil, Roger (2003) "Progress report: Urban political ecology," *Urban Geography*.

Pulido, Laura (1996) *Environmentalism and Economic Justice: Two Chicano Struggles in the Southwest*, Tucson: University of Arizona.

Pulido, Laura (2000) "Rethinking environmental racism: White privilege and urban development in Southern California," *Annals of the Association of American Geographers* 90, 1: 12–40.

Pulido, Laura (2002) "Reflections on a white discipline," *Professional Geographer* 54, 1: 42–9.

Swyngedouw, E. and Heynen, N. (2003) "Urban political ecology, justice and the politics of scale," *Antipode: A Journal of Radical Geography* 35, 5: 898–918.

United Church of Christ, Commission for Racial Justice (1987) *Toxic Waste and Race in the United States*, New York: Author.

United States Environmental Protection Agency (1994) *1987–1992 Toxic Release Inventory*, EPA 749/C-94-001.

Wolch, J., Pincetl, S., and Pulido, L. (2002) "Urban nature and the nature of urbanism," in M.J. Dear (ed.) *From Chicago to L.A.: Making Sense of Urban Theory*, Thousand Oaks, CA: Sage.

"On the Links between Home and Work: Family-Household Strategies in a Buoyant Labour Market"

from *International Journal of Urban and Regional Research* (1991)

Geraldine Pratt and Susan Hanson

Editors' Introduction

For nearly a decade, beginning in 1986, feminist geographers Susan Hanson (Clark University) and Geraldine Pratt (University of British Columbia) examined gender and the geography of home and work in Worcester, Massachusetts. Their rich empirical analysis reveals the extent to which employers and employees both intentionally and unintentionally create "extremely local labour markets" within the metropolitan area (1995). Such markets, they demonstrate, effectively enable *and* impose specific family and gender relations, while they reinforce class and racial identities in different parts of the city. For instance, women with the heaviest domestic workloads were the most likely to find work close to home; therefore power relations in the home played an important role in producing bounded labor markets. The sexual division of labor enmeshed in this home/work myth, as Pratt and Hanson explain, defines a woman's paid work as secondary to her primary reproductive role in the household and establishes a man's reproductive role as secondary to his primary productive one.

In the following reading, produced as part of their Worcester study, Hanson and Pratt introduce their findings on gendered responses to the home/work dynamic within dual-headed households and the narrow range of strategies employed by the women and men within them. Sequential scheduling of work hours for both parents is a remarkably common response, allowing the family to accommodate dual careers while caring for their children themselves. In response to this domestic strategy, employers advertise "mothers' hours," offering part-time positions in female-dominated occupations that require employees to work the hours children are typically in school.

By examining such household/workplace strategies, Pratt and Hanson highlight particular theoretical concerns that critique both feminist and geographic literature on the home/work dynamic of the family household and thus open what has thus far been the unexplored "black box" of household decision-making. Focusing on this under-theorized unit of analysis, they explore arrangements made within the family to mediate relations with the formal economy, noting that this unit of analysis is "not simply a legitimate and interesting object of study, but rather an absolutely central one for those studying economic processes." By producing a careful, empirical study of family-household arrangements that underscores the complex links between domestic and employment responsibilities, Pratt and Hanson critique the functionalist inclinations of certain feminist scholars' assumptions about the family's relationship to patriarchy and capitalism while they urge a greater tolerance for ambiguity in theorizing the complex responses of the family household. They also challenge the neoclassical economic influenced assumptions about the household that underlie the bulk of urban geographic research.

Their conclusion is reminiscent of Linda McDowell's earlier insistence that urban theory must focus on the interrelationship of production and reproduction as a single process by examining relations during various periods and circumstances. As Linda McDowell (1983) argues, conventional urban theory contains unexamined and habitual distinctions between the public and private sectors, work and home, and production and reproduction. The second half of each of these dualisms tends to be associated with women. Pratt and Hanson's work draws on McDowell's argument to demonstrate the interrelationship between each half of these dualisms showing production and reproduction as part of a single process, underscoring the expanded range of links that should be drawn between home–work relations, thus exposing the partial and often inaccurate assumptions associated with conventional urban geographic analyses. Furthermore, their choice to study the strong (albeit uneven) local economy of Worcester complements the urban restructuring literature that was previously limited by its focus on family-household arrangements in declining economies.

Since completion of the Worcester study, both Hanson and Pratt have continued to focus on gender in their research. Susan Hanson's presidential address to the Association of American Geographers reflected upon the opportunities for geography and feminism to transform each other as areas of inquiry (Hanson 1992). She noted in particular the importance of gender as an organizing principle of social life and the grounded nature of geography's empirical research. Geraldine Pratt acknowledges that the findings of the Worcester study may be unpopular among those who emphasize the mobile nature of gender identities, but she argues that their narrative, constructed around stabilized identities and bounded places, is neither natural nor static. It does, in fact, "reveal the processes and power relations that produce bounded areas and the implications of these for those who are contained and enact identities within them" (Pratt 1998).

For an interesting example of this as it relates to super-skilled, male-dominated employment in the restructured economy, see Doreen Massey's work on the contemporary "Cambridge Phenomenon" among British high-tech employees (1996). Associated with extremely long working hours and the consequent worker demand for very high degrees of both temporal and spatial flexibility, this male-dominated field ties into the dualistic thinking dividing work and home life that constructs a culture glorifying rationality, competition, and a life focused on career as opposed to family life. Massey observes that the strength of this embedded culture draws participants in the high-technology industry towards these "traits of masculinity" and, through their valorization of work, makes these people more easily exploitable by capital. In her conclusion, Massey reflects on the effects of the restructured economy that creates an extreme of de-skilled and super-skilled positions. Noting the irony that after years of criticizing de-skilling within industry she criticizes work that is too absorbing, she supports this conclusion by underscoring its impact on daily practice and "the pressure towards what can only be called a competitive workaholism and the inability to keep things under control" (Massey 1996: 123).

In the traditional urban literature – especially in the models of residential and employment location – arrangements within the household have received very little explicit attention. A few key assumptions are made about the family and the household and then the analysis proceeds; household arrangements remain 'exogenous' to the model and, as a consequence, are left virtually untheorized. In more recent work, household arrangements are recognized as being considerably more interesting, both empirically and theoretically, than the earlier treatment implies. Household relationships do not belong to a private sphere, standing outside of and apart from the important business of society: labour markets, capitalist production relations, the state. Relationships within the household are the critical arrangements through which these societal processes are mediated, experienced, and in part shaped by individuals.

... [The] household has been 'rediscovered' in large part by feminist scholars and by those studying the restructuring of capitalist economies. Feminist theory has exposed the family and the household as an object of inquiry and debate, forcefully arguing that neither women's nor men's work, within the household or in the labour force, can be conceptualized adequately without a clear understanding of household arrangements.

[. . .]

Although feminist writers have often made a case for looking within the family and the household, their theoretical arguments have developed in near isolation from careful empirical studies designed to expose the specific arrangements that households develop for accomplishing domestic and employment responsibilities. The consequence, in our view, is a body of theory that has tended to oversimplify social relations within the family-household and to portray them as overly functional to the maintenance of patriarchy and of capitalism.

The urban restructuring literature also has limitations for a comprehensive understanding of the family-household, in so far as studies that highlight family-household arrangements tend to have been done in declining local economies (e.g., MacKenzie, 1987; Pahl, 1985). Given fundamental and widespread transformations in the economies and occupational structures of most industrialized countries, transformations that include a massive increase in the number of female workers and traditionally female jobs, there is a need to explore more fully the arrangements that family-households develop to accommodate various forms of work and to sustain particular values and goals (such as family-based care of children).

[. . .]

The family-household in conventional urban analyses

Standard urban models (e.g., Alonso, 1964; Muth, 1969) rest on the implicit assumptions that all households have one worker, that all workers commute to jobs in the centre of the city, and that all households have adopted the strategy of sending one (male) adult into the labour force while the other (female) adult works in the home. Moreover, the decision process behind the selection of the household's residential location is assumed to be that of trading off the one worker's commuting distance/cost against housing lot size/cost.

For a number of reasons these assumptions about the household are inadequate. First, these models ignore the processes that structure the household arrangements upon which the models are based. . . . [T]he larger societal and local contexts that contribute in important ways to the shaping of family-household strategies and decisions are noticeably absent *from* these models. Further, by building the existing domestic division of labour into the models in the *form* of an assumption, existing relations between men and women are not open to scrutiny and tend therefore to be naturalized.

Second, the presumption made about the incidence of the assumed household maximizing strategy mentioned above is seriously biased. Even in the 1950s and 1960s, when these urban models were first formulated, the dual-headed, male-breadwinner/female-homemaker household was undoubtedly more characteristic of the white middle class than it was of other ethnic or class groups. Many working-class ethnic minority households certainly had more complex household strategies *vis-à-vis* work–home arrangements, because a large proportion of married women worked, and links within the extended family persisted. For example, in 1960 in the United States, while only 36.5 per cent of white females were employed outside the home, 48.2 per cent of black women had paid employment (Smith, 1987: 423). . . .

Behavioural modellers attempt to meet a third criticism, namely that certain assumptions underlying the traditional urban land use models (for example, complete information, utility maximization, one central employment site, housing space and commute cost as the sole bases of residential choice, and single-earner households) unnecessarily oversimplify the locational decision-making process. . . . Attempts to integrate the knowledge that a significant proportion of contemporary households have two wage-earners into models of residential location decisions are symptomatic of the unexamined nature, or at least overly mechanistic understanding, of the household (Curran *et al.*, 1982) which simply add together the wages of person 1 and person 2, weighted by the hours each works in paid employment, in order to determine the income and time constraints that a household faces. They assume that each person's time/income constraints then enter equally into the residential location decision. . . .

The fourth criticism of the assumptions underlying traditional urban models has come largely from feminists, who have pointed to the lack of attention paid to power imbalances and to potential conflicts within the households. . . .

FEMINIST PERSPECTIVES ON THE FAMILY-HOUSEHOLD

Feminists have refused the conceptual separation between home and work, insisting that the household is a site of labour, that relations within the home structure relations in the workplace and public realm, and that social relations in the workplace affect those within the family-household. They draw causal connections between the nature of the family and household, the sex segregation of the labour market and, in some cases, the development of capitalist society. . . . How do feminist theorists see the family-household functioning to maintain male dominance and (in the case of socialist feminists) to sustain capitalist relations? There are numerous strands to this argument, as well as very real disagreements among feminists of different theoretical persuasions. . . . Many feminists have developed the links between power relations in the home and women's marginalization in the paid labour force. Responsibility for home and child-care leads to somewhat more discontinuous work histories . . . or to temporary and/or part-time paid employment. Responsibility for domestic work and sexual stereotypes developed and reproduced in the home feed into conceptions of what types of paid employment are suitable for women (for example, 'caring' professions, jobs in the textile industry, 'fine' work). Presumptions about the significance of a women's wage for household maintenance may be used to legitimize lower wage rates for women (Beechey, 1977) and a 'last hired, first fired' policy *vis-à-vis* female workers (Connelly, 1978).

Although feminist theories have without doubt highlighted some absolutely central connections between the family-household and the economy, we wish to draw out some of their limitations. First, despite an ambitious theoretical programme aimed at uncovering the interconnections between 'the private' and 'the public', an unfortunate conceptual fragmentation has developed within some branches of feminist theory. Numerous commentators have noted the tendency within relatively early

feminist theory to treat housework and waged work as separate topics of investigation, dividing them into two separate and somewhat self-contained areas of study and debate: the domestic labour debate on the one hand and studies of occupational segregation in the labour market on the other. . . . An ironic side effect of this conceptual separation is that the full complexity of family and household relations has sometimes been ignored. . . .

[. . .]

A second line of criticism follows from a tendency towards functional explanation. Feminists have tended to be critical of the nuclear family-household, viewing it as an institution that is functional to patriarchal and (in the case of socialist feminists) capitalist systems of domination. . . . If many theorists now find it unlikely that capitalism functions as a totality or system in the way implied by structural-functionalist theory, it seems even less likely that patriarchy functions as a closed system. The term 'patriarchy' has a controversial history within recent feminist theory, and even a theorist such as Walby (1989), who wishes to retain the notion of a patriarchal mode of production, acknowledges the multiplicity of institutional features that constitute a system of patriarchy, the specific content of which varies across time and space.

Another strand of the criticism of functionalist explanation plays on the reification of social processes and the tendency to ignore the active role of individuals in reproducing social relations through everyday practices. Again, this criticism seems particularly meaningful for feminists; it is difficult to see women as anything other than victims or to see real possibilities for political action and social change if current practices are considered to result from the functional requirements of a capitalist or patriarchal system.

. . . [The] nuclear family and gender relations within it may be reproduced in different ways and the possibility exists that small but significant changes may take place over time. Aligned with this is the possibility that not all family practices and arrangements are functional to capitalism and/or work to maintain male dominance. Some feminist theorists have wished to acknowledge this: Humphries (1977) argues that the working-class family was a supportive institution in early industrial society, both by providing aid to individuals who could not obtain a wage (e.g. the old, the sick) and by controlling the supply of labour in order to maintain wage levels; hooks (1984) argues that black women have valued the family as a nurturing environment (a considerably less alienating one than the world of paid employment) and as a site for political mobilization against racism. . . .

In our view, feminists must strive for a greater tolerance for ambiguity in their theorizing about the family-household than has typically been the case if the full implications of the criticisms of functional explanation are accepted. A careful look within the family-household is likely to reveal a more complex picture of household interactions than that painted by many theorists thus far. . . .

The theoretical subtlety that we call for requires, as a complement, a closer empirical study of actual household arrangements and the linkages between these 'private' accommodations and economic processes. There is, for example, a need to focus carefully, in very concrete ways, on relationships within the household in order to explain the continuing occupational segregation of women. Although employer discrimination and exclusion by (white) male trade unionists offer partial explanations for the occupational segregation of women, a full explanation undoubtedly pushes the analysis back into the family and household (Barrett, 1980; Barrett and McIntosh, 1982). . . . It is necessary to extend this analysis of the concrete ways in which domestic work structures waged work. In this paper we intend to highlight one set of linkages that has hitherto received little attention, that is the sequential scheduling of work shifts among family-household members.

[. . .]

THE FAMILY-HOUSEHOLD AND ECONOMIC RESTRUCTURING

At present we know little about household work arrangements in . . . [local economies] that have been stable, or even growing. . . . [S]table or growing local economies have also been influenced by economic restructuring, including the decline of traditional manufacturing jobs and the expansion of service industries and occupations. Indeed, our study area lies in the heart of 'Bluestone and Harrison (1982) country': traditional manufacturing jobs in mills and factories have declined, to be replaced by jobs in high-tech industries and the service sector. . . . The new expanding industries are characterized by a dual-market structure, having a good proportion of poorly paid jobs with limited career mobility alongside high-paying professional, technical and managerial jobs. Even in healthy local economies, therefore, one expects a decline in real wages and a reorganization of household arrangements involving the massive entry of women into the labour force to take up the bottom tier of newly created jobs and to compensate for the decline in real wages. Smith notes that the increase in female labour force participation between 1960 and 1982 in the United States occurred disproportionately among single *and married* women . . . leading her to conclude that: 'The great increase in the participation of married women in the wage labor force was a response by white working-class families to the decline in the value of their wage. The wages of all working-class wives, whether white or black, became the crucial ingredient in keeping families out of poverty' (1987: 424–5).

[. . .]

. . . During the early and mid-1980s Worcester County enjoyed an economic boom, brought on by the growth of high-technology firms along the Route 128 and Interstate 495 corridors, which lie to the east of Worcester County, between Worcester and Boston. The Worcester Metropolitan Statistical Area (MSA) unemployment rate was 3.3 per cent in October 1987, slightly lower than that (3.4 per cent) for the state of Massachusetts as a whole (Herwitz, 1987: 4–6).[1] Worcester is an old industrial centre, well known for the production of industrial abrasives, wire, leather products and textiles. Recent economic growth has occurred within the service sector, however; to take 1985 as an example, Worcester and its surrounding suburbs gained 6,389 service sector jobs while simultaneously losing 2,486 manufacturing ones.

Almost one-third of Worcester's service sector increases in 1985 came in the retail trades; it is thus not surprising to find that women participate in the labour force at high rates. Sixty per cent of women in New England were employed in the labour force in 1987 – a jump from 55.7 per cent in 1982 which compares favourably to a national average of 56.0 per cent in 1987 (French, 1988: 22).

In Massachusetts the 1987 unemployment rate was lower for women than for men, at 3 per cent and 3.4 per cent respectively (French, 1988: 22). These figures reflect the large and, to some extent, unmet demand for entry-level service workers. . . .

For assessing household strategies in Worcester, we draw upon survey data that we collected in the summer and autumn of 1987. We conducted personal interviews with at least one, and in some cases two, adults (both women and men) in each of 620 households resident throughout the Worcester MSA. The households were chosen via an area-based, spatially stratified random sample of census blocks from the MSA's 95 census tracts, so as to constitute a representative sample of the area's working-age (21–65 years) population. The interviews were lengthy, averaging about 75 minutes each, and consisted of both closed and open-ended questions designed to elicit information on how people decide where to live and where to work, how they come to do the work they do, how they value different job attributes, how they handle household and family responsibilities, and how they view their current job situations.

HOUSEHOLD STRATEGIES IN WORCESTER, MASSACHUSETTS

[. . .]

In the analysis that follows we focus exclusively on dual-headed households and a narrow range of strategies employed by these households. We have selected these households and these strategies because they allow us to make the theoretical points that we are concerned to highlight.

Domestic work and paid employment – family strategies and social reproduction

Feminists have long recognized that household and child-care responsibilities structure women's paid employment. Women may choose to work full-time within the household, a decision mediated by the spatial constraints set up by the separation between home and work. Many women opt to work part-time (this is the case for 35 per cent of employed women in our Worcester sample) as a way of managing the twin obligations of caring for a

family and contributing to the family income. That this is an employment strategy directly linked to child-care responsibilities is suggested by the fact that Worcester women with pre-school children are much more likely to work part-time (as opposed to full-time) than are women without pre-school children. Roughly 50 per cent of women with pre-school children who work outside of the home, work part-time. Only 28 per cent of women without pre-school children who work outside of the home work part-time hours. . . . Further, cycles of child-bearing often interrupt a woman's employment history. Of the employed women in our Worcester sample, 63 per cent had experienced at least one break in their labour force participation in the last 10 years; 80 per cent of these breaks were related to childbearing or rearing.

Child-care responsibilities may structure paid employment opportunities in a more subtle way as well; women who work full-time may severely curtail the spatial extent of their job search in order to work close to home so as to be readily available in case of family emergencies. This may partially explain the well-established fact that women work closer to home than do men. That women care more about the proximity between home and work is clear from the Worcester interviews. . . . If women 'choose' to work closer to home, in part because of domestic responsibilities, it is clear that they are heavily dependent on a relatively restricted range of local employment opportunities, to an extent that men are not.

We have found that temporal and spatial constraints come together in a quite striking way within many Worcester households, taking the form of the sequential scheduling of paid employment so that one adult can always be in the home to care for children. This is a family-household strategy that has received very little attention in the literature. We point to one further way, then, that family power relations reproduce sex segregation in the labour force. One example will give a sense of this household strategy and offer a point of entry for our discussion of the contradictory links between the family and women's paid employment.

The household under consideration consists of four members: a 25-year-old woman, her 26-year-old husband and their two children, aged two and six. The couple married in January 1981, before she completed high school. Though their first child was born in 1981, this was a busy year in terms of their employment history. The woman obtained a job in February, monitoring alarms for a security company from 4 p.m. until midnight. After graduating from high school she added a second job, putting boxes of screws and nuts into packages from 7 a.m. until 3 p.m. She was able to work from 7 a.m. until midnight because she and her husband were living in her parents' home and presumably her mother and sisters could be relied upon for child-care. In September 1981 she left both of these jobs, one because her employer would not give her a rise, the other because 'it stunk: I didn't like it'. She went back to a part-time job as a cashier in a supermarket, a job that she had done on and off since October 1978. In 1982 there were a number of changes in their work/home arrangements. In February they moved out of her parents' home into an apartment of their own, at which point child-care became a greater responsibility. In the spring of 1982, her husband graduated from college. Now able to rely on her husband for child-care, the woman took a full-time waitressing job at a pub. In September her husband began working for the City of Worcester as a computer operator. They worked for these same employers when interviewed in July 1987 and their schedule took the following form: she worked a 13-hour shift on Friday, an 11-hour shift on Saturday, a 13-hour shift on Sunday and a 6-hour shift one other (rotating) day of the week. The City allowed her husband to compress his 40-hour working week into three weekdays by working from 7 a.m. until 7 p.m. Arranging their schedules in this way, they were each able to work full-time (she a 50-hour week) without relying upon paid child-care. . . .

The case outlined above is typical in the complex scheduling of work and jobs within the household. . . .

We have been struck by the remarkable frequency of these sequential scheduling strategies. Sixty-six couples in our sample and 29 per cent of all dual-earner households with children under 13 years of age arranged their work in this manner. This is a strategy that is used by all racial/ethnic groups: of those arranging their schedules sequentially, 88 per cent were identified as 'white', 3 per cent as 'black', 1.4 per cent as 'Hispanic' and 7 per cent as 'other'. These proportions roughly parallel those in the sample as a whole.

In the majority (75 per cent) of the cases when households arrange their paid employment sequentially, it is done for the purpose of child-care so that the family does not have to rely on an outside agent for this task. It reflects, for the most part, a distrust of non-familial child-care, and not simply the expense or inadequacy of child-care options in Worcester. In some cases, the subordination of the relationship between the two adults in the household to the family ideal is quite striking, as in one case where a man and woman worked at the same place, located 40 minutes away from home. She worked an 8 a.m. to 4.30 p.m. shift; he from 2 p.m. to 10 p.m. He then went to a second job, where he worked until 2:30 a.m. 'My wife works mornings. I work nights, so someone's home with the kids all of the time.' They had maintained this schedule for 11 years and continued to do so for the sake of the children, despite the fact that their children were now aged 14 and 17. Serial scheduling is, therefore, a household strategy that is adopted in a very conscious way as a means of reconciling domestic and paid work.

. . . It is significant that in the majority of cases it is the woman who takes the less optimal or less 'conventional' time slot. In the first example cited above, the woman worked weekends, the male a more conventional, though compressed, Monday to Thursday working week. Among households that have a daily sequential strategy, it is typically the woman who works the second or third shift; this was the case for 66 per cent of the households whose members schedule their work sequentially. More than one-third (22 per cent) of the women in households using this strategy spontaneously mentioned the effect that this schedule had on the type of job they could get, mentioning a compromise in the type of work or an inability to advance into management or administration. One woman bartender said: 'I can't be a secretary and work 9 to 5. My kids come first so I have to work around their schedules'. . . . As one further example, a woman who works from 3 p.m. to 11:30 p.m. as a medical technologist reported that she had to turn down a promotion because the hours of the higher-level job involved day work. Analysts have focused on the large numbers of women who do part-time work and the implications that this holds for the type of work that women do, but these data suggest that the gender characteristics of the second and third shifts also warrant attention because they have implications for the type of paid employment women do as well as women's participation in workplace organization.

Our survey data indicate that sequential scheduling arrangements not only push many women into working in irregular time slots, but also drive women into female-dominated occupations: 65 per cent of women in households in which wife and husband arrange paid employment sequentially work in female-dominated jobs; this compares to 51 per cent of women in households that do not use this sequential scheduling strategy. . . . This greater propensity to work in female-dominated jobs may well reflect the fact that women in households that use sequential scheduling are also more likely than are other women to work part-time: 49 per cent of the former group, but only 33 per cent of the latter group, work part-time. . . . Clearly, the sequential scheduling strategy – largely through the hours and timing of the job – affects the type of work many women do, their prospects for internal mobility and their capacity to organize.

Sequential scheduling strategies influenced not only type of work but the location of paid employment as well. Almost one-third (32 per cent) of the women interviewed in the households using these strategies spontaneously mentioned that their schedules led them to work close to home. In some cases this is because they were reluctant to travel long distances at night. In others it is because the household work schedules are very closely timed, precluding lengthy work trips. . . . This highlights the implausibility of behavioural geographers' assumptions that in two-earner households equal weight is placed on the work-trip time/distance of each partner in the residential location decision. Sequential work schedules place special temporal and hence spatial constraints on the journeys to work of each parent, but it is most often the woman's work trip that is the more constrained of the two.

It is readily arguable that we see patriarchal relations played out through these serial scheduling arrangements. Women tend to arrange their paid employment around the schedules of their husbands and their children. In the words of one woman: 'I had to arrange my own job around everyone else's schedule.' Women's subordination to the schedules of other members in the household clearly has some implications for their subordination within the

paid labour force, as when preferences about type of paid employment are totally subordinated to the suitability of hours. In many cases, career ambitions are set aside, at least for a time. The spatial limits of many individuals' labour markets are shrunk to a narrow circumference surrounding the home . . .

. . . [T]hus far we have emphasized the ways in which interactions within the household serve to constrain women's employment options. The analysis of the family has to be deepened, however, to include a recognition that the family functions as a resource as well. The men in these households are taking sole responsibility for child-care for a portion of each day. In some (admittedly only nine) cases it is the male who takes the second or third shift so that he can be with the children during the day. In the words of one male who had voluntarily worked a night shift (11 p.m. to 7 a.m.) as a supervisor at a supermarket for the past 10 years: 'I don't trust anyone else to raise them . . . not even a relative . . . especially for the moral training.' In other words, men are contributing a considerable amount of domestic labour (and, in some cases, accommodating their paid employment to the family's needs) so that women can take employment outside the home. Even within a buoyant labour market, therefore, there is evidence for a readjustment of the gender division of labour within the household to allow for two wage-earners.

[. . .]

Household and family strategies that allow women to enter the labour force can be interpreted in terms of their functionality to the economic system. [I]n the . . . Worcester context, there is an extreme shortage of clerical, hospital and service workers, precisely the types of jobs that women tend to take up (especially if they are constrained to work the second or third shift due to domestic responsibilities). Accommodations made within the family allow women to fill these jobs. It is particularly working-class families who make these accommodations; considering the occupations of the male heads in households using a serial scheduling strategy, 32 per cent can be classified as non-skilled manual, 9 per cent as non-skilled non-manual, 29 per cent as skilled manual, 23 per cent as skilled non-manual and only 5 per cent as managerial/professional. . . . If one looks at the incomes of male household heads, it is clear that sequential scheduling strategies are used most in families in which the male incomes are lowest. Of those households that schedule paid employment sequentially, 57.1 per cent of male heads had personal incomes below $25,000 per annum, while only 40.7 per cent of male heads in other households had incomes below this level. This fits nicely with Smith's argument that a decline in real wages has forced the increased participation of married women in the paid labour force, and with an argument that sees sequential scheduling as a strategy to compensate for the erosion of the family wage without threatening the continuation of home- and family-based childrearing.

[. . .]

CONCLUSIONS

We have tried to demonstrate the significance of the family-household for understanding individual's lives, as they determine where to work and the type of paid employment to pursue. This is not the type of analysis that can be 'tacked on' to studies of occupational structures and labour processes. An understanding of where and why men and women work and the types of jobs they do demands a close look at the power dynamics within the family-household. This point has been developed by many feminist scholars. Our analysis contributes to this feminist literature by drawing attention to yet another way in which decisions within the realm of the family-household, in this case about the timing of paid employment for individual members, reproduces sex segregation in the labour market.

In trying to understand how families cope in communities that have suffered employment losses due to industrial restructuring, researchers have turned their attention to the household to examine coping strategies. We have found quite elaborate family strategies in a buoyant labour market, most probably reflecting the fact that wage levels necessitate two wage-earners in lower- and middle-income families. . . . When one comes to abstract the implications of these strategies for the maintenance of existing social relations, male and class dominance, more subtlety is required than has often been exhibited by feminist theorists. On the one hand, the way in which individual family members arrange their work shift sequentially does seem to reflect male dominance, and certainly the

view that women should fit their paid employment around their domestic responsibilities and partner's jobs. This then reinforces the existing gender division of labour in so far as it tends to lead women into female-dominated occupations. In a seemingly perfect fit, these are also the occupations with the shortest labour supply; the arrangement seems quite functional to the capitalist 'system'.

And yet, should the analysis stop here? At the very least there is a tension between the two systems of domination, a point that Hartmann (1987) has been concerned to highlight. The lowering of real wages so as to necessitate two wage-earners in a household is forcing at least some men to take a more active role in child-care (or alternatively, two wages allow them to do so). Families also have and provide resources: the help of extended kin allows women to take paid employment and the intergenerational transference of domestic property provides some measure of independence from the labour market. Without wishing to press this point beyond the credible, there is a complexity and layering of possibilities within these household arrangements that extends beyond the needs of capital (or men) and allows the possibilities for individual freedoms, choices and alternative gender relations. We have sought to emphasize the very active process of social reproduction and the potential for counter-hegemonic tendencies within the household.

Our discovery that some family-households employ work strategies that require men to take responsibility for child-care and that others use residential strategies that loosen the family-household's commitment to waged labour lends support to a point that many feminists are now keen to acknowledge: that there is a diversity of experience for women in different circumstances. Feminist theory on the family has tended to be overly reductionist, often working with clumsy stereotypes of the nuclear family. Recognizing the diversity of family-household arrangements is not only more accurate in empirical terms; it also admits the possibility of incremental but radical changes in social relations within the family.

NOTE

1 If not otherwise noted, data in this section come from Herwitz, 1987.

REFERENCES FROM THE READING

Alonso, W. (1964) *Location and Land Use*, Cambridge, MA.: Harvard University Press.

Barrett, M. (1980) *Women's Oppression Today*, London: Verso.

Barrett, M. and M. MacIntosh (1982) *The Antisocial Family*, London: Verso.

Beechey, V. (1977) 'Some notes on female wage labor in capitalist production,' *Capital and Class* 3: 45–66.

Connelly, P. (1978) *Last Hired, First Fired: Women and the Canadian Workforce*, Toronto: The Women's Press.

Curran, C., L.A. Carlson and D.A. Ford (1982) 'A theory of residential location decisions of two worker house-holds,' *Journal of Urban Economics* 12: 102–14.

French, D. (1988) '60 per cent of N.E. women work,' *Boston Globe*, 23 May: 22.

Hartmann, H. (1987) 'The family as the locus of gender, class and political struggle: the example of housework,' In S. Harding (ed.) *Feminism and Methodology*, Bloomington and Indianapolis: Indiana University Press.

Herwitz, E. (1987) 'Area's labor pool reflects a mismatch of needs and resources,' *Business Digest*, February 1987: B-55.

hooks, b. (1984) *Feminist Theory: From Margin to Center*, Boston: South End Press.

Humphries, J. (1977) 'Class struggle and the persistence of the working-class family,' *Cambridge Journal of Economics* 1: 241–58.

MacKenzie, S. (1987) 'Neglected spaces in peripheral places: homeworkers and the creation of a new economic center,' *Cahiers de Geographie du Quebec* 31: 247–60.

Muth, R.F. (1969) *Cities and Housing*, Chicago: University of Chicago Press.

Pahl, R.E. (1985) 'The restructuring of capital, the local political economy and household work strategies,' in D. Gregory and J. Urry (eds) *Social Relations and Spatial Structures*, New York: St Martin's Press.

Pahl, R.E. and C. Wallace (1985) 'Household work strategies in economic recession,' in N. Redclift and E. Mingione (eds) *Beyond Employment: Household, Gender and Subsistence*, Oxford: Blackwell.

Smith, J. (1987) 'Transforming households: working-class women and economic crisis,' *Social Problems* 24: 416–36.

Walby, S. (1989) 'Theorizing patriarchy,' *Sociology* 23: 213–34.

Editors' references and suggestions for further reading

Hanson, S. (1992) "Geography and feminism: worlds in collision?," *Annals of the Association of American Geographers* 82, 4: 569–86.

Hanson, S. and Pratt, G. (1995) *Gender, Work and Space*, London and New York: Routledge.

McDowell, L. (1983) "Towards an understanding of the gender division of urban space," *Environment and Planning D: Society and Space* 1: 59–72.

Massey, D. (1996) "Masculinity, dualism and high technology," in N. Duncan (ed.) *Body Space*, London: Routledge.

Pratt, G. (1998) "Grids of difference," in R. Fincher and J. Jacobs (eds) *Cities of Difference*, New York: Guilford Press.

"Gender, Class, and Gentrification: Enriching the Debate"

from *Environment and Planning D: Society and Space* (1999)

Liz Bondi

Editors' Introduction

Gentrification, a term coined in 1964 to describe class transformations in central city neighborhoods (Glass 1964), continues to figure prominently in discussions of the restructuring of urban geographical space. Scholars' focus on this process of residential change rests, in part, because gentrification makes visible in the urban landscape the creation and constitution of distinctive social groups. Neil Smith (p. 128) and David Ley (p. 304) provide examples of this concern in their examination of the politics of development with the "return" of white-collar residents to cities such as New York and Vancouver, BC.

Since the 1980s, however, several scholars commented specifically on the prominent role that women as members of a new urban middle class play in the process of gentrification. These studies suggested that an increasing number of these women chose to postpone marriage and child-bearing in favor of careers, and that an increasing number of women experienced success in professional and managerial classes, thus gaining access to "decent-paying jobs" (Beauregard 1986; Markusen 1981: 32). Speculating on the potential social restructuring caused by this economic shift, a growing number of geographers asked questions about gender's constitutive role in the gentrification process (Bondi 1991; Rose 1984, 1989; Smith 1987). At least one such examination of gender and gentrification suggested that a common factor underpinning various forms of gentrification might be career-oriented women who drive the process, whether in affluent dual-career households or non-traditional households, headed by lone women (Warde 1991). Another, however, argued that "the proposition linking women and gentrification has remained a general affirmation with little documentation of actual empirical trends" (Smith 1987: 156).

Feminist geographer Liz Bondi at the University of Edinburgh concurred with this assessment as recently as 1999, when she noted that despite the frequent attention given to gender and gentrification, empirical studies of the relationship remained rare. In "Gender, Class, and Gentrification: Enriching the Debate," she seeks to fill that gap in the scholarly debates by examining specific gender and class practices in several Edinburgh neighborhoods. As one of the first to offer such a critique of the treatment of gender issues within gentrification studies (1991), Bondi's empirical study allows her to rework her own previous analyses while enriching the literature by drawing attention to the complex ways in which gender, class and life course are interwoven in various neighborhoods of a city not previously considered in gentrification studies.

Much of the gentrification research depends upon the experiences of centers of the global economy such as New York as exemplified in Neil Smith's research (p. 128; 1987). However, Bondi introduces her study of gentrification in the Scottish capital by noting that her choice of location allows for analysis of the process outside a "world city," yet within an urban center with a strong information services economy and its prestigious white-collar occupations. She then takes up several important issues in the Edinburgh research, including: (1) the significance of financially independent middle-class women whose lifestyles and outlooks are similar to professional middle-class men; (2) the role of local contexts in the interpretation of the gentrification process; and (3) the influence of a woman's life course on residential decision-making. By comparing two distinct examples of gentrification in the central city neighborhoods of Stockbridge and Leith, she is able to ask "who are the gentrifiers?" In answering this question, Bondi concludes: "That gentrification has proven to be a resilient term despite its elusive and sometimes contradictory qualities suggests that it remains important to 'unpack' its characteristics" (1999: 279).

While Bondi's focus on feminist issues begins to "unpack" the forces of gentrification, others have contributed to this literature by focusing on ethnicity and sexuality. Loretta Lees provides a useful overview of the late twentieth-century gentrification literature in which she suggests new directions in its examination (2000). With an emphasis on the issues of social restructuring and geographies of difference, Lees notes that class and gender studies of gentrification far outweigh studies of ethnicity and race, and therefore places ethnic minority gentrification on the agenda for needed future studies. Lees also notes, however, that a more detailed investigation of masculinity and space would add to the gender and gentrification literature (see also Knopp (1990), Lauria and Knopp (1985), and Rothenberg (1995) for studies that address gay and lesbian gentrification and the extent to which sexuality influences the making of urban social space).

A founding editor of the journal *Gender, Place, Culture: A Feminist Journal of Geography*, Bondi describes her research interests as feminism and gender issues in relation to identities, subjectivities and the urban. Recently, her career and research interests have shifted to the spatialities, situated histories and identity politics associated with psychotherapies and counseling as cultural practices.

■ ■ ■ ■ ■ ■

The relationship between gender and gentrification has attracted a good deal of attention during the last few years. . . . However, empirical studies of the relationship remain rare, and, as I will elaborate in due course, address the issues at stake only partially. My aim in this paper is to contribute to the development of understandings of gender and gentrification with reference to case-study research conducted in Edinburgh.

[. . .]

GENDER AND GENTRIFICATION

That gentrification is a multifaceted phenomenon is widely accepted. It involves processes of consumption *and* production; it depends upon individual human agency and broader structural processes; it is driven by, and expresses, cultural *and* economic imperatives. . . . Moreover, despite what has been described as 'the chaos and complexity' of gentrification (Beauregard 1986), the term has

demonstrated considerable resilience: it continues to fulfill a powerful discursive function despite difficulties in defining its limits precisely. . . .

In coining the term to describe aspects of change in postwar London, Ruth Glass (1964) powerfully evoked issues of middle-class identity in relation to transformations in the class composition of previously working-class inner-urban neighbourhoods. In the decades since Glass's observations about London, one strand of research on gentrification has explored processes of class definition in some detail (Caulfield 1994; Jager 1986; Ley 1994, 1997; Ley and Mills 1993; Mills 1988; Smith 1987, 1992; Williams 1986; Zukin 1982). Some of these researchers have observed that women play a prominent part within the urban middle class, and, gradually, the significance of gender has been examined more systematically.

Several feminist urban researchers have explored ways in which women are *disadvantaged* in cities, whether through the operation of the market, public policies, male violence, and/or discursive practices (e.g. Bondi and Christie 1997; Booth *et al.*, 1996; Garber and Turner 1995; Massey 1994; McDowell 1983, 1991; Pain 1991; Roberts 1991). Others have shown that poor women, including many lone mothers and many lone elderly women, are disproportionately concentrated in deprived inner-urban neighbourhoods (e.g. Holcomb 1986; Winchester 1990; Winchester and White 1988). Against this, Elizabeth Wilson (1991) has argued that cities provide (and have long provided) emancipatory opportunities for women. Wilson's argument that this applies as much to working-class as to middle-class women has been contested (Bondi 1998; Ravetz 1996), but evidence of connections between gentrification and changes in gender practices amongst strands of the urban middle class is beginning to accumulate.

For example, by using data for Montreal from the Canadian census, Damaris Rose (1989) demonstrated an association between public-sector professions dominated by women (especially the education, health, and social services sectors), households other than conventional nuclear family units (including lone women and lone mothers), and residence in inner-urban areas undergoing gentrification. She argued that 'changes in women's employment situation, mediated by other aspects of gender relations, may often be *constitutive* of gentrification processes' (p. 133).

Rose (1984, 1989) also drew attention to gentrification as a housing strategy adopted by less affluent sectors of the urban middle classes, such as those in public-sector professions. This interpretation is also evident in Tamar Rothenberg's (1995) account of the development of a lesbian community in Park Slope, Brooklyn. In developing Rose's theme, Alan Warde (1991) has suggested that the common factor encompassing different forms of gentrification might be that they are all strategies driven principally by career-oriented women, whether in affluent dual-career households (often without children), which are associated with the most expensive commercially led forms of gentrification, or in 'nontraditional' households, including many headed by lone women [such as those on whom Rose (1989) and Rothenberg (1995) focus] whose participation in gentrification depends upon 'sweat equity'.

[. . .]

RESEARCHING GENDER AND GENTRIFICATION: AN EDINBURGH STUDY

A number of researchers have argued that gentrification relates closely to the emergence of a small group of cities that serve as 'command and control' centres in a global economy (Hamnett 1991; Sassen 1991, 1994). London is, of course, a prominent member of this group. But there is plenty of evidence to suggest that gentrification is not limited to 'global cities': Larry Knopp, for example, has documented examples of gentrification in which gay men have played a key role in several regional centres in the USA (Knopp 1987, 1990a, 1990b; Lauria and Knopp 1985) and David Ley (1986, 1988, 1997) has identified neighbourhoods subject to gentrification in several regional Canadian cities. What is evident, however, is a strong link between gentrification and employment in prestigious white-collar occupations in the service sector, especially in business and financial services. 'Global cities' may contain a large proportion of all such positions but they do not have a monopoly on them.

Although a European capital, Edinburgh cannot claim to be a global city. Nevertheless, it has a number of characteristics that make areas within it strong candidates for gentrification. First, in terms of sources of demand, although Edinburgh is dwarfed by London in terms of the total volume of service-sector jobs, its employment structure is remarkably similar, with the large percentage employed in the so-called 'information services' (i.e. banking, financial, insurance, and business services) being particularly significant. By using data from the 1981 Census, Anthony King (1990) observed that, among British cities, Edinburgh was the only one that came within five percentage points of London in terms of the relative significance of employment in services in general and 'information services' in particular. Edinburgh owes its substantial professional and administrative middle class to its historic role as a capital city, and as a centre of learning and culture (Gordon 1986; Richardson *et al.* 1975). But it is also worth noting that the percentage growth in 'information services' during the period 1971 to 1981 exceeded that in London, and between 1981 and 1991 fell only very slightly behind. Thus, given the socioeconomic characteristics of its working population, Edinburgh is likely to have a plentiful supply of potential consumers of gentrified property.

Second, on the supply side, the urban fabric of Edinburgh is also well suited to gentrification. Exceptional among British cities in retaining a residential core, central Edinburgh is also unusually well preserved. The mediaeval origins of the city remain in evidence, and large tracts of Georgian and Victorian buildings have survived various twentieth-century ravages more or less intact. At the same time, disinvestment – the precursor of gentrification – has had a major impact on particular neighbourhoods during particular periods, generating 'rent gaps' and therefore property and/or land suitable for gentrification (Clark 1995; Smith 1979). The most recent and dramatic example is in the port of Leith, two miles from the city centre, where the decline of dockyard and related industries has resulted in disinvestment continuing through several decades, with the opportunities to profit from reinvestment eventually being taken up in the 1980s and 1990s.

More generally, some parts of Edinburgh are best characterised as long-term high-status neighbourhoods: the New Town, for example, has never suffered a mass exodus by the middle classes (although individual streets have experienced a decline in status for limited periods). Other neighbourhoods, notably some parts of the Old Town, have been renovated and rehabilitated largely by local government and housing associations, through which a wholesale transfer from working-class to middle-class occupancy has at least been impeded. But there are also several neighbourhoods that have undergone or are undergoing a process of gentrification, in which a marked change in social composition is accompanied by a substantial reinvestment in the physical condition of the urban fabric.

[...]

The research reported here is based on detailed examination of changes in two neighbourhoods subject to gentrification. In the first, Stockbridge, gentrification can be traced to the 1960s, when individual owner-occupiers began to buy and upgrade the largely Victorian housing stock (see Kersley 1974). By the 1980s the neighbourhood could be described as one of 'mature' gentrification. In the second neighbourhood, the Leith waterfront,

gentrification occurred much more recently and through different mechanisms: from the early 1980s developers have built on gap sites or have converted industrial buildings for owner-occupation. In this neighbourhood the gentrification process remains incomplete and continuing.

[...]

... [The] areas contained between 3,000 and 3,500 households, and all had experienced population growth between 1981 and 1991.... Within each of the ... neighbourhoods, a group of adjacent streets was selected, consisting of between 350 and 500 housing units trading in similar price ranges. ... Altogether, 62 households were contacted in Stockbridge, [and] 84 in Leith.... In the end, fourteen interviews were conducted in the Stockbridge study area, [and] sixteen in the Leith study area. ... All the interviews took place during 1991. This small and nonrepresentative sample does not provide a basis for making generalisations about the people moving into neighbourhoods subject to gentrification. However, it provides some useful qualitative evidence....

CHANGING PLACES; THE NEIGHBOURHOODS

Stockbridge: From red lights to flower boxes

The earliest example in Edinburgh of a process similar to what Glass (1964) observed in London when she coined the term 'gentrification' appears to have occurred in the Stockbridge neighbourhood. Adjacent to the New Town and lying on the potentially picturesque Water of Leith, this neighbourhood was ideal for middle-glass gentrifiers. Their first target was an area of terraced artisan dwellings built on a cooperative basis between 1861 and 1911 for working-class owner-occupancy. By the mid-1960s many were in a rundown condition and still lacked standard amenities. At this point, middle-class professionals began to purchase the houses from their largely working-class owners, and to refurbish them with the benefit of local authority improvement grants (Kersley 1974). Much of this property came on the market after the death of elderly long-term owners and without any overt class conflict. Indeed it is difficult to find any evidence of resistance of any kind to the gentrification of Stockbridge.

The upgrading of these dwellings by their new middle-class owners led to considerable increases in value with spin-off effects for property in other parts of Stockbridge. Much of the housing stock in Stockbridge consists of Victorian tenements originally built for Edinburgh's expanding middle classes. In addition there are some mews flats in buildings originally owned by affluent Victorian households living in substantial town-houses nearby. The mews were originally stables, coach-houses and living accommodation for coachmen. Some of the stables were converted into garages, and as the neighbourhood grew in popularity, other parts, and even the garages, were converted into well-appointed flats. The narrow cobbled streets with unusual building frontages provided imagery most amenable to gentrification (Bondi 1998). Upgrading was undertaken by individual owner-occupiers: through the 1970s and into the 1980s countless flats were described by selling agents as 'full of potential', 'in need of renovation', and so on. By the mid-1980s, which marks the beginning of the period on which this study focuses in detail, these kinds of descriptions had become rare, which suggests that the gentrification of Stockbridge had reached saturation point. The outward appearance of residential streets in the neighbourhood is neat and tidy, and the numerous flower boxes suggest that considerable care and pride are invested in local property.

During this period Stockbridge became increasingly dominated by nonfamily households in the economically active age range, so that at the time of the 1991 Population Census only 11 per cent of households in the neighbourhood included dependent children whereas 34 per cent consisted of lone, nonpensioner adults (compared to 22 per cent and 18 per cent respectively for the city as a whole). Socioeconomically, the 1991 Population Census indicated that the neighbourhood was then home to substantial numbers of those employed in professional and managerial grades (35 per cent of the economically active, compared to 22 per cent for the city as a whole). Unemployment rates were below the city average. Although the great majority of households living in Stockbridge in 1991 were in owner-occupied housing (72 per cent), a substantial proportion (20 per cent) were in privately

rented property. In both 1981 and 1991 about one-fifth of households reported that they had moved during the 12 months prior to each census. Combined with the relative prominence of the private rental sector this suggests that the neighbourhood has for some time served as a place in which households tend to reside on a relatively short-term basis.

Stockbridge has a slightly bohemian and trendy reputation. It is home to a community theatre, its own annual festival, a small award-winning council estate, and some relics of the numerous watermills that once flourished along the Water of Leith. In addition, the existence until the mid-1970s of thriving brothels is very much part of the neighbourhood's credentials. It is very close to city-centre galleries, museums, theatres, and filmhouses, and Stockbridge itself is renowned throughout Edinburgh for its range of high-quality specialist shops. Although the neighbourhood includes some properties that command high prices likely to be within the price range only of the most affluent, at the time of the study there were also many more modestly priced one- and two-bedroom flats likely to be attractive to single first-time buyers.

Leith: From industrial wasteland to industrial heritage

. . . Leith flourished in the eighteenth and nineteenth centuries on the strength of its port and related industries. Many fine Victorian buildings continue to testify to the affluence of its merchants and industrialists. During the twentieth century, however, Leith's industrial fortunes have experienced an accelerating decline, culminating in the closure of the last shipyard in the late 1970s. Although the population fell progressively through the century, overcrowding continued to be a problem into the 1960s. Many of the working-class tenements built in the nineteenth and early twentieth centuries were chronically neglected by their landlord owners. Some were demolished to make way for council flats, others have been rehabilitated with the assistance of local authority repair grants. In the early 1990s, when this study was undertaken, some remained in a derelict state.

The gentrification of Leith began in the early 1980s. The Leith Project, under the combined guidance and finance of the Scottish Development Agency, Lothian Regional Council, and Edinburgh District Council, supported private investment through environmental improvements and grants to small businesses. This provided the context for developers who saw the potential for lucrative residential projects. These include conversions of former industrial buildings and newly built complexes on or close to the waterfront. Between 1981 and 1991, Leith became more similar to Stockbridge demographically: the proportion of households with dependent children fell from 24 per cent to 18 per cent, whereas the proportion of households consisting of lone adults below pensionable age rose from 14 per cent to 28 per cent. Socioeconomically there were related trends, with the percentage of economically active residents in professional and managerial employment increasing from 7 per cent to 11 per cent. However, in 1991 the neighbourhood remained numerically dominated by the working classes, those in manual work accounting for 43 per cent of the economically active (compared to 32 per cent for the city as a whole and 28 per cent in Stockbridge). Unemployment rates were well above the city average. The decade between 1981 and 1991 witnessed a sharp reduction in the proportion of the population living in local authority housing throughout the United Kingdom and this was reflected to some degree in tenurial change in Leith, where public sector rental declined from 20 per cent to 15 per cent of households. Although this was accompanied by a rise in owner-occupation in the neighbourhood, a significant component was produced by transfers from the local authority to housing associations. As in Stockbridge, the private rental sector has remained fairly substantial (16 per cent of households) and rates of in-migration increased between 1981 and 1991 with the percentage of households moving into the area during the year preceding the two censuses increasing from 12 per cent to 18 per cent.

These changes took place with relatively little overt class conflict. As gentrification occurred in spaces and buildings already abandoned, it did not precipitate direct displacement of people and jobs. Indeed, the total population of Leith rose by about 20 per cent between 1981 and 1991 whereas that for the city as a whole fell slightly. It is harder to detect the incidence of indirect displacement, which results from increases in the rents and sale prices of residential property together with changes in employment opportunities, consumer demand, and so on (Marcuse 1986). However, the continuing presence of local authority and housing association rentals indicates impediments to the wholesale exclusion of existing working-class residents. Indeed, despite the dominance of commercially led gentrification, an image of urban pioneering is prominent in Leith, expressed in part through visual references to Leith's maritime past, which abound in the gentrified areas (see Bondi 1998).

At the time the research reported in this paper was conducted, the gentrification of Leith was confined to a relatively small area close to the mouth of the Water of Leith. However, the process has been continuing since then, further stimulated by the opening of a major office development in the area. But whether the neighbourhood will ever support the saturation gentrification evident in Stockbridge remains an open question . . .

[. . .]

HOUSE PURCHASERS AND RESIDENTS: WHO ARE THE GENTRIFIERS

. . . [V]ery different proportions of couples and single people were buying in the inner-urban areas and in the suburban area during the period 1985 to 1990 . . . [I]n Leith and most especially in Stockbridge, single people greatly outnumbered couples. Among the single people buying in both inner-urban neighbourhoods men outnumbered women but not by very many, with the smaller 'gender gap' in Stockbridge.

To assess the full significance of these findings it is useful to compare the figures with national patterns. Some of the owner-occupiers . . . bought their homes outright (without mortgages) although the majority did not. Consequently, these house purchasers are similar to (but not identical with) those taking out mortgages during the same period. Using evidence collected by the Council of Mortgage Lenders, Fionnuala Early and Michelle Mulholland (1995) estimate that between 1983 and 1993, lone women as a percentage of all borrowers in the United Kingdom increased from 8 per cent to 17 per cent, with lone men increasing from

17 per cent to 21 per cent and male and female couples decreasing from 75 per cent to 62 per cent. Relative to these patterns, [several] things are clear about the Edinburgh evidence. . . . [I]n both of the inner-urban areas the proportion of lone purchasers of either sex was much higher than national patterns. . . . [A]nd most significant, the proportion of lone women purchasing property in both of the inner-urban areas, and most especially in Stockbridge, was very much greater than in the United Kingdom as a whole.

In each of the study areas, a small number of transactions named two individuals of the same sex. . . . Although these might include gay and lesbian households, some were siblings who had inherited the property. . . . [T]he number of such transactions was too small to suggest that substantial numbers of gay or lesbian couples were buying in any of the three areas, although the data at least hint at higher rates in the two inner-city areas than the suburban area.

It is not possible to differentiate between households any further using the data collected. . . . In particular there is no evidence of the age of house purchasers, of the presence or absence of children, or of the occupations of household members. To enrich the picture available from this source I therefore turn again to the UK Population Census.

Those participating in gentrification have often been characterised as career-oriented young adults in a phase of their lives between the completion of their education and the commencement of child-rearing, which, at least for women, is 'postponed' relative to dominant norms, sometimes permanently (e.g. Bondi 1991; Markusen 1981). This image was challenged by Rose's (1984, 1989) argument concerning the significance of 'nontraditional' households including lone parents. Several subsequent studies have found significant proportions of households supported by two economically active adults living with dependent children in neighbourhoods subject to gentrification. In the Edinburgh study, however, both of the inner-urban neighbourhoods contained significantly smaller proportions of households with dependent children than the city as a whole, together with significantly greater proportions of lone adults of working age. With respect to lone parents too, the 1991 census indicates that there were relatively low numbers living in any of the . . . areas and only the

Leith neighbourhood had experienced a marked rise during the 1980s. As the increase in Leith was associated with a particularly sharp decline in the percentage of female lone parents who were economically active, it probably had more to do with patterns among working-class residents than incoming gentrifiers. Thus, there is no evidence here to suggest that Stockbridge has become a neighbourhood of particular significance to middle-class women raising children in nontraditional households, and only a weak suggestion that the Leith neighbourhood might have done so.

In order to explore further the context in which women are purchasing property in their own right in Stockbridge and in Leith, it is necessary to examine patterns of occupational status more closely. . . . When the professional and managerial categories are examined separately the particular significance of the former becomes clear. Across the city as a whole, in 1991, 14 per cent of the economically active population were employed in managerial grades and 8 per cent in professional grades. . . . In Leith the rates were approximately half the city-wide averages (8 per cent and 3 per cent respectively). But in Stockbridge, although managerial employment was four percentage points above the city-wide rate, professional employment was more than twice as great (18 per cent and 17 per cent respectively). This suggests that Stockbridge is a residential location particularly attractive to those in professional employment. It might be objected that the difference between Leith and Stockbridge is exaggerated by the fact that the spatial units concerned extend beyond the limits of the areas subject to gentrification. However, the relevant rates for a more tightly defined area centred on the gentrified property on the Leith waterfront were very similar to those for the larger area. In Stockbridge the proportion of those in professional employment was even higher in that part of the ward furthest from the long-time middle-class stronghold of the neighbouring New Town.

These data strongly suggest that the gentrification of Stockbridge is closely linked with processes of class formation within the professional strand of the middle class. However, they also suggest that class processes operating in Leith are different: people resident in those parts of Leith affected by gentrification did not, in 1991, include many in professional occupations. . . .

What part do women play in these socioeconomic profiles? . . . [W]omen formed the majority of those in occupations classified as intermediate nonmanual (dominated by the so-called 'semi-professions' such as nursing, teaching, and so on) and junior nonmanual (clerical work), but a minority in all the remaining categories. So far as professional employment is concerned women accounted for only about one quarter of the total, with little variation between the [two] neighbourhoods. These rates demonstrate the persistence of gender differences in occupational status, reflecting both the gendered character of the system of classifying occupations and gender divisions in employment (e.g. Beechey 1986; Crompton 1995; Murgatroyd 1984). More specifically, although women resident in Stockbridge in 1991 were much more likely to be in professional careers than women residents in Leith . . . they continued to lag as far behind men in occupational status as in the other . . . neighbourhood. In fact it is within the managerial category that women in Stockbridge and Leith appear to fare better, relative to men, than in the city as a whole. So, although Stockbridge may be an area particularly attractive to other professional middle-class, it is no more attractive to professional women than to professional men. Furthermore, with some caution in view of the limited sample sizes, if Stockbridge attracts a disproportionate number of middle-class women it would appear to be those in the managerial strand.

Savage *et al.* (1992) emphasise the importance of educational qualifications in mediating the relationship between gender and occupational status. . . . [I]n 1991 nearly half the adult population of Stockbridge had some form of advanced educational qualifications, and roughly a third held at least first degrees or the equivalent. Like figures for professional employment, these rates were more than double those found in the city as a whole, whereas rates for Leith were barely half the city-wide rates. Although male residents in each of the . . . areas are more likely to hold first degrees than female residents, the gender gaps in educational qualifications at this level are certainly much smaller than the gender gaps in rates of professional employment. . . . Advanced educational qualifications below first degree level are dominated by forms of vocational accreditation associated with the 'semi-professions'. When these are added to the higher level qualifications, the gender gaps close further, or, in the case of Leith, are reversed. These data confirm the link between Stockbridge and the professional middle class, and illustrate that persistent gender inequalities in occupational status cannot be attributed to gender differences in educational qualifications.

In summary, the data examined in this section suggest complex and variable class and gender dimensions of gentrification, with three points of particular significance. First, in two rather different areas subject to gentrification, large numbers of lone women have been purchasing property in their own right. Second, although house prices are similar, these two areas appear to attract different socioeconomic groups, with men and women in professional careers much more strongly represented in the neighbourhood of 'mature' gentrification. Third, despite small gender gaps in educational qualifications, gender inequalities in socioeconomic status persist in roughly equal measure in all neighbourhoods, inner-urban and suburban. Taken together, these points suggest that career-orientation among women within the professional middle class may play a part in gentrification in some areas but that other processes are also at work.

CONCLUDING DISCUSSION

[. . .]

Evidence from both Stockbridge and Leith lends support to Butler and Hamnett's (1994) argument that gentrification is associated with a professional strand of the middle class, within which a substantial proportion of women have become *strongly* career-oriented. I would, however, define this strand in slightly broader terms than Butler and Hamnett. In the Edinburgh study women did not form a larger proportion of those in professional employment than elsewhere in the neighbourhoods subject to gentrification, although they did constitute a very substantial proportion of those purchasing residential property. This points to the significance of financially independent middle-class women whose occupations are not classified as 'professions' as frequently as their male counterparts but whose lifestyles and outlooks are broadly similar to those of professional middle-class men. . . . It must also be noted, however, that the association between gentrification and the professional middle class is not an exclusive one: just as some professional households choose residential locations other than inner-urban neighbourhoods, so too particular neighbourhoods attract groups other than those in well-established professional careers, as exemplified in the case of Leith. . . .

Evidence from the Edinburgh study lends weight to Lees's (1994) argument that the context in which gentrification occurs must inform its interpretation. . . . In the case of Stockbridge, what emerged was the significance attaching to this neighbourhood for a professional middle-class group rooted in and oriented towards a particular city. Leith held no such attractions for this group but served instead as a location of choice for a more footloose professional middle-class group relatively new to the city. Studies of gentrification, particularly those in global cities, have tended to emphasise the geographical mobility of those pursuing professional careers. In this paper, I have illustrated other possibilities, and in so doing further illuminated some of the complexity of gentrification. The differences between Stockbridge and Leith may in part reflect the different periods during which gentrification began, but subdivisions within the professional middle class are also of undoubted importance. These subdivisions relate to variations in geographical mobility produced by the existence of different kinds of professional careers, some requiring participation in a UK-wide or international labour market and therefore a very high degree of mobility, others linked closely to Edinburgh's position as the administrative capital of Scotland, which has created a locally based and locally oriented professional middle class. An issue meriting further research in other cities concerns the question of whether those who participate in the gentrification of neighbourhoods adjacent to long-term high-status enclaves tend to share with the Stockbridge interviewees this more locally based career orientation when compared with those who participate in the gentrification of other inner-urban neighbourhoods.

Local specificity is also crucial to understanding why different studies of gentrification produce such strongly contrasting evidence concerning the importance of households with dependent children. In a study based on interviews with people working in the financial sector in London, McDowell (1997: 2071) found that women with children 'felt completely unable to combine the demands of careers and motherhood unless they lived relatively close to work'. For her interviewees this meant living in inner London. . . . [I]n the case of Edinburgh its relatively small size (a population of about half a million) together with its compactness may have contributed to the views expressed by several women interviewees that it might well be possible to combine motherhood with commuting from suburban or ex-urban locations. Whether these women find this possible in practice is, of course, another question, but the small proportion of households with dependent children in Leith and especially in Stockbridge does confirm that relatively few women remain in these neighbourhoods to raise children. It is certainly the case that studies of gentrification in London figure prominently among those reporting the presence of significant numbers of households with dependent children in neighbourhoods subject to gentrification (Butler 1997; Butler and Hamnett 1994; Lyons 1996; Munt 1987). The study presented here serves as a reminder that the geographical mobility of career-oriented women and men may vary widely between (as well as within) cities on a daily basis as well as through their working lives. Again, further research on this, which compares different cities, would be welcome.

[. . .]

REFERENCES FROM THE READING

Beauregard, R. (1986) 'The chaos and complexity of gentrification', in N. Smith and P. Williams (eds) *Gentrification of the City*, Allen and Unwin, Boston, MA, pp. 35–55.

Beechey, V. (1986) 'Women's employment in contemporary Britain', in V. Beechy and E. Whitelegg (eds) *Women in Britain Today*, Open University Press, Milton Keynes, pp. 77–131.

Bondi, L. (1991) 'Gender divisions and gentrification', *Transactions of the Institute of British Geographers: New Series* 16: 190–8.

Bondi, L. (1998) 'Gender, class and urban space', *Urban Geography* 19: 160–85.

Bondi, L. and Christie, H. (1997) 'Gender', in M. Pacione (ed.) *Britain's Cities*, Routledge, New York, pp. 300–16.

Bondi, L. and Domosh, M. (1998) 'On the contours of public space: a tale of three women', *Antipode* 30: 270–89.

Booth, C., Darke, J. and Yeandle, S. (1996) *Changing Places: Women's Lives in the City*, Paul Chapman, London.

Butler, T. (1997) 'Gentrification and the urban middle classes', in T. Butler and M. Savage (eds) *Social Change and the Middle Classes*, UCL Press, London, pp. 188–204.

Butler, T. and Hamnett, C. (1994) 'Gentrification, class, and gender: some comments on Warde's "Gentrification as consumption"', *Environment and Planning D: Society and Space* 12: 477–94.

Caulfield, J. (1994) *City Form and Everyday Life. Toronto's Gentrification and Critical Social Practice*, University of Toronto Press, Toronto.

Clark, E. (1995) 'The rent-gap re-examined', *Urban Studies* 32: 1489–503.

Crompton, R. (1995) 'Women's employment and the "middle class"', in T. Butler and M. Savage (eds) *Social Change and the Middle Classes*, UCL Press, London, pp. 58–75.

Early, F. and Mulholland, M. (1995) 'Women and mortgages', *Housing Finance* 25: 21–7.

Garber, J. and Turner, R. (eds) (1995) *Gender in Urban Research*, Sage, Thousand Oaks, CA.

Glass, R. (1964) *London: Aspects of Change*, MacGibbon & Kee, London.

Gordon, G. (1986) 'Capital and regional city', in G. Gordon (ed.) *Regional Cities in the UK, 1890–1980*, Harper & Row, New York, pp. 149–70.

Hamnett, C. (1991) 'The blind men and the elephant: the explanation of gentrification', *Transactions of the Institute of British Geographers: New Series* 16: 259–79.

Holcomb, B. (1986) 'Geography and urban women', *Urban Geography* 7: 448–56.

Jager, M. (1986) 'Class definition and the aesthetics of gentrification: Victoriana in Melbourne', in N. Smith and P. Williams (eds) *Gentrification of the City*, Allen & Unwin, London, pp. 78–91.

Kersley, S. (1974) 'Improvement grants: their contribution to the process of gentrification', B.Sc. dissertation, Department of Housing and Planning, Heriot-Watt University, Edinburgh.

King, A. (1990) *World Cities*, Routledge, London.

Knopp, L. (1987) 'Social theory, social movements and public policy: recent accomplishments of gay and lesbian movements in Minneapolis', *International Journal of Urban and Regional Research* 11: 243–61.

Knopp, L. (1990a) 'Exploiting the rent gap: the theoretical significance of using illegal appraisal schemes to encourage gentrification in New Orleans', *Urban Geography* 11: 48–64.

Knopp, L. (1990b) 'Some theoretical implications of gay involvement in an urban landmarket', *Political Geography Quarterly* 9: 337–52.

Lauria, M. and Knopp, L. (1985) 'Toward an analysis of the role of gay communities in the urban renaissance', *Urban Geography* 6: 152–69.

Lees, L. (1994) 'Gentrification in London and New York: an Atlantic gap?', *Housing Studies* 9: 199–219.

Ley, D. (1986) 'Alternative explanation for inner-city gentrification', *Annals of the Association of American Geographers* 76: 512–35.

Ley, D. (1988) 'Social upgrading in six Canadian inner cities', *Canadian Geographer* 32: 31–45.

Ley, D. (1994) 'Gentrification and the politics of the new middle class', *Environment and Planning D: Society and Space* 12: 53–74.

Ley, D. (1997) *The New Middle Class and the Remaking of Central Cities*, University Press, Oxford.

Ley, D. and Mills, C. (1993) 'Can there be a postmodernism of resistance in the urban landscape?', in P. Knox (ed.) *The Restless Urban Landscape*, Prentice-Hall, Englewood Cliffs, NJ, pp. 255–78.

Lyons, M. (1996) 'Employment, feminisation, and gentrification in London, 1981–93', *Environment and Planning A* 28: 341–56.

Marcuse, P. (1986) 'Abandonment, gentrification and displacement: the linkages in New York City', in N. Smith and P. Williams (eds) *Gentrification of the City*, Allen & Unwin, Boston, MA, pp. 153–77.

Markusen, A. (1981) 'City spatial structure, women's household work, and national policy', in C.R. Stimpson, E. Dixler, M.J. Nelso and K.B. Yatrakis (eds) *Women and the American City*, University of Chicago Press, Chicago, IL, pp. 20–41.

Massey, D. (1994) *Space, Place and Gender*, Polity Press, Cambridge.

McDowell, L. (1983) 'Towards an understanding of the gender division of urban space', *Environment and Planning D: Society and Space* 1: 59–72.

McDowell, L. (1991) 'Life without father and Ford: the new gender order of post-Fordism', *Transactions of the Institute of British Geographers: New Series* 16: 400–19.

McDowell, L. (1997) 'The new service class: housing, consumption, and lifestyle among London bankers in the 1990s', *Environment and Planning A* 29: 2061–78.

Mills, C. (1988) '"Life on the upslope": the post-modern landscape of gentrification', *Environment and Planning D: Society and Space* 6: 169–89.

Munt, I. (1987) 'Economic restructuring, culture, and gentrification: a case study in Battersea', *Environment and Planning A* 19: 1175–97.

Murgatroyd, L. (1984) 'Women, men and the social grading of occupations', *British Journal of Sociology* 35: 473–97.

Pain, R. (1991) 'Space, sexual violence and social control: integrating geographical and feminist analyses of women's fear of crime', *Progress in Human Geography* 15: 415–31.

Ravetz, A. (1996) 'Revaluation, "The Sphinx in the City"', *City* 112: 155–61.

Richardson, H.W., Vipond, J. and Furbey, R. (1975) *Housing and Urban Spatial Structure: A Case Study*, Saxon House, Farnborough, Hants.

Roberts, M. (1991) *Living in a Man-made World*, Routledge, London.

Rose, D. (1984) 'Rethinking gentrification: beyond the uneven development of Marxist urban theory', *Environment and Planning D: Society and Space* 2: 47–74.

Rose, D. (1989) 'A feminist perspective of employment restructuring and gentrification: the case of Montreal', in J. Wolch and M. Dear (eds) *The Power of Geography*, Unwin Hyman, Boston, MA, pp. 118–38.

Rothenberg, T. (1995) '"And she told two friends": lesbians creating urban social space', in D. Bell and G. Valentine (eds) *Mapping Desire*, Routledge, London, pp. 165–81.

Sassen, S. (1991) *The Global City: London, New York and Tokyo*, Princeton University Press, Princeton, NJ.

Sassen, S. (1994) *Cities in a World Economy*, Pine Forge Press, Thousand Oaks, CA.

Savage, M., Barlow, J., Dickens, P. and Fielding, A. (1992) *Property, Bureaucracy and Culture: Middle Class Formation in Contemporary Britain*, Routledge Chapman and Hall, New York.

Smith, N. (1979) 'Toward a theory of gentrification: a back to the city movement by capital not people', *Journal of the American Planners Association* 45: 538–48.

Smith, N. (1987) 'Of yuppies and housing: gentrification, social restructuring, and the urban dream', *Environment and Planning D: Society and Space* 5: 151–72.

Smith, N. (1992) 'New city, new frontier: the Lower East Side as Wild West', in M. Sorkin (ed.) *Variations on a Theme Park: The New American City and the End of Public Space*, Hill & Wang, New York, pp. 61–93.

Warde, A. (1991) 'Gentrification as consumption: issues of class and gender', *Environment and Planning D: Society and Space* 9: 223–32.

Williams, P. (1986) 'Class constitution through spatial reconstruction? A re-evaluation of gentrification in Australia, Britain and the United States', in N. Smith and P. Williams (eds) *Gentrification of the City*, Allen & Unwin, Boston, MA, pp. 56–77.

Wilson, E. (1991) *The Sphinx in the City*, Virago, London.

Winchester, H. (1990) 'Women and children last: the poverty and marginalisation of one-parent families', *Transactions of the Institute of British Geographers: New Series* 15: 70–86.

Winchester, H. and White, P. (1988) 'The location of marginalised groups in the inner city', *Environment and Planning D: Society and Space* 6: 37–54.

Zukin, S. (1982) *Loft Living: Culture and Capital in Urban Change*, Hutchinson, London.

Editors' references and suggestions for further reading

Beauregard, R. (1986) "The chaos and complexity of gentrification," in N. Smith and P. Williams (eds) *Gentrification in the City*, Boston, MA: Allen & Unwin.

Bondi, L. (1991) "Gender divisions and gentrification," *Transactions of the Institute of British Geographers: New Series* 16: 190–8.

Glass, R. (1964) *London: Aspects of Change*, London: MacGibbon & Kee.

Knopp, L. (1990) "Some theoretical implications of gay involvement in an urban landmarket," *Political Geography Quarterly* 9: 337–52.

Lauria, M. and Knopp, L. (1985) "Toward an analysis of the role of gay communities in the urban renaissance," *Urban Geography* 6: 152–69.

Lees, L. (2000) "A reappraisal of gentrification: towards a 'geography of gentrification,'" *Progress in Human Geography* 24, 3: 389–408.

Ley, D. (1997) *The New Middle Class and the Remaking of Central Cities*, Oxford: Oxford University Press.

Markusen, A. (1981) "City spatial structure, women's household work, and national urban policy," in C.R. Stimpson, E. Dixler, M.J. Nelson and K.B. Yatrakis (eds) *Women and the City*, Chicago, IL: University of Chicago Press.

Rose, D. (1984) "Rethinking gentrification: beyond the uneven development of Marxist urban theory," *Environment and Planning D: Society and Space* 2: 47–74.

Rose, D. (1989) "A feminist perspective of employment restructuring and gentrification: the case of Montreal," in J. Wolch and M. Dear (eds) *The Power of Geography*, Boston, MA: Unwin Hyman.

Rothenberg, T. (1995) "'And she told two friends': lesbians creating urban social space," in D. Bell and G. Valentine (eds) *Mapping Desires*, London: Routledge.

Smith, N. (1987) "Of yuppies and housing: gentrification, social restructuring, and the urban dream," *Environment and Planning D: Society and Space* 5: 151–72.

Smith, N. (1996) *The New Urban Frontier*, London: Routledge.

Warde, A. (1991) "Gentrification as consumption: issues of class and gender," *Environment and Planning D: Society and Space* 9: 223–32.

"(Re)Negotiating the 'Heterosexual Street': Lesbian Production of Space"

from Nancy Duncan (ed.), *Bodyspace: Destabilizing Geographies of Gender and Sexuality* (1996)

Gill Valentine

Editors' Introduction

Gill Valentine, Professor of Geography at the University of Sheffield, uses the term "the street" in place of the increasingly problematic phrase "public space," and challenges those who assume that the street is asexual. She argues that the space of the street is naturalized by dominant society as "authentically" heterosexual. Valentine's contribution to our understanding of how dominant forms of heterosexuality are inscribed on urban space places her work among those in the forefront of urban geographic research related to social identities, sexuality and exclusion. The following article addresses lesbian territoriality while drawing attention to society's discomfort with (and often violent reactions to) those who do not conform to dominant expectations of gender identity and sexual behavior. Furthermore, it contributes to a growing literature that challenges our conceptualization of public space.

Valentine explains that she has chosen the term "the street" to characterize "everyday publicly accessible places" because the common phrase "public space" is inconsistent with contemporary experience for three reasons. First, increasingly, so-called "public" spaces are privately owned, controlled and managed. Second, these spaces often exclude people based on age, class, race, sexuality, and so on. And third, the term "public" deflects attention from the fact "that many so-called 'private' relationships, such as sexualities, are actually part of 'public' space" (p. 155 in original). As convention determines the heterosexuality of sidewalks, shops and cafes, "sexual dissidents" are only permitted "to be gay in specific spaces and places," and gays and lesbians are "relegated to the margins of the 'ghetto' ... and preferably, the closeted or private space of the 'home'" (p. 147). Valentine explores this public coding of sexual space by examining the manner in which lesbians often silently produce their own relational space – "using heterosexual space against the grain" (p. 147) – and/or contest the heterosexism of public space quite openly.

Increasingly, the hegemony of heterosexual social relations in everyday environments draws geographers' research attention. David Bell and Gill Valentine outline the relatively recent history of geographers' concern for "putting sexualities on the map" in their introduction to *Mapping Desire: Geographies of Sexualities* (1995). Starting with an initial focus on "locating" homosexual regions and neighborhoods in the 1970s and continuing with research into the impact which gay communities have on the urban fabric through such processes of change as gentrification in the 1980s (Knopp 1990; Lauria and Knopp 1985; Lyod and Rowntree 1978), the issue of sexuality continued to garner serious attention alongside class, gender, race and ethnicity as a research theme in the 1990s. For example, Liz Bondi's study of gentrification, gender and class in Edinburgh neighborhoods (p. 251) provides an interesting contrast with her analysis of the same neighborhoods, since she opens up distinctions among sex, gender and sexuality as concepts in interpretation of the gentrification process (1998). Although we might question Thrift and Johnston's assertion (1993) that sexuality would be to the discipline of geography in the 1990s what class and gender were in the 1980s, important contributions *have* been made to the examination of sexuality in spatial terms that illuminate not only the "margins" of society but also the center. Yet, as suggested in Larry Knopp's theoretical framework for examining the relationship between sexualities and aspects of urbanization in contemporary Western societies (1995), there is much more to be explored.

Gill Valentine's current research continues to address issues of social identity, lifestyles and exclusion (2001) with a major project on processes of marginalization and resistance among lesbian, gay men and the Deaf. Other recent research projects address children's safety in public spaces (Valentine and Holloway 2000, 2003).

See Don Mitchell's "Public space and the city," introducing two special issues of *Urban Geography* (1996), for an overview of the scholarship that problematizes "publicly accessible places." For other examples of research related specifically to sexuality and territoriality, see Adler and Brenner (1992) and Rothenberg (1995). Both studies challenge the conclusion made by Manuel Castells (1983) that lesbians tend not to concentrate in particular areas because they lack the resources needed for residential ownership and thus tend to establish aspatial networks instead. Adler and Brenner's work focuses on an anonymous American city and a "neighborhood (which) has a quasi-underground character (that) . . . does not have its own public subculture and territory" (p. 31). Tamar Rothenberg's study of the Park Slope neighborhood in Brooklyn further suggests that although lesbian neighborhoods do exist, they form through a casual network that operates by "word-of-mouth."

THE HETEROSEXUAL STREET

In November 1991 a lesbian couple made the headlines in the British gay press when they were thrown out of a supermarket in Nottingham for kissing in the store (*Scene Out* 1991). What their experience demonstrates is that the street – and I mean this to include not only the pavement/sidewalk, but also the places, such as shops and cafés, which the street contains – is not an asexual space. Rather, it is commonly assumed to be 'naturally' or 'authentically' heterosexual (Bell *et al.* 1994). Whilst couples of the opposite sex are free to embrace over the supermarket trolley, the lesbian kiss caused panic because 'images of selves trouble as they cut into spaces where they don't "belong"' (Probyn 1992: 505).

Judith Butler has famously argued in her book *Gender Trouble: Feminism and the Subversion of Identity* that: 'gender is the repeated stylization of the body, a set of repeated acts within a highly rigid regulatory framework that congeal over time to produce the appearance of substance, of a natural sort of being' (1990: 33). In the same way

the heterosexing of space is a performative act naturalized through repetition and regulation (Bell *et al.* 1994; Bell and Valentine 1995). This repetition takes the form of many acts: from heterosexual couples kissing and holding hands as they make their way down the street, to advertisements and window displays which present images of contented 'nuclear' families; and from heterosexualized conversations that permeate queues at bus stops and banks, to the piped music articulating heterosexual desires that fill shops, bars and restaurants (Valentine 1993). These acts produce 'a host of assumptions embedded in the practices of public life about what constitutes proper behaviour' (Weeks 1992: 5) and which congeal over time to give the appearance of a 'proper' or 'normal' production of space.

Whilst heterosexuals have the freedom to perform their heterosexuality in the street – because the street is presumed to be a heterosexual space – sexual dissidents, as the Nottingham lesbians found out, are only allowed 'to be gay in specific spaces and places' (Bristow 1989: 74). Whilst the space of the centre – the street – is produced as heterosexual, the production of 'authentic' lesbian and gay space is relegated to the margins of the 'ghetto' and the back street bar and preferably, the closeted or private space of the 'home' (although even this is not always acceptable, as David Bell (1995) argues concerning the complex ways that the state regulates sexual citizenship). Thus the London Lesbian Offensive Group claim that 'Heterosexual privilege is about having, and assuming, the right to be more "normal" in both public and private. (Public not meaning outside your home, but in absolutely all dealings with the everyday world)' (London Lesbian Offensive Group 1984: 257).

But the production of heterosexual space is not only tied up with the performance of heterosexual desire but also with the performance of gender identities. Despite Gayle Rubin's claim that 'it is essential to separate gender and sexuality analytically to more accurately reflect their separate social existence' (1984: 308), it is hard to escape the role gender identities play in the production of heterosexual space. On the one hand, gender and sexuality are *not* the same thing but on the other hand they are certainly closely related, mutually constitutive perhaps, as Vron Ware (1992) has argued about gender and race. Certainly, Judith Butler makes a convincing case for the argument that binary gender identities only make sense within a heterosexual framework or matrix. She writes:

> The institution of a compulsory and naturalized heterosexuality requires and regulates gender as a binary relation in which the masculine term is differentiated from a feminine term, and this differentiation is accomplished through the practices of heterosexual desire. The act of differentiating the two oppositional moments of the binary results in a consolidation of each term, the respective internal coherence of sex, gender and desire. (Butler 1990: 22–3)

The specific 'feminine' 'shape' and 'look' that is perceived as heterosexually desirable and that is (re)presented and (re)produced through the biopower channels fashion, health, diet, fitness and so on (Evans 1993) may change over space and time but essentially being a woman is about performing a gender identity that is perceived to maintain the unity or coherence of gender, sex, desire by articulating a discrete asymmetrical opposition between the 'feminine' self and 'masculine' other (so that, to paraphrase Rosalind Coward (1984: 42) 'sexuality pervades our bodies almost *in spite of ourselves*').

Lesbians and gay men have historically been assumed to have 'twisted' gender identities, so that gay men are labelled as effeminate and lesbians as butch just as effeminate men and masculine women are perceived to be gay. This despite plenty of evidence to the contrary; for example, lipstick lesbians are the embodiment of femininity and 'many heterosexuals are not respectively masculine and feminine, or not in certain respects all the time' (Sinfield 1993: 22). Thus repetitive performances of hegemonic asymmetrical gender identities, like repetitive performances of heterosexualities, also produce a host of assumptions about what constitutes 'proper' behaviour/dress in everyday spaces which congeal over time to produce the appearance of 'proper', i.e. heterosexual, space. In this way 'the lesbian subject is always a doubled subject caught up in the doubling of being a woman and a lesbian' (Probyn 1995: 81). For example, Linda McDowell's (1995) study of merchant bankers demonstrates the way that working in the city involves the construction of an embodied gender performance, in which attributes of masculinity and femininity, including a more or less authentic presentation of sexual identity, are not only an integral part of doing business but also produce the bank as a heterosexual space.

But sex, gender and desire do not necessarily map coherently onto each other to maintain the logic of heterosexuality. As Alan Sinfield argues 'ideological categories fail to contain the confusions that they must release in the attempt to achieve control. That is why we observe heterosexuality plunged into inconsistency and anxiety; it is aggressive because it is insecure' (Sinfield 1993: 22). This insecurity often manifests itself in the form of regulatory regimes that constrain the possible performances of gender and sexual identities, in order to maintain the 'naturalness' of heterosexuality. These are regimes which take the form of multiple processes, of different origin and scattered location, regulating the most intimate and minute elements of the construction of space, time, desire and embodiment' (Foucault 1979: 138).

One such process, as the Nottingham lesbians trying to shop for groceries found out, is the simple removal of those who cut into and disrupt the 'normality' of heterosexual space by performing their desires in a way that produces (an)other space. In the UK, for example, a number of statutory and common laws can be and often are, used to criminalize public displays of same-sex desire on the streets. Although these laws do not explicitly single out lesbians and gays, they are often interpreted and applied in an extremely discriminatory way against sexual dissidents (Foley 1994). Public order laws, for example, have been invoked to threaten a couple walking hand in hand with prosecution, and have been used to obtain a conviction for 'insulting behaviour' against two sexual dissidents seen kissing at a bus stop in the early hours of the morning (Foley 1994).

Often, however, anxious straight citizens don't wait for the police or private security forces to step in and stabilize the heterosexuality of the street, rather they actively regulate it through aggression (Herek 1988). As Virginia Apuzzo, former executive director of the National Gay and Lesbian Task Force has pointed out, 'To be gay or lesbian in America is to live in the shadow of violence' (Comstock 1991: 54). For example, levels of victimization reported by lesbians in a survey in Philadelphia were twice as high as those recorded for women in the general urban population (Aurand *et al.* 1985; Comstock 1989, 1991). Dr Stewart Flemming describes the type of injuries sustained by sexual dissidents who are brought into his San Francisco medical centre for treatment as:

> vicious in scope and the intent is to kill and maim. . . . Weapons include knives, guns, brass knuckles, tire irons, baseball bats, broken bottles, metal chains, and metal pipes. Injuries include severe lacerations requiring extensive plastic surgery; head injuries, at times requiring surgery; puncture wounds of the chest, requiring insertion of chest tubes; removal of the spleen for traumatic rupture; multiple fractures of the extremities, jaws, ribs and facial bones; severe eye injuries, in two cases resulting in permanent loss of vision; as well as severe psychological trauma the level of which would be difficult to measure. (Comstock 1991: 46)

Whilst gay men are primarily attacked by other men, the perpetrators of violence against lesbians include not only men (alone and in groups with other men and/or women), but also women (alone and in groups). There are also gender differences in the geography of homophobic assaults – lesbians report more violent encounters in the 'heterosexual street' than gay men, who are more likely to be victimized in cruising areas, or gay-identified neighbourhoods (Berrill 1992; Comstock 1991; Valentine 1993).

In many cases these incidents are not 'provoked' because lesbians are articulating their sexuality, for example, by kissing or cuddling, but rather because they are not performing their gender identity in an 'appropriate' heterosexually identified way. Similarly, many women who identify as heterosexual but who do not perform their gender in a way that can be read as differentiated from the opposite sex in a heterosexually desirable way also encounter harassment in the form of anti-lesbian abuse (Bunch 1991), whilst very 'feminine' lesbians can be taken for heterosexual. In these cases you don't have to be 'one' just to *look like* 'one' to be seen as a threat to the heterosexuality of the street. This just goes to show how the identity of those present in a space, and thus the identity of the space being produced, can sometimes be constructed by the gaze of others present rather than the performers.

Not all the processes at work maintaining the heterosexuality of the street directly involve violence and aggression. There are many other more subtle omnipresent regulatory regimes constraining performative possibilities which pass unnoticed by those not subject to their pressures. Heterosexual looks of disapproval, whispers and stares are used to spread discomfort and make lesbians feel 'out of place' in everyday spaces. These in turn pressurize many women into policing their own desires and hence reinforce the appearance that 'normal' space is straight space (Valentine 1993). In this way, whilst sexual dissidents are constantly aware of the performative nature of identities and spaces, heterosexuals are often completely oblivious to this because they rarely have to be conscious of or examine their own performativity. They can take the street for granted as a 'commonsense' heterosexual space precisely because they take for granted their freedom to perform their own identities. In contrast many dykes exercise constant self-vigilance, policing their own dress, behaviour and desires to avoid confrontation. . . .

NOW YOU SEE US, NOW YOU DON'T: (RE)NEGOTIATIONS OF THE HETEROSEXUAL STREET

The identity of spaces and places, like the identities of individuals are 'frequently riven with internal tensions and conflicts' (Massey 1991: 276). Spaces are rarely being produced in a singular, uniform way as heterosexual (even though this is usually the hegemonic performance of space). Rather, as the quote above demonstrates, there are usually 'others' present who are producing their own relational spaces, or who are reading 'heterosexual' space against the grain – experiencing it differently.

'Lesbian desires and manners of being can restructure space' (Probyn 1995: 81) in many different ways. One of these is through dress. Subtle signifiers of lesbian identity, such as pinkie rings, labris earrings and rainbow ribbons; or lesbian dress codes such as butch-femme style, articulate subtly different spaces. Dress and body language (such as gestures, a glance, an independent or confident manner) also help dykes to 'spot' each other, to recognize a sense of sameness – she's like me – even though no words may be spoken. 'These features are not perceived or interpreted as indicators of lesbianism by straight women, because identifying lesbians is not relevant or necessary for them' (Painter 1981: 77). . . .

Often it may be not so much what is there but what is missing, the wedding ring for example, that marks out (an)other identity. Thus through these fleeting glances or cruising, dykes can produce 'gay(ze) space' (Walker 1995: 75). Or as Probyn has argued 'space is sexed through the relational movements of one lesbian body to another' (Probyn 1995: 81). Sometimes the object of the gay(ze) doesn't reciprocate 'the look', rather a lesbian reading is imposed upon her, more in hope than anticipation. But the voyeur can still momentarily imagine the space as her own, producing a small fissure in hegemonic heterosexual space.

Lesbian spaces are also mobilized through linguistic structures of meaning. By 'dropping pins', for example, by referring to lesbian cultural icons or appropriated films, books or music, dykes can establish contact with other gay women, subverting heterosexual spaces with their own meaning.

Lesbian social knowledge is used to interpret verbal and non-verbal behaviour [of others] such that the reality of one's lesbianism is not tied to external acts, but instead to the unquestioned and unquestionable propositions of the community itself. The woman, whether she perceives herself as a lesbian or not, is verbally constituted as a lesbian through the indexical use of members' talk. (Painter 1981: 72)

Hayes (1976) even goes so far as to suggest that there is a gay speech 'community' and gay lexicon – 'gayspeak'. Whilst language is undoubtedly important in constructing lesbian meanings, however, all sexual dissidents do not speak with one voice but are polyvocal.

[. . .]

Like language, music also has the power to produce space. The music of artists such as kd lang and Melissa Etheridge, becomes infused with their lesbian sexuality, even though their lyrics and the sounds they make may have no explicit lesbian content and the artists themselves may resist this reading of their work (Valentine 1995). A lang track playing in street space, like a shop or a bar, can therefore facilitate the materialization of a lesbian space by causing two women to catch each other's eye and establish fleeting contact or even long-term friendships (Bradby 1993; Valentine 1995). 'Do you like kd lang?' is, after all, Cherry Smyth argues [1992], *the* litmus test of a woman's sexuality.

Music also has the power to act as a vehicle that can transport the listener to another place and time. This use of music with fantasy allows women to use Walkmans to escape the street into imaginary lesbian spaces. Similarly, by playing music that has lesbian meaning publicly on tapedecks in shops or bars, or by using 'boom boxes' on the street, women can subvert straight space (Valentine 1995).

What these examples show is that:

Lesbian identity constructed in the temporal and linguistic mobilization of space, and as we move *through* space we imprint utopian and dystopian moments upon urban life. Our bodies are vital signs of this temporality and intersubjective location. In an instant, a freeze-frame, a lesbian is occupying space as it occupies her. Space teems with . . . 'possibilities, positions, intersections, passages, detours, u-turns, dead-ends [and] one-way streets', it is never still. (Munt 1995: 125)

However, these subtle possibilities and singular productions of relational sexed and gendered spaces often pass unnoticed by heterosexuals, either because overwhelming repetitive performances of heterosexuality swamp out these articulations of difference or because these subtle signifiers of 'otherness' are not read or understood by a heterosexual audience. Some lesbians therefore are actively using more 'in your face' tactics to challenge the stability of heterosexual productions of space.

'IN YOUR FACE': (RE)NEGOTIATIONS OF THE HETEROSEXUAL STREET

Don Mitchell has argued that public spaces are

very importantly, *spaces for representation*. That is, public space is a place within which a political movement can stake out the space that allows it to be *seen*. In public space, political organisations can represent themselves to a larger population. By claiming space in public, by creating public spaces, social groups themselves become public. (Mitchell 1995: 115)

Lesbian and gay pride marches, held annually in Europe and the UK, are one example of sexual dissidents being seen. By numerically appropriating the streets (and surrounding transport system, car parks, pubs, parks, shops, McDonalds and so on) and filling them with lesbian and gay meaning for one day, marchers pierce the complacency of heterosexual space. Sally Munt describes this spectacle as 'fifty thousand homosexuals parading through the city streets, of every type, presenting the Other of heterosexuality, from Gay Bankers to Gay Men's Chorus singing "It's Raining Men", a carnival image of space being permeated by its antithesis' (Munt 1995: 123). But as this quote implies, Pride marches also achieve more than just visibility, they also challenge the production of everyday spaces as heterosexual. The importance of space is something that has particularly been seized on by queer activists.

In the early 1990s, impatient with lesbian and gay assimilationist tactics and inspired by aggressive unapologetic political tactics of AIDS activists such as ACT UP (AIDS Coalition To Unleash Power), a group met in New York to discuss how to resist the growing number of assaults against lesbians and gay men in East Village. Under the slogan of 'queers bash back', Queer Nation was born (Smyth 1992). This shift in politics (that quickly spread to Europe) away from integration and equality issues towards an 'in your face' confrontational tactics has also brought with it a recognition, to paraphrase Tim Davis, that heterosexism is a spatially constituted discourse that can be interrupted and undermined (Davis 1995: 287). Rather than merely trespassing in heterosexual public space with the political intention of standing out or gaining a share of it, queer is also about confronting and contesting the very production of public space.

[. . .]

One group that has set out not only to be seen, but also to bend and queer space, are the Lesbian Avengers. The notion of a lesbian direct action group was the brainchild of Sarah Schulman, an American writer and activist, who developed Lesbian Avengers with five friends, with the intention of prioritizing lesbians' needs and making lesbians visible (in both heterosexual and queer spaces). They began by recruiting activists at the New York Lesbian and Gay Pride where they circulated flyers stating: 'We want revenge and we want it now!' and with the invitation to 'Imagine what your lives could be. Aren't you ready to make it happen?' (Hopkins 1994: 18). This was rapidly followed by a series of actions to challenge heterosexism in everyday places, in which the Avengers sometimes joined forces with other groups, such as Las Buenas Amigas and African Ancestral Lesbians for Societal Change, to target specific organizations or places for protests (Meono-Picado 1995).

These tactics quickly spilled over into the UK. In July 1994 the birth of a UK version of the Lesbian Avengers was announced in the magazine *Rouge*: 'A new dyke direct action group has arrived, bringing with it style from AIDS activism, theatricality from queer and, above all, determination and bravado from suffragists, Greenham women and a long tradition of anger and action' (Hopkins 1994: 18).

The first action of the British group was to target the memorial to Queen Victoria (who famously denied that lesbians existed) near tourist focal point, Buckingham Palace. The statue was circled by over fifty women, chanting and banging drums and carrying banners such as 'Lie back and think of Lesbians' and 'The Lesbians are not amused' (*Diva* 1994). In September they followed up this action by taking over another space colonized with heterosexual meaning – the shop window (Reekie 1993). Invading window displays in one of London's major shopping streets, Oxford Street, lesbians posed next to the mannequins with labels such as 'Designer Dyke', 'Lesbian Boy' and 'Funky Femme'.

These 'in your face', aggressive tactics both raise the visibility of lesbians but also rupture the taken-for-granted heterosexuality of these spaces by disrupting the repetitive performances of the mall and the shopping street as heterosexual places and (re)imagining/(re)producing them as queer sites. Implicit in these 'other' performances of the street is a recognition that control over the way that space is produced is fundamental to heterosexuals' ability to reproduce their hegemony. The insecurity and anxiety of straights in the face of this challenge is evident in the heteropatriarchal jeers of passers-by and attempts to police such actions.

Disruptive performances of dissident sexualities on the street are therefore about empowerment and being 'in control'. These actions are also not only transgressive, in that they trespass on territory that is taken for granted as heterosexual, but also transformative, in that they publicly articulate sexualities (both lesbian and, by exposing its taken-for-granted presence in everyday spaces, heterosexuality) that are assumed to be 'private' (and in the case of lesbians also invisible) and thus change the way we understand space by exposing its performative nature and the artifice of the public/private dichotomy.

CONCLUSION: A KISS IS NOT JUST A KISS

The experience of the two Nottingham lesbians which opened this chapter demonstrates that a kiss is not just a kiss when it is performed by a same-sex couple in an everyday location. Rather, repetitive performances of hegemonic asymmetrical gender identities and heterosexual desires congeal over time to produce the appearance that street spaces (such as shops, parks and bars) are normally or naturally heterosexual spaces. The heterosexuality of the street is, however, as the response of the store manager to the lesbian embrace demonstrates, an insecure appearance that has to be maintained by regulatory regimes (from harassment and violence, to jeers and stares). It is insecure because space teems with many other possibilities. As Elspeth Probyn has argued 'we need to think about how space presses upon bodies differently; to realize the singularities of space that are produced as bodies press against space' (Probyn 1995: 83).

[I have] tried to highlight two contrasting ways that lesbian bodies (re)produce space; first, by picking out some of the subtle ways women use lesbian manners and styles to fleetingly produce relational sexed and gendered spaces; second, by examining the more 'in your face' efforts of lesbian activists to visibly replace heterosexual space with other performances. These spaces are 'multiple and intersecting, provisional and shifting, and they require ever more intricate skills in cartography' (Rose 1993: 55) if we as geographers are to begin to try to map them.

REFERENCES FROM THE READING

Aurand, S., Addessa, R. and Bush, C. (1985) *Violence and Discrimination against Philadelphia Lesbian and Gay People: A Study by the Philadelphia Lesbian and Gay Task Force*, Philadelphia: Philadelphia Lesbian and Gay Task Force.

Bell, D. (1995) 'Perverse dynamics, sexual citizenship and the transformation of intimacy', in D. Bell and G. Valentine (eds) *Mapping Desire: Geographies of Sexualities*, London: Routledge.

Bell, D. and Valentine, G. (eds) (1995) *Mapping Desire: Geographies of Sexualities*, London: Routledge.

Bell, D., Binnie, J., Cream, J. and Valentine, G. (1994) 'All hyped up and no place to go', *Gender, Place and Culture* 1: 37–47.

Berrill, K. (1992) 'Anti-gay violence and victimisation in the United States: an overview', in G. Herek and K. Berrill (eds) *Hate Crimes: Confronting Violence against Lesbians and Gay Men*, London: Sage.

F
I
V
E

Bradby, B. (1993) 'Lesbians and popular music: does it matter who is singing?', in G. Griffin (ed.) *Outwrite: Lesbianism and Popular Culture*, London: Pluto Press.

Bristow, J. (1989) 'Being gay: politics, pleasure, identity', *New Formations* 9: 61–81.

Bunch, C. (1991) 'Not for lesbians only', in S. Gunew (ed.) *A Reader in Feminist Knowledge*, London: Routledge.

Butler, J. (1990) *Gender Trouble: Feminism and the Subversion of Identity*, London: Routledge.

Comstock, G.D. (1989) 'Victims of anti-gay/lesbian violence', *Journal of Interpersonal Violence* 4: 101–6.

Comstock, G.D. (1991) *Violence Against Lesbians and Gay Men*, New York: Colombia University Press.

Coward, R. (1984) *Female Desire: Women's Sexuality Today*, London: Paladin Books.

Davis, T. (1995) 'The diversity of Queer Politics and the redefinition of sexual identity and community in urban spaces', in D. Bell and G. Valentine (eds) *Mapping Desire: Geographies of Sexualities*, London: Routledge.

Diva (1994) 'Lesbian avengers campaign kicks off with series of zaps', 9 October.

Evans, D. (1993) *Sexual Citizenship: The Material Constructions of Sexualities*, London: Routledge.

Foley, C. (1994) *Sexuality and the State: Human Rights Violations against Lesbians, Gays, Bisexuals and Transgendered People*, London: National Council for Civil Liberties.

Foucault, M. (1979) *Discipline and Punish*, New York: Vintage Press.

Hayes, J. (1976) 'Gayspeak', *Quarterly Journal of Speech* 62: 256–66.

Herek, G. (1988) 'Heterosexuals' attitudes toward lesbians and gay men: correlates and gender differences', *Journal of Sex Research* 25, 4: 451–77.

Hopkins, R. (1994) 'Sisterly subversion', *Rouge* 18: 22–3.

London Lesbian Offensive Group (1984) 'Anti-lesbianism in the Women's Liberation Movement', in H. Kanter, S. Lefanu, S. Shah and C. Spedding (eds) *Sweeping Statements: Writings from the Women's Liberation Movement 1981–83*, London: The Women's Press.

Massey, D. (1991) 'The political place of locality studies', *Environment and Planning A* 23: 267–81.

McDowell, L. (1995) 'Bodywork: heterosexual gender performances in city workplaces', in D. Bell and G. Valentine (eds) *Mapping Desire: Geographies of Sexualities*, London: Routledge.

Meono-Picardo, P. (1995) 'Redefining the barricades: Latina lesbian politics and the appropriation of public space', paper presented at the New Horizons in Feminist Geography Conference, Kentucky, USA. Available from the author at the School of Geography, Clark University, USA.

Mitchell, D. (1995) 'The end of public space? People's Park, definitions of the public and democracy', *Annals of the Association of American Geographers* 85, 1: 108–33.

Munt, S. (1995) 'The lesbian flaneur', in D. Bell and G. Valentine (eds) *Mapping Desire: Geographies of Sexualities*, London: Routledge.

Painter, D.S. (1981) 'Recognition among lesbians in straight settings', in J.W. Chesbro (ed.) *Gayspeak*, New York: Pilgrim Press.

Probyn, E. (1992) 'Technologising the self: a future anterior for cultural studies', in L. Grossberg, C. Nelson and P. Treichler (eds) *Cultural Studies*, London: Routledge.

Probyn, E. (1995) 'Lesbians in space. Gender, sex and the structure of missing', *Gender, Place and Culture* 2, 1: 77–84.

Reekie, G. (1993) *Temptations: Sex, Selling and the Department Store*, St Leonards, NSW, Australia: Allen and Unwin.

Rose, G. (1993) *Feminism and Geography: The Limits of Geographical Knowledge*, Cambridge: Polity Press.

Rubin, G. (1984) 'Thinking sex: notes towards a radical theory of the politics of sexuality', in C. Vance (ed.) *Pleasure and Danger: Exploring Female Sexuality*, London: Routledge.

Scene Out (1991) 'It "Asda" be perv-ect . . .', December 33: 9.

Sinfield, A. (1993) 'Should there be lesbian and gay intellectuals?', in J. Bristow and A. Wilson (eds) *Activating Theory: Lesbian, Gay and Bisexual Politics*, London: Lawrence and Wishart.

Smyth, C. (1992) *Lesbians Talk Queer Notions*, London: Scarlet Press.

Valentine, G. (1993) '(Hetero)sexing space: lesbian perceptions and experience of everyday spaces', *Environment and Planning D: Society and Space* 11: 359–413.

Valentine, G. (1995) 'Creating transgressive space: the music of kd lang', *Transactions of the Institute of British Geography* 20: 474–85.

Walker, L. (1995) 'More than just skin-deep: fe(me)inity and the subversion of identity', *Gender, Place and Culture* 2: 71–6.

Ware, V. (1992) *Beyond the Pale: White Women, Racism and History*, London: Verso.

Weeks, J. (1992) 'Changing sexual and personal values in the age of AIDS', paper presented at the Forum on Sexuality Conference, Sexual Cultures in Europe, Amsterdam, June.

Editors' references and suggestions for further reading

Adler, S. and Brenner, J. (1992) "Gender and space: lesbians and gay men in the city," *International Journal of Urban and Regional Research* 16: 24–34.

Bell, D. and Valentine, G. (eds) (1995) *Mapping Desire: Geographies of Sexualities*, London: Routledge.

Bondi, L. (1998) "Sexing the city," in R. Fincher and J. Jacobs (eds) *Cities of Difference*, New York: Guilford Press.

Castells, M. (1983) *The City and the Grassroots*, Berkeley: University of California Press.

Duncan, N. (ed.) (1996) *Body Space: Destabilizing Geographies of Gender and Sexuality*, London: Routledge.

Knopp, L. (1990) "Some theoretical implications of gay involvement in an urban land market," *Political Geography Quarterly* 9: 337–52.

Knopp, L. (1995) "Sexuality and urban space: a framework for analysis," in D. Bell and G. Valentine (eds) *Mapping Desire: Geographies of Sexualities*, London: Routledge.

Lauria, M. and Knopp, L. (1985) "Toward an analysis of the role of gay communities in the urban renaissance," *Urban Geography* 6: 152–69.

Lyod, B. and Rowntree, L. (1978) "Radical feminists and gay men in San Francisco: social space in dispersed communities," in D. Lanegran and R. Palm (eds) *An Invitation to Geography*, New York: McGraw-Hill.

Mitchell, D. (1996) "Public space and the city," Special Issue: Public Space and the City I – *Urban Geography* 17: 127–31.

Rothenberg, T. (1995) "'And she told two friends': lesbians creating urban social space," in D. Bell and G. Valentine (eds) *Mapping Desire: Geographies of Sexualities*, London: Routledge.

Thrift, N. and Johnston, R.J. (1993) "The futures of *Environment and Planning A,*" *Environment and Planning A, Anniversary Issue*: 83–102.

Valentine, G. (2001) *Social Geographies: Society and Space*, Harlow: Longman.

Valentine, G. and Holloway, S. (eds) (2000) *Children's Geographies: Living, Playing, Learning*, London: Routledge.

Valentine, G. and Holloway, S. (2003) *Cyberkids: Children in the Information Age*, London: FalmerRoutledge.

PART SIX

Form and symbolism

Concord Pacific Place on the north shore of False Creek, Vancouver BC. (Courtesy of Elvin Wyly)

INTRODUCTION TO PART SIX

Architectural critic Michael Sorkin asserts provocatively that the phone and modem have produced the "ageographical city" and rendered the street irrelevant as time and space become obsolete. Particularly advanced in the United States, he argues, the ageographical postmodern city is "visible in clumps of skyscrapers rising from well-wired fields next to the Interstate; in huge shopping malls . . . surrounded by swarms of cars; in hermetically sealed atrium hotels cloned from coast to coast; in uniform 'historic' gentrifications and festive markets; in the disaggregated sprawl of endless new suburbs" (Sorkin 1993: xi). Ending his description of this built environment as a product of the electronic age and the global economy, Sorkin comments on the "antenna bristle of a hundred million rooftops" and "clouds of satellite dishes" and then makes the connection – the structure of this city is very similar to television. The comparison of the current city's form with TV programming rests on the "packaging" of discrete urban landscapes which lack the integration that might otherwise be provided by meaningful public space. Without the space for these necessary public connections, he warns, an integrated social experience is lost.

Sorkin is not alone in his concern for the twenty-first-century city with its landscape of consumption rather than production, its privatization of public space, and its simulation of meaning through "ersatz architecture." He colorfully conjures up an image of such a postmodern city in his description of new urban elements scattered across a vast metropolitan area grid of difference, not unlike Michael Dear's model of postmodern urban form (p. 138). Many would object, however, to Sorkin's notion of the ageographical city, arguing instead that we must work to understand the new geographies of the contemporary Western city in terms of both material form and symbolic content. In the decade since Sorkin collected various authors' views on and concerns regarding the postmodern city in *Variations on a Theme Park: The New American City and the End of Public Space* (1993), many geographers have contributed in theorizing public space, landscapes of consumption and the politics of meaning within contemporary social life as manifested in the built environment.

Up to this point, we have used "postmodern" without question but, as indicated in the volume's Introduction and elsewhere in our comments, this term is fraught with controversy as to whether it should be associated with a break from previous social and economic processes or whether it is best understood as an historic period. Certainly the period of postmodernity (post-1973) is associated with the increasing significance of style and decoration in urban discourse as well as the recognition of the central role that the culture industries play within the informational and service-based advanced capitalist economies. While "the postmodern" may not be specifically addressed, each of the following readings, as they assess various theoretical and methodological approaches for the examination of the built environment, share a concern for the complex relationship between the symbolic and material form of contemporary Western cities.

In the first article of this section, Paul Knox describes the Washington, DC metropolitan area, an area that resembles Sorkin's architectural critique of postmodern metropolitan form and landscape elements. Citing David Harvey's observations on the changing character of the landscape of capitalism, Knox indicates that the "restless(ness)" may be seen as the outcome of the stress and contradiction associated with capitalism's perpetual struggle to create a social and physical landscape "in its own image" and necessary to its own need. Ranked as a major world city, Washington, DC serves particularly well as a case study of urbanization influenced by global capitalism. Knox's overview of the metropolitan form of Washington, DC is primarily descriptive, but he also outlines the connections between economic and social change, the culture of consumption influencing the "new bourgeoisie," and the postmodern style that characterizes the new built environment. The postmodern style, he argues, is linked to globalizing capitalism and the legitimatization of the new social, professional, commercial and financial elites. Consumers acquire social status through cultural practices that involve the exercise of taste or judgment and thus tastes, as aesthetic judgments, function as markers of class and elite affiliation.

Knox's article emphasizes the extent to which cultural consistency permeates the material possessions of the new class fractions and consequently provide cultural reinforcement of urban spatial structures. Differences of taste and values among members of the elites create diverse market niches for residential real estate ranging from gentrified neighborhoods in the central city to master-planned communities in exurbia. Neil Smith (p. 128) and Liz Bondi (p. 251) offer further insight on the process of gentrification. At the other extreme, decentralization has occurred to such an extent that the term "rural gentrification" has been coined to address those members of the new economy who have "colonized" areas previously inaccessible to professionals and thus changed rural areas in Britain and the USA (see Cloke and Thrift 1987; Ghose 2004; Nelson 2000).

Knox lends fresh insight by examining the new metropolitan form as a response to the supply and demand for built structures with related consequences for: (1) the organization of developers and construction companies, and (2) the changing roles and orientation of architects and planners. To further define the political economy of the contemporary development process, he argues for a theoretical framework that foregrounds the reciprocal relationships among individuals, the built environment, and society at large (Knox 1987). Such a research agenda requires the study of consumers' perceptions and attitudes as well as the motivations and professional practices of developers, architects and planners.

Jon Goss' work on the "Magic of the Mall," the second entry in this section, shares with Knox's analysis a general interest in the culture of consumption and its influence on the built environment with a specific emphasis on the role of developers and designers who construct "ideal spaces" for shopping according to consumers' preferences, thereby maximizing profit for retailers and the developers themselves. Given the significance of retail activity in contemporary urban (re)development schemes, local governments also perceive the retail landscape as an integral part of the city's economic vitality. However, rather than focus on public/private partnerships and strategies for revitalization from the perspective of local government, which is a common theme in literature addressing the political economy of late twentieth-century capitalist cities (see Clarke and Gaile, p. 159), Goss considers how the symbolic value of commodities interweaves with the social activity in specialized places (in this case "the mall") to create profit for the retailer and produce new forms of identity for the consumer. Goss' framework for analysis attends to both the developer's manipulation of space, which materially influences the shopper's behavior, and the manipulation of symbols, which symbolically creates a positive marketplace theme. He stresses that an understanding of how these strategies work in the retail built environment can create an avenue of resistance.

During the 1990s there were two themes that inspired important research in human geography: (1) the geography of consumption (Jackson and Thrift 1995); and (2) the privatization of public space (Aitken et al. 2003). As noted in the volume Introduction, the shift from the modern era's economy of mass production for mass consumption to the postmodern era's fragmentation of production and creation of market niches explains in part the current fascination with the consumer. In an era defined arguably by the culture of consumption, empirical studies by geographers have made important contributions by grounding research in the social sciences and humanities on "shopping, place, and identity" (Miller et al. 1998). New and distinctive retail landscape elements, such as the megastructures, festival settings, and theme parks for shopping, have produced concern about the "malling of America" and death of the street. Increasingly, activities once associated with the outdoor public space have been brought inside into the privately controlled space of the mall or, even in many streets, public space is being increasingly controlled by private interests. In his work on festival settings, however, Jon Goss contemplates the ambiguity of public space noting the extent to which social control is exercised *and* new public behaviors are displayed to unsettle conventions (Goss 1996; see also Valentine, p. 263). Other researchers comment on the extent to which public surveillance is reshaping the experience of public space. See, for instance, Fyfe and Bannister's (p. 364) examination of the growing use of technology for surveillance and the public's response.

Related to this concern for the potential loss of public space is the apprehension that the design of contemporary Western cities does not support community. In "Styles of the Times," the third article in this section, David Ley examines two landscapes developed in Vancouver, BC during the 1980s, the first reflecting modernist architectural styles and planning theory, and the second postmodernist. He argues that each landscape reflects the cultural politics of neo-conservatism and liberalism respectively. The small-scale, place-specific design, and liberal goals associated with the postmodern development on the south shore of False Creek correspond to its municipal reform proponents' belief in the ability to plan for diversity, community and sense of place. In contrast, Ley argues, the north shore BC Place development offers spectacle for mass consumption, emphasizing a "vision for the future" rather than a commitment to cultural values rooted in place. Ley's wider argument considers the extent to which the cultural is political and emphasizes that cultural distinctions are rarely made on purely aesthetic grounds but frequently involve conflicting material interests.

Ley also raises for consideration the power of architecture to influence social life. The archetypical modernist LeCorbusier stated: "Architecture or revolution. – Revolution can be avoided." Confidence in the ability of design to produce change of such a magnitude receives little critical support today. Yet the advocates of New Urbanism, a planning movement which endorses the values Ley describes as postmodern, argue that correcting the failings of modernism with traditional urban design principles would enhance community and benefit the environment. Critical analyses by geographers, evaluating New Urbanist developments from a political economy perspective and/or examining its ideology, have concluded that New Urbanism represents a market niche (McCann 1994) and, in its celebration of traditional community, a rejection of diversity (Till 1994). A case study of the environmental benefits of a specific New Urbanist project also produced skepticism surrounding the claims (Zimmerman 2001). Knox (1991) and Harvey (1989) also dismiss postmodernist design, describing architectural postmodernism as a superficial, popular style associated with global capitalism. In response, David Ley cautions against turning postmodernism into a "discursive monologue" incapable of offering any form of resistance to the political economy of global capitalism (2003).

The first three readings in this section treat postmodernism as an epoch or as an architectural style associated with that epoch. In the fourth article, "Portland's Comprehensive Plan as Text," Judith Kenny borrows from the postmodern critique which challenges the belief that language provides direct access to reality. While avoiding the radical extreme of relativism connected with postmodernism, she states that language and discourses (frameworks through which social meanings are produced) are socially constructed and historically specific (Duncan 2003). Examining a planning document as a product of discourses within the discursive field of liberal planning, she identifies contradictory values and goals that coexist within this liberal ideology, and demonstrates how these conflicts can result in competing interpretations of a single text. In her Portland study, various groups challenge each other with their different readings of a single planning document, each group arguing that the meanings they found in the text were "the" legitimate ones. Controversy surrounding this particular land use case indicates the extent to which "reading" a document can become a political act with material consequences for a city's built environment.

Such study of texts and discourses has shaped contributions to the analysis of urban policy and planning since the early 1990s. For example, Kathryne Mitchell's "Visions of Vancouver" explores the texts of local planning reports and the debates surrounding them to critique the urban development ideology naturalized in local politics (Mitchell 1996). Robert Beauregard's *Voices of Decline* (1993) confronts the influence of anti-urban discourses on public policy since World War II. In a similar use of discourse analysis, David Wilson explores the influence of representations of Black-on-Black violence (2004) and efforts to stimulate growth in America's "Rust Belt" cities (Wilson and Wouters 2003). Glenda Laws summed up the significance of applying discourse analysis to urban plans and policy when she wrote: "The material form of urban policies and plans often is textual, in the more conventional sense of a written text. To understand the ideological power of these texts we must indeed read them closely" (Laws 1994).

The final reading in this section offers another form of text that influences popular images of the city. Larry Ford's "Sunshine and Shadow" addresses conventions of color and light employed in film to represent environmental qualities. His analysis suggests the extent to which films expose a popular audience to complex representations of the modern urban experience. The Chicago School sociologist Robert Park famously commented, "the city is a state of mind." That state of mind has significant impact when translated to the movie screen, drawing an audience into a narrative set against a backdrop of real and imagined places given the power of visual images. Film, it might be argued, naturalizes dominant representations of the city even more effectively than does literature. As such, cinematic representations offer significant insight into popular views of the urban and suggest the extent to which they might be challenged in alternative narratives drawn from different perspectives. Although progress in this area of geographic research and film study has been relatively limited, we need only return to Sorkin's image of the ageographical city. However dubious that phrase might be, consider the ubiquitous and unrelenting influence that they have when picturing the "antenna bristle of a hundred million rooftops" and the "clouds of satellite dishes" that blanket Western urban space.

References and suggestions for further reading

Aitken, S., Mitchell, D. and Staeheli, L. (2003) "Urban Geography," in G. Gaile and C. Willmott (eds) *Geography at the Dawn of the 21st Century*. Oxford: Oxford University Press.

Beauregard, R. (1993) *Voices of Decline: The Postwar Fate of U.S. Cities*. Oxford: Blackwell.

Cloke, P. and Thrift, N. (1987) "Intra-class conflict in rural areas," *Journal of Rural Studies* 3(4): 321–33.

Duncan, J. (2003) "Discourse," in R. Johnson *et al.* (eds) *Dictionary of Human Geography*. Oxford: Blackwell.

Ghose, R. (forthcoming, 2004) "Big sky or big sprawl: rural gentrification and the changing cultural landscape of Missoula, Montana," *Urban Geography*.

Goss, J. (1996) "Disquiet on the waterfront: reflections on nostalgia and utopia, the urban archetypes of festival marketplaces," *Urban Geography* 17: 221–47.

Harvey, D. (1989) *The Condition of Postmodernity*. Oxford: Blackwell.

Jackson, P. and Thrift, N. (1995) "Geographies of consumption," in D. Miller (ed.) *Acknowledging Consumption: A Review of New Studies*. London: Routledge.

Knox, P. (1987) "The social production of the built environment: architects, architecture and the post modern city," *Progress in Human Geography* 11: 354–77.

Knox, P. (ed.) (1993) *The Restless Urban Landscape*. Englewood Cliffs, NJ: Prentice Hall.

Laws, G. (1994) "Urban policy and planning as discursive practices," *Urban Geography* 15(6): 592–600.

Ley, D. (2003) "Forgetting postmodernism? Recuperating a social history of local knowledge," *Progress in Human Geography* 27: 537–60.

McCann, E. (1994) "Neotraditional development: the anatomy of a new urban form," *Urban Geography* 13: 210–33.

Miller, D., Jackson, P., Thrift, N., Holbrooke, B. and Rowlands, M. (eds) (1998) *Shopping, Place and Identity*. London: Routledge.

Mitchell, K. (1996) "Visions of Vancouver: ideology, demography, and the future of urban development," *Urban Geography* 17: 478–501.

Nelson, P. (2000) "Rural restructuring in the American west: land use, family and class discourses," *Journal of Rural Studies* 16: 1–13.

Sorkin, M. (ed.) (1993) *Variations on a Theme Park: The New American City and the End of Public Space*. New York: Hill and Wang.

Thrift, N. (2000) "Commodity" and "Geography of consumption," in R. Johnson *et al.* (eds) *Dictionary of human geography*. Oxford: Blackwell.

Till, K. (1994) "Neotraditional towns and urban villages: the cultural production of a geography of 'otherness'," *Environment and Planning D: Society and Space* 11: 709–32.

Wilson, D. (2004) *Inventing Black-On-Black Violence: Discourse, Space, Representation*. Syracuse: Syracuse University (in press).

Wilson, D. and Wouters, J. (2003) "Spatiality and growth discourse: the restructuring of America's Rust Belt cities," *Journal of Urban Affairs* 25(2): 123–39.

Zimmerman, J. (2001) "The 'nature' of urbanism on the New Urbanist frontier: sustainable development, or defense of the suburban dream," *Urban Geography* 22: 249–67.

S
I
X

"The Restless Urban Landscape: Economic and Sociocultural Change and the Transformation of Metropolitan Washington, DC"

from *Annals of the Association of American Geographers* (1991)

Paul Knox

Editors' Introduction

Paul Knox borrows David Harvey's description of the "restless formation and reformation of geographical landscapes" for his article's title, and then undertakes a detailed description of the radical changes that have influenced the transformation of Washington, DC since the mid-1970s. He introduces this descriptive work by focusing on the challenging task of understanding the relationship between class, culture and space in our contemporary urban landscape as the economy has transitioned from "Fordist" industrial capitalism to advanced capitalism. The edited pages that follow cover the first half of his original article, where Knox gives a well-structured explanation of the current trends in the production and consumption of the built environment. Those wishing to read additional details on new elements of the built environment and to benefit from a rich description of the particular case of Washington, DC must read the original full text.

Knox's case study of Washington, DC brings together topics addressed by several other entries in this Reader, as he bridges the literature on economic restructuring and on the culture of consumption. Just as Dear and Flusty (p. 138) argue that contemporary cities demonstrate a radical break with earlier models of urban development such as that outlined in Burgess' concentric zone model (p. 19), Knox explores the influence of global restructuring on the built environment. Dear and Flusty privilege the periphery in their examination of the "proto-postmodern" urban process, while Neil Smith (p. 128) focuses on gentrification and the creation of "bourgeois playgrounds" in the inner city. Paul Knox addresses the processes of both decentralization and gentrification in defining six elements that influence the "restless urban landscape" of advanced capitalism. These include: gentrification; historic preservation; postmodern architecture; mixed-use development (MXD) and multiple-use development (MUD); high-tech corridors; and master planned suburban and exurban development.

Knox's article also shares a similar focus with David Ley (p. 304) and Jon Goss (p. 293), who each use the theme of architecture as *Zeitgeist* to link the influence of the culture of consumption to new construction practices and the built environment. However, while Ley assesses liberal and conservative perspectives that influence the cultural politics of development, and Goss considers the particular strategies used by those developing retail landscapes in order to capitalize on the culture of shopping, Knox provides an overview of the development industry's response to the consumption patterns of the new bourgeoisie and petite bourgeoisie. Skillfully drawing together a sociological literature on the culture of consumption, Knox highlights such concepts as the "Diderot effect" and Bordieu's notion of habitus. Each suggests that consumers strive to maintain a cultural consistency in their ensemble of material possessions, thus developing new material cultures associated with new class fractions at the aggregate level. Niche marketing of the urban landscapes further reflects the creation of products in response to fragmentation of classes.

As he relates the development industry's use of postmodern architectural styles to contemporary cultures, Knox describes postmodernism as the "cultural clothing" of advanced capitalism. Despite the apparent potential for a postmodernism of resistance in the built environment (Ley and Mills 1993), he argues that the cultural expression of postmodernity in terms of architecture and design styles is primarily a postmodernism of reaction – a superficial response to modernism that merely replaces postmodern aesthetics with modern.

Paul Knox is currently the Dean of the School of Architecture and Urban Planning at Virginia Polytechnic Institute. His body of work sheds light on the changing role of design professionals during an era of corporate concentration in the development industry, and he has written specifically on the social production of the built environment, particularly the role of design professionals in this process (1987; Knox and Ozolins 2000). The particular article reproduced in this reader, however, also reflects Knox's record of scholarship dealing with the changing global economy (Knox and Taylor 1995) and his political economy approach. His examination of the restructuring of the development industry in the production of our contemporary urban landscapes offers an interesting contrast with others' work on the built environment – such as Edward Relph's (1987) and Larry Ford's (1994) discussion of architects, architecture and consumers.

■ ■ ■ ■ ■ ■

The purpose of this study is to probe the radical changes that have occurred within American cities since the mid-1970s as a result of the interplay of new economic, social, political and cultural forces associated with the transition from "Fordist" industrial capitalism to advanced capitalism. The American city has not only "lost its neat social patches of the 1950s and witnessed a resorting that is reminiscent of the period of ecological competition in the first decades of this century" (Kirby 1989: 16), but has also experienced a significant transformation of the built environment as part of what Harvey (1985: 150) has described as the "restless formation and reformation of geographical landscapes." For Harvey, this restlessness is to be seen principally as an outcome of the tensions and contradictions associated with capitalism's perpetual struggle to create a social and physical landscape "in its own image" and requisite to its own needs at a particular point in time. Yet, as Soja (1989) insistently points out, the spatial imprint of capitalism is not a smooth and automatic process in which the "needs" of capital are stamped, without resistance or constraint, onto the landscape. Rather, we have to contend with a "sociospatial

dialectic" (Soja 1980) in which the urban landscape is, in Meinig's words (1979), "mold and mirror" of our economy, culture and society.

The central premise is that the built environment is both the product of, and the mediator between, social relations. It is central to the sociospatial dialectic of human geographies that are simultaneously contingent and conditioning, outcome and medium, product and premise. Harvey (1987: 263) has argued persuasively that there is an "intimate connection" between aesthetic and cultural movements and the changing nature of the urban experience, while David Ley has shown that urban landscapes may be read interpretively as a text, "as a product which expresses a distinctive culture of ideas and practices, of often oppositional social groups and political relationships" (1985: 419).

. . . I seek to contribute to an understanding of the built environment as part of the sociospatial dialectic, as mirror and mold, first by exploring current trends in the production and consumption of the built environment, and, second, by attempting to read the new urban landscapes resulting from these trends.

ECONOMIC AND SOCIOCULTURAL CHANGE

While there is little doubt as to the onset of an epochal change in the dynamics of contemporary capitalism, there is little consensus on the labels and concepts – let alone the theories – most appropriate to their understanding. Concepts of advanced capitalism, flexible accumulation, post-Fordism, and postmodernism are somewhat tentative and exploratory, the subject of active discourse, and inevitably, therefore, of overstatement and confusion (deliberate or otherwise). Without wishing to be drawn too far into this discourse, I consider it important, nevertheless to stake out the principal dimensions of the trends that constitute the context for the production of the contemporary built environment. For present purposes, this context is best described as the confluence of two processes of change.

The first stems from the restructuring of the economic world that has followed the (partial) eclipse of the Fordist economic regime of mass production and mass consumption. It is characterized by (a) the pursuit of economies of scope rather than scale; (b) corporate concentration, centralization and reorganization; (c) an internationalization and decentralization of productive processes; (d) an increasing flexibility in the organization of production (in order to deal with and exploit increasingly segmented markets and the time–space compression introduced by new transport and communications technologies and the globalization of consumer markets); and (e) an intensification of the role of finance capital, a weakening of the role of organized labor, a shift in occupational structure, and an increasingly instrumental partnership between the public and private sectors of the economy (Moulaert and Swyngedouw 1989; Schoenberger 1988).

The second process involves a philosophical, cultural and attitudinal shift away from modernism towards postmodernism. Whereas modernism is paradigmatic, universalistic, purposive, hierarchical, synthetic, selective, and concerned with master codes and metanarratives, postmodernism is syntagmatic, playful, anarchical, antithetical, combinatorial, ideolectal, localistic and anti-narrative (Foster 1985; Hassan 1987; Hebdige 1989). The most tangible expressions of postmodernity are found in art, architecture, literature, music, design and commodity aesthetics. It should be noted, however, that the *cultural* expression of postmodernity does not always echo the *philosophical* underpinnings of postmodernity. Thus Foster (1985) makes the distinction between a postmodernism of resistance (which seeks to deconstruct modernism and undermine the status quo) and a postmodernism of reaction (which is a more superficial response to modernism, seeking merely to replace one style, fashion, or system of practices with another). Postmodernism in architecture and urban design is overwhelmingly a postmodernism of reaction, and it owes little to the intellectual discourse and cultural roots of a postmodernism of resistance.

Many writers have nevertheless suggested that postmodernism can be seen in more general terms as a product of the *Zeitgeist*, or spirit of the age. There is widespread agreement that the emergence of postmodern architecture, culture and philosophy in Western society has proceeded in tandem with the emergence of globalized, more flexible forms of capitalist enterprise. More specifically, Harvey (1987) argues that, while this transition and the critique of modernism had been under way for some time, it was not until the international economic crisis of 1973 that the relationship between art and society was sufficiently shaken to allow postmodernism to become both accepted and institutionalized as the cultural clothing of advanced capitalism. It should be emphasized that this does not mean that the major part of contemporary culture is to be seen as postmodern. Rather, postmodernism is an increasingly important part of contemporary urban culture, and it articulates with certain trends in economic organization in generating new urban landscapes. In this context, these new landscapes can be seen as the product of changes associated with, on the one hand, patterns of demand and consumption and, on the other, patterns of supply and production.

CONSUMPTION: COMMODITY AESTHETICS AND THE NEW BOURGEOISIE

One of the most important preconditions for the creation of a sufficient audience and market for postmodern objects and settings has been the emergence of new class fractions under advanced capitalism. Bourdieu (1984) equates this new audience with the "new bourgeoisie" (members of the professions, public administrators, scientists, and private-sector executives – especially those involved in non-material products: financial analysts, economists, management consultants, personnel experts, designers, marketing experts, purchasers, etc.) and the "new petite bourgeoisie" (junior commercial executives, engineers, medical and social service personnel and those directly involved in cultural production and organization – authors, editors, radio and TV producers and presenters, magazine journalists, etc.). In the U.S., such occupations expanded dramatically in the 1970s. Whereas total employment in the U.S. grew by 27.5 per cent between 1970 and 1980, the number of accountants and auditors grew by 56 per cent, lawyers by 83 per cent, architects 98 per cent, and social scientists and urban planners 97 per cent. Among the new petite bourgeoisie, counselors grew in number by 69 per cent, health technologists and technicians by 81 per cent, therapists by 149 per cent, designers by 44 per cent, and public relations specialists by 47 per cent.

Yet, as Neil Smith (1987) points out, the transformation of occupational structures does not in itself establish the emergence of new class fractions. It is spending power and patterns of consumption that, in addition to occupational status, are the defining characteristics of the new bourgeoisie/petite bourgeoisie. Several factors seem to be at work here. The relative affluence of the new bourgeoisie/petite bourgeoisie coincided with the maturing of a cohort – the baby boom generation, raised under the "forced democratization and egalitarianism of taste" (Harvey 1989: 80) of a modernist aesthetic. Their repressed demand for variety and symbolic ornamentation, suggests Harvey, was an important precondition for the blossoming of postmodern culture. Furthermore, this same generation, as it came of age in the late 1960s, spawned a variety of anti-modernist, counter-cultural movements that led to the exploration of self-realization through "new" politics and iconoclastic habits in music, clothing and lifestyles (Ehrenreich 1989).

As Sack (1988) points out, people's actions of consumption (whether of "heritage" settings, new homes, automobiles or magazines) are among the most powerful and pervasive processes within the

sociospatial dialectic. Consumption, he argues, molds people's consciousness of place, helps them to construct real places, connects the realms of nature, social relations and meaning, and reveals how geographical settings are constitutive of contemporary tensions and paradoxes. Consumption is also epigrammatic, having as a major role the transfer of meaning from the object to the consumer. It is, moreover, not only the material objects themselves that are epigrammatic, but also the settings in which they are purchased. "Implicated in [a] purchase, be it of gourmet ice cream, a nouvelle cuisine meal, or a dance lesson, is the status of being at that shop in that neighborhood and buying that particular brand" (Beauregard 1986: 44). Merely to be in attendance in upmarket retail settings – "the court of commodities" (Shields 1989) – is an important ritual in establishing individual and group identity.

. . . Thus we move from the "vulgar functionalism" of modernist mass consumption to the "aestheticized commodity" of postmodernity. . . . Superimposed on this general shift is the so-called "Diderot effect," a force that encourages individuals to maintain a cultural consistency in their *ensemble* of material possessions. "Before yuppies," as McCracken observes (1988: 121), "there was no compelling connection between the Rolex and the BMW." The point is that once insinuated into a person's possessions through one or two key "departure" purchases, aestheticized commodities will tend to exert a kind of gravitational pull that, cumulatively, rewrites the inventory of their possessions. At the aggregate level, the Diderot effect will thus tend to revise the material culture of entire subgroups or class fractions.

Also relevant here is Bourdieu's notion of habitus – the values, cognitive structures, and orienting practices of individual class fractions. . . . The habitus of a particular subgroup is a collective perceptual and evaluative schemata that derives from its members' everyday experience and operates at a subconscious level through commonplace daily practices, dress codes, use of language, comportment, and patterns of consumption. The result is a distinctive pattern in which "each dimension of lifestyle symbolises with the others" (Bourdieu 1984: 173). According to Bourdieu, each class fraction will seek to sustain and extend its habitus (and each *new* class fraction – the new bourgeoisie and new petite bourgeoisie, for example – will seek to *establish* its habitus) through the appropriation of "symbolic capital." The latter is defined as "luxury goods attesting to the taste and distinction of the owner." There are clear echoes here of Veblen's theory of the leisure class and Hirsch's (1977) ideas on "positional goods.". . .

[. . .]

This brings us back to the new bourgeoisie and new petite bourgeoisie, for it is these groups that seem, on the one hand, to have been most successfully targeted by the development industry and, on the other, to have been strikingly successful in strengthening and extending their habitus, largely through their sponsorship of new (postmodern) cultural goods and residential settings. Gentrification, loft living, and the appropriation of areas of historic preservation have already been shown to be pivotal to the spatial practices of the "new" middle classes (Goodey 1985; Jager 1986; Mills 1988; Smith 1987; Zukin 1988), and I argue that some new suburban residential settings are also the product of their role within the sociospatial dialectic. . . .

PRODUCTION: THE DEVELOPMENT INDUSTRY

As Ventre (1988) has noted, the development industry has always been oriented towards flexible production systems. Although Fordist mass production techniques certainly found expression in suburban Levittowns and in modular, factory-built inner-urban renewal projects, architects, builders, construction engineers and developers have for a long time been geared to small-batch production and a high level of subcontracting, delivering customized products based on specific user requirements. Nevertheless, new initiatives and a new balance of forces are apparent within the development industry, largely as a result of broader economic shifts and exigencies. Like many other industries, the development industry has consolidated in the wake of the system-shock of the 1970s and is now more highly integrated vertically, with finance capital having assumed greater control of both design and construction (Healey and Nabarro 1990; King 1988). Between 1985 and 1987 the top 25 per cent of diversified U.S. development companies (those engaged in developing offices, shopping centers, and hotels) increased their market share from 64 per cent to 73 per cent (Survey of Diversified Developers 1988). The housebuilding industry, traditionally decentralized (with more than 110,000 firms nationally), has also exhibited significant consolidation: more than 15 per cent of the residential market is now accounted for by 0.09 per cent of the firms (Schwanke 1989). Similarly, the real estate industry has come to be dominated by national companies like Coldwell Banker. At the same time, an increased involvement in real estate and construction on the part of conglomerate corporations and financial institutions (as investment switched into secondary circuit of the built environment) has made investment capital more readily available, particularly for large-scale, highly visible projects (King 1989; Schwanke 1987). This, in turn, has raised the stakes: a typical mixed-use development (MXD) or multi-use development (MUD) of about 1 million square feet now requires an investment of about $200 million, spread over the best part of ten years, before any return can be made. To fully utilize the skilled labor force that has to be assembled in order to follow through on such projects, many of the larger development companies have moved into asset and property management, financing and brokerage, thus further accelerating corporate concentration (Schwanke 1989).

Another result of the system-shock of the 1970s has been that firms throughout the development industry have been forced to develop still more flexible strategies in order to exploit more effectively the tastes and preferences of distinctive subgroups and, in particular, the affluent employees of the more buoyant advanced services sector. . . .

In the residential sector, both larger firms and local "niche" developers have repositioned themselves to build for the growing move-up market, which accounted for two out of every three homes built in 1988 and in which the most marketable packages include community amenities, expensive-looking materials, dramatic master bedroom/bathroom suites, and integrated but distinctive design based on traditional and vernacular styling (Suchman 1989). Large, privately planned communities have become increasingly popular with developers during the 1980s because, as the Urban Land Institute's (1989) review of real estate markets puts it: "They permit developers more flexibility in design and product type . . . [and] enable developers to respond quickly to changing market demand" (Suchman 1989: 38). In the commercial sector, specialization and product differentiation have been reflected in the emergence of so-called "power centers": community shopping centers located near regional shopping malls and dominated by specialty retailers offering brand-name products at discount prices. Another significant new product line consists of specialty settings such as medical malls, designed to provide busy, affluent consumers with one-stop shopping (physicians, counselors, therapists, medical laboratories, outpatient facilities, fitness centers, drug stores, health food stores and cafés) in suitably upscale settings (Martin and Hughes 1989). In hotel construction, "The industry answer to overbuilt markets is continued segmentation" (Beyard 1989: 52). Among the new products to emerge from this segmentation are all-suite/limited service hotels, luxury/full service hotels, limited service/economy hotels, extended-stay hotels (featuring kitchenettes and washer-dryers), and executive conference resorts. Mixed-use and multi-use developments, themselves a product of the industry's restructuring, also reflect the increasing importance of product differentiation and management integration. The latest generation of MXDs is smaller (just over 1 million square feet on average, compared to just under 2 million square feet in the 1970s) and more specialized, with a greater use of postmodern architecture and of renovation of older structures, an infusion of residential space, and a more widespread provision of CARE (culture, amusement, recreation and entertainment) packages (Schwanke 1987). Similarly, developers of business and industrial parks (now called planned corporate environments in the language of the trade) have increasingly sought "to enhance the attractiveness and value of projects by including daycare services and facilities, fitness programs and centers, jogging trails, restaurants, and convenience retail facilities . . . marble and brass architectural trim in public lobbies, intensive interior and exterior landscaping" (Beyard 1989: 45, 47).

In the course of all this repositioning and restructuring, a state of flux has emerged in the professional domains attached to the development industry. Competition between architects, builders, engineers, construction managers, interior designers and project consultants has become acute and traditional divisions of labor have become blurred. Architects, in general, seem to have been the major losers in this process. With few exceptions, they are no longer able to assume the traditional role of master builder, with patrons supporting their creative acts while requiring little direct involvement in the design process. Now, patronage has turned to clientage: developers impose tight design specifications on architects, who must increasingly hustle for work, and when they get it, must cede more and more of the process to other professionals (Gutman 1987, 1988). As a result, professional discourse within architecture has shifted a long way from the visionary social ideals inspired by early modernists to an overwhelming emphasis on style and visual effect, with no immediate concern for the social dimensions of urban design (Amendola 1989; King 1988; Montgomery 1989). . . .

The role of the planning profession has also changed significantly. Having lost a good deal of its moral authority as a result of its inability to deliver the utopian goods, it has become increasingly cooperative and instrumental in its relationships with the development industry. . . . As such, it is increasingly geared to the needs and wants of specific producers and consumers rather than to overarching notions of rationality or public good. The new emphasis is on pluralistic, organic approaches aimed at producing a collage of highly differentiated spaces and settings. As Boyer observes,

> fragmented elements of the city whole are planned or redeveloped as autonomous elements, with little relationship to the whole and with direct concern only for adjacent elements. In other words fragments of the city are regulated by special district or contextual zoning, Historic Preservation Controls, TDRs [Transfer of Development Rights] off of historic structures, and even the dictates of an EIS [Environmental Impact Statement], all of which pay attention to the *artful fragment* but say nothing about the city as a whole. (1987: 6, emphasis added)

One of the most important changes in planning practice from the point of view of the development industry has been the relaxation of single-purpose zoning in downtown areas. Encouraged by post-modern urban design theorist/practitioners such as Krier (1985), who see single-purpose zoning as wasteful, monotonous and anti-ecological, mixed-use zoning has come to be seen as attractive by planning agencies and planning committees because of the potential role of mixed-use and multi-use developments in enhancing a city's tax base, initiating urban revitalization, and increasing ridership on public transit systems (Lassar 1989). Increasingly, mixed-use downtown zoning is being combined with incentive systems in which developers are awarded additional height or density allowances in exchange for specified building features (Seattle, for example, offers density bonuses for ornate building cornices and rooflines) or facilities such as day care centers, residential space, or space for services that might help restore variety and vitality to downtown districts. A similar mix of postmodern orientations and fiscal mercantilism has fostered radical changes in suburban zoning practice. The key mechanism here has been cluster zoning and planned unit development (PUD) zoning, whereby regulations are applied to an entire parcel of land (in contrast to traditional lot-by-lot regulation). With PUD zoning, densities can be aggregated and calculated on a project basis, allowing the clustering of buildings to create open spaces or preserve attractive site features, and facilitating a mixture of residential and nonresidential elements and a mixture of housing types. For developers, PUD zoning offers economies of scale plus scope for product diversity and flexibility within a predictable regulatory environment. The attraction for planners and planning committees is that PUD zoning offers the prospect of high-quality development with a minimum of public expenditure on services and amenities.

NEW URBAN LANDSCAPES: THE EXAMPLE OF THE WASHINGTON, DC METROPOLITAN AREA

Given the context provided by the conjunction of these trends in the production and consumption of the built environment, one can attempt to read the inchoate landscapes of the postmodern metropolis. The example discussed is the Washington, DC

metropolitan area which, although by no means typical of large American metropolitan areas (because of the relative absence of manufacturing activity and the relative importance of federal civilian employment in its economic structure), does exhibit the full imprint of advanced capitalist urban development, with a social geography that has become an exemplar of the congested, fragmented and polarized urbanization of the postmodern metropolis (Fuller 1989; Gale 1987; Knox 1987). As in Los Angeles (Soja 1989), the internationalization of the *economy* and the imprint of advanced capitalism have prompted a striking recentralization of economic activity that has been expressed in a downtown renaissance and the emergence of "edge cities" (Garreau 1988) within the metropolitan periphery. It contains a population of three million which, in overall terms, is distinctively affluent and well educated. Nearly one-third of all adults are college graduates, compared with around one-fifth in metropolitan New York, Houston and Los Angeles. In 1985, when the national average household income was $28,400, the average for the Washington metropolitan area was $42,000. In suburban counties, affluence is particularly marked, with commensurately high levels of consumer spending power. In terms of projected growth of median household incomes, Fairfax County, Virginia, and Montgomery County, Maryland, rank highest in the conterminous U.S. Marketing studies have shown suburban Washington to contain a disproportionately high incidence (four or five times the national average) of "blue blood" and "furs-and-station wagon" neighborhoods, where, among other things, ownership of Jaguar and BMW 5-series automobiles, subscriptions to *Architectural Digest* and *Town and Country* magazines, and sales of imported wines (by the case) are important marketing yardsticks. These studies also show that the District, despite its relatively low overall levels of spending power contains a disproportionately high incidence of "money and brains" and "bohemian mix" neighborhoods, both distinctive in marketing terms for their indulgence in relatively upscale consumption (Weiss 1988). While federal jobs account for some of this distinctiveness, much of it is attributable to the expansion of middle- and upper-middle-income jobs in the private advanced services sector. Business associations, educational services, legal services, investment trusts, mortgage banking, professional associations, services to buildings and real estate leasing have all expanded rapidly (Howland 1989); the area is now home to a growing number of Fortune 500 companies.... In 1990, the Washington metropolitan area was listed by *Inc.* magazine as the second most entrepreneurial metropolitan region in the U.S. (after Las Vegas, Nevada, and ahead of Orlando and Tallahassee, Florida, and San Jose, California), as measured by business startups, job generation, and young companies with high growth rates (Case 1990). The net result has been a pronounced growth in both the new bourgeoisie and the new petite bourgeoisie.

The expansion of the advanced services sector has not only modified the composition of class fractions in Washington but has also fueled a major development boom. In downtown Washington, this has been geared to office space for the "Four A's": accountants, analysts, associations and attorneys. Of the almost 300 million square feet of office space in the metropolitan area, almost 20 per cent is located in the District's downtown. In 1988 more than 5 million square feet of office space were added to the CBD, and it is estimated that space under construction or planned will add another 18 million square feet by 1992 (Boselovic 1989). In addition, for every square foot of office space built in the District in recent years, there have been two square feet of new MXD space. Retailing space has also increased, including the only full-sized freestanding department store ... to be built in a U.S. downtown since 1945. It should be noted that this boom has in turn been supported by public agencies. The District of Columbia was one of the first central cities to relax single-purpose zoning when it established mixed-use zoning for the West End in 1974, along with incentives and bonuses for residential, retail and service space, pedestrian and cycling areas, and the preservation or enhancement of historic structures. The Pennsylvania Avenue Development Corporation has assembled land (more than 110 acres) and established a planning framework for a number of large-scale development projects, including prominent MXDs.... The District has also enacted an ordinance requiring developers in some downtown areas to devote at least 20 per cent of the space in office projects to retailing and to conform to design standards aimed at promoting variety and vitality.

In the suburbs, office and MXD projects each account for about 40 per cent of recent construction activity, with retailing accounting for the remaining 20 per cent (Boselovic 1989). Perhaps the most striking single outcome of this activity is the edge city that has emerged around Tyson's Corner. The two malls at its center ... account for $1 billion in annual sales; surrounded by more than 20 million square feet of office space, they now constitute a commercial core larger than downtown Miami (Downey 1989). Similar but smaller edge cities are emerging at Bethesda, Rosslyn/Ballston, Reston, Crystal City/Pentagon City, Fairfax/Fair Oaks, Shady Grove, Rockville, Merrifield, Dulles, Gaithersburg/Germantown, Lanham/Landover, and Silver Spring (Figure 1). The spatial distribution of these edge cities reflects the broad socioeconomic structure of the metropolitan area; all but one are in three suburban counties, Arlington, Fairfax and Montgomery, with the highest incomes.

... The "influence industry," with its emphasis on the iconography of prestige and image, has been particularly prominent in these changes. As a result, the transformation of Washington from a federal town to a major control point for advanced services has been accompanied by some striking changes in its built environment. In addition to the suffusion of postmodern architectural styles throughout much of the metropolitan area, there are several kinds of new settings. These include buildings and districts subject to historic preservation, gentrified neighborhoods, master-planned communities, mixed-use and multi-use developments, festival settings, and high-tech corridors. These categories are not mutually exclusive, and they are not exhaustive. They do, however, provide cogent illustrations of the new urban landscapes produced as a result of the confluence of recent economic and sociocultural change.

[...]

CONCLUSIONS: NEW URBAN LANDSCAPES AND A NEW URBAN GEOGRAPHY

I have argued that a series of distinctive new urban landscapes is emerging from a sociospatial dialectic dominated by the effects of the reconfiguration of

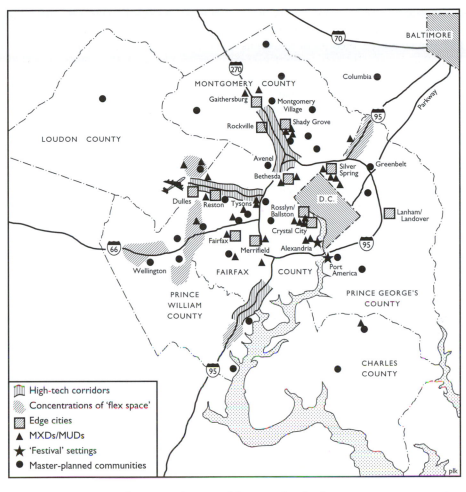

Figure 1 The Washington metropolitan area: elements of the new urban landscape.

economic and cultural life. I have also attempted to show that the emergence of these new urban landscapes provides a rich source of conceptual and empirical issues that link urban geography with economic, social and cultural geography. The particular approach adopted here clearly lends itself to a theoretical framework rooted in historical materialism, in which the built environment is seen as part of the superstructure that is not only produced by but also helps to sustain the dominant relations of production. Postmodern design, along with postmodern culture and philosophy, thus becomes the latest, incipient, dialectical expression of the transformation from rationalist/Modernist/Fordist capitalism to an emergent, globalizing, advanced capitalism. According to this perspective, the new settings described here can be interpreted

as sustaining the transformation of capitalism because of their capacity to legitimize new social, professional, commercial and financial elites and their capacity to enhance the circulation and accumulation of capital within an emergent global political economy. More specifically, they can be held to reflect the ideology and values of elements of the new social formations that have emerged with this transformation and be implicated in some of the consequent tensions and conflicts between them. Similarly, they can be interpreted as reflecting and reinforcing the broader fragmentation and polarization of urban space. . . .

The examination of new urban landscapes in the Washington metropolitan area lends support to the notion of a new urban geography, with a radically different form and ecology from that of the classic American city depicted by factorial ecologies and explained, with varying degrees of success, by bid-rent theory, theories of residential mobility, Weberian theory, and neomarxist theories (see, for example, Cadwallader 1988; Cooke 1990; Davis 1990). The spatial patterns associated with the landscape elements described in this essay do not fit comfortably within the sectors, zones or mosaic patterns that have been the focus of academic debate surrounding the classic American city. Fragmentation, multinodality, fluidity, plurality, and diffusion are more in evidence than homogeneity, nodality and hierarchy (Gottdiener and Gephart 1991). Vance recognized these qualities some time ago in his survey of Bay Area urbanization (1972: 205), suggesting that "We are witnessing the birth of a new complex urbanism in which the specialized social districts have begun to replace a synoptic pattern (of land rent) in shaping the morphology of settlement." Peirce Lewis coined the term "galactic metropolis" to capture a new urban geography in which "The residential subdivision, the shopping centers, the industrial parks seem to float in space; seen together, they resemble a galaxy of stars and planets, held together by mutual gravitational attraction, but with large empty areas between clusters" (1983: 35). This new urban geography, as some geographers have pointed out (Fonseca 1976; Wood 1988), is characterized by landscapes that are semiurban: landscapes of mixed densities and unexpected juxtapositions of forms and functions. All this should not imply, of course, that the "old" urban geography has been completely overwritten or that the restlessness of the built environment is somehow linked to an attenuation of polarization and segregation within the city. Rather, we have to contend with new economic and sociocultural trends, a changing framework for metropolitan growth, changing urban landscapes and, probably, changing patterns of economic, social, and cultural differentiation. Most of this, clearly, has yet to be mapped out and set within appropriate theoretical frameworks. It remains to be seen whether the new urban landscapes described in this essay will become paradigmatic of the new urban geography: the symbolic landscapes of the postmodern, *fin-de-siècle* American metropolis, the successors to Meinig's (1979) Main Street and Hummon's (1990) Commonplaces.

REFERENCES FROM THE READING

Amendola, G. (1989) "Postmodern architects' people," in R. Ellis and D. Cuff (eds) *Architects' People*, Oxford University Press, New York, pp. 239–59.

Beauregard, R. (1986) "The chaos and complexity of gentrification," in N. Smith and P. Williams (eds) *Gentrification of the City*, Allen & Unwin, Boston, MA, pp. 35–55.

Beyard, M. (1989) "Commercial and industrial development," in D. Schwanke (ed.) *Development Trends*, pp. 42–56.

Boselovic, I. (1989) "Washington," *National Real Estate Investor* (June): 117–30.

Bourdieu, P. (1984) *Distinction: A Social Critique of the Judgment of Taste*, Routledge & Kegan Paul, London.

Boyer, C. (1987) "The return of the aesthetic to city planning: Future theory as a departure from the Past," Paper presented at Rutgers University Center for Urban Policy Research Conference on Planning Theory in the 1990s, Washington, DC.

Cadwallader, M.T. (1988) "Urban geography and social theory," *Urban Geography* 9: 227–51.

Case, J. (1990) "The most entrepreneurial cities in America," *Inc.* (March): 41–8.

Cooke, P. (1990) "Modern urban theory in question," *Transactions, Institute of British Geographers* 15: 331–43.

Davis, M. (1990) *City of Quartz: Excavating the Future of Los Angeles*, Verso, New York.

Downey, K. (1989) "Emerging cities," *Washington Post* (April 3): K1–K5.

Ehrenreich, B. (1989) *Fear of Falling*, Pantheon, New York.

Fonseca, J.W. (1976) "The semi-urban landscape," *Landscape* 21: 23–5.

Foster, H. (ed.) (1985) *Postmodern Culture*, Pluto Press, London.

Fuller, S. (1989) "The internationalization of the Washington, DC area economy," in R.V. Knight and G. Gappert (eds) *Cities in a Global Society*, Sage, Newbury Park, CA, 108–19.

Gale, D.C. (1987) *Washington, D.C: Inner City Revitalization and Minority Suburbanization*, Temple University Press, Philadelphia, PA.

Garreau, J. (1988) "Edge cities," *Landscape Architecture* 78: 48–55.

Goodey, B. (1985) "Urban designs: Context, style and history in the post-modern era," *Planning History Bulletin* 7: 46–52.

Gottdiener, M. and Gephart, G. (1991) "The multinucleated metropolitan region: A comparative analysis," in R. Kling, S. Olin and M. Poster (eds) *Postsuburban California*, University of California Press, Berkeley, pp. 31–54.

Gutman, R. (1987) "Architects in the homebuilding industry," in J. Pipkin, M. LaGory and J. Blau (eds) *Professionals and Urban Form*, SUNY Press, Albany, NY, pp. 208–13.

Gutman, R. (1988) *Architectural Practice*, Princeton University Press, Princeton, NJ.

Harvey, D.W. (1985) *The Urbanization of Capital*, Blackwell, Oxford.

Harvey, D.W. (1987) "Flexible accumulation through urbanization: Reflections on 'postmodernism' in the American city," *Antipode* 19: 260–86.

Harvey, D.W. (1989) *The Condition of Postmodernity*, Blackwell, Oxford.

Hassan, I. (1987) *The Postmodern Turn: Essays in Postmodern Theory and Culture*, Ohio State University Press, Columbus.

Healey, P. and Nabarro, R. (eds) (1990) *Land and Property Development in a Changing Context*, Gower Press, Brookfield, VT.

Hebdige, R. (1989) "After the masses," *Marxism Today* (January): 48–55.

Hirsch, F. (1977) *The Social Limits to Growth*, Routledge & Kegan Paul, London.

Howland, M. (1989) "The growth and location of services in Washington, D.C.," Paper presented to the Association of Collegiate Schools of Planning, Portland, OR.

Hummon, D.M. (1990) *Commonplaces*, SUNY Press, Albany, NY.

Jager, A. (1986) "Class definition and the aesthetics of gentrification: Victoriana in Melbourne," in N. Smith and P. Williams (eds) *Gentrification of the City*, Allen & Unwin, Boston, MA, pp. 78–91.

King, R.J. (1988) "Urban design in capitalist society," *Society and Space* 6: 445–74.

King, R.J. (1989) "Capital switching and the role of groundrent: Theoretical problems," *Environment and Planning A* 21: 445–62.

Kirby, A. (1989) "Time, space, and collective action: Political space/political geography," Discussion paper #89-1, Department of Geography, Tucson.

Knox, P.L. (1987) "City profile: The Washington metropolitan area," *Cities* 4: 290–8.

Krier, L. (1985) "Visionary architecture: The completion of Washington, D.C.," *Architecture and Urbanism* 11: 91–100.

Lassar, T.J. (1989) *Carrots and Sticks: New Zoning Downtown*, Urban Land Institute, Washington.

Lewis, P.F. (1983) "The galactic metropolis," in R.H. Platt and G. Macinko (eds) *Beyond the Urban Fringe*, University of Minnesota Press, Minneapolis, pp. 23–49.

Ley, D. (1985) "Cultural/humanistic geography," *Progress in Human Geography* 9: 415–23.

McCracken, G. (1988) *Culture and Consumption: New Approaches to the Symbolic Character of Consumer Goods and Activities*, Indiana University Press, Bloomington.

Martin, D. and Hughes, S. (1989) "The real estate implications of changes in the health care industry," *Urban Land* (August): 16–20.

Meinig, D.W. (1979) "Symbolic landscapes," in D.W. Meinig *et al.* (eds) *The Interpretation of Ordinary Landscapes*, Oxford University Press, New York, pp. 164–92.

Mills, C.A. (1988) "Life on the upslope: The post modern landscape of gentrification," *Society and Space* 6: 169–89.

Montgomery, R. (1989) "Architecture invents new people," in R. Ellis and D. Cuff (eds) *Architects' People*, Oxford University Press, New York, pp. 261–81.

Moulaert, F. and Swyngedouw, E.A. (1989) "A regulation approach to the geography of flexible production systems," *Society and Space* 7: 327–45.

Sack, R.D. (1988) "The consumer's world: Place as context," *Annals of the Association of American Geographers* 78: 642–65.

Schoenberger, E. (1988) "From Fordism to flexible accumulation: Technology, competitive strategies and international location," *Society and Space* 6: 339–65.

Schwanke, D. (1987) *Mixed-use Development Handbook*, Urban Land Institute, Washington, DC.

Schwanke, D. (1989) *Development Trends*, Urban Land Institute, Washington, DC.

Shields, R. (1989) "Social spatialization and the built environment: The West Edmonton Mall," *Society and Space* 7: 147–64.

Smith, N. (1987) "Of yuppies and housing: Gentrification, social restructuring and the urban dream," *Society and Space* 5: 151–72.

Soja, E. (1980) "The sociospatial dialectic," *Annals of the Association of American Geographers* 70: 207–25.

Soja, E. (1989) *Postmodern Geographies*, Verso, New York.

Suchman, D.R. (1989) "Housing and community development," in D. Schwanke (ed.) *Development Trends*, Urban Land Institute, Washington, DC, pp. 30–41.

Survey of Diversified Developers (1988) *Building Design and Construction* 34: 34–7.

Vance, J.E. Jr. (1972) "California and the search for the ideal," *Annals of the Association of American Geographers* 62: 182–210.

Ventre, F. (1988) "Myth and paradox in the building enterprise," in P. L. Know (ed.) *The Design Professions and the Built Environment*, Nichols, New York, pp. 147–74.

Weiss, M.J. (1988) *The Clustering of America*, Harper & Row, New York.

Williams, P. (ed.) (1986) *Gentrification of the City*, Allen & Unwin, Boston, MA.

Wood, J.S. (1988) "Suburbanization of center city," *Geographical Review* 78: 325–9.

Zukin, S. (1988) *Loft Living, Culture and Capital in Urban Change*, Radius Books, New York.

Editors' references and suggestions for further reading

Ford, L.R. (1994) *Cities and Buildings: Skyscrapers, Skid Rows, and Suburbs*, The Johns Hopkins University Press, Baltimore, MD.

Knox, P.L. (1987) "The social production of the built environment: architects, architecture, and the post-Modern city," *Progress in Human Geography* 11: 354–77.

Knox, P.L. (1993) *The Restless Urban Landscape*, Prentice-Hall, Englewood Cliffs, NJ.

Knox, P.L. and Ozolins, P. (eds) (2000) *Design Professionals and the Built Environment: An Introduction*, John Wiley, Chichester.

Knox, P.L. and Taylor, P.J. (1995) *World Cities in a World System*, Cambridge University Press, Cambridge.

Ley, D. and Mills, C. (1993) "Can there be a postmodernism of resistance in the urban landscape?," in P. Knox (ed.) *The Restless Urban Landscape*, Prentice-Hall, Englewood Cliffs, NJ, pp. 255–71.

Relph, E. (1987) *The Modern Urban Landscape*, Johns Hopkins University Press, Baltimore, MD.

"The 'Magic of the Mall': An Analysis of Form, Function, and Meaning in the Contemporary Retail Built Environment"

from *Annals of the Association of American Geographers* (1993)

Jon Goss

Editors' Introduction

To truly understand the contemporary shopping center, cultural geographer Jon Goss argues that we must closely examine the connection between environmental design and the "consciousness industry" since they are both "media of mass communication, employing rhetorical devices to effect hidden persuasions" (1993, p. 21). Advertisers, the "captains of consciousness" (Ewen 1976), work to link a commodity with common cultural symbols, drawing upon the consumer's ability to interpret subtle (as well as not so subtle) references to shared meanings. They hope to create not only demand for the item but also identification with it. Thus shopping is not simply a means of satisfying material needs, it is also a way for individuals to shape their identities. Advertisers work to underscore this message – "You are what you buy."

If "you are what you buy," you are also defined by where you buy. The advertising drive to blend the material and symbolic in an object is reinforced by the retailers'/developers' ability to design settings that encourage the consumer to shop. Drawing shoppers to a particular mall and keeping them there, after all, is a key part of a retailer's and a developer's profit strategy. In his analysis of the environmental design and image-making of contemporary shopping centers, Goss uses the professional literature of the retail development industry to demonstrate that the mall is produced as a "machine for shopping" to be consumed as a "retail resort." Furthermore, the "magic of the mall" manipulates shoppers' behavior through the configuration of space and construction of symbolic landscapes that, ironically, allow for a "fantasized dissociation" from the act of shopping. Goss argues that to legitimize shopping as an activity that may seem otherwise frivolous the developer seeks to create a setting with positive associations.

In the following article, Goss explores the design of the shopping mall, linking it to the spatial archetypes of traditional public space, marketplace, and festival settings, and the cultural symbols of idealized, historical streetscapes and market experiences. Nostalgia for a shared public life pervades the design of postmodern retail environments. By embracing the past in the design elements of a contemporary shopping mall, however, the developer creates *pseudoplaces* – a normative term developed to indicate "places made over to be something they were never meant to be" (Wood 1985). That is, they create profoundly contradictory spaces which articulate a desire for "genuine" urban life with an expectation of a public accessibility while providing opportunities for social control.

Goss extends his interpretation of contemporary retail environments in a later article on festival market-places – or themed, open-air retail center (1996). His most recent article on landscapes of shopping provides an "unofficial guide to malls of America" (1999). By choosing the largest retail and entertainment complex in the USA for examination, Goss again addresses the blend of shopping and tourism in popular culture, the masking of private and public space, and the illusions inherent in both the commodities for sale and the environmental design of shopping areas.

For additional examples of the changing landscape of retail in this volume, see David Ley's (p. 304) comments on the festival/marketplace setting of Vancouver's False Creek development, and Paul Knox's (p. 281) overview of metropolitan Washington, DC.

At least one urban geographer has complained that geographers tend to study "housing but not houses, retailing but not department stores" (Ford 1994). Goss' research begins to correct that failing, and in the process he demonstrates the complexity involved as one comes to terms with an examination of architectural form and meaning (1988, 1993). See Loretta Lees' article, "Towards a critical geography of architecture" (2001), for a more recent review of the challenges of developing an architectural geography.

■ ■ ■ ■ ■ ■ ■

Shopping is the second most important leisure activity in North America, and although watching television is indisputably the first, much of its programming actually promotes shopping, through both advertising and the depiction of model consumer lifestyles. The existential significance of shopping is proclaimed in popular slogans such as: "Born to Shop," "Shop 'Til You Drop," and "I Shop Therefore I Am." An advertisement for Tyson's Corner, Virginia, asks: "The joy of cooking? The joy of sex? What's left?" and the answer provided is, of course, "The joy of shopping"! As Tyson's obviously knows, recent market research shows that many Americans prefer shopping to sex (Levine 1990).

Despite increases in catalog sales, shopping remains essentially a spatial activity – we still "go" shopping – and the shopping center is its chosen place. By 1990, there were 36,650 shopping centers in the U.S., providing 4.2 billion square feet (151 square miles!) of gross leasable area and accounting for more than $725 billion of sales, or 55 per cent of retail sales excluding automobile sales ("Retail Uses" 1991: 23). The time spent in shopping centers by North Americans follows only that spent at home and at work/school. Centers have already become tourist destinations, complete with tour guides and souvenirs, and some include hotels so that vacationers and conferees need not leave the premises during their stay. Downtown retail complexes often include condominia, and residential development above the suburban mall is predicted to be an inevitable new trend ("The PUD Market Guarantee" 1991). Their residents can literally shop without leaving home (or be at home without leaving the shops?). Moreover, planned retail space is colonizing other privately owned public spaces such as hotels, railway stations, airports, office buildings and hospitals, as shopping has become the dominant mode of contemporary public life.

Nevertheless, there persists a high-cultural disdain for conspicuous mass-consumption resulting from the legacy of a puritanical fear of the moral corruption inherent in commercialism and materialism, and sustained by a modern intellectual contempt for consumer society. This latter critique condemns the system of correspondences between material possessions and social worth (Boorstin 1973; Veblen 1953), the homogenization of culture and alienation of the individual (Adorno and Horkhiemer 1969; Marcuse 1964) and the distortion of human needs through the manipulation of desire (Haug 1986). The contemporary shopper, while taking pleasure in consumption, cannot but be aware of this authoritative censure, and is therefore, like the tourist (Frow 1991), driven by a simultaneous desire and self-contempt, constantly alternating between assertion and denial of identity. . . .

This paper argues that developers have sought to assuage this collective guilt over conspicuous consumption by designing into the retail built environment the means for a fantasized dissociation from the act of shopping. That is, in recognition of the culturally perceived emptiness of the activity for which they provide the main social space, designers manufacture the illusion that something else other than mere shopping is going on, while also mediating the materialist relations of mass consumption and disguising the identity and rootedness of the shopping center in the contemporary capitalist social order.

The analysis proceeds in several parts, elaborating upon a conception of the built environment developed elsewhere (Goss 1988) and employing the professional literature of the retail development industry as well as empirical observation of shopping centers. . . . [First], I examine the retail built environment as an object of value; that is, a private, instrumental space designed for the efficient circulation of commodities which is itself a commodity produced for profit. [Second], I discuss the means by which developers have obscured this logic by constructing shopping centers as idealized representations of past or distant public spaces. [Finally], I describe the operation of the shopping center as a spatial system structuring opportunities and constraints for movement and social interaction.

This account is necessarily limited to the workings of the design, that is the assumptions made about the retail built environment and its users, and the intent of the developers as inferred from a reading of their professional literature and of the landscape itself. This requires some care lest we fall into the same trap that compromises the modernist critique of consumption, a critique which holds much intellectual force but little political potential. This is to conceive of the consumer as cultural dupe and helpless object of technical control, exactly as the (mostly) male middle-class designers imagine them. . . . [T]he shopper is not merely the object of a technical and patriarchal discourse and design, but is also a subject who may interpret the design aberrantly or intentionally appropriate meaning for her/his own purposes.

[. . .]

THE MAKING OF THE MALL

Of course, it must be kept in mind that architects do not design malls for architects; they design them for developers and retailers that are interested in creating malls and other shopping centers to attract consumers and keep them coming back. (Richards 1990: 23)

The developer's profit accrues from the construction and sale of shopping centers, lease rent, and deductions from retail revenues. Unlike other forms of real estate, where markets have been rapidly saturated and are dependent upon urban and regional economic fortunes, shopping center construction has been a relatively secure investment, whether in the suburbs, always provided a big name department store could be enticed to sign an agreement (Frieden and Sagalyn 1989: 79), or downtown, provided subsidies could be negotiated from cooperative municipal governments. Recently however, there has been a marked slowdown in the speculative development observed in the 1970s and early 1980s. This trend is attributed to a variety of factors: the combination of a shortage of suitable greenfield sites; escalating costs of land assembly, construction, and operation; tightened developmental controls; declining federal government programs that provide infrastructure and capital incentives (Frieden and Sagalyn 1989); organized resistance from local communities ("Building despite the obstacles" 1990); the financial vulnerability of highly leveraged retail chains (Hazel 1989; Reynolds 1990) changing market demographics; and the segmentation of the retail industry (Goss 1992: 168). As a result, many regions are effectively saturated and intercenter competition is intense. An extreme example is Dallas, where three megacenters (Galleria, Prestonwood, and Valley View) are within two miles of each other. . . . Profit increasingly depends, therefore, upon image making and the creative management of shopping centers.

It is important at the outset to realize the scale and detail of the conception. Shopping centers are typically produced by huge corporations or ad hoc coalitions of finance, construction, and commercial capital (typically pension funds, developers, and department stores), and are meticulously planned. They usually involve state agencies and teams of market researchers, geo-demographers, account-ants, asset managers, lawyers, engineers, architects, landscape artists, interior designers, traffic analysts, security consultants, and leasing agents. Develop-ment, therefore, involves the coordination of a complex of concerns, although always overdeter-mined by the goals of retail profit.

The shopping centers profit from an inter-nalization of externalities; that is, by ensuring strict complementarity of retail and service functions through an appropriate tenant mix (Goss 1992). Leasing agents plan the mix of tenants and their locations within the center, inevitably excluding repair shops, laundromats, or thrift stores that might remind the consumer of the materiality of the commodity and attract those whose presence might challenge the normality of consumption. Where resale shops are found, they conventionally indicate difficulty in attracting more desirable tenants (Ricks 1991).

While individual retailers may pursue their own strategies for profit within limited bounds, the center operates as a whole to maximize "foot traffic" by attracting the target consumers and keeping them on the premises for as long as possible. . . .

The task begins with the manufacture and mar-keting of an appropriate sense of place (Richards 1990: 24), an attractive place image that will entice people from their suburban homes and downtown offices, keep them contentedly on the premises, and encourage them to return. This occurs in an increasingly competitive retail market resulting from the "overmalling of America" and in response to consumer loyalties shifting from name-retailers to specific shopping centers, the personality of the center is critical (McDermott 1990: 2–3).

IMAG(IN)ING THE MALL

The sense of place is also a political fact. What can be done to the look of a locality depends on who controls it. . . . People can be excluded, awed, confused, made acquiescent, or kept ignorant by what they see and hear. So the sense of the environment has always been a matter of moment to any ruling class. (Lynch 1976: 72–3)

In constructing an attractive place image for the shopping center, developers have, with remark-able persistence, exploited a modernist nostalgia for authentic community, perceived to exist only in past and distant places, and have promoted the con-ceit of the shopping center as an alternative focus for modern community life. Shopping districts of the early years of this century, for example, were based on traditional market towns and villages, and a strong sense of place was evoked using stylized historical architecture and landscaping (typically evoking the village green). They were built on a modest scale, functionally and spatially integrated into local communities, in order to pro-vide an idyllic context for consumption by the new gentry (Rowe 1991). The picturesque Country Club Plaza in Kansas City, Missouri, built in 1922, is a prototypic example. With the contemporary postmodernist penchant for the vernacular, this original form is undergoing a renaissance in the specialty center, a collection of high-end outlets that pursue a particular retail and architectural theme. Typically these are also idealizations of villages and small towns, chock-full of historical and regional details to convince the consumer of their authenticity (Goss 1992: 172). Examples include Pickering Wharf in Salem, Massachusetts (a New England village), the Borgata in Scottsdale, Arizona (a thirteenth-century Italian village), the Pruneyard in San Jose, California (a Spanish-American hacienda), the Mercado in Phoenix, Arizona (a Mexican hillside village).

In contrast, the modern regional shopping center was built on a large scale with regular, unified architecture. Its harsh exterior modernism and automobile-focused landscaping refused any com-promise with the rustic aesthetic. As Relph (1987: 215) notes, however, "modernism . . . never wholly succeeded in the landscape of retailing," and the interior contained pedestrian walkways, courts, fountains and statuary that referred reassuringly to the traditional urbanism of southern Europe, Victorian Britain or New England. According to Victor Gruen, the acknowledged pioneer of the modern mall, his "shopping towns" would be not only pleasant places to shop, but also centers of cultural enrichment, education, and relaxation, a suburban alternative to the decaying downtown (Gruen and Smith 1960).

Gruen's shopping centers proved phenomenally successful, and he later argued that by applying the lessons of environmental design learned in the suburbs to downtown, "we can restore the lost sense of commitment and belonging; we can counteract the phenomenon of alienation, isolation and loneliness and achieve a sense of identity" (Gruen 1973: 11). James Rouse, effectively heir to Gruen and heralded as "the savior of downtown America" (Sawicki 1989: 347), similarly argued that shopping centers "will help dignify and uplift the families who use them . . . promote friendly contact among the people of the community . . . [and] expose the community to art, music, crafts and culture" (Rouse 1962: 105). . . . The key, Rouse argues, is not so much the design features of the shopping mall, but centralized retail management (CRM) and leasing strategies (cited in Stokvis and Cloar 1991), which would include levels of security and maintenance well beyond that provided by municipal authorities, market research, cooperative advertising, common business hours, common covenants, and a regulated tenant mix (Cloar 1990). Downtown is now "learning from the mall": as the director of the National Mainstreet Center, an organization established by the National Trust for Historic Preservation, argues, "shopping centers . . . are well-planned, well-funded, and well-organized. . . . Main streets need management like that" (Huffman 1989: 95).

The new downtown retail built environment has taken two essential forms, which in practice may be mixed. First is the commercial gentrification of decaying historical business and waterfront districts, pioneered by James Rouse with Quincy Market in Boston. Its opening in 1979 supposedly marked "the day the urban renaissance began" (Rouse, cited in Teaford 1990: 253) and subsequently no self-respecting city seems complete without its own festival marketplace, replicating more or less the original formula. Historical landmarks and "water exposure" (Scott 1989: 185) are critical features, as this retail environment is consciously reminiscent of the commercial world city, with its quaysides and urban produce markets replete with open stalls, colorful awnings, costermonger barrows, and nautical paraphernalia liberally scattered around.

A second form is the galleria, the historic referent of which is the Victorian shopping arcade and especially the famous Galleria Vittorio Emanuele II in Milan. After Cesar Pelli pioneered the galleried arcade in the early 1970s . . . glazed gallery and atria became standard feature in downtown mixed-use developments. . . . Enclosed streetscapes refer to the idealized, historic middle-American Main Street or to exotic streets of faraway cities, including Parisian boulevards, Mexican paseos, and Arabic souks or casbahs, if only because the contemporary North American street invokes fear and loathing in the middle classes. They reclaim, for the middle-class imagination, "The Street" – an idealized social space free, by virtue of private property, planning, and strict control, from the inconvenience of the weather and the danger and pollution of the automobile, but most important from the terror of crime associated with today's urban environment.

The malling of downtown could not work, however, without the legislative and financial support of the local state. These developments exploit historic preservation laws and federal and municipal funds to subsidize commercial development. Newport Center in Jersey City, for example, is the recipient of the largest-ever Urban Development Action Grant (Osborne 1988). Frieden and Sagalyn (1989) provide a particularly incisive analysis of the coalitions of private capital and municipal government necessary to the successful development of the new urban retail built environment.

In creating these spaces, developers and public officials articulate an ideology of nostalgia, a reactionary modernism that expresses the "dis-ease" of the present (see Stewart 1984: 23), a lament on the perceived loss of the moral conviction, authenticity, spontaneity, and community of the past; a profound disillusionment with contemporary society and fear of the future. More specifically, we collectively miss a public space organized on a pedestrian scale, that is, a setting for free personal expression and association, for collective cultural expression and transgression, and for unencumbered human interaction and material transaction. Such spaces no longer exist in the city, where open spaces are windswept tunnels between towering buildings, abandoned in fear to marginal populations; nor were they found after all in the suburb, which is subdivided and segregated, dominated by the automobile, and repressively predictable and safe. Such spaces only exist intact in our *musees imaginaire*, but their forms can now be expertly reproduced for us in the retail built environment. Below, I discuss the form and the contradictions inherent in the reproduction of such spaces as conceived in their idealized civic, liminal and transactional forms.

THE SHOPPING CENTER AS CIVIC SPACE

By virtue of their scale, design, and function, shopping centers appear to be public spaces, more or less open to anyone and relatively sanitary and safe. This appearance is important to their success for they aim to offer to middle-Americans a third place beyond home and work/school, a venue where people, old and young, can congregate, commune, and "see and be seen" (Oldenburg 1989: 17). Several strategies enhance the appearance of vital public space, and foremost is the metaphor of the urban street sustained by street signs, streetlamps, benches, shrubbery, and statuary – all well-kept and protected from vandalism. Also like the ideal, benign civic government, shopping centers are extremely sensitive to the needs of the shopper.... They may house post offices, satellite municipal halls, automated government services, and public libraries; space is sometimes provided for public meetings or religious services. They stage events not only to directly promote consumption (fashion and car shows), but also for public edification (educational exhibits and musical recitals). Many open their doors early to provide a safe, sheltered space for morning constitutionals – mall-walking – and some have public exercise stations with health and fitness programs sponsored by the American Heart Association and YMCAs (Jacobs 1988). Some even offer adult literacy classes and university courses.... Public services not consistent with the context of consumption are omitted or only reluctantly provided, often inadequate to actual needs and relegated to the periphery. This includes, for example: drinking fountains, which would reduce soft drink sales; restrooms, which are costly to maintain and which attract activities such as drug dealing and sex that are offensive to the legitimate patrons of the mall (Hazel 1992); and public telephones, which may be monopolized

by teenagers or drug dealers. As a result, telephones in some malls only allow outgoing calls (Hazel 1992).

The idealized public street is a relatively democratic space with all citizens enjoying access with participatory entertainment and opportunities for social mixing; and the shopping center re-presents a similarly liberal vision of consumption, in which credit-card citizenship allows all to buy an identity and vicariously experience preferred lifestyles, without principles of exclusion based on accumulated wealth or cultural capital (Zukin 1990). It is, however, a strongly bounded or purified social space (Sibley 1988) that excludes a significant minority of the population and so protects patrons from the moral confusion that a confrontation with social difference might provoke (see also Lewis 1990).... There have been several court cases claiming that shopping centers actively discriminate against potential minority tenants, employees, and mall users. Copley Place in Boston, for example, has been charged with excluding minority tenants ("Race is not the issue" 1990: 32); a Columbia, South Carolina mall was accused of discriminatory hiring practices ("NAACP in hiring pact..." 1991: A20): and security personnel have been widely suspected of harassing minority teenagers.... [S]ome managers have even tried to regulate hours during which teenagers can shop without adult supervision ("Retailers use bans..." 1990), and passed ordinances and erected barricades in parking lots to prevent "unnecessary and repetitive driving" ("Suburbs rain on teens'..." 1990). "Street people" are harassed because their appearance, panhandling, and inappropriate use of bathrooms (Pawlak *et al.* 1985) offend the sensibility of shoppers, their presence subverting the normality of conspicuous consumption and perverting the pleasure of consumption by challenging our righteous possession of commodities. Even the Salvation Army may be excluded from making its traditional Christmas collections, perhaps because they remind the consumer of the existence of less-privileged populations and so diminish the joy of buying.

Developers must of course protect their property and guard themselves against liability (Hazel 1992), but the key to successful security apparently lies more in an overt security presence that reassures preferred customers that the unseemly and seamy side of the real public world will be excluded from the mall. It is argued that the image of security is more important than its substance:

> Perception is perhaps even more important than reality. In a business that is as dependent as film or theater on appearances, the illusion of safety is as vital, or even more so, than its reality. (Hazel 1992: 28)

In extreme cases, however, overt and pervasive security may itself be part of the attraction, and this applies particularly to the "defensible commercial zones" (Titus 1990: 3) which reclaim part of the decaying inner city for the display of cultural capital and lifestyles of the middle classes.... Such pan-optical presence has been enhanced in some cases by donating mall space for local police. Vermont-Slauson Shopping Center, Capital Center in Trenton, New Jersey and Chicago's Grand Boulevard Plaza, all contain police substations, while Crenshaw Plaza has a 200-officer police station on its premises....

Finally, the politics of exclusion involves the exclusion of politics, and there is an ongoing struggle by political and civil liberties organizations to require shopping centers to permit hand billing, picketing, and demonstrations on their premises, on the grounds that they cannot pretend to be public spaces without assuming the responsibility of such, including recognition of freedom of expression and assembly. Courts have generally found in favor of free speech in shopping centers by virtue of their scale and similarity to public places, provided that the activities do not seriously impair their commercial function (Peterson 1985). The Supreme Court, however, has ruled that it is up to individual states to decide (Kowinski 1985), and in a recent case, an anti-war group was successfully banned from leafletting in New Jersey malls ("Judge bars group..." 1991).

THE SHOPPING CENTER AS LIMINAL SPACE

The market, standing between the sacred and secular, the mundane and exotic, and the local and global, has always been a place of liminality; that is, according to Turner (1982), a state between social stations, a transitional moment in which established rules and norms are temporarily suspended. The marketplace is a liminoid zone, a place where potentiality and transgression are engendered by the exciting diversity of humanity, the mystique of exotic objects, the intoxicating energy of the crowd channeled within the confined public space, the prospects of fortunes to be made and lost in trade, the possibility of unplanned meetings and spontaneous adventures, and the continuous assertion of collective rights and freedoms or *communitas* (Bahktin 1984). The market thrives on the possibility of "letting yourself go," "treating yourself," and of "trying it on" without risk of moral censure, and free from institutional surveillance.

Places traditionally associated with liminoid experiences are liberally quoted in the contemporary retail built environment, including most notably seaports and exotic tropical tourist destinations, and Greek agora, Italian piazzas, and other traditional marketplaces. Colorful banners, balloons and flags, clowns and street theater, games and fun rides, are evocative of a permanent carnival or festival. Lavish expenditure on state-of-the-art entertainment and historic reconstruction, and the explosion of apparent liminality is perfectly consistent with the logic of the shopping center, for it is designed explicitly to attract shoppers and keep them on the premises for as long as possible.

This strategy reaches its contemporary apotheosis in the monster malls that contrive to combine with retailing the experiences of carnival, festival, and tourism in a single, total environment. This includes, most famous, the West Edmonton Mall (WEM), Canada, which has already become a special concern of contemporary culture studies ("Special issue on the WEM" 1991; Blomeyer 1988; Hopkins 1990; Shields 1989; Wiebe and Wiebe 1989), and others inspired by its extravagant excess: Franklin Hills in Philadelphia, River Falls in Clarksville, Tennessee; the controversial new Mall of the Americas in Bloomington, Minnesota; Meadowhall in Sheffield and Metrocentre in Gateshead in England; and Lotte World in Seoul, South Korea. The shopping center has become hedonopolis (Sommer 1975). Shopping centers have become tourist resorts in their own right, recreating the archetypical modern liminal zone by providing the multiple attractions, accommodations, guided tours, and souvenirs essential to the mass touristic experience, *all* under a single roof.

WEM, which receives 15 million visitors a year (and is responsible for more than 1 per cent of all retail sales in Canada [Jones and Simmons 1987: 77]), claims that:

> Tourists will no longer have to travel to Disneyland, Miami Beach, the Epcot Park . . . New Orleans . . . California Sea World, the San Diego Zoo, the Grand Canyon. . . . It's all here at the WEM. Everything you've wanted in a lifetime and more. (Winter City Showcase cited in Hopkins 1990: 13)

There are necessarily strict limits to any experience of liminality in these environments. Developers are well aware of the "more unsavory trappings of carnival life" (McCloud 1989: 35), and order must be preserved. . . . Liminality is thus experienced in the nostalgic mode, without the inherent danger of the real thing: the fairground is recreated without the threat to the social order that the itinerant, marginal population and the libidinal temptations that traveling shows might bring, while the revitalized waterfronts lack the itinerant sailors, the red lights, the threatening presence of foreign travelers and shiphands. The contrived retail carnival denies the potentiality for disorder and collective social transgression of the liminal zone at the same time that it celebrates its form.

THE SHOPPING CENTER AS TRANSACTIONAL SPACE

Regardless of the location and scale of the development, a constant theme in contemporary retail space is a nostalgia for the traditional public marketplace, or what we might call *agoraphilia*. In the idealized traditional marketplace, there is an immediate relationship between producer and consumer, and both apply knowledge and skill to judge quality and negotiate price. Vendors ideally sell their own product and have direct responsibility for its quality. They are also in competition with other traders so presentation and service are important, and they acquire considerable interpersonal skill and extensive knowledge of their customers. Such commitment and initiative is not to be expected among the retail staff of the increasingly large, centralized retail corporations, but in response to the perceived deterioration of

service, mall management may organize training sessions to improve sales techniques ("Developers of big shopping malls . . ." 1991), while on-site research is constantly conducted to discover the special desires and problems of customers and the ways in which staff might meet them (1991: 32). Competition for customer service awards motivates personnel, and plain-clothes shopping police, or undercover shoppers, watch for "testy cashiers and inattentive managers" (Levine 1990: 187).

To further solve the problem of indolent and insolent attendants, contemporary retailing has learned from the theater, and particularly the total theater of North American theme parks. For example . . . sales staff in a Fred Meyer megastore in appropriately named Hollywood West in Portland, Oregon, enter through the Stage Door and are admonished to "get into character" ("Fred Meyer megastore . . ." 1990). The Disney Store sales staff are "cast members" and customers are "guests," while the staff at Ringling Brothers and Barnum and Bailey Circus stores are educated in Clown College.

[. . .]

THE SHOPPING CENTER *IS* INSTRUMENTAL SPACE

Most shoppers know that the shopping center is a contrived and highly controlled space, and we all probably complain about design features such as the escalators that alternate in order to prevent the shopper moving quickly between floors without maximum exposure to shopfronts, or the difficulty finding restrooms. Some of us are also disquieted by the constant reminders of surveillance in the sweep of the cameras and the patrols of security personnel. Yet those of us for whom it is designed are willing to suspend the privileges of public urban space to its relatively benevolent authority, for our desire is such that we will readily accept nostalgia as a substitute for experience, absence for presence, and representation for authenticity. We overlook the fact that the shopping center is a contrived, dominated space that seeks only to resemble a spontaneous, social space. Perhaps also, we are simply ignorant of the extent to which there is a will to deceive us. The professional literature is revealing. *Urban Land*, for example, congratulates the Paseo Nuevo project in Santa Barbara for its deception:

it "*appears* to be a longstanding part of downtown" (when it isn't) and is a "*seemingly* random arrangement of shops, tree-shaded courtyards, splashing fountains, and sunny terraces" when it is a carefully designed stage for "*choreographing* pedestrian movement" ("Fitting a shopping center . . ." 1991: 28, emphasis added). In this professional literature, the consumer is characterized as an object to be mechanistically manipulated – to be drawn, pulled, pushed, and led to flow magnets, anchors, generators, and attractions; or as a naive dupe to be deceived, persuaded, induced, tempted, and seduced by ploys, ruses, tricks, strategies, and games of the design. Adopting a relatively vulgar psychogeography, designers seek to environmentally condition emotional and behavioral response from those whom they see as their malleable customers.

The ultimate conceit of the developers, however, lies in their attempt to recapture the essence of tradition through modern technology, to harness abstract space and exchange value in order to retrieve the essence of use value of social space (Lefebvre 1971). The original intention may have been more noble, but the contradiction soon became apparent, and the dream of community and public place was subordinated to the logic of private profit. Victor Gruen himself returned to his home city of Vienna disillusioned and disgusted at the greed of developers (Gillette 1985), while James Rouse formed a nonprofit organization engaged in urban renewal! The contemporary generation of developers may still express the modernist faith in the capacity of environmental design to realize social goals, but one somehow doubts that Nader Ghermezian, one of the developers of the monstrous WEM, is genuine when he claims their goal is "to serve as a community, social, entertainment, and recreation center" (cited in Davis 1991: 4).

REFERENCES FROM THE READING

Adorno, T. and Horkheimer, M. (1969) *The Dialectic of Enlightenment*, Continuum, New York.

Bahktin, M.M. (1984) *Rabelais and His World*, Indiana University Press, Bloomington.

Blomeyer, G. (1988) "Myths of malls and men," *The Architect's Journal*, May: 38–45.

Boorstin, D.J. (1973) *The Americans: The Democratic Experience*, Vintage Books, New York.

Building despite the obstacles: Anti-growth sentiment, local restrictions slow retail development (1990) *Chain Store Age Executive*, June: 27–32.

Cloar, J.A. (1990) *Centralized Retail Management: New Strategies for Downtown*, Urban Land Institute, Washington.

Davis, T.C. (1991) "Theatrical antecedents of the mall that ate downtown," *Journal of Popular Culture* 24, 4: 1–15.

Developers of big shopping malls tutor faltering tenants in retail techniques (1991) *Wall Street Journal*, April 24: B. 1.

Fitting a shopping center to downtown (1991) *Urban Land*, July: 28–9.

Fred Meyer megastore goes Hollywood (1990) *Chain Store Age Executive*, March: 76–8.

Frieden, B.J. and Sagalyn, L.B. (1989) *Downtown, Inc.: How America Rebuilds Cities*, MIT Press, Cambridge.

Frow, J. (1991) *Tourism and the Semiotics of Nostalgia*, October, 57: 123–51.

Gillette, H. (1985) "The evolution of the planned shopping center in suburb and city," *Journal of the American Planning Association*, 51, 4: 449–60.

Goss, J.D. (1992) "Modernity and postmodernity in the retail built environment," in F. Gayle and K. Anderson (eds) *Ways of Seeing the World*, Unwin Hyman, London.

Goss, J.D. (1998) "The built environment and social theory: Towards an architectural geography," *The Professional Geographer* 40: 392–403.

Gruen, V. (1973) *Centers for the Urban Environment: Survival of the Cities*, Van Nostrand Reinhold, New York.

Gruen, V. and Smith, L. (1960) *Shopping Towns USA: The Planning of Shopping Centers*, Van Nostrand Reinhold, New York.

Haug, W.F. (1986) *Critique of Commodity Aesthetics*, University of Minnesota Press, Minneapolis.

Hazel, D. (1989) "After the fall: Lessons from L.J. Hooker," *Shopping Center Age*, September: 27–30.

Hazel, D. (1992) "Crime in the malls: A new and growing concern," *Chain Store Age Executive*, February: 27–9.

Heison, S. (1989) *The Heritage Industry*, Methuen, London.

Hopkins, J.S.P. (1990) "West Edmonton Mall: Landscape of myths and elsewhereness," *Canadian Geographer* 34,1: 2–17.

Huffman, F. (1989) "Mall Street, USA," *Entrepreneur*, August: 95–9.

Jacobs, J. (1988) *The Mall: An Attempted Escape from Everyday Life*, Waveland Press, Prospect Heights, IL.

Jones, K. and Simmons, J. (1987) *Location, Location, Location: Analyzing the Retail Environment*, Methuen, London.

Judge bars group from leafleting in malls (1991) *New York Times*, July 28, 1: 31.

Kowinski, W.S. (1985) *The Malling of America: An Inside Look at the Great Consumer Paradise*, William Morrow, New York.

Lefebvre, H. (1971) *Everyday Life in the Modern World*, Harper and Row, New York.

Levine, J. (1990) "Lessons from Tysons Corner," *Forbes*, April 30: 186–7.

Lewis, G.H. (1990) "Community through exclusion and illusion: The creation of social worlds in an American shopping center," *Journal of Popular Culture* 24: 121–36.

Lynch, K. (1976) *Managing the Sense of Region*, MIT Press, Cambridge.

McCloud, J. (1989) "Fun and games is serious business," *Shopping Center World*, July: 28–35.

McDermott, M.J. (1990) "Too many malls are chasing a shrinking supply of customers," *Adweek's Marketing week*, February 5: 2–3.

Marcuse, H. (1964) *One-dimensional Man*, Beacon Press, Boston.

NAACP in hiring pact with South Carolina Mall (1991) *New York Times*, March 7: A. 20.

Oldenburg, R. (1989) *The Great Good Life*, Paragon House, New York.

Osborne, T. (1988) "Revolutionizing the retail landscape," *Marketing Communications*, October: 17–25.

Pawlak, E.J. *et al.* (1985) "A view of the mall," *Social Service Review*, June: 305–17.

Peterson, E.C. (1985) "Diverse special interest groups may have access to center property," *Shopping Center World*, May: 85.

The PUD Market Guarantee (1991) *Chain Store Age Executive*, April: 31–2.

Race is not the issue, Copley Place says (1990) *Boston Globe*, August 16: 32.

Relph, E. (1987) *The Modern Urban Landscape*, Johns Hopkins University Press, Baltimore.

Retail Uses (1991) *Urban Land*, March: 22–6.

Retailers use bans, guards and ploys to curb teen sport of mall-mauling (1990) *Wall Street Journal*, August 7: B. 1.

Reynolds, M. (1990) "Revamps on the rise," *Stores*, July: 34–7.

Richards, G. (1990) "Atmosphere key to mall design," *Shopping Center World*, August: 23–9.

Ricks, R.B. (1991) "Shopping center rules misapplied to older adults," *Shopping Center World*, May: 52–6.

Rouse, J.W. (1962) "Must shopping centers be humane?" *Architectural Forum*, June: 105–7, 196.

Rowe, P.G. (1991) *Making a Middle Landscape*, MIT Press, Cambridge.

Sawicki, D.S. (1989) "The festival marketplace as public policy: Guidelines for future policy decisions," *American Planning Association Journal*, summer: 347–61.

Scott, N.K. (1989) *Shopping Centre Design*, Van Nostrand Reinhold, London.

Shields, R. (1989) "Social spatialization and the built environment: West Edmonton Mall," *Environment and Planning D: Society and Space* 7: 147–64.

Sibley, D. (1988) "Survey 13: The purification of space," *Environment and Planning D: Society and Space* 6: 409–21.

Sommer, J.W. (1975) "Fat city and hedonopolis: The American urban future," in R. Abler *et al.* (ed.) *Human Geography in a Shrinking World*, Duxbury Press, North Scituate, MA, pp. 132–48.

Special issue on the West Edmonton Mall (1991) *Canadian Geographer* 35, 3.

Stewart, S. (1984) *On Longing: Narratives of the Miniature, the Gigantic, the Souvenir, the Collection*, Johns Hopkins University Press, Baltimore.

Stokvis, J.R. and Cloar, J.A. (1991) "CRM: Applying shopping center techniques to downtown retailing," *Urban Land*, April: 7–11.

Suburbs rain on teens' "1 Big Hormone" parade (1990) *Chicago Tribune*, August 5: 2C, 1.

Teaford, J.C. (1990) *The Rough Road to Renaissance: Urban Revitalization in America, 1940–1985*, Johns Hopkins University Press, Baltimore.

Titus, R.M. (1990) "Security works," *Urban Land*, January: 2–3.

Turner, V. (1982) *From Ritual to Theater*, Performing Arts Publications, New York.

Veblen, T. (1953) *The Theory of the Leisure Class*, Mentor Books, New York.

Wiebe, R. and Wiebe, C. (1989) "Mall," *Alberta* 2, 1: 81–90.

Zukin, S. (1990) "Socio-spatial prototypes of a new organization of consumption: The role of real cultural capital," *Sociology* 24, 1: 37–56.

Editors' references and suggestions for further reading

Ewen, S. (1976) *Captains of Consciousness: Advertising and the Social Roots of Consumer Culture*, McGraw-Hill, New York.

Ford, L. (1994) *Cities and Buildings: Skyscrapers, Skid Rows, and Suburbs*, Johns Hopkins University Press, Baltimore, MD.

Goss, J. (1988) "The built environment and social theory: towards an architectural geography," *Urban Geography* 17, 3: 221–47.

Goss, J. (1996) "Disquiet on the waterfront: reflections on nostalgia and utopia in the urban archetypes of festival marketplaces," *Urban Geography* 17, 3: 221–47.

Goss, J. (1999) "Once-upon-a-time in the commodity world: an unofficial guide to malls of America," *Annals of the Association of American Geographers* 89, 1: 45–75.

Lees, L. (2001) "Towards a critical geography of architecture: The case of an ersatz Colosseum," *Ecumene* 8, 1: 51–86.

Wood, J. (1985) "Nothing should stand for something that never existed," *Places* 2, 2: 81–7.

SIX

"Styles of the Times: Liberal and Neo-conservative Landscapes in Inner Vancouver, 1968–1986"

from *Journal of Historical Geography* (1987)

David Ley

Editors' Introduction

Just west of the Chinatown analyzed by Kay Anderson (p. 219), two redeveloped industrial areas lining both shores of False Creek inlet in Vancouver, BC offer city residents and tourists alike lively landscapes of picture-postcard beauty. Amid the natural backdrop of mountains and water, False Creek's new marinas, residences, city parks and shops epitomize the urban renaissance that many cities' leaders have strived to achieve since the 1960s. These are the neighborhoods that inspired David Ley's extended research project on gentrification and the "new middle-class" role in remaking the central city (1996). In his analysis, Ley, a social and cultural geographer at the University of British Columbia, emphasizes the changing lifestyle choices that have fueled the transformation of inner-city, industrial areas into landscapes of consumption. He reminds us that there is a geography to gentrification that varies not only between cities such as Vancouver and Winnipeg, but also among neighborhoods within a single city. Just as Liz Bondi explores the different gentrification processes in two Edinburgh neighborhoods (p. 251), Ley evaluates in his monograph *The New Middle Class and the Remaking of the Central City* those attributes of the natural and built environment that attract potential gentrifiers.

In "Styles of the Times," however, Ley compares two False Creek neighborhoods, which he describes as "perhaps glaring at each other" on either side of the inlet. The hostility he imagines stems from their contrasting architectural styles and the diametrically opposed cultural politics that produced them. Inspired by the work of Bernice Martin (1981), Ley argues that these two neighborhoods also reflect the competition of two major ideologies that have "run intermittently through western culture over the past two hundred years" – the rational (or instrumental) versus the romantic (or expressive). The modern movement as it influenced urban planning and architectural theory stressed functionality and universality in organizing space; whereas, he argues, contemporary postmodern currents in design promote communitarian values, contextuality and place-making.

Ley offers False Creek Southside as an example of a reaction against the modernist influence in the city. Begun in the mid-1970s as part of a municipal reform group's efforts to create a "liveable city," the development reflects the liberal professionals members' commitment to the construction of a medium-density, human-scale landscape, expressive of local urban subcultures. In contrast, six years later, a conservative provincial government returned to modernist values in their proposed development of the high-rise, high-density British Columbia Place on the north side of False Creek, which included the site of Expo '86 and other mega-structures. We should note that since the publication of Ley's article and the end of Expo '86, the land associated with BC Place was sold to a Hong Kong developer, ending the provincial government's direct control of the project. The Concord Pacific development claims to be the largest development scheme in North America (2003).

By comparing the different political, social and cultural values involved in the construction of these two contemporary landscapes – evidence of the potential influence of these competing ideologies – Ley undertakes the sensitive task of interpreting the built environment as a landscape text linked to other systems of meaning. To explore the epistemological foundations of his textual analysis, see the introduction to *Place/Culture/Representation* by the co-editors James S. Duncan and David Ley (1993), particularly the section on hermeneutics which underscores the influence anthropologist Clifford Geertz has had on the new cultural geographers (e.g. Geertz 1973).

Other analyses that examine the manner in which "building elites" inscribe their visions of social order and social life on a city's landscape include Mona Domosh's study of nineteenth-century New York and Boston (1996) and David Harvey's study of the symbolic significance inherent in the construction of Paris' Sacre-Coeur (1979). As the unedited version of "The Style of the Times" recommends, Carl Schorske's study of late nineteenth-century critical responses to Vienna's Ringstrasse is also worth examining, since Schorske provides a masterful analysis of the competing themes of expressivism/romanticism and instrumentalism/rationalism in modern thought as well as their influence on the built environment (1981). To explore the links between Vancouver's future urban development and global market forces, see Kathryne Mitchell's article on the "Visions of Vancouver" (1996).

Architecture is the will of the epoch translated into space. (Mies van der Rohe, 1926)[1]

A common theme in interpretation of the modern movement in twentieth-century culture has been the interconnection of developments in the arts, architecture and philosophy not only with each other but also with broader trends in social, economic and political life. This is not to impose a historical *zeitgeist* upon reality, with its tendencies to reification, but to recognize that the identification of both influence and independence among different domains in society as they confront common problems and opportunities, negates a reductionism which mechanistically reads off the imprint of one domain upon another.[2] The relations between domains are manifest in Le Corbusier, perhaps the central figure in modern architecture, who was not only an architect in the morning and a painter and sculptor in the afternoon, but also a regular contributor to syndicalist journals.[3] At the same time, Le Corbusier's machine-age aesthetic was closely attuned to dominant social and economic realities of his time. A design he completed with his cousin of an economy car for mass production, and his characterization of a house as a machine for living in,[4] locate him unambiguously in the technological

thrall of an early twentieth-century society convinced of the capacity of rationalism in all its manifestations to propel society into a brave new world. The strength of the "engineering aesthetic" in art and design repeated the exalted position of the engineer and his blueprint in society. The Hegelian sensitivity of Mies van der Rohe simply made the internal relations between domains more self-evident; for this master of modern architecture, to believe that technology constituted the cultural arsenal of modern man was to sustain a building programme of minimalist, geometric skyscrapers which became the standard downtown urban form in the post-war years.

[. . .]

If the act of interpretation is perhaps more nuanced than the Hegelian outlook of Mies van der Rohe would suggest, none the less landscape style is intimately related to the historic swirl of culture, politics, economics and personality in a particular place at a particular time. . . . [I]n an analysis of contemporary culture, Bernice Martin has counterposed two . . . ideologies which have run intermittently through western culture over the past 200 years.[5] . . . [T]he *expressive dimension* represents a romantic theme treasuring the subjective, the interpersonal and the aesthetic. . . .

[T]he *instrumental dimension* is associated with the world-view of modernism: functional, technological and sharing the purposive rational values of bourgeois society.

Martin interprets the 1960s counterculture movement as a reappearance of the romantic expressive style, a style which passed beyond the youth culture to find a receptive sanctuary in the liberal middle class, particularly in the arts and such "soft" professions as teaching, the media and the caring professions. For a period it dislodged the pre-existing instrumentalism of growth-oriented society, but though some of its attributes have become absorbed in mainstream culture, it has receded before the rise of neo-conservative politics in the 1980s. Expressivism and instrumentalism are not free-floating spirits but the ideologies of discrete social groups who emerge in particular places at particular times when, according to the extent of their prominence, they may become significant cultural architects, moulding a repertoire of symbols and forms, including the built environment.

This paper offers an interpretation of two distinctive inner-city renewal landscapes in Vancouver, authored by social groups adhering to expressive and instrumental values. Though particular and local, these landscapes also manifest broader political and cultural processes. Begun only six years apart, and facing (or perhaps glaring at) each other across False Creek, a narrow body of water, these projects offer two widely divergent texts. The earlier redevelopment, False Creek Southside, started in the mid-1970s, incorporates current post-modern planning themes, with a studied attempt to construct a picturesque, medium density, human-scale landscape, with mixed residential, commercial and leisure uses, and expressive of a cross-section of local urban subcultures. It is a product of the ideology of liberal reform, emerging from the social innovation of the 1960s and reacting against the rationalism of an earlier generation of municipal politicians, planners and designers. The later project, British Columbia Place, incorporating the site of Expo 86, was begun in 1981 and is a more conventional outcome of the rational, corporate mind, a high-density residential and commercial district to be built around a sports stadium and the site of a world's fair, and conceived as an instrument in the consolidation of a mass society. It displays the ideology of neo-conservatism, the counter-reaction

of the 1980s against the perceived fuzzy-mindedness of the welfare state and its departure from the discipline of accountability through the market place. Publicly initiated redevelopment of both shores of False Creek has been a substantial undertaking; the Southside project, now in its third phase, has been under development for ten years, while the Northside project, British Columbia Place, was conceived in 1980 and is advertised as the largest current renewal project in North America, with a twenty- to thirty-year development horizon along its three-kilometre site (Figure 1). As we shall see, False Creek presents a landscape text which, when interrogated, reveals the evolving intersection around the built environment of political and cultural processes and the opposition of social classes that have been adherents to the broader and conflicting ideologies we have outlined.

EXPRESSIVISM: DESIGN FOR COMMUNITY AND THE PROBLEM OF PLACELESSNESS

By the late 1960s the ideals of the modernist prophets of the 1920s had grown increasingly hollow in North American cities. Mies van der Rohe's concern for spirit and value in architecture in the 1920s had by the 1950s evolved into the "symmetrical monumentality . . . of a highly rationalized building method"[6] and had been appropriated as the distinctive urban signature of large business corporations. Le Corbusier's utopian radiant city of highways and high-density tower blocks became a widely imitated formula for post-war residential development by public and private sector builders. A corporate urban landscape, the product of an increasingly corporate society, became the legacy of the modern movement, and through the 1960s and 1970s a critique emerged that the planning and design of the modern city was a blueprint for placelessness, of anonymous, impersonal spaces, massive structures and automobile throughways. . . . A philosophical reorientation has been influential in the social sciences, in planning and architecture, and in urban politics, as a critical ideology concerned with the reconstitution of meaning, with a respect for human subjectivity and the private realms of everyday life. In planning and architecture the criticism is directed against a functionalist

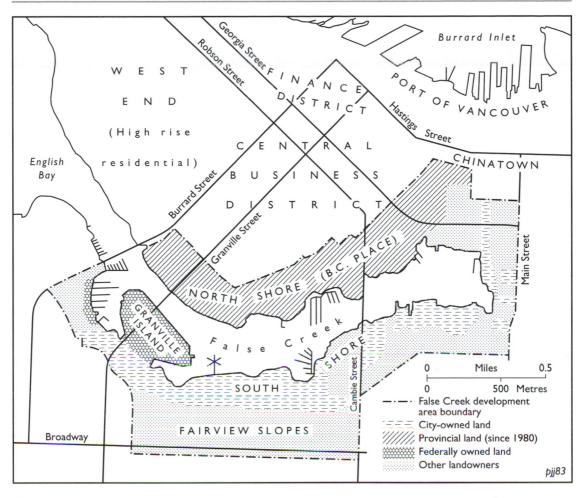

Figure 1 Land holdings in the False Creek basin (1982).

landscape. . . . Against the uniformity of the modern movement is a renewed interest in the specificity of regional and historical styles and a respect for the diversity of urban subcultures.

Perhaps the most celebrated of the early critics of modern design and planning was Jane Jacobs . . . who sought to demonstrate the destructive effects of centralized rational planning upon urban life. Her influential book, an "attack on current city planning and rebuilding",[7] identified a faulty logic in detached technical plans for urban development. . . . What was at stake, Jacobs argued, was the preservation of community and the personalizing of space. "People places", vital, human, liveable space, was being destroyed by detached, rational planning; the antidote was human-scale planning, a built environment which in its landmarks, folk allusions and meeting places would sustain personal associations, and what geographers have called a sense of place. In opposition to the segregated land uses of current plans, diversity of people and land uses was the cardinal principle: "Planning for vitality must stimulate and catalyze the greatest possible range and quantity of diversity among users and among people throughout each district of a big city."[8] . . . [N]ot the least of the innovativeness of Jacobs' approach was epistemological, in its distrust of the detached expert, in its appreciation of folk and personal knowledge, and in its method which established direct communication between everyday life and planning practice.

An identical development may be seen in current reactions to modernism in architecture where, in the plural styles of what has become known as post-modern architecture,[9] there is a more complex attention to theories of space. . . . [T]he objective is the construction of forms which suggest and evoke symbolic associations, "sensitive urban place-making".[10] Post-modern architecture, in short, aims to communicate recognizable intersubjective themes and to acknowledge local history and culture. In reaction to the large structures of the modern movement, it attempts to create smaller units, seeks to break down a corporate society to urban villages and to maintain historical associations through renovation and recycling. Jencks sees in post-modern architecture "the return of the missing body", an attempt to restore meaning, rootedness and human proportions to place in an era dominated by depersonalizing standardization.[11]

[. . .]

FALSE CREEK SOUTHSIDE: THE EXPRESSIVE LANDSCAPE OF LIBERAL REFORM

Redevelopment of the southside of False Creek was marked by an intention to construct a novel environment in conscious opposition to existing urban landscapes. The project was envisaged as an antithesis to both private and public sector development of the city core, sanctioned and encouraged by a conservative pro-business City Council in the 1950s and 1960s. By the mid-1960s, Vancouver seemed on course for a destiny as a high-rise, freeway city, another fulfilment of the 1920s blueprint of the machine-age metropolis. With symptomatic ambivalence, Arthur Erickson, a late modern Vancouver architect who was a consultant for the freeway system, found in such urban cores "a desolating impersonality as if no creature but a machine had made them. But this was also their vitality . . . purely utilitarian space."[12]

However, the social and landscape consequences of the machine-age aesthetic contained few redeeming virtues for other residents. In False Creek . . . there was "a clear philosophical rejection of 'modernist' approaches to urban design and architecture. This was a common theme of the urban reform movement in general."[13] Beginning with opposition to the freeway system and the demolition of Chinatown-Strathcona through urban renewal, a loose coalition spearheaded by professionals developed into a social movement that rapidly led to the founding of a liberal municipal reform party

in 1968. This party, The Electors Action Movement (TEAM), was dominated by young professionals; over a quarter of party members were architects, teachers, university professors and lawyers. Its momentum owed not a little to the spirit of the times in North America, an era of protest and liberal social movements, a period when the rights of minorities and individual conscience were set against the impersonal control of precisely the same centralized rational society which had appeared to be progressive to an earlier generation. In Canada was added the spring fever of "Trudeaumania" in 1968, the sometimes euphoric sense of a new beginning in society, politics and, not least, lifestyle. A more distant influence was the North American and French student riots the same year, with their celebration of self-expression, creativity and openness, and their central motif of "exhilarating, joyous festival".[14] TEAM was born into the midst of the "expressive revolution."[15]

[. . .]

The reform approach to False Creek was instructive. A report in 1968 had described the Creek as "a garbage dump, a sewer outlet for the City of Vancouver. . . . It is first and foremost, the industrial heart of the city."[16] Its dominant land use was noxious industry, lumber mills, smelters and engineering works, which in the 1950s employed over 10,000 people, though by the late 1960s a number had become economically marginal. But industrial land use continued to be advocated and in 1967 the conservative City Council passed a motion that "it is Council's policy at this time, that the land abutting False Creek be retained as an industrial area."[17] There followed a period of political mobilization about the built environment which culminated with the election of a TEAM majority to the Council in 1972, and the setting in motion of an altogether different future for False Creek.

Reform politics and social planning

Design and planning criteria for development were adopted in late 1973 by a Council with a strong reform majority. In its 1972 campaign material, TEAM had promoted a future False Creek as offering "inner city living at its best", while "the waterfront should be a continuous system of parks and marinas for all the people to enjoy".[18] . . . Liberal advocates of human-scale planning and community development such as Jane Jacobs and Herbert Gans appear prominently in the planning documents, while Christopher Alexander's pattern language was employed. . . . Consultation with the public was ongoing. . . .

The outcome of this process was an innovative urban landscape. . . . Following closely Jane Jacobs' philosophy, diversity was a major theme of the development. Social mix was deliberately fostered with income levels representative of the larger city, a goal achieved according to the 1981 Census which shows an average income for False Creek residents almost identical with that throughout Vancouver. Tenure forms and lifestyles were also heterogeneous, including self-owned condominiums, rental apartments, housing cooperatives, and non-profit associations representing such groups as the elderly, the handicapped, and a floating homes society. . . .

The underpinning ideology of development represented the fusion of aesthetics, ecology and social justice. This substitution of liveability for the highest and best use meant a considerable reduction in taxable property values for the City. The 1970 concept plans had identified taxable assessments in the False Creek basin of up to $500 million with a residential population of 60,000. By 1972 densities had been reduced, however, with a new target population of 30,000 residents, and with half the housing on city-owned land to incorporate some form of social assistance. For its own land, the City aimed to cover its costs, but no more, through land leases; use value rather than exchange value guided its policy-making.

Post-modern landscape, post-modern lifestyle

The design guidelines for False Creek set out by reform politicians identified the importance of careful urban place-making. The planners noted critically that "our cities are becoming more and more uniform", but promised "In the False Creek Development we are all looking for something new."[19] They would "avoid apartment slabs with cell-like windows," the stereotype of modern architecture, and build to a density half that of the adjacent, and much criticized, West End district.

A number of post-modern design themes figure prominently in the self-consciously constructed landscape. First, particularly in Phase 1, the built environment assumes human proportions, with its medium densities and three-storey family townhouses. Secondly, circulation is largely limited to pedestrian and bicycle traffic; a fundamental premise was that False Creek would be "a non-automobile oriented environment."[20] Thirdly, in place of uniformity, the housing enclaves are differentiated by colour, materials and design, representing a deliberate attempt both to foster pluralism but also to personalize the built form, "to support rather than destroy the vast mosaic of subcultures".[21] Fourthly, historical, regional and vernacular allusions, important elements of post-modern style, are conspicuous. Historical associations are retained in street names reminiscent of the Creek's industrial past, and the use of an industrial vernacular style for new buildings. . . . Historical references have been fitted into the evolving landscape of Granville Island in the deliberate and almost patronizing preservation of old-established industrial workshops, including a chain company and a small nail factory. New "cottage industries" including a small brewing company have opened on the site. Other industrial plants, such as a factory formerly fabricating millwork equipment, have been recycled into commercial, recreational and, in one instance, residential, units. Regional symbolism includes a planted forest grove, and a garden of native British Columbia species. A fifth post-modern feature, acknowledgement of the local environmental context, is reflected in the attention given to Ian McHarg's view on ecological harmony in design.[22] The result is a landscape which veers towards the romantic in its open space: a waterfall, stream and lake in a sixteen-acre park with ecological niches for Canada geese, and a pedestrian overpass, connecting the site with an adjacent neighbourhood designed as a landscaped winding trail sixty feet wide, broader than a normal residential street. A sixth post-modern feature, an orientation towards the picturesque, was inevitable in a process which began with the mandate that False Creek should be "a beautiful addition to Canada's most beautiful city."[23] The aesthetic character of the project is paramount: a sensuous landscape has been constructed, maximizing views of water, the downtown skyline, and the enveloping mountain rim; extensive design features suppress noise pollution; while the range of colours, textures and materials add visual diversity to the built form. According to design peers, the visual landscape is "too good to be true and in this has some of the character of a film set . . .".[24] The post-modern influence of Granville Island offers "an oasis in the architectural desert . . . a perfect antidote to the massive boxes" of the downtown core.[25]

Granville Island contains important clues revealing the social interests implicated in a post-modern landscape. It represents a mix of industrial, commercial, cultural and leisure uses. The most important commercial use is a farmer's market, offering fresh produce and seafood, but also a bazaar of small booths selling ethnic and craft products. The emphasis here is on style differentiation, goods and experience which are personalized not mass produced, the result (or so it would seem) of "folk" rather than standardized commodity production. . . . [I]n certain respects, the landscape in its complexity and surprises is symbolically overloaded. In its proportions, style and facade materials, the new Granville Island Hotel has the appearance of a small factory, and the art college that of a warehouse. What does this almost patronizing allusion to the industrial past convey? The taming of the industrial city? Or, more than this, even a confirmation of its passing? . . . There are tendencies towards both elitism and indulgence in this landscape and the lifestyles it both reflects and reinforces. As with all styles and lifestyles there is a narrow line dividing statement from overstatement, authenticity from ersatz and even parody.

False Creek, and especially Granville Island, is a landscape for a lifestyle, the lifestyle of the new middle class, particularly those in the expressive professions. The Island contains three theatres, an art college, and studios for artists, architects and designers. To what extent is the cultural new class both patron and client of post-modernism? And to what extent is the post-industrial centre city its natural habitat?

INSTRUMENTALISM: RATIONAL DESIGN AND THE PROBLEM OF SOCIAL ORDER

Nikolaus Pevsner has observed that "the twentieth century is the century of the masses . . . science, technology, mass locomotion, mass production and consumption, mass communication".[26] The transition from folk society to mass society introduced a double challenge, for not only were the self-regulating control mechanisms of rural communities severely eroded, but also formerly separate social and cultural groups were thrown together in the maelstrom of urban life. The problem of social order in a mass society is a repetitive theme in the literature on the new industrial city. . . . Inevitably the appropriate model for urban administration was perceived to be the rational private corporation, to be imitated in an attempt "to introduce into city government the standardization and scientific management already found in industry".[27] . . .

The technological thrall in early city planning and urban administration during the Progressive era resonated with the slogans of modern movements in architecture. A functional aesthetic created forms which were honest to the spirit of the times, a technological, mass society.[28] . . .

The cultural agnosticism of the modern movement has continued to the present. . . . Following Keynesian economics with its reorientation of state intervention to fuel the engine of mass consumption, and with the related rise of mass advertising in the inter-war period, a new public culture has emerged in western nations, the culture of consumption.[29] Mass consumption is a critical instrument of the modern state for assuring social and political order; increasingly it provides government with its legitimacy.[30] The culture of mass consumption gives rise to its own distinctive icons, its own characteristic landscapes. The current late phase of modern architecture finds its most complete expression in the megabuildings of a corporate society. Late-modernism, Jencks and Chaitkin have noted, is oriented to "create extravagant forms meant to impress by their grand, overpowering sweep. One finds this sort of architecture in World's Fairs, airports, stadia, city halls, and architectural schools."[31] As we shall see, the populism evoked by the megastructure has been a potent electoral instrument of neo-conservative politicians.

BRITISH COLUMBIA PLACE: THE INSTRUMENTAL LANDSCAPE OF NEO-CONSERVATISM

The north shore of False Creek, which includes the site of the 1986 World's Fair, reveals an emerging landscape which symbolizes a very different politics of consumption. If the south shore presents an expressive and personalized landscape, sired by liberal professionals, the north shore promises to be a more instrumental landscape for a mass society, the product of neo-conservative political priorities.

The politics of a megastructure

. . . [t]he birth of British Columbia Place . . . coincided with a public crisis in confidence in the provincial government, a crisis of political and, later, social order. The ruling Social Credit Party, a populist right-wing group, had traditionally resorted to the megastructure as both a policy and a political tool. With uncharacteristic haste, a report was commissioned on the construction of a new sports stadium for Vancouver, and less than five months later, in January 1980, Premier Bennett announced his Cabinet's decision. To the sports stadium was now to be added a world transportation fair on the north shore of False Creek. . . . In the Premier's words: "This is a commitment to a vision for the future, a vision to build a great meeting place for all of our people that we would call British Columbia Place."[32] Far more than a sports stadium, it would be "the focal point of our great province", "the great central showplace on False Creek", "the chance for a celebration that will leave a lasting legacy". . . .

The desire for a sports stadium and a fair, encouraged by a crisis in political support and endorsed by a non-technical twenty-two page report, has provided the wedge around which is gathering a $2 billion redevelopment project to include also up to 10,000 housing units, up to seven million square feet of office space, other commercial and public buildings, and a billion dollar transit line. But the choice of a sports stadium was an appropriate one, for the spectacle occupies a central position in the contemporary culture of mass consumption. According to John Alt, "The spectacle requires and fosters a social milieu of masses" a social order which channels "the emotional needs of consumers in instrumental ways".[33]

The mandate of the new right

Once the provincial government had decided to purchase False Creek land from the Canadian Pacific Railway and assume a multifaceted programme of urban redevelopment, it determined that BC Place should not only break even in cost terms (the objective for False Creek Southside) but that it should be profitable, with its profits redistributed to similar projects in smaller urban centres. As we shall see, a number of design themes and conflicts flowed from this Cabinet decision. There is a strong private sector orientation to the project. Two of the initial executive officers were former land developers, the chairman and former president are eminent corporate businessmen, and the project planners see themselves "as a new breed of planners who get things done. We don't regard ourselves as civil servants – we're privately employed."[34]

The market orientation requires high densities and, despite early protestations to the contrary, it is now evident that residential densities will approximate those of the West End. The market mandate also requires that the liberal social philosophy of social mixing will not be repeated. Socially assisted units will account for a smaller proportion of residential construction: family units will also have a diminished presence, in part because a child-oriented environment would require infrastructure, such as a school and open space, which would be expensive in opportunity costs in comparison to development for childless households. The City of Vancouver has strongly resisted the density and social mix proposals tendered by BC Place.

Conflict has arisen over the implementation of other components of the political philosophy of the new right. The Expo Board, comprising corporate business leaders and Social Credit politicians and appointees, awarded construction contracts to non-union companies, despite the high level of unionization in the building trades, and set up a wage structure with these companies independent of an agreement with the unions.[35] With building contracts valued at between $600 and $900 million, the stakes were high for the unions and labour unrest including work stoppages and demonstrations almost led to the Fair's cancellation.

But labour unrest was only one part of Expo's problems. Like megaprojects elsewhere the financial accountability of the project has been slender. . . . Uncertain attendance estimates and rising costs from an initial figure of $78 million, have created increasing deficit estimates from $12 million in 1980 to over $75 million in 1982. Six months after a private report by economists suggested a loss of $500 million, the Expo Board in January 1985 admitted to a new projected deficit of $311 million.[36] When pressed about the costs of his megaproject in 1980, Premier Bennett responded, "It's more important to proceed than worry about how much it might cost."[37] By 1984–85 his personal identification with the Fair had become ever closer, as he described himself as "the father of Expo suffering through its birth".[38] The projection of not only ideology, but also personality, on to the landscape is being etched ever clearer.

Design for mass consumption

While planning and design guidelines for BC Place are at present fairly generalized, there are few signs of the post-modern influences of the south shore. Symptomatic was the dispute over the C.P.R. Roundhouse, a facility for locomotive storage and maintenance and part of the railyards on the BC Place site. Sections of the Roundhouse dated back to 1888, two years after the railway was completed and the City of Vancouver was incorporated. The Roundhouse thus held significant symbolic value in its local cultural and historical associations, but it became clear that to the BC Place Corporation the structure was an impediment to redevelopment and would be cleared. . . . A similar problem has emerged in the interface between the BC Place site and Yaletown, an area of old warehouses, some renovated, with heritage value. The maintenance of continuity, of historical and cultural symbols, so central to post-modernism, is not a priority in a project which seems to share in its "commitment to a vision for the future" a view of progress and an antipathy to historicism widely held by the modern school.

This assessment is supported by the initial selection of Arthur Erickson, Canada's foremost corporate architect, as a project designer. Erickson's successful career has been built upon his mega-buildings in Vancouver, Toronto, Los Angeles and elsewhere. A common critique of these structures is that they were built as spectacles, to impress the visitor rather than to accommodate users, "designed for the visitor . . . who had come to see it as a *thing*, an object set in a landscape."[39] In this respect Erickson is an archetypal designer for modern society, for a mobile consumer who encounters the built environment as an outsider, in motion. Erickson's design concept for BC Place called for an extension of the high-use apartments in the West End along the north shore of False Creek. The rationale for this proposal is "a funny imbalance" which the architect found between densities in the two districts, requiring high densities in BC Place to accomplish a "balancing" with the West End. This aesthetic asymmetry may not be evident on the ground but "aerial photographs show it well."[40] True to modernism's fascination with speed and locomotion almost a quarter of Le Corbusier's *Towards a New Architecture* is given over to the study of airplanes, steamships and automobiles. It is the aesthetic of the mobile outsider, in this instance the view from the airplane, which is used to justify a design solution. In this design perspective users largely disappear; in a view from the air they are invisible. And so the West End to Erickson offered a landscape to be emulated; for the Southside planners it was a landscape to be avoided. . . . The emancipatory mass culture anticipated by Le Corbusier and Frank Lloyd Wright has diverged from their utopian goals to a fully developed culture of consumption.

The process of centralized control

In May 1980, the British Columbia Place Act passed through the provincial legislature and established BC Place Ltd as a crown corporation. This legal status conferred upon the developer sweeping and centralized powers. It has the authority to raise and disperse its own funds, the power of expropriation, and may override all city by-laws and zoning regulations, not to mention planning policy. As a crown corporation, BC Place Ltd has the powers of a private corporation, additional privileges and fewer legal restrictions. It represents an extension of the centralized administrative model advocated by Progressive reformers early in this century, and is a vivid illustration of the argument of Weber, Habermas and others concerning the range of social control exercised by the modern bureaucracy. The corporate model gives maximum discretion and minimum disturbance to technical specialists acting as the agents of a central executive. As one corporate planner observed, "The Cabinet are our shareholders."[41]

[. . .]

BC Place offers an effective example of what has been labelled the "authoritarian centralism" of the new right in North America and Britain.[42] The paradox is that pre-election promises of less government become in practice a more entrenched corporatism of centralized control. Also . . . advocates of free market accountability do not engage in self-critical cost-benefit studies themselves. BC Place represents one illustration of the broader ideology of the Social Credit government to "centralize power in an overwhelmingly dominant executive".[43] This ideology is reflected throughout social and economic policy. . . . [T]he mayor of Vancouver has complained that the BC Place Corporation wishes to resurrect the highway, high-rise urban vision of the 1950s; other writers have extended this critique to a broader range of provincial social and economic policy.[44] But BC Place does not simply represent in miniature the provincial government of British Columbia; it also provides a text of the script being written more generally by new right legislators. The Southside development remains a document to be read *for its cues to an earlier era of liberal design and planning*, dominated by the expressive values of a cultural new class. In this manner the hermeneutics of the built environment return us to Mies van der Rohe's claim that "Architecture is the will of the epoch translated into space", where landscape is interpreted in the context of a broader political culture, and where style is seen as one element in an interrelated domain of ideas and interests in social life.

NOTES

1 Frampton, K. (1985) *Modern Architecture: A Critical History*, London, p. 231.

2 LaCapra, D. (1983) "Rethinking intellectual history and reading texts," in D. LaCapra (ed.) *Rethinking Intellectual History: Texts, Contexts, Language*, Cornell, pp. 23–71.

3 Michel, J. (1984) "Le Corbusier exhibition: the architect as painter," *Manchester Guardian Weekly*, 10 June; and Frampton, *op. cit.*, pp. 184–5.

4 One of his standardized designs was called Maison Citrohan, a pun on the car manufacturer: Frampton, *op. cit.*, pp. 153–4.

5 Martin, B. (1981) *A Sociology of Contemporary Cultural Change*, Oxford.

6 Frampton, *op. cit.*, p. 231.

7 Jacobs, J. (1961) *The Death and Life of Great American Cities*, New York, p. 161.

8 *Ibid.*, p. 408.

9 Jencks has been the chief publicist of this movement in architecture. See Jencks, C. (1984) *The Language of Post-modern Architecture*, New York. See also Stern, R. (1980) "The doubles of post-modern," *Harvard Architecture Review* 1: 75–87; Portoghesi, P. (1982) *Postmodern*, New York.

10 Jencks, *op. cit.*, p. 82.

11 Jencks, C. and Chaitkin, W. (1982) *Architecture Today*, New York, p. 217.

12 Iglauer, E. (1981) *Seven Stones: A Portrait of Arthur Erickson, Architect*, Madeira Park, BC, p. 93.

13 Julchanski, D. (1984) *St. Lawrence and False Creek: A Review of the Planning and Development of Two New Inner City Neighbourhoods*, School of Community and Regional Planning, University of British Columbia, Planning Papers 10, p. 41.

14 Poster, M. (1975) *Existential Marxism in Postwar France*, Princeton, p. 373.

15 Martin, *op. cit.*; Ley, D. (1980) "Liberal ideology and the postindustrial city annals," *Association of American Geographers* 70: 238–58.

16 Fukui, J. (1968) *A Background Report on False Creek for the Vancouver Board of Trade*, Vancouver.

17 Elligott, F. (1977) "The planning decision-making of Vancouver's False Creek: a case study 1968–74," unpublished M.A. thesis, University of British Columbia, p. 45.

18 *Vancouver Province*, 5 October 1972.

19 False Creek Study Group (1972) *Reports 4–5*, Vancouver.

20 Rodger, R. (1976) *Creating a Livable Inner City Community*, Ottawa and Vancouver.

21 False Creek Study Group (1972) *Report 3*, Vancouver.

22 Anon. (1972) "False Creek basin offers opportunity of the decade," *BC Business Journal* 4 (April): 38–9.

23 False Creek Development Group (1977) *False Creek South Shore*, Vancouver.

24 Anon. (1980) "Vancouver then and now," *Architectural Review* 167: 322–3.

25 Kemble, R. (1980) "Granville Island," *Canadian Architect* 25: 16–27.

26 Pevsner, N. (1981) *The Sources of Modern Architecture and Design*, New York, p. 7.

27 Zorbaugh, H. (1929) *The Gold Coast and the Slum*, Chicago, p. 272.

28 Ley, D. (1986) "Making space for people: the theory and practice of a humane urban design," Paper presented to the inaugural symposium of the Center for Advanced Studies in Humanist Social Science, University of Waterloo, May.

29 Jackson Lears, T. (1983) "From salvation to self-realization: advertising and the therapeutic roots of the consumer culture, 1880–1930," in R. Wightman Fox and T. Jackson Lears (eds) *The Culture of Consumption: Critical Essays in American History, 1880–1930*, New York, pp. 3–38.

30 Habermas, J. (1975) *Legitimation Crisis*, Boston; Ley, D. (1983) *A Social Geography of the City*, New York.

31 Jencks and Chaitkin, *op. cit.*, p. 21.

32 Bennett, W. (1980) "Speaking notes for Premier Bennett," Four Seasons Hotel, Vancouver, 29 January.

33 Alt, J. (1983) "Sport and cultural reification: from ritual to mass consumption," *Theory, Culture and Society* 1: 93–107.

34 Personal communication, 1982.

35 *Vancouver Sun* (1984) 13 July; see also *Globe and Mail* (1984), 30 April.

36 Blackorby, C., Donaldson, G. and Slade, M. (1985) *Expo 86: an economic impact analysis*, BC Economic Policy Institute, University of British Columbia, Paper No. 84,11; *Vancouver Sun* (1985), 25 January.

37 *Globe and Mail* (1984), 13 April.

38 *Globe and Mail* (1984), 12 April.

39 Gustein, D. (1974) "Arthur Erickson: the corporate artist-architect," *City Magazine* 1: 6–15.

40 *Vancouver Sun* (1983), 27 March.

41 Personal communication, 1982.

42 Elliott, B. and McCrone, D. (1982) *The City: Patterns of Domination and Conflict*, London.

43 Dobell, A.R. (1983) "What's the BC spirit? Recent experiences in the management of restraint," Institute of Policy Analysis, University of Victoria.

44 Woodcock, G. (1983) "Bill Bennett wants to take British Columbia back to the 1950s," *Saturday Night* 98: 11–14.

Editors' references and suggestions for further reading

Domosh, M. (1996) *Invented Cities: The Creation of Landscape in Nineteenth-Century New York and Boston*, Yale University Press, New Haven.

Duncan, J.S. and Ley, D. (eds) (1993) *Place/Culture/Representation*, Routledge, London.

Geertz, C. (1973) "Thick description," in C. Geertz (ed.) *The Interpretation of Cultures*, Basic Books, New York.

Harvey, D. (1979) "Monument and myth," *Annals of the Association of American Geographers* 69: 362–81.

Ley, D. (1996) *The New Middle Class and the Remaking of the Central City*, Oxford University Press, Oxford.

Martin, B. (1981) *A Sociology of Contemporary Cultural Change*, Blackwell, Oxford.

Mitchell, K. (1996) "Visions of Vancouver, ideology, demography, and the future of urban development," *Urban Geography* 17: 478–501.

Schorske, C. (1981) *Fin-de-Siècle Vienna: Politics and Culture*, Vintage Books, New York.

"Portland's Comprehensive Plan as Text: The Fred Meyer Case and the Politics of Reading"

from T. Barnes and J. Duncan (eds),
Writing Worlds: Discourse, Text and Metaphor in the Representation of Landscape (1992)

Judith Kenny

Editors' Introduction

Whether or not to approve a zoning change that would allow construction of a 153,000 square foot retail center would not be a momentous decision for many cities. However, in 1985 in the City of Portland, Oregon, this requested change that required a revision to the city's comprehensive plan sparked a heated debate. The scale of the proposed development held less significance than the challenge that the "Fred Meyer Inc. Case" posed to community leaders' ability to resolve land use disputes rationally through the technical skills of the professional planner and the guidance of an involved citizenry. Those who had taken an active role in writing the comprehensive plan assumed authorship of the legal document and, therefore, the ability to interpret it in a singular manner. That neighborhood activists and land-use legal experts lost this contentious land-use case, even after appealing it to the State Supreme Court, continues to reverberate in a city and state known for pioneering liberal planning prescriptions and concern for the conservation of a valued environment.

Many asked what went "wrong"; that is, how could an active, informed neighborhood group supported by specialists lose what originally appeared to be a clear-cut case in defense of "the public's interest"? As a Portland transportation planner during the early 1980s, urban geographer Judith Kenny (University of Wisconsin-Milwaukee) observed the land-use battle firsthand, and her observations inspired the following article. By critically examining the tradition of liberal planning and its relation to the modern city, Kenny uses this conflict to mine the ideology implicit in planning, an ideology that attempts to manage society for the "public good" while supporting the rights of individuals. Her study reveals that this ideology may be used to support contradictory positions, since it contains contradictory goals that embody multiple meanings and intentions. Furthermore, competing interpretations of Portland's planning documents are shaped by external economic and political conditions as well as by the particular goals of the interpreter. Consequently, defining the public good inevitably involves the "politics of reading." Ultimately, power relations play a role in how those plans will be translated into the urban landscape.

David Ley's analysis of Vancouver's liberal politics and the landscape text of False Creek (p. 304) comments upon planning activities that also occurred during the 1980s. A remarkable similarity may be noted between the liberal project of neighborhood activists described in the Portland study and those involved in the redevelopment of Vancouver's False Creek Southside project. Thus, both of these readings might fuel a discussion of postmodernism and landscape interpretation as well. While Ley's work emphasizes postmodernism as an architectural style with ties to the cultural and political values of the times, Kenny draws upon the epistemological critique associated with postmodernism to analyze the ways in which discursive practices shape social experience. Specifically, she deconstructs the liberal foundation of contemporary planning. While drawing from the critical spirit of poststructural discourse theory, as outlined in the following article, she distinguishes her analysis from the most anarchic versions of poststructuralism and postmodernism to incorporate its insights into a reconstructed modernist geography.

Kenny examines how a written plan becomes a concrete use of land by focusing on the concept of a discursive field, the range of competing discourses relevant to a particular realm of social practices. Her attention to the cultural politics of development proceeds intertextually by drawing connections among the City of Portland's Comprehensive Plan, the landscape values encoded in the plan, broader sets of political writings and the specific texts produced by competing interests in Portland's controversial land-use case. This focus on a planning document as a text for interpretation is an approach rarely taken by geographers. Yet Kenny argues that examining the comprehensive plan provides a powerful means of analyzing the social and political ideologies that are woven into the material fabric of the city.

As noted by Trevor Barnes and James Duncan, editors of the volume in which Kenny's study first appeared, geographers often exhibit some confusion as to the meaning of postmodernism since it is used to refer to: (1) an architectural style, (2) a critical method, and, (3) an epoch. Paul Knox's writings (p. 281) on the changing landscape of Washington, DC emphasize the architectural style, as does David Ley's article on Vancouver, BC (p. 304). In terms of the critical method discussed in this article, ambivalence surrounds the approach since radical relativism results when it is taken to an extreme, and that radical relativism makes it impossible to position oneself politically. To explore these concerns further, see the introductory comments and afterword in *Writing Worlds* (Barnes and Duncan 1992). In addition, essays by Mona Domosh (1993) and Glenda Laws (1994) examine work by geographers and planners which address the discourses that underpin urban policy and local planning. David Wilson and Jared Wouter (2003) analyze growth discourses in America's Rust Belt cities and offer another opportunity to consider the discursive framework influencing specific urban policies as they illustrate the theoretical and methodological approach. As to the use of postmodernism to define an epoch – more accurately called postmodernity – David Harvey's *The Conditions of Postmodernity* exemplifies that work.

INTRODUCTION

Summarizing the dominant view of its citizens, the City of Portland *Comprehensive Plan* (1976) begins: 'Portland is more than a geographic area – it is a way of life.' . . . Associated with [a] proud description of the city and its quality of life, the plan expresses a concern that without care [this way of life] might be lost. Addressing this possibility in conservative tones, the document indicates that 'the task facing us is to retain the most important characteristics of our city in the face of change we cannot control'. In apparently simple terms, it is accepted that the city can be defined by its landscape, since it exhibits qualities that are 'Portland'. Beyond any 'sense of place' that might be attributed

to this appreciation are values, attitudes and beliefs which define what is good and what should be reproduced in Portland's landscape.

In its broadest terms, the purpose of this chapter is to demonstrate the value of textual analysis as a means of interpreting the landscape. A planning document, possibly more than any other written text, articulates the ideology of dominant groups in the production of the built environment. Consequently an examination of a document such as the Portland *Comprehensive Plan* provides a means of analysing the social and political ideologies that are to be transformed into physical form. In the quotations cited above a liberal belief in the role of government and a commitment to certain qualities in the city's landscape are clearly indicated.

The authors of the plan have created a vision of Portland's future informed by local values and an interpretation of national trends. The resulting document represents layers of meaning that are drawn from the multiple discourses within American liberal ideology.

Thus, more narrowly defined, I will evaluate the relationship between the *Comprehensive Plan*, liberal planning practice, and the landscape of Portland. This is far from being a straightforward task. The work of several geographers warns against the assumption of a simple relationship between landscapes, ideology and social practice (Cosgrove 1984; Duncan and Duncan 1988). James and Nancy Duncan argue that while 'virtually any landscape can be analyzed as a text in which social relations are inscribed . . . naive readings commit the error of "naturalizing" these social relations' (1988: 123). Too easily the landscape's political and social underpinnings are masked by a 'taken-for-granted' world.

Conflict over the plan and the landscape to which it refers, however, effectively denaturalizes the 'taken-for-granted'. In such disputes the political nature of the interpretation and the influence of economic and social conditions are emphasized. Although the planning process and the plan's legal status imply a single meaning for the document, conflict in land use cases highlights not only competing interests but also, at a more abstract level, tensions between various discourses of liberal thought. These tensions can allow competing material interests to produce opposing textual readings. This is demonstrated in what has been described as Portland's most controversial land use decision, the Fred Meyer Inc. *Comprehensive Plan* amendment request. The Fred Meyer Inc. case is noted not only for the level of controversy surrounding the decision, but also for the company's traditional hostility to planning (MacColl 1979). Despite this reputation Fred Meyer Inc.'s representatives were able to appropriate the text by focusing debate on contradictions inherent in it. A review of this case will demonstrate the impact that the politics of reading can have on the production of landscape.

Before interpreting the conflict, however, I will outline both the textual and the contextual bases of the study. First I will address the interpretive method used and the concepts of text . . . discourse and discursive field. Then I will discuss the liberal foundations of Oregon's and, more specifically, Portland's planning programme. Since Portland provides one of the few examples in which nationally accepted prescriptions for planning are being tested (Abbott 1983: 8), an effort will be made to relate this to national trends in liberal planning theory as well as to the specific values of Portland's textual communities.

TEXT AND DISCOURSE

To call the Portland *Comprehensive Plan* a text may evoke fewer questions than applying the same term to describe a landscape. The meaning of written texts is generally thought to be accessible to any member of the 'reading public'. However, it is acknowledged that there may be cases where the meaning, being obscured by the language of planning, must be elucidated by various interpreters, such as city planners, planning consultants and lawyers employed to translate its message. In these terms the *Comprehensive Plan* would appear to fit the traditional definition of a text as an 'entity which always remains the same from one moment to the next' (Hirsch, quoted in Fish 1980: vii).

Challenging this view, however, literary critics and more recently social scientists have removed texts from the realm of objective meaning. As redefined, a text is 'the structure of meaning that is obvious and inescapable from the perspective of whatever interpretive assumptions happen to be in force' (Fish 1980: vii). This move away from understanding a text as a single stable meaning is a relatively recent development in poststructuralist literary criticism. In its most radical form, reception theory transfers authority over meaning in a text from the author to the readers. Authors' intentions are lost in the act of reading where meaning is created by a reader's beliefs and expectations.

This theory can deteriorate into radical individualism unless we recognize the social and historical bases of the reader's belief system. I would argue that without acknowledgement of the social construction of meaning, this theory becomes of questionable use to social scientists and presumably of limited value to literary critics as well. Eagleton (1983: 87) warns that a radical reader response theory is 'a simple fantasy bred in the minds of those who have spent too long in the classroom'. His argument proceeds with the reminder that: '. . . such texts belong to a language as a whole, have intricate relations to other linguistic practices, however much they might also subvert and violate them; and language is not in fact something we are free to do what we like with.' Although there may be multiple readings of a text, readings are not in any important sense unique to an individual. Meaning is contained within the limits of language, as Eagleton describes, but also has an element of stability based on the social and historical context of interpretation or discourse.

I will argue that the *Comprehensive Plan* text was produced reflecting discourses within the discursive field of liberal planning ideology. Inherent in definitions of discourse and discursive field is the politics of interpretation. As used in this essay, a discursive field is a range of competing discourses that are relevant to a particular realm of social practice. Discourses can be defined as social frameworks that enable and limit ways of thinking and acting. The definition of discourse followed in this essay is not the deterministic one of Michel Foucault (1967), which assumes that discourses are incommensurable or indisputable from the outside. Instead, following the interpretation of James Duncan, it is assumed that while some discourses are hegemonic others are contestatory. As Duncan (1990: 16) states:

> Whereas words may have different meanings within different discourses as Foucault has argued, I do not assume the impossibility of translation between discourses, nor do I reject the possibility of real, resolvable conflict between those subscribing to the terms of different discourses. Of course, I would acknowledge that the difference between the discourses may be based in real and irreconcilable material interests, and thus resolutions may often be the product of an unequal power struggle within which one group loses its voice.

Since a 'strong' definition of discourse is not employed, it is also possible to assume that a stable discursive order may exist in which competing discourses coexist in an uneasy alliance. I will argue that this is the case within the discursive field of liberal planning.

[. . .]

In the case of the Portland *Comprehensive Plan* the text is produced . . . based on its interpretation of a larger set of texts both written and otherwise inscribed (in the landscape, for example). This interpretation is the product of social discourses reflecting the hegemonic value system of dominant groups within Oregon and more specifically Portland. Environmental interests and control of land use are significant aspects of this value system. . . . [The land use laws] represent a complex interpretation of liberal planning ideology, reflecting a concern for both the rights of individuals and the rational management of society.

DISCOURSES OF LIBERALISM

Social order and individual choice

Within liberal ideology there are multiple strands of meaning which have varied in significance historically. Contrasting American and British liberalism, Gordon Clark and Michael Dear have identified two primary strands: the concern for social order and the concern for an individual's natural rights (1984: 10, 169). In the one, government's role is to act as a social engineer to redress socioeconomic imbalances and maintain fairness for disadvantaged groups; the other is concerned with individualism and the guarantee of the potential for individual choice.

The American tradition, primarily, is one of Lockean liberalism. The conception of natural rights is central to Locke's notion of individualism and justice. Throughout the Revolutionary period an individual's natural rights of life, liberty and property were proclaimed. During most of the nineteenth century liberals continued to assert that government's exercise of power diminished the liberty of individual Americans (Volkmer 1969: 3). Decentralizing power to the local level was viewed as the most advantageous means of governing, since local control was associated with a greater responsiveness to the will of the people.

The other major strand of liberalism is associated with Bentham and the utilitarians, who defined the state in terms of the need to respond to the 'greatest happiness of the greatest number' (Clark and Dear 1984: 10). This was to be accomplished with scientific rationality and state involvement in managing the problems of society. Although Clark and Dear identify these beliefs with the British tradition of liberalism, twentieth-century conditions resulted in revisions of American liberalism which increased government's role based on a similar rationale. Replacing the older fear of government, liberalism accepted the need for a relatively strong government which would protect the welfare of weaker members of society and thus create greater equality.... The basic philosophy of twentieth-century liberalism is pragmatic and secular, placing reliance on humans and their capacity to find solutions to political, social and economic problems. 'Positive freedom' for each member of society is viewed as the goal of a liberal government.

While the view of government's role has evolved, concern for the welfare of the individual has persisted as the central concept of liberalism. The traditional liberal belief in the individual's right to life, liberty and property stresses the individual. This belief distinguishes the liberal from the conservative, who is likely to emphasize the fundamental nature of property rights over the needs of society. Such a fine distinction can result in misconception of liberal philosophy. There is no hostility to property rights within liberalism, nor is there opposition to capitalism.

As Richard Walker and Douglas Greenburg argue, a common error is to view liberal reform movements as diametrically opposed to business interests (1982). Liberal ideology does not so much question the capitalist social order as suggest rational social management as a means of obtaining more equitable 'positive freedom' for society's members. Walker and Greenburg describe this rationalist ideology of scientific management as the basis of the contemporary environmental movement as well as of much planning theory (1982: 23).

Liberal planning theory

Oregon's land use laws were introduced in the late 1960s when confidence in the economic climate and planning solutions was strong. Describing the decade as a time of remarkable ferment in planning theory, Michael Dear has labelled the 1960s the period of 'new scientism and the rise of popular planning' (1986: 377). The first philosophy stressed technical solutions and rational management for the public's good, while the second emphasized citizen participation in the definition of the public good. The emphasis on popular planning here reflects the more individualistic discourse deriving from classical liberalism, whereas the emphasis on scientific management reflects the 'public good' discourse within contemporary liberal ideology. Although these two different discourses fit within the larger discursive field of liberal planning ideology, they interact uneasily. There are tensions between them which are sometimes glossed over by planners anxious to serve as many different interests as possible, but which can quickly erupt into obvious controversies. Planners attempting to juggle these two competing discourses are known to use the phrase 'a win-win situation', which summarizes the optimism of liberal planning. Conflict in land use decisions could be eliminated, it was assumed, through the technical skills of the professional planner and the guidance of an active citizenry. The implied assumptions are that goals are uniform among members of a community and that there are no contradictions between various goals.

Oregon would serve as one of the first laboratories for these new planning prescriptions.

Sprawled development and its pressure on agricultural land was a concern shared by many in the rural areas of western Oregon. The impulse for state-wide planning came as much from these areas as it did from the new environmentalists associated with the cities.

Ten thousand citizens participated in public meetings where state goals were developed, linking urban and rural concerns. With an unusual combination of liberalism and conservative language, the goals were written to direct growth while maintaining 'what Oregon is'. At this time the phrase the 'Oregon tradition' was introduced and cited frequently. That tradition was described in terms of Oregon's history as a 'pioneer' rather than a 'boom' state. Various historical references were made, citing the efforts of nineteenth-century residents and their 'pains taken to build up an orderly and enlightened society' (in contrast to California, where 'the land systems and titles were in a state of chaos', Lyman 1903: 38). Its twentieth-century land use and environmental efforts were felt to be a modern pioneering effort, and Oregon's landscape became synonymous with the Oregon tradition. An example is provided in a book of scenic photographs. The brief introduction provided by the editor instructs: 'If you move there be prepared to love it and protect it. That's become an Oregon tradition' (*Oregon* 1985).

Public choice theory, with its emphasis on citizen participation in planning, was translated easily into another Oregon tradition, that of local control. By creating a plan through the efforts of the community, it was assumed, the interests of members of that community would best be met. At the same time, planners at the state level were responsible for reviewing local plans for consistency with overall state requirements. This two-level process was intended to acknowledge both the interests of local control and the necessary collective action required for scientific and rational planning in the interests of the state population.

[. . .]

While the collective need is used as the standard of decision-making, there is a lengthy amendment appeal process that focuses on the rights of the individual. The amendment request can be heard by up to six official 'interpreters' of the text before the process is exhausted . . . [ending with] the State Supreme Court. This quasi-judicial process can be quite time-consuming and expensive, but it is viewed as planning's system of checks and balances.

As a legal document with the necessary support to maintain its enforceability, the Comprehensive Plan is a specific written text with a perceivable impact on the landscape of Portland. The basis of this planning apparatus is found in the 1960s debates of planning theory, inspired by the strands of liberal philosophy. Other reform groups of the 1960s and 1970s clearly shared similar influences and expressed similar goals (Ley 1980). Portland's plan and its implementation, however, must be 'read' as a local interpretation of culturally and historically specific discourses. The 'Oregon tradition' legitimizes certain values and beliefs, while at the same time the phrase is related to, if not created by, contemporary academic and popular views of American society.

Those who have authored the plan have relied on the texts of planning theory and the larger 'text' they call 'tradition'. The future landscape of Portland, however, will not simply be structured by authors' intentions. Various competing interpretations of the plan, influenced by economic and social conditions, will be based in different political discourses within the larger discursive field of liberal planning ideology, and it will be this competition which will determine what aspects of the larger ideology predominate in the 'text' on the ground. Ultimately the relationship between plan and landscape is an issue of power – whose interpretation will prevail in the 'politics of reading'?

POLITICS OF READING

Authors' intentions

During the major battles that ensued over the construction of a large Fred Meyer shopping facility in inner north-east Portland, various residents repeatedly voiced belief in the planner's ability to resolve issues of public interest. One resident expressed these views by saying:

> Sullivan's Gulch has a record of working in a positive fashion in land use issues. We are proud of the work done on the Comprehensive Plan. Many volunteer hours were put into developing a strong, viable, workable plan to take us through the next twenty years of growth. . . . (Public hearing, November 1985)

In effect, neighbourhood representatives claimed authorship of the plan and assumed an interpretation necessarily consistent with those who shared in its creation: 'I believe the City's decision regarding this proposal will affect the future integrity of the *Comprehensive Plan* and that the future of the Northeast neighborhoods lies in the balance. . . . The City and State are responsible for upholding these policies just as the police are responsible for upholding our laws.'

This knowledge of, and commitment to, the planning process on the part of neighbourhood activists contributed to the length and significance of the Fred Meyer land use battle. Portland's mayor, Bud Clark, described the Fred Meyer *Comprehensive Plan* amendment case as the most difficult decision he had faced as mayor. Given the other issues dealt with by Portland's leaders, including racial tension between the police and Portland's black community and a threatened police strike, one might wonder at such a sense of priorities. An explanation might be provided in the mayor's background as a former neighbourhood activist and a small businessman. . . . The controversy surrounding the plan amendment was undeniably a challenge to notions of a singular meaning for the text.

The controversy began in 1985 when Fred Meyer Inc. presented a request to the City of Portland for an amendment to the *Comprehensive Plan* map and zone change on a 13.4 acre parcel of property in north-east Portland. Formerly the site of a manufacturing operation, it was necessary to amend the property's industrial use designation before construction of the proposed 153,000 square foot shopping facility would be allowed. As the applicant for the amendment, the Fred Meyer company was required to demonstrate that the plan map revision was not only appropriate but necessary. The argument provided was that the new shopping centre would fill unmet market demand in the area. Citing the *Comprehensive Plan*'s economic development goal, Fred Meyer asked the city to demonstrate that it truly was 'open for business'. Given the poor economic climate and the city's campaign to let firms know that Portland was hoping to grow, this posed a more significant challenge than it might have when the plan was adopted in the 1970s. . . . In the early 1980s, the impact of recession on Oregon's timber economy resulted in a decline in Portland's and the state's population.

The proposed development site is located in the midst of several older neighbourhoods. It has been said that these older residential sections within five miles of the downtown give Portland its recognizable character. Although there has been residential change in Portland's inner-city neighbourhoods, this change is not characterized as gentrification so much as a transition from an older to a younger population. An older housing stock that belongs to working-class and upper middle-income occupants has been maintained in these residential areas. In each, a strong neighbourhood commitment has been demonstrated in planning and civic activities. The eight organized neighbourhood associations that represented the inner north-east voiced a strong negative reaction when the Fred Meyer proposal was made public. Objections were based on the proposed development's traffic impact on the neighbourhood and the anticipated adverse impact on the existing businesses in the area. . . . Citing its violation of the neighbourhood, transport and economic development goals of the *Comprehensive Plan*, the neighbourhood organizations launched a coordinated opposition to the plan change and the proposed development.

The organized protest against the proposed revision demonstrated the well articulated belief system of the active residents. Consistently, residents cited their commitment to the urban character of their neighbourhood and explained this on the basis of both their interest in the particular environment of the residential area and their concern for a 'healthy' Portland.

[. . .]

The neighbourhood associations and individual residents who were opposed to the proposed land use amendment supported a reading of the plan that they felt was consistent with its liberal text. Rather than calling for no growth, they argued for 'controlled growth'. Economic development and neighbourhood goals were weighed equally in their arguments. The proposed shopping centre, with its potentially negative impact on both residential and business areas, was viewed as a proposal that 'simply does not fit the place or time' (public hearing, November 1985). City planning staff and the Hearings Officer also recommended denial of the requested plan map amendment.

Contested readings

Fred Meyer Inc. appealed against this negative decision to the City Council, while at the same time a Fred Meyer official appealed to public opinion by describing frustration with the planning process. . . . In a more private forum, Fred Meyer management called for the help of Portland business people:

The business climate of Oregon is *the* topic of discussion in board rooms, over lunch, on street corners, in every political utterance before any gathering and even during football games. We strive daily to improve it for our companies, ourselves and all Oregonians. . . . Now we must ask for your support, help and suggestions.

Our request for this land use change has been heralded as the 'acid test' of Oregon's land use laws; as the land use decision of the 'century', the 'true' test of Oregon's quest for economic development; and the test of the cities [*sic*] responsiveness to Neighborhood Associations. (Fred Meyer letter, 30 October 1985)

The founder of Fred Meyer Inc., Fred Meyer, had been known for his opposition to public planning and his belief that 'people will act in their own best interests and society will benefit accordingly' (MacColl 1979: 630). Seven years after his death, Fred Meyer's successors were presenting similar arguments to the business community and the public at large. In the land use hearings, however, a different sentiment was required for a successful campaign.

Using the language of the plan text, the Fred Meyer consultants, lawyers and management staff attempted to demonstrate that there was an interest other than their own to be served by constructing the proposed facility. The public would benefit because the facility would serve an unmet market demand and create new jobs. Those opposing the development accused the Fred Meyer consultants of 'magical thinking' in their efforts to develop a case for a new land use that would not create any change in the surrounding area. Minimizing any potential negative impact, the Fred Meyer consultants argued that the new facility would neither bring in significant amounts of new traffic nor drain business away from the existing commercial centres. Community organizations, according to the Fred Meyer representatives, were overreacting to the potential negative impact of the facility based on miscalculations. . . . The proposal was argued to be consistent with the goals of the city as expressed in the *Comprehensive Plan*. In effect, it was said, this proposal offered a 'win-win' decision for the city even if it was not recognized as such by the neighbourhood groups that would most directly 'benefit' from it.

In addition to the testimony of experts, Fred Meyer built its case on the interest expressed by individuals within the area, conducting several large surveys of residents. Significant support for the store was documented among area residents. While Fred Meyer was reminded by members of the neighbourhood association and the City Council that planning was not a democratic process, their figures did challenge the neighbourhood organizations' ability to represent the position of area residents. With the *Comprehensive Plan*'s liberal emphasis on citizen involvement, certain City Council members questioned the role of the neighbourhood organization versus the individual residents who responded in support (City Council Record, January 1986). The neighbourhood organizations, it was implied, were elitist and not capable of addressing the issue of public need.

The definition of citizen participation in the planning process became problematic. Although the neighbourhood organizations argued that their position represented informed opinion, as a result of their participation in the development of the *Comprehensive Plan*, a cloud was cast on the issue of public choice. A resident in opposition to the development responded to this debate by saying:

Land use planning means never having to say you blew it. No amount of public opinion will make our streets safe again after several hundred to several thousand vehicles a day are added to every block in my immediate neighborhood. No amount of postcard returns will bring back the ambiance to this inner-city neighborhood after it's been rendered commonplace.

Fred Meyer's arguments, however, implied that the neighbourhood group represented self-interest rather than informed opinion. From their position close to the proposed development site, it was stated, the neighbourhood activists had lost sight of the public good.

Highlighting the economic development goal, the Fred Meyer position was focused on an ability to achieve rational management and find technical solutions to achieve the public good. It was this belief that rational scientific management could remedy development problems that resulted in a City Council decision in favour of the shopping complex's construction. Two of the council members voted against the amendment in the belief that the project would have a detrimental impact on the area's neighbourhoods and business district. Two voted in support of the amendment on the basis of its perceived economic benefits. Mayor Clark, who had been a neighbourhood activist himself, was the crucial vote. Citing the need for economic development and the potential for mitigation of any use-related problems, he voted in favour of the proposed plan amendment.

The land use case did not end with that decision. Members of the neighbourhood organizations and their legal representative . . . appealed against the council's decision. During the early stages of the appeal process there was a certain confidence on the part of the appellants. The intent of the *Comprehensive Plan*, it was felt, was clearly at odds with the decision on behalf of Fred Meyer.

In March 1988, nearly three years after the case was introduced, the State Supreme Court upheld the City Council's decision approving the plan amendment. At each level of the appeal it was decided that no procedural error had been made and that the amendment was consistent with goals expressed in the plan. The official interpreters of the text gave approval to the council's alternative reading.

CONCLUSION

The Fred Meyer case became a paper war over interpretation of a planning document. Behind what appeared to be simply battling interests were the 'politics of reading'. Different textual communities found within the legal document alternative meanings and, consequently, multiple readings. Duelling technical consultants and issues of public participation and representation filled hours of public hearings. I would argue that these tensions and ambiguities are inherent within the discursive field of liberal ideology, with its attempt to manage society for the public good while supporting the rights of individuals. The difficulty of achieving this balance is highlighted by the varied interests within a community. Defining the public good becomes a matter of power. In a poor business climate, confidence in Portland's ability to manage growth had been shaken and, consequently, a new interpretation of the city's goals gained political favour. The availability of jobs and the concern with economic health were more conspicuous in people's evaluation of the quality of life in the 1980s. The changing economic conditions provided a new context for evaluating issues of planning theory and practice. As Terry Eagleton describes, the act of reading provides a 'frame of reference within which to interpret what comes next' (1983: 87).

REFERENCES FROM THE READING

Abbot, C. (1983) *Portland: Planning, Politics and Growth in a Twentieth Century City*, University of Nebraska Press, Lincoln, Neb.

City of Portland (1976) *Comprehensive Plan*, City Planning Department, Portland, Oregon.

Clark, G. and Dear, M. (1984) *State Apparatus: Structures and Language of Legitimacy*, Allen and Unwin, Boston, Mass.

Cosgrove, D. (1984) *Social Formations and Symbolic Landscapes*, Croom Helm, London.

Dear, M. (1986) 'Postmodernism and planning', *Environment and Planning D: Society and Space* 4: 367–84.

Duncan, J.S. (1990) *The City as Text: The Politics of Landscape Interpretation in the Kandyan Kingdom*, Cambridge University Press, Cambridge.

Duncan, J.S. and Duncan, N. (1988) '(Re)reading the landscape', *Environment and Planning D, Society and Space*, 6: 117–26.

Eagleton, T. (1983) *Literary Theory: An Introduction*, University of Minnesota Press, Minneapolis.

Eagleton, T. (1986) *Against the Grain*, Verso, London.

Fish, S. (1980) *Is there a Text in this Class? The Authority of Interpretive Communities*, Harvard University Press, Cambridge, Mass.

Foucault, M. (1967) *Madness and Civilization: A History of Insanity in the Age of Reason*, trans. R. Howard, Tavistock, London.

Ley, D. (1980) 'Liberal ideology and the postindustrial city', *Annals of the Association of American Geographers* 70: 238–58.

Lyman, H. (1903) *History of Oregon: The Growth of an American State 4*, North Pacific, New York.

MacColl, E. Kimbark (1979) *The Growth of a City: Power and Politics in Portland*, Georgian Press, Portland, Oregon.

Oregon: Pictures to Remember Her by (1985) Crescent Books, New York.

Portland City Council (1986) Hearing transcript, 29 January, Portland, Oregon.

Public testimony before the Portland Planning Hearings Officer: Fred Meyer Inc. Comprehensive Plan Amendment Request (November 1985) Portland, Oregon.

Volkmer, W. (1969) *The Liberal Tradition in American Thought*, Putnam, New York.

Walker, R. (1989) 'What's left to do? Some principles to live by', *Antipode* 21, 2: 133–65.

Walker, R. and Greenburg, D. (1982) 'Post-industrialism and political reform in the city: critique', *Antipode* 14, 1: 17–36.

Editors' references and suggestions for further reading

Barnes, T. and Duncan, J. (eds) (1992) *Writing Worlds: Discourse, Text and Metaphor in the Representation of Landscape*, Routledge, London.

Domosh, M. (1993) "Imagining our cities: utopic visions and contested realities," *Urban Geography* 14, 6: 568–73.

Harvey, D. (1989) *The Conditions of Postmodernity: An Enquiry Into the Origins of Cultural Change*, Oxford University Press, Oxford.

Laws, G. (1994) "Urban policy and planning as discursive practice," *Urban Geography* 15, 6: 592–600.

Wilson, D. and Wouter, J. (2003) "Spatiality and growth discourse: restructuring of America's rust belt cities," *Journal of Urban Affairs* 25, 2: 123–39.

"Sunshine and Shadow: Lighting and Color in the Depiction of Cities on Film"

from S. Aitken and L. Zonn (eds), *Place, Power, Situation, and Spectacle* (1994)

Larry Ford

Editors' Introduction

How does the depiction of cities in popular media influence our response to urban issues and problems? What understanding might be achieved by paying closer attention to the cinematic images that entertain movie viewers? Although the need for dramatic effect influences the scenarios and characterizations constructed by filmmakers, the resulting images draw on a larger public's collective knowledge of urbanism. The complex social, political and ideological forces that shape representations of urban spaces in film warrant examination. Thus, geographers could usefully pay much closer attention to the "city of cinema," as these images are part of a complex cultural process in which meanings are produced and consumed.

In the introduction to their anthology on the geography of film, Stuart Aitken and Leo Zonn (1994) note that this idea of film as a medium that shapes our understanding of place had been left largely unexplored. Exceptions such as Burgess and Gold's (1985) as well as Zonn's (1990) collections on geography and the media were offered alongside their book as evidence of what might be accomplished as they called for a new research direction focused on the cultural texts embodied in films, and the manner in which they produce and communicate meaning.

Larry Ford, an urban cultural geographer at San Diego State University, contributed the following reading concerning the changing role of cities in American and European cinema to Aitken and Zonn's book. Ford focuses on the narrative conventions which filmmakers use in order to examine the intertextual and multi-vocal representations that constitute their versions of urban reality. Although "the city" began as simply a background for the action in the earliest American movies, it gradually assumed a major role in narrative by the 1940s with the development of *film noir*. Lighting, the sunshine and shadow of Ford's title, was used in a blatantly allegorical manner in the already black-and-white film depiction of the *noir* city. This particular genre, inspired by the pulp crime novels of such authors as Raymond Chandler, portrayed the city as a shadowy, dystopian world of fear and desire. Raymond Chandler himself summarized, "the streets were dark with something more than night" (cited in Krutnik 1997, p. 83). Although *film noir* lost prominence in the 1950s and even greater significance with the advent of color-film movies, its use of light carried over into the films of later decades through parodies and transference of light and dark. Ford suggests the impact of *film noir* lighting conditions on more recent films such as *Bladerunner* (1982) and *Grand Canyon* (1991), and, through a reversal of the light and dark narrative conventions, in the satire of suburbia portrayed in *Edward Scissorhands* (1990).

Directors use other twists on color and light to tell their urban tales. For example, Woody Allen has reversed lighting conventions to make night-time New York heroic and an over-exposed Los Angeles villainous, thus underscoring alternative possibilities for sunshine (bright color) and shadow (dark color). Drawing on a large filmography, Ford skillfully interprets the manner in which "sunshine and shadow" characterize the city as the director tells a story.

Although a decade has elapsed since the publication of Ford's essay, geographers' attention to representations of the city in the visual arts remains limited. David Clarke's *The Cinematic City* (1997), a collection by geographers and film studies specialists that suggests the "central but uncredited role" which urban space plays in a number of films, is a notable exception. Other important analyses of spatial and temporal articulations in films include David Harvey's consideration of *Bladerunner* in "Time and space in the postmodern cinema" (1989) and Natter and Jones' interpretation of *Roger and Me* (1993). For further reading on the pedagogical use and scholarly value of critical readings of urban images on film, see Leigh and Kenny's article "The city of cinema" (1996).

. . . . The purpose of this chapter is to trace the ways in which cities have been depicted in films over time with a special focus on the roles of lighting and color. More specifically, I hope to affirm with this focus some recent theories that suggest that the role of cities in film gradually changed over time from serving as mere background scenery to acting as the equivalent of major characters in many stories. My contention is that lighting and color are of major importance in this development. Second, I hope to show that in recent years, cities have been illuminated in increasingly complex and often contradictory ways in films and that by examining this topic, we may add an additional layer of understanding to our knowledge of place-making and place representation. Sense of place is of vital concern to geographers, yet it remains a nebulous concept. Films provide controlled and replicable visual experiences that can enable us to study the ways in which people and places interact as stories unfold.

THE CITY AS A BACKDROP

In the early days of American silent films during the 1920s, cities were used as random, often unidentified, stages for action. City scenes appeared in many of the early film comedies. Memorable images include Harold Lloyd hanging from a clock tower (*Safety Last*, 1923) and Laurel and Hardy pushing a piano up a steep hillside stairway (*The Music Box*, 1932), but the city was a stage, not a player. It did not influence unduly the psyches of the human participants in the stories. There were tall buildings, traffic jams, busy sidewalks, and so forth, but Harold Lloyd and Laurel and Hardy took them all in stride. The city was just "there." For technical reasons, most films had to be shot in bright daylight and so the mood of the films was not much affected by changing levels of lighting. In America, the focus of early films was on the actors. A strong tradition of stage comedy and Vaudeville meant that famous stars could transfer their antics to film and be presented to an even wider and more appreciative audience. One could argue that America was naive and unsophisticated in its demand for entertainment compared to European *haute couture* and that Charlie Chaplin doing slapstick comedy was what people wanted to see. City scenes were used most often because they were handy. Much invaluable footage of early Los Angeles is found in these films, but the city as a purposeful character is not fully developed. . . .

This casual concern for the role of place and setting was less prevalent in Europe where early films were seen as part of artistic movements rather than as mass entertainment. During the early 1920s, for example, German Expressionist filmmakers were creating urban worlds (on artificial sets) that were carefully contrived to contribute to a sense of mood and to enhance the emotional valence of films. In many, cities are depicted as brooding, tension-filled places that actually participate in the character's descent into nightmarish predicaments. Dark shadows are often contrasted with tense, malevolent light. The city was depicted as having an uneasy ambience that contributed mightily to human despair. In films such as *The Cabinet of Dr. Caligari* (1919) and *The Blue Angel* (1930), the claustrophobic, isolating settings are important in creating the mood of the stories. In Fritz Lang's *Metropolis* (1926), the city itself becomes a monster villain contributing directly to the misery of the human characters. . . . In Europe, films were often made for a relatively small audience by directors who sought to author a piece of art. In America, the goal was more often to create a blockbuster that would make lots of money and showcase a star.

The chaos in Europe during the late 1920s and early 1930s brought an end to the creative period in German filmmaking, and many German directors and actors sought refuge in America. By the mid-1930s, Hollywood was the undisputed film capital of the world, but by then everyone was preoccupied with the advent of sound, and experimentation with lighting took a back seat to finding actors who could speak in sentences and perhaps even sing. Early talking pictures involved a great deal of talking. Many films look like live theater captured on celluloid. Actors were recruited heavily from the theaters of Broadway, and fast-paced stagelike dialogues dominated the screen. Dazzling, larger-than-life musicals became very important money-makers in Depression America. Interior scenes came to dominate exterior settings as films moved inside, at least in "city films" (less so in westerns, etc.). Fred Astaire and Ginger Rogers dance around ballrooms, and Chicago gangsters argue and fight in hotel rooms. The city itself was still largely a backdrop – something to be seen from restaurant windows or speeding cars. Outdoor urban scenes often feature the decks of luxury liners or penthouse patios. During the 1930s, big cities were seen as fantasy lands full of tuxedo-clad dancers and cigar-chomping gangsters but the presence of the city was lightly felt. *Forty-Second Street* (1933) was not really a place as much as a state of mind. There were, of course, some exceptions. When King Kong (*King Kong*, 1933) meets his fate climbing the newly completed Empire State Building, the symbolism is powerful – the best and biggest of the modern world defeats the best and biggest of the traditional world. Four decades later when King Kong once again visits New York, in 1976, it takes TWO skyscrapers to defeat him.

NIGHT AND THE CITY: GERMAN EXPRESSIONISM AND FILM NOIR

By the 1940s, carefully contrived city scenes were becoming more important in many American films. World War II shortages made big, expensive films difficult to produce and actors and directors turned to lower budget dramas to fill the theaters. It may also be that audiences were simply ready for a change. The endless Broadway extravaganzas full of tap dancers with feather boas were becoming somewhat predictable and the novelty of sound and music was wearing off. In the absence of big orchestras and chorus lines, many directors sought to create a stronger film atmosphere through creative lighting and camera angles. Directors such as Billy Wilder, Otto Preminger and Fritz Lang had been trained in the German Expressionist tradition of the 1920s and were ready to show what they could accomplish with low-key lighting, a few dark alleys, and rain-slick streets. These dark and brooding urban dramas were later dubbed "Film Noir."

It is difficult to say just when Film Noir (and the Film Noir city) first emerged. The term was coined after the fact in France during the late 1940s and did not come into wide usage until the mid-1950s. The earliest date often suggested is 1941 when *The Maltese Falcon* and *Casablanca* [1942] appeared. The latter in particular developed the theme of people watching their lives and their worlds go slowly out of control. Others would put the beginning closer to the mid-1940s with the production of such classics as *Double Indemnity* (1944) and *Scarlet Street* (1946). The latter view holds that fully developed Film Noir was too bleak and morally ambiguous to be accepted by the wartime public. At any rate, the heyday of the genre was the late 1940s, a time when new philosophies of the individual and concern with anomie came to the fore. Film Noir and existentialism went hand in hand. By the late 1940s, audiences knew that the world was a complex and often evil place and that some discussion of this was in order. Tap dancers could sugarcoat the Depression for a while, but Nazis and Iron Curtains eventually wore them down. It was time to examine the depths of human nature.

Isolation and anomie in the midst of the city was a favorite theme, although different directors and different national contexts sometimes suggested other ways of depicting the increasing concern for these existential effects. British filmmakers tended to focus on the importance of social class, often with neo-Marxist overtones. In *The Loneliness of the Long-Distance Runner* (1962), for example, a reform school inmate becomes a working-class hero to the Establishment because of his ability to win races against "better" schools. In the end, he recognizes the idiocy of his role (and society?) and refuses to finish a race he can easily win.

[. . .]

Deciding when Film Noir began may be easier than deciding just what Film Noir is and just what films belong in the genre. In its purist form, however, the type is fairly easy to describe: Film Noir usually features a psychological drama in which seemingly normal people are drawn ever deeper into a very personal, isolating nightmare. Mistakes are made, crimes are committed, and characters gradually lose touch with their normal lives and friends. There is moral and ethical ambiguity in the sense that everyday people may gradually become criminals as if by fate. Many of the actors associated with the genre were selected to epitomize ambiguity in contrast to the swashbuckling heroics of the films of both the 1930s and the late 1950s. . . .

Even strong characters like Humphrey Bogart and Robert Mitchum exhibited a sort of tight-lipped resignation as they were drawn into their respective fates. Why fight it when night comes to the city? Night scenes, usually very dark and shadowy, dominate the films. Low-key lighting and strong black and white contrasts, often provided by streetlights shining through venetian blinds, are also used to create the dark psychological moods that led to the term Film Noir. The city plays an important role in the development of a nightmare atmosphere. In the best films, however, the role is subtle. The city itself is not depicted as horrible or nightmarish but rather as a setting that gradually contributes to the development of such feelings.

The Film Noir city is a tense, brooding, lonely, isolating place that tends to push people over the edge. Typical titles were *Night and the City* (1950), *The Naked City* (1948), *Side Street* (1950), *The Street with No Name* (1948), *The Asphalt Jungle* (1950), *The Dark Corner* (1946), *Cry of the City* (1948), and *Nightmare Alley* (1947). . . . Film techniques borrowed heavily from the German Expressionists but the urban images were also derived from American Realist artists such as George Bellows, Reginald Marsh, and Edward Hopper. City scenes were presented as strangely stark and aloof. Still photography may have also had an influence – especially the stark urban scenes of artists such as Alfred Steiglitz.

In *Scarlet Street* (1946), Chris Cross (Edward G. Robinson), a solidly middle-class clerk leading a humdrum life in a small apartment with a dominating wife, chooses to walk home alone after an award dinner. The night is dark and rainy and he becomes disoriented and confused in the mazelike streets of Greenwich Village. He witnesses a man hitting a woman, and in coming to her rescue he begins his descent into mayhem and, finally, murder. In such a setting, he is helpless to resist. "It's the city that's getting to us." He enters a shadowy, claustrophobic world with no light at the end of the tunnel. The atmosphere is not all-pervasive evil as in later terror films, but rather gives the feeling of ambiguous tension and uneasiness. I contend that this mood has stuck with us and that American cities have been associated with ambiguous tension for the past fifty years. But that is getting ahead of the story.

Like *Scarlet Street*, most of the early Film Noir dramas were filmed in soundstages rather than real city streets. The sense of isolation could thus be complete with designed-in emptiness and foreboding. Lighting was often blatantly allegorical with happy times bathed in the noonday sun while the worst moments occurred in almost total darkness with perhaps only a flashing neon sign to illuminate the scene. The fact that everything was artificial and could be controlled contributed to the nightmarish emptiness in Film Noir atmosphere. Back and side lighting could create just the look of uneasy weariness on the face of the hapless victim of life in the big city.

By the late 1940s, however, the stage set was beginning to lose some of its appeal and the trend was toward films made "on location," especially if the location was nearby. . . . By the end of the war, documentaries had become popular in Europe as liberation revived the possibility of real news and the accurate filming of rapidly unfolding events. In Europe, artistic directors were also intrigued by the documentary approach, partly because elaborate soundstages were simply not available and also because the Nazi era had made people suspicious of contrived and unauthentic art. There was a need to film real people in the real world in order to establish credibility. . . . [N]owhere was the need for "real world representation" more vital than in Italy where many postwar films were seen to represent a movement dubbed Italian neorealism. The influence of Italian neorealism was being felt in America by 1948 as directors began to seek a more documentary approach in urban crime dramas. Capturing, but at the same time controlling, the real city became the challenge of the day. . . .

REALISM AND THE CITY

Postwar Italian neorealism films such as *The Bicycle Thief* (1948) and *Open City* (1945) demonstrated to American directors that real urban scenes could be just as conducive to the creation of an atmosphere of lonely uneasiness as the stage set had been. In addition, the use of real settings could lend a sort of exposé/documentary quality to the dramas that could take advantage of the public's familiarity with the newsreels being shown widely in theaters. Location films such as *The Naked City* (1948), *Criss Cross* (1948), and *Kiss of Death* (1947) used neorealist techniques to create a city that is more than a merely neutral and uninflected backdrop. The city is molded into a powerful neurotic element in the story. According to Hirsch (1981, p. 17), for example, in the brilliant *Night and the City* (1950), a real London, oozing with slime and enshrouded with fog, becomes a maze of crooked alleyways, narrow cobbled streets and waterfront dens; a place of pestilential enclosure.

[. . .]

THE CITY OF MONSTERS AND THE DEATH OF A FILM NOIR

Classic Film Noir was done in black and white, and whether true Film Noir can be done in color is a hotly debated point. Color changed everything. To a very real degree, however, Film Noir was on its way out well before color films took over. While the heyday of the genre was the late 1940s and early 1950s, variations on the Film Noir theme continued until at least 1960 when *Psycho* [1960] and *The Hustler* (1961) were released. By that time, however, such films were becoming few and far between. Black and white films remained predominant through the 1950s but Film Noir gradually gave way to other types of films as the decade progressed. Westerns were particularly popular, as were films featuring an assortment of oversized monsters. Perhaps inspired by concerns over atomic war and possible associated mutations, moviegoers were treated to visitations from giant ants, spiders, space creatures, and blobs. The city (once again an artificial city) became something to be either stepped on or eaten. The depiction of the city as a major character diminished as more scenes were designed to simply blow up in a picturesque and exciting manner. Lighting still played a role in creating desired effects, but artistic subtlety waned.

Film Noir died in the early 1960s but vestiges of the Film Noir city live on. The remainder of this chapter focuses on the ways in which new color and lighting technologies have been used to reproduce, redefine, modify, and occasionally reverse the image of the city as created in Film Noir.

THE TECHNICOLOR CITY

Directors experimented with color almost as soon as films were invented. Some tried hand coloring but that proved to be too expensive and tedious. Others tried tints such as blue for night and amber for fireside romances. For the most part, however, filmmakers stuck with black and white until color film was invented in the 1930s. . . . It was not until the late 1960s that color films became the norm. It was also not until the 1960s that color films looked truly accurate as Technicolor brightness gave way to more varied and sophisticated experiments with hue and tone.

[. . .]

Primary colors gave way to more somber and subtle greens and grays. A neo-Film Noir quality appeared along with a new emphasis on realism in *Midnight Cowboy* (1969). A darker New York of nighttime scenes and somber grays replaced the pervasive brightness. . . .

Perhaps the final and most pervasive attempt to create a Film Noir city occurred in the making of

Blade Runner in 1982. Set in Los Angeles in the year 2020, the dark, oppressive cityscape plays such a powerful role in the film that it all but overwhelms the characters. The sky is always dark and polluted, the streets are always wet and foggy, the alleys always narrow and claustrophobic, and the lighting is strong and focused. In addition, there are the ingredients of moral and ethical ambiguity, despair, and isolation. The connection between *Blade Runner* and its Film Noir heritage is solidified by the use of Los Angeles' historic Bradbury Building, an oft-used setting in the detective stories of the 1940s. The role of the city is so rich and powerful in *Blade Runner* that any further attempts to recreate the Film Noir city as a major dramatic element may seem trite and underwhelming. The otherwise effective setting used in *Brazil* (1985), for example, seems to have less impact than it might have as a result of the impact of *Blade Runner*. New images of the city were needed. It was no longer enough to pit darkness against light. The Film Noir city was in danger of becoming hackneyed.

WOODY ALLEN'S *ANNIE HALL*: A REVERSAL OF IMAGES

Woody Allen, according to Woody Allen, likes cities and dislikes the countryside, likes New York and dislikes Los Angeles, likes somber colors and dislikes bright colors. This all comes together in his 1977 film, *Annie Hall*. The main character, Alvy Singer, loves a New York that is filmed almost entirely in dark and subtle greens and grays. New York is seen as visually calm with few strong contrasts or bright accents. Gray-green trees and green-gray buildings provide a comfortable background for chameleonlike, gray-green-clad Alvy Singer. Nighttime is also depicted as comfortable – a time for strolling and chatting. The city has a role in the film but it is a muted, subtle role like a favorite old sweater. The cityscape contributes to a mood of serenity.

Bright lights and colors appear chiefly when there is conflict. Arguments between the characters often occur in association with the appearance of bright colors such as yellow taxis or bright red doorways. In the scene where Annie and Alvy finally break up, both characters are (uncharacteristically) wearing red. When the story shifts to the West Coast, it is in Los Angeles that the city as ambiguous tension reappears.

Los Angeles is awash with blazing sunshine in *Annie Hall*. The antipathy that Alvy Singer has for California is symbolically represented by a pervasive, glaring sunlight. Sunlight is reflected off bland buildings, automobile windshields, and residential patios. Characters are often backlit as they stand in front of windows or patio doorways. An interesting reversal of Film Noir lighting occurs as the characters appear darkened while a halo of bright light surrounds them. At other times, the actors are depicted as sort of fading away and/or washing out into the inescapable sunshine. Most of the characters wear white or a variety of bright clothing that makes the brightness even more pervasive (*Film Soleil*?). At one point, a character dons a space suit in order to safely drive a convertible into the blazing California sun as Alvy Singer asks, "What are we, driving through plutonium?" . . .

Reflected sunlight seems to represent Allen's view of the lack of depth of Southern California culture as if the glaring sun limits access to subtle meaning and deeper truths just as it limits access to subtle colors and shading. Everything is surface/superficial. In "shallow" Los Angeles, the characters wear bright colors and romp in the sunshine but the setting is somehow less serene than subtle, gray-green New York. The roles of light and dark are not only expanded but reversed.

The films of Woody Allen represent a purposeful reorienting of our image of the (traditional) city. He admittedly tries very hard to select scenes that will portray Manhattan in a positive light and to emphasize the pleasures of urban life. In a sense, he is reflecting an emerging pro-urban ideology associated with the pervasive gentrification and yuppification of many cities during the 1970s. Woody Allen films such as *Annie Hall*, *Manhattan* (1979), and *Hannah and Her Sisters* (1986) are not so much blatant "boosterism" films for the city as illustrations that normal people can live good, interesting lives in the midst of comfortable but very urban landscapes. The city itself is more a calming influence than a menacing one. When interpersonal tension arises, there is always a friendly coffeehouse or a good street to walk down. The city is still a major player in the film but in a more subtle, complex way. It is backdrop but it is not mere backdrop.

Woody Allen has continued to experiment with lighting and color. In 1979, he presented Manhattan in black and white (by now a daring thing to do). In the opening credits, Allen makes it clear that he feels New York City is BEST seen in black and white. "He adored New York City . . . to him, no matter what the season was, this was still a town that existed in black and white and pulsated to the great tunes of George Gershwin" (Allen 1980, p. 181). Once again, the shadowy city is depicted as a friendly place. A decade later, in *Scenes From a Mall* (1991), he represents the (for him) polar opposite Los Angeles as a brightly lit, almost totally white shopping mall, which is nothing more than a bland backdrop for the trials and tribulations of the main characters. Woody Allen and Bette Midler are in the mall but somehow they cannot interact with it. It is all surface. It is as though "whiteness" and sunshine now contribute to the isolation and anomie that were once the forte of dark and rainy nights. The characters can interact comfortably with green-gray New York but the white glare of Los Angeles is isolating. In a sense, Allen perceived the increasingly negative image of Los Angeles and perhaps California as a whole, which is now pervasive.

UNEASY SUNSHINE, A CONTINUING TRADITION?

Over the past decade, there have been a number of films that have helped to perpetuate the reversal of day and night and light and dark in the creation of a disturbing urban atmosphere. In *Choose Me* (1984), for example, the characters seem at home in the friendly confines of a dark and dreary Skid Row but in the light of day, they seem awkward and disoriented. The city at night seems warm and comforting whereas sunshine seems strangely oppressive. The light and dark reversal here is perhaps related to character development in that the people in the story have night jobs and are more used to night living, but there is more to it than that. Dark and rainy "claustrophobic" alleys just do not threaten anymore. White heat does.

[. . .]

In *Edward Scissorhands* (1990), an overly colorful, sunny suburban landscape is seen as sterile and superficial. The people there are generally amiable but their lives are bland and empty. Edward Scissorhands, a Frankenstein-like creation, adds depth and meaning to life in suburbia. His home is a classic Psychoesque abandoned Victorian on the top of a hill. Significantly, the castlelike compound is depicted as dark and nearly devoid of color. The images of good and evil (or at least good and not-so-good) are completely reversed.

In other films the reversal is not quite so blatant. In *Grand Canyon* (1991), for example, Los Angeles is a very disturbing city but there is no difference between night and day. Some of the most obvious criminal encounters take place in classic Film Noir settings but bright, sunny days bring little respite from the tension. They almost seem worse simply because crime is supposed to happen at night. We are so used to the allegorical use of light and dark that tension on a pleasant, sunny street can be particularly disturbing, especially when people are gunned down in their neighborhoods. The same undiminishing uneasiness is masterfully accomplished in *Boyz in the Hood* (1991). Once again the sun plays a role in the creation of a "tense and brooding" city and provides an escape from the evils of the night.

[. . .]

Now that sufficient time has passed since the heyday of Film Noir, some of the settings from the Film Noir city can be revived safely and even parodied in ways that can appeal to both new audiences and those who remember the originals. In *The Player* (1992), for example, a film set in an otherwise sun-filled Los Angeles, the murder takes place in a dark, wet, gloomy, brick-filled alley. Film Noir strikes again. . . .

NEW USES FOR COLOR AND LIGHT: THE INFLUENCE OF THEATER AND COMIC STRIPS

[. . .]

THE INCREASING VARIETY AND COMPLEXITY OF CITY IMAGES IN FILM

Cities are now depicted in a variety of ways in films. . . . The golden age of Film Noir lasted only a decade or so, from the late 1940s to the late

1950s, but its influence lingers on. While the classic Film Noir city still appears occasionally in many American productions, it is now as likely to be parodied as emulated. . . .

In recent years, the city of bright sunshine has begun to compete successfully with the Film Noir city in the area of brooding urban tension. The "city of uneasy sunshine" has been perfected so that there is no longer a consensus way to create an atmosphere of urban tension or, conversely, an atmosphere of jolly goodwill. In films such as *Grand Canyon* (1991) and *Boyz in the Hood*, the city by day offers little respite from the tensions of the night. Cities can now be depicted with a complex combination of color coding as well as with combinations of stereotypical and reversed images. Relatively muted colors have been developed to the extent that lighting can sometimes dominate color to a degree unknown since the black and white films of the 1950s.

Over the past eighty years, films have been one of the most important sources for images of the city and urban life. It is important, even imperative, that we geographers begin to examine the roles that films have played in shaping our understanding of and attitudes toward the city. While the role of the urban scene as a character in films may be less obvious and stereotyped today than it was at one time, it is still useful and fun to ponder how the city is being depicted and portrayed. We will never really understand the ways in which Americans perceive cities unless we pay attention to the roles that cities have played in films.

REFERENCES FROM THE READING

Allen, W. (1980) *Four Films of Woody Allen*, Random House, New York.

Hirsch, F. (1981) *Film Noir: The Dark Side of the Screen*, Da Capo Press, New York.

FILMOGRAPHY

Annie Hall (1977) United Artists/Jack-Rollins-Charles H. Joffe. Director, Allen, W.

The Asphalt Jungle (1950) M.G.M. Director, Huston, J.

The Bicycle Thief (1948) Italy: P.D.S.-E.N.1.C. Director, de Sica, V.

Blade Runner (1982) Warner Bros. Director, Scott, R.

The Blue Angel (1930) Germany: U.F.A. Director, Pommer, E.

Boyz in the Hood (1991) Columbia Pictures. Director, Lee, S.

Brazil (1985) Universal Studios. Director, Gilliam, T.

The Cabinet of Dr. Caligari (1919) Deda-Bioscop Productions. Director, Wiene, R.

Casablanca (1942) Warner Bros. Director, Curtiz, M.

Choose Me (1984) Island Alive. Director, Rudolph, A.

Criss Cross (1948) U-I. Director, Siodmak, R.

Cry of the City (1948) T.C.F. Director, Siodmak, R.

The Dark Corner (1946) T.C.F. Director, Hathaway, H.

Double Indemnity (1944) Paramount. Director, Wilder, D.

Edward Scissorhands (1990) Twentieth Century Fox. Director, Burton, T.

Forty-Second Street (1933) Warner Bros. Director, Bacon, L.

Grand Canyon (1991) T.C.F. Director, Kasdan, L.

Hannah and Her Sisters (1986) Orion. Director, Allen, W.

The Hustler (1961) T.C.F. Robert Rossen. Director, Rossen, R.

King Kong (1933) RKO. Directors, Cooper, M.C., and Schoedsack, E.

King Kong (1976) Dino de Laurentis. Director, Guillermin.

Kiss of Death (1947) T.C.F. Director, Hathaway, H.

The Loneliness of the Long-Distance Runner (1962) British Lion/Bryanston/Wood Fall. Director, Richardson, T.

The Maltese Falcon (1941) Warner Bros. Director, Huston, J.

Manhattan (1979) United Artists. Director, Allen, W.

Metropolis (1926) Germany: U.F.A. Director, Lang, F.

Midnight Cowboy (1969) United Artists/Jerome Hellman. Director, Schlesinger.

The Music Box (1932) M.G.M. Director, Roach, H.

The Naked City (1948) Universal. Director, Dassin.

Night and the City (1950) United Kingdom: T.C.F. Director, Dassin.

Nightmare Alley (1947) T.C.F. Director, Goulding, E.

Open City (1945) Italy: Minerva. Director, Rossellini, R.

The Player (1992) Hollywood. Director, Altman, R.

Psycho (1960) Shamley/Alfred Hitchcock. Director, Hitchcock, A.

Safety Last (1923) Harold Lloyd. Directors, Taylor, S. and Newmeyer, F.

Scarlet Street (1946) Universal/Diana. Director, Lang, F.

Scenes From a Mall (1991) Touchstone Pictures. Director, Allen, W.

Side Street (1950) M.G.M. Director, Mann, A.

The Street with No Name (1948) T.C.F. Director, Keighley, W.

The Window (1949) R.K.O. Director, Tetzlaff, T.

Editors' references and suggestions for further reading

Aitken, S. and Zonn, L. (eds) (1994) *Place, Power, Situation, and Spectacle*, Rowman and Littlefield, Lanham, MD.

Burgess, J. and Gold, J.R. (eds) (1985) *Geography, the Media, and Popular Culture*, St. Martin's Press, New York.

Clarke, D. (ed.) (1997) *The Cinematic City*, Routledge, London.

Harvey, D. (1989) "Time and space in the postmodern cinema," *The Condition of Postmodernity*, Blackwell, Oxford, pp. 308–26.

Krutnik, F. (1997) "Something more than night: tales of the *noir* city," in D. Clarke (ed.) *The Cinematic City*, Routledge, London, pp. 83–109.

Leigh, N.G. and Kenny, J. (1996) "The city of cinema: interpreting urban images on film," *Journal of Planning Education and Research* 16: 51–5.

Natter, W. and Jones, J.P. (1993) "Pets or meat: class, ideology, and space, in *Roger and Me*," *Antipode* 25, 2: 140–58.

Zonn, L. (ed.) (1990) *Place Images in the Media: Portrayal, Meaning and Experience*, Rowman and Littlefield, Savage, MD.

S
I
X

PART SEVEN

Technologies

Surveillance camera in Glasgow, Scotland. (Courtesy of Elvin Wyly)

INTRODUCTION TO PART SEVEN

In an article entitled "Megalopolis and antipolis: the telephone and the structure of the city," the geographer Jean Gottman (1977) reflected on the impact of the telephone on urban form. "[T]he telephone," he concluded, "has helped to make the city better, bigger, more efficient, more exciting, providing, where needed, quasi-immediate verbal communication between all elements of an expanding metropolis at minimum cost" (p. 312). Gottman's interest in the interplay between the telephone, urban structure and the urban experience fits into a long tradition of work concerned with technology and the city (Konvitz *et al.* 1990), that has ranged from speculation over the role played by agricultural technology in origins of urbanism (Goodwin and Chant 1999) to the potential of new information technologies for building urban "e-topias" (Mitchell 2000). Yet many of the established perspectives on the interplay between technology and urban development adopt a narrow, deterministic approach. Technological innovations are viewed as a kind of autonomous, "independent external force acting upon cities" (Knox and Pinch 2000: 23) which lead inexorably to changes in urban form and development. The contributors to this section, however, clearly distance themselves from this simplistic separation of "technology" and "society." Instead, in different ways and in relation to different technologies, they all insist on the interdependence of the "social" and the "technical": they all view the city as a "sociotechnical process" (Graham and Marvin 2001).

The first reading, by Kaika and Swyngedouw, reflects on the technological networks required to supply water to European cities during the period stretching from the mid-nineteenth to the mid-twentieth centuries. "The city," as Swyngedouw (1997: 390) has observed elsewhere, "cannot survive without capturing, transforming and transporting nature's water," so the construction of complex arrangements of pipes, tunnels and towers was a crucial element in what Gandy (1999: 35) calls "the progressive rationalization of urban space." The impetus behind these infrastructural developments, along with similar networks for energy and light, was a combination of nineteenth-century ideological beliefs in the positive transformative powers of science and networked technologies, and the rise of urban planning as a mechanism for managing the problems of disease and disorder associated with the rapidly growing cities of industrial capitalism (Graham and Marvin 2001: 43). Yet the resulting infrastructures of water and sanitation systems, electricity grids, gas pipelines and telephone connections that developed over the next hundred years and created "the modern networked city" (Graham and Marvin 2001) have received remarkably little analytical attention. Explanation for this neglect partly reflects the way networked infrastructure forms a "hidden city" of wires, ducts, pipes and tunnels that lie beneath the streets. When this relative invisibility is combined with the fact that geography has tended to give a "privileged role to the visual as a means of accessing landscape" (Rose 1993: 70), the lack of research on technological networks is more understandable. But as Kaika and Swyngedouw reveal in the first article, the technological networks of cities have not always been so opaque. Providing an intriguing link back to the Form and Symbolism section, part of their article focuses on the early nineteenth century when "urban networks and their connecting iconic landmarks [like water towers] were prominently visual and present," symbolizing "progress" towards a modern society through the control of nature by technology. Drawing inspiration from Marxist analysis, Kaika and Swyngedouw also emphasize how, under the pressures of capitalist urbanization, infrastructural networks became "fetishized" in the nineteenth century. By representing new urban technological infrastructure of water management as the "embodiment of progress" and as providing freedom from the "tyranny of nature," the capitalist relations of production and asymmetries in social power that underpinned this infrastructure were effectively hidden. Only in the early part of the twentieth century, as the promises of a better society were largely unfulfilled, did the contradictions and tensions of capitalist modernization reveal that "technological innovation and progress were profiting . . . the ones who had control over the means of production" (see also Harvey's article in Part 3, Restructuring).

The second reading shifts from water pipes and towers to the "gigantic invisible cobwebs" of optic fibre, copper cable, wireless, microwave and satellite communications networks (Hillman 1991: 1, quoted in Graham and Marvin 1996: 3). These infrastructural elements provide the framework for telecommunications technologies and telematics, the latter referring to the services and infrastructures which link computer and digital media equipment over telecommunications links (Graham and Marvin 1996: 2–3). With their capacity to transcend spatial barriers instantaneously, telecommunications raise fundamental questions about the future form of urban development. From the perspective of the technological determinists, for example, the assumption has been that new telecommunications technologies will result in the radical decentralization of cities. As Graham observes, "there are widespread predictions that concentrated urban areas will lose their spatial glue in some wholesale shift towards reliance on broadband, multimedia communications grids" (1998: 168). One such prediction, made twenty-five years ago by the "futurist" Alvin Toffler (1981), even envisaged the dissolution of cities and the rise of a home-centred society based around a rural world of "electronic cottages." Such thinking, however, has been roundly dismissed as an "infantile utopia" (Robins and Hepworth 1988) which wildly exaggerates the degree to which information technology can substitute for place-based, face-to-face interaction.

A more sophisticated perspective on the interplay between telecommunications and the city emerges from work grounded in the political economy tradition (see Part 3), and it is this perspective which informs Warf's examination of the geographies of knowledge transmission. An important theme running through this reading is how telecommunications systems are deeply implicated in the production of new economic geographies associated with global restructuring (see Part 2, Globalization). Warf is particularly interested in how, on the one hand, telecommunications are reinforcing the growth and development of metropolitan centers, particularly world cities, while, on the other hand, information technology is also facilitating the radical decentralization of routine producer and consumer service functions. The former is clearly evident in the fact that "In terms of hard infrastructural investment [in telecommunications], demand for services and rates of innovation, the largest globally orientated metropolitan areas are clearly maintaining their dominance" (Graham 1998: 173). With respect to the latter (the decentralization of routine functions) there is strong evidence that the more information-intensive organizations become, the more they will seek to minimize their information costs through the so-called "back-officing of routine functions" (Hepworth 1992). As Warf's analysis indicates, this is occurring at a range of geographical scales. In some instances, the split is between the developed and developing world, while in other cases it may involve a division within developed countries between core and peripheral regions, metropolitan and non-metropolitan areas, or between city centers and suburbs. Warf's article therefore provides a robust counter-argument to the view of the technological determinists. Far from ushering in a world of greater homogeneity and absolute mobility, telecommunications are producing "new landscapes of production, consumption and distribution" underpinned by established asymmetries of power as transnational corporations use information technologies to engineer new divisions of labour at a range of geographical scales (Graham 1998: 175).

Telecommunications are not just important to economic restructuring. The capabilities of this technology for control and surveillance in the economic sphere can also be extended into the social realm and used to address problems of crime and disorder. This is a theme taken up in the third reading by Fyfe and Bannister. As increasingly "form follows fear" (Ellin 1996) in the contemporary city and the "fortress impulse" underpins urban design (see Dear and Flusty, p. 138), there has been a proliferation in the use of electronic surveillance technologies, such as Closed Circuit Television (CCTV) cameras, alongside the more traditional "target hardening" measures of gates, bars and locks (see Davis 1990). CCTV technology is, of course, not new. Cameras have been used in private spaces such as banks or sports stadia for many years. But the extension of this technology into public space is clearly more controversial and significantly increases the "surveillance capacity" in contemporary cities. CCTV cameras now watch and record people on transportation systems, in shops and offices, and on the streets, such that "quite literally, a person going about his or her daily routine may be under watch virtually the entire time spent outside the house" (Squires 1994: 396). In understanding the background to the introduction of public space CCTV, we return to some of the themes raised in Part 3. Throughout the 1980s, de-industrialization and the rapid growth of out-of-town retail and business parks led to the decline of many downtown areas, and growing fears about the economic and social consequences of this spatial reorganization of economic activity. In an effort to reverse this decline, many local authorities have attempted to create a "downtown as mall" (Christopherson 1994: 418), recognizing that improvements in security are one of the keys to maintaining property values in gentrified enclaves (see Smith, p. 128) and profits in shopping malls, restaurants and cultural centres (see Goss, p. 293). This economic agenda is, as Fyfe and Bannister illustrate, crucial to understanding the introduction of CCTV in central Glasgow. But CCTV also raises vital social and political questions. Far from being a neutral, value-free technology, the monitoring of, and response to, CCTV images depends on "normative ecologies," assumptions about who "belongs" in particular spaces at particular times. Of particular concern in this respect are claims that CCTV is being used to preserve "the public spaces of our town centres . . . for the consumer citizen, while those whose spending power is low . . . are effectively excluded" (Williams *et al.* 2000: 184). In addition, and intersecting with issues raised in Part 5, Difference, there is already evidence of racism and sexism by those responsible for the day-to-day operation of CCTV systems. As one study concludes, "the gaze of the camera does not fall equally on all users of the street but those who are stereotypically predefined as potentially deviant or, through appearance and demeanour, are singled out by operators as unrespectful" (Norris and Armstrong 1999: 264).

These themes of surveillance and exclusion resurface in the final reading but in the context of a rather different form of technology: Geographical Information Systems (GIS). GIS bring together technological advances in computer cartography and data collection for the analysis of spatial data. First developed in the 1960s, GIS had, by the 1980s, become a widely used tool by geographers, a "facilitating and application-led technology, which transparently assesses the importance of space, and as such . . . [is] central to our geographical understanding of the world" (Longley 2000: 157). Urban geographers were not slow to recognize the role GIS could play in studying the city. Its uses have include detailed accounts of urban spatial structure (Waddell and Shukla 1993), examining the spatial implications of changes in transport policy (Shaw 1993), generating urban population densities (Batty and Kim 1992) and investigating the effects of socio-economic status and ethnic heterogeneity on the spatial pattern of crime (Taylor and McDonald 1989). Against this background, one "progress" report on the use of GIS in urban geography concluded that GIS add "immensely to urban geographers' analytical capabilities in urban empirical research . . . [and provide] a new vehicle for urban geographers to close the gap between theory and practice" (Sui 1994: 263).

Nevertheless, as GIS gained in disciplinary prominence, so too questions have been asked about the social and political implications of this technology. Specifically, debate has focused on whether GIS is a tool for community empowerment or marginalization (Sieber 2003: 52; Pickles 1995; Craig *et al.* 2002). It is this which provides the background to Elwood's reading. Proponents of GIS, as Sieber (2003: 51) observes, "hope that the technology will allow communities to better understand and advocate for their concerns, promote the geographical visions of previously unheard people and provide greater entrée into policy-making because communities use the tools and data of the policy-makers." Set against this, however, "Detractors fear that by submitting to the quantification and resource requirements of GIS, organizations will lose their sense of place and become tools of corporate concerns" (ibid.). Elwood's article is a strategically important intervention into this highly polarized debate. Drawing on a case study of a community organization using GIS in a neighbourhood regeneration initiative, Elwood introduces some conceptual clarity into the debate by disentangling different meanings of "empowerment." But she also recognizes the profound ambiguities and ambivalences that surround the use of GIS within communities. For Elwood there is strong evidence that GIS is a tool of both empowerment *and* marginalization. Individuals within a community who are more technically proficient at GIS may be better able to voice their opinions, but in doing so open up asymmetries in power relations which can undermine notions of internal democracy. Elwood's study, like the others contained in this section, therefore demonstrates that technology is not "natural or neutral" (Curry 1998: 2); nor does it "impact" linearly on cities and urban life (Graham and Marvin 2001: 21). Rather, its role in the making of urban landscapes and mediating the urban experience is a fundamentally economic, political and socio-cultural process.

References and suggestions for further reading

Batty, M. and Kim, K.S. (1992) "Form follows function: reformulating urban population density functions," *Urban Studies* 29: 1043–70.

Christopherson, S. (1994) "The fortress city: privatized spaces, consumer citizenship," in A. Amin (ed.) *Post-Fordism: A Reader*, Oxford: Blackwell, pp. 409–27.

Craig, W.J., Harris, T.M. and Weiner, D. (eds) (2002) *Community Participation and Geographical Information Systems*, London: Routledge.

Curry, M. (1998) *Digital Places: Living with Geographic Information Technologies*, London: Routledge.

Davis, M. (1990) *City of Quartz: Excavating the Future in Los Angeles*, London: Verso.

Ellin, N. (1996) *Postmodern Urbanism*, Oxford: Blackwell.

Gandy, M. (1999) "The Paris sewers and the rationalization of urban space," *Transactions of the Institute of British Geographers* 24: 23–44.

Goodwin, C. and Chant, D. (eds) (1999) *Pre-industrial Cities and Technology*, London: Routledge.

Gottman, J. (1977) "Megalopolis and antipolis: the telephone and the structure of the city," in I. de Sola Pool (ed.) *The Social Impact of the Telephone*, Cambridge, MA: MIT Press, pp. 303–17.

Graham, S. (1998) "The end of geography or the explosion of place? Conceptualizing space, place and information technology," *Progress in Human Geography* 22: 165–85.

Graham, S. and Marvin, S. (1996) *Telecommunications and the City: Electronic Spaces, Urban Places*, London: Routledge.

Graham, S. and Marvin, S. (2001) *Splintering Urbanism: Networked Infrastructures, Technological Mobilities, and the Urban Condition*, London: Routledge.

Hepworth, M. (1992) "Telecommunications and the future of London," *Policy Studies* 13, 2: 31–45.

Hillman, J. (1991) *Revolution or Evolution? The Impact of Information and Communication Technologies on Buildings and Places*, London: Royal Institute of Chartered Surveyors.

Knox, P. and Pinch S. (2000) *Urban Social Geography: An Introduction*, Harlow: Longman.

Konvitz, J., Rose, M. and Tarr, J. (1990) "Technology and the city," *Technology and Culture* 31, 2: 284–95.

Longley, P.A. (2000) "Spatial analysis in the new millennium," *Annals of the Association of American Geographers* 90: 157–65.

Mitchell, W.J. (2000) *E-topia: "Urban Life, Jim – But Not As We Know It*," Cambridge, MA: MIT Press.

Norris, C. and Armstrong G. (1999) *The Maximum Surveillance Society: The Rise of CCTV*, Oxford: Berg.

Pickles, J. (1995) *Ground Truth: The Social Implications of Geographic Information Systems*, New York: Guilford Press.

Robins, K. and Hepworth, M. (1988) "Electronic spaces: new technologies and the future of cities," *Futures*, April, pp. 155–76.

Rose, G. (1993) "Some notes towards thinking about spaces of the future," in J. Bird, B. Curtis, T. Putnam, G. Robertson and L. Tickner (eds) *Mapping the Futures: Local Cultures, Global Change*, London: Routledge, pp. 70–86.

Shaw, S.L. (1993) "Hub structures of major U.S. passenger airlines," *Journal of Transport Geography* 1: 47–58.

Sieber, R.E. (2003) "Public participation geographic information systems across borders," *Canadian Geographer* 47: 50–61.

Squires, J. (1994) "Private lives, secluded spaces: privacy as a political possibility," *Environment and Planning D: Society and Space* 12: 387–401.

Sui, D.Z. (1994) "GIS and urban studies: positivism, post-positivism, and beyond," *Urban Geography* 15: 258–78.

Swyngedouw, E. (1997) "Power, nature and the city: the conquest of water and the political ecology of urbanization in Guayaquil, Ecuador: 1880–1990," *Environment and Planning A* 29, 2: 311–22.

Taylor, R.W. and McDonald, D. (1989) "GIS techniques applied to spatial analysis of crime in the city of Dallas: preliminary findings," *GIS/LIS'89* 1: 171–80.

Toffler, A. (1981) *The Third Wave*, New York: Morrow.

Waddell, P. and Shukla, V. (1993) "Employment dynamics, spatial structuring, and the business cycle," *Geographical Analysis* 25: 35–52.

Williams, K.S., Johnstone, C. and Goodwin, M. (2000) "CCTV surveillance in urban Britain: beyond the rhetoric of crime prevention," in J.R. Gold and G. Revill (eds) *Landscapes of Defence*, London: Prentice Hall, pp. 168–87.

SEVEN

"Fetishizing the Modern City: The Phantasmagoria of Urban Technological Networks"

from *International Journal of Urban and Regional Research* (2000)

Maria Kaika and Erik Swyngedouw

Editors' Introduction

Despite their importance to the functioning of cities, the complex technological infrastructures along which water, energy, waste and information flow have often been, quite literally, overlooked by urban geographers. The "hidden city" of wires, tunnels and ducts that exists below the streets has "long been relegated to the sidelines of urban analysis, eliciting yawns of boredom from those who view it through the lens of technological determinism as an inert web of transportation and communication lines devoid of politics" (Warf 2003: 246). Yet far from simply being examples of some neutral, value-free technology, the infrastructure networks of cities are intrinsically political and economic phenomena. Decisions about where and when to make capital investments in the wires, ducts and tunnels that are embedded within cities and about who has access to vital utilities, such as power and water, depend crucially on uneven and changing distributions of economic and political power. In particular places and at particular times, networks may provide some degree of equality in access to basic infrastructural services; however, in other places and at other times, there may be significant social and spatial disparities in access. This inequality of access in turn has important implications for understanding different experiences of the city and making sense of variations in the quality of urban life for different socio-economic groups (Graham 2000).

For urban geographers, then, the largely hidden and taken-for-granted technological infrastructures and networks that exist within (and between) cities raise a range of challenging social, political and economic questions. In this reading, two Oxford University geographers, Maria Kaika and Eric Swyngedouw, take up some of these questions by examining "the shifting meanings of urban technological networks" in European cities. Focusing initially on the development of water supply and sewerage systems in the mid-nineteenth century, Kaika and Swyngedouw describe how the first water towers, purification plants, dams and reservoirs were often prominent features in the urban landscape. The high visibility of these structures was not unimportant because they were potent symbols of "progress" towards a better society through the "mastering and taming of nature" by technology. The desire to be connected to the rapidly developing network of pipes was therefore about more than simply acquiring water; it meant "connecting to progress, 'betterment' and emancipation." This fascination with the technological infrastructure of water supply also had the effect, Kaika and Swyngedouw suggest, of transforming the networks into what Marxists would describe as "fetishized products" in which the underlying inequalities in power involved in the production and use of this new infra-structure were obscured. "Progress" was made to appear as simply a process of technological innovation and connection rather than being enmeshed in social power relations.

In the period after World War I, however, doubts began to surface about the assumption that technological progress and social progress went hand in hand. Kaika and Swyngedouw describe how the old urban networks and infrastructures begin to disappear, quite literally, from public view, becoming "an underground city that veiled the failure of modernization to create a better society." In their place came an emphasis on the "sanitized" and "clean" city in which pipes, wires and ducts are completely hidden and "new phantasmagoric networked constructions" replace them. Urban and regional highways with their promise of speed and mobility in linking "the modern networked and private home to the places of consumption, leisure and work" are one example of the phantasmagoric.constructions. As Kaika and Swyngedouw conclude, however, no matter how sanitized and clean cities have attempted to become, burst pipes, polluted water and accumulated waste provide stubborn reminders of the "materiality of the networked city."

By focusing on water supply, this article therefore illuminates the vital but often hidden interplay between, on the one hand, the economic and functional role of urban networks and infrastructures and, on the other hand, the social, cultural and political significance of these taken-for-granted technologies. Other urban technological networks can, of course, be subject to similar forms of analysis. In *Splintering Urbanism: Networked Infrastructures, Technological Mobilities and the Urban Condition*, Graham and Marvin (2001) examine a wide range of networks, including water, electricity, telecommunications, gas and transportation, to show how the increasing shift towards market-based provision of these services is contributing to growing social and spatial inequalities in access to urban infrastructures. Water does, however, have a particular significance for urban geographers, because it provides one example of the complex interrelationships that exist between "nature" and the "city" (see Harvey 1996; Swyngedouw and Kaika 2000). The transformation of nature is inextricably connected to the processes of urbanization, and the water towers, pumping stations and purification plants that appeared in the urban landscape in the mid-nineteenth century were an important visual reminder of this transformation. As Kaika and Swyngedouw discuss, however, the disappearance of this infrastructure underground and the "miraculous" appearance of water at the end of a tap "coming from nowhere in particular and from everywhere" appeared to sever the connection between nature and the city. The technological control of water nevertheless remains a crucial factor both in the process of urbanization and in the experience of the city, as further research by Swyngedouw (1997, 1999) in Ecuador and Spain, and Gandy (1999, 2002) in Paris and New York vividly illustrate.

INTRODUCTION

Technological networks (water, gas, electricity, information etc.) are constitutive parts of the urban. They are the mediators through which the perpetual process of transformation of nature into city takes place (Russell *et al.* 1997). Technological networks are the material mediators between nature and the city; they carry the flow and the process of transformation of one into the other. The city is a space of flows, of flux, of translocation. The urban fabric and the technological networks that carry the flows are a nexus of entry-exit points of a myriad of interconnected circuits and conduits. . . . However, urban networks in the contemporary city are largely hidden, opaque, invisible, disappearing underground, locked into pipes, cables, conduits, tubes, passages and electronic waves. It is exactly this hidden form that renders the tense relationship between nature and the city blurred, that contributes to severing the process of social transformation of nature from the process of urbanization. Perhaps more importantly, the hidden flows and their technological framing render occult the social relations and power mechanisms that are scripted in and enacted through these flows.

However, urban networks have not always been opaque. Along with their 'urban dowry' – water towers, dams, pumping stations, power plants, gas stations etc. – they have undergone important historical changes in their visual role and their material importance in the cityscape. In particular, during the early stages of nineteenth-century modernization, urban networks and their connecting iconic landmarks were prominently visual and present (Portaliou 1998). When the urban became constructed as agglomerated use values that turned the city into a theatre of accumulation and economic growth, urban networks became the iconic embodiments of and shrines to a technologically scripted image and practice of progress. Once completed, the networks became buried underground, invisible, rendered banal and relegated to an apparently marginal, subterranean urban underworld.

If we consider the city as a process of transformed nature, as the metabolic and social transformation of nature through human labour, the city turns into a 'hybrid' of the natural and the cultural, the environmental and the social (Latour 1993; Swyngedouw 1996; Swyngedouw and Kaika 2000). Entering the city posits the city as a flow, a flux and a movement, and suggests social, material and symbolic transformations and permutations. It also puts the focus on the process of commodification. Indeed, commodification, understood fundamentally as a social and cultural process of inserting socially metabolized goods into commodity or market relations, becomes in the modern city the form and medium through which 'nature' is turned into urbanity and the production of an urban environment. Of course, as we shall argue below, a process of fetishization parallels commodification. Fetishization is exactly the process through which the commodity form becomes *the* form of existence, severed from its historical and geographical (hence social) process of production; a process that is, of course, full of ambiguities and contradictions.

In this paper, we shall take water as the emblematic example to excavate the shifting meanings of urban technological networks. Indeed, as water becomes commodified and fetishized, nature itself becomes re-invented in its urban form (aesthetic, moral, cultural codings of hygiene, purity, cleanliness etc.) and severed from the grey, 'muddy', kaleidoscopic meanings and uses of water as a mere use value. Burying the flow of water via subterranean and often distant pinpointed technological mediations (dams, purification plants, pumping stations) facilitates and contributes to masking the social relations through which the metabolic urbanization of water takes place. The veiled subterranean networking of water facilitates severing the intimate bond between use value, exchange value and social power. The discrete technologies themselves become enshrined as the sources of all the wonders

of the city's water. . . . Dams, water towers, sewage systems and the like were celebrated as glorious icons, carefully designed, ornamented, and prominently located in the city, celebrating the modern promise of progress. During the twentieth century, the symbolic and material shrines of progress started to lose their mobilizing powers and began to disappear from the cityscape. Water towers, dams and plants became mere engineering constructs, often abandoned and dilapidated, while the water flows disappeared underground and in-house. They also disappeared from the urban imagination.

[. . .]

Commodity fetishism is the entry into our excavation of the dialectics between the economic/political and the cultural/ideological role of networks, and the hide/show of their material existence in the urban. 'Commodity fetishism' uniquely permits the bringing together of economics, politics and culture. Urban networks became 'urban fetishes' during early modernity, 'compulsively' admired and marvelled at, materially and culturally supporting and enacting an ideology of progress. The subsequent failure of this 'ideology of progress' is paralleled by their underground disappearance during high modernity, while the abandonment of their urban dowry announced a recasting of modernity in new ways.

ON COMMODITY FETISHISM

[. . .]

All goods necessary to sustain human lives are produced. Water, food, clothing, housing, even air, undergo a production process involving the extraction of raw materials and their subsequent transformation through human labour (Cronon 1991; Swyngedouw 1996). Under market exchange conditions and capitalist relations of production, these goods enter the social and urban fabric as commodities. The particular use values of goods that satisfy the wants and desires of individuals and social groups become combined with the distinct, universal and homogenized characteristic of their exchange value (Marx 1976). The exchange value acquired by commodities is based precisely on the fact that they are produced under specific social relations of production (Lefebvre 1991). This production process presupposes the transformation of nature through human labour. Although the natural foundation of this socioenvironmental metabolism that we call the labour process is an essential mechanism in the creation of exchange value of commodities, the link between nature and the final product (commodities) is severed and the socioeconomic conditions of their production are obscured. Commodities begin to appear as mere embodiments or containers of exchange value. In this way commodities become 'naturalized', and the qualitative relations involved in their production process become quantified. Blurring the socioenvironmental process of their production by foregrounding their character as universally exchangeable for anything else becomes an amazingly powerful mechanism. Severing materially and symbolically the connection between producing exchange and use values contributes to masking the qualitative social and environmental relations of production. Acquiring exchange value, without revealing at the same time the social power relations of their production, permits commodities to be presented as exceptional, as outside and over the thing that really makes them exceptional (i.e. the social metabolism of nature). The special character of commodities, the thing that makes them desirable and makes consumers want to pay the price, comes from presenting the commodity as an autonomous entity, as having a life of its own and a value in itself. In short, commodification turns the commodity into a fetish (Pietz 1993).

THE FETISHIZATION OF THE URBAN: TECHNOLOGICAL NETWORKS AND THEIR 'URBAN DOWRY'

Like other commodities in a market economy, the urban environment (roads, parks, buildings, networks) is also produced and commodified through the same process of transformation of nature by human labour (Harvey and Chatterjee 1973; Davis 1990; Lefebvre 1991). However, although the urban is part and parcel of our everyday experience, the human labour and social power relations involved in the process of its production are forgotten. The production of the urban remains, therefore, unquestioned and the urban becomes 'naturalized', as if it had always been there on the one hand,

and as distinct and separate from nature on the other. Yet, the urban undergoes the same process of production/commodification and fetishization, similar to other commodities.

Technological networks are a constitutive part of the urban as a collective means of consumption (Lojkine 1976; Castells 1977). The latter adds a particular twist to the process of their commodification and fetishization. The use value of networks dwells exactly in their capacity of and role in facilitating the process of socioenvironmental transformation and metabolization; the networks permit exactly the urbanization of nature *and* the fetishization of the commodities it carries. Water supply networks, for example, are the means of transforming H_2O (a natural element) into potable, clean, translucent water (a socially produced commodity embodying powerful cultural and social meanings) (Illich 1986; Swyngedouw 1997). Water enters one end of the network as H_2O and subsequently undergoes a chemical and social transformation to end up at the other end (the tap) as potable water, as a commodity properly priced and treated. Networks express through their material existence the socioeconomic process and the material flow of this transformation of nature into commodities. They represent, as they cut their way through and underneath the urban, the production process.

[. . .]

EARLY MODERNITY (MID-NINETEENTH CENTURY–1914): MYTHS OF MODERNIZATION AND PROGRESS

[. . .]

The expansion and consolidation of free trade, the establishment of a global monetary system and mass movements of goods and people from the mid-nineteenth century onwards went hand in glove with the need to connect the world via the proliferation of all kinds of networks. Technology, in the form of railways, steamships and the telegraph were becoming part of daily life (Burchell 1970). Being 'connected' became an icon and expression of progress. Technological networks and constructions, apparently mastering and taming nature, were among the most prominent material expressions of this practice and ideology in the urban sphere. Because of their significant role in the functioning of the modern capitalist city, networks of technology became *the* embodiment of progress during early modernity. The more the urban environment was filled with networks, the closer humankind would appear to approach the final goal of emancipation and freedom from the 'tyranny of nature'. Being connected to technology meant in itself emancipation, was in itself a way of participating in the new society. Being excluded from the technological networks, on the contrary, symbolized exclusion from the spheres of the powerful. Hence, the connection to the electricity or water networks of the city, or, similarly, the connection of one's home to a network of highways became a symbol of prestige and authority on the one hand and a terrain of controversies and power struggles on the other.

Of course, as many would soon discover, this process did not necessarily lead to emancipation. It also harboured practices of exclusion and segregation. Hence, despite the inevitable fascination with the new and the unknown, 'people were torn between a sense of euphoria at the progress and a romantic blurring of the past' (Gympel 1996: 72). The wonders of technology entering everyday urban life created both awe and fear. This combination of anxiety and admiration was also related to the rapidly deteriorating living and working conditions of the working class, despite the promises enshrined in technological progress. Social unrest intensified, culminating in, among others, the Paris Commune in 1871 and the Dock Strike of 1880 in the UK. Along with other urban 'miasma', the living conditions in slums in combination with lack of sanitary conditions resulted in the growth of epidemic diseases such as cholera, typhus etc. Two epidemics of cholera hit London in 1831 and 1848–49 (Coley 1989). During the 1840s, the sanitation movement emerged and Chadwick embarked on his mission to link cleanliness with water supply.

[. . .]

By the late nineteenth century, 'social reformers' worked hand in hand with engineers to construct a better and sanitized world (Goubert 1989). The early threat imposed by the introduction of technology gradually begins to give way to a 'new deal' between man and technology, a deal based on an assumed mutual benefit between technology and the conditions of life. The process of familiarization with the expressions of technology in everyday life bears fruit and technological achievements begin to be marvelled at for what they really are: crude expressions of the power of progress. Technology and new materials are not only accepted but start being aestheticized in a new way. The Eiffel Tower, for example, 'assembled out of steel girders, illuminated with electric lights, powered with dynamos and petrol engines, linked by copper wires' (Burchell 1970: 39), became the classic statement of this trend. For Eiffel, the tower expressed 'its own unique beauty' (Gympel 1996: 76) and, despite the original decision to tear it down after the exhibition, it remained dominating the Parisian landscape as a reminder of the continuing fascination with technology and its own special aesthetics.

The late nineteenth-century fascination with technological myths and urban utopias was paralleled by an increasing importance of the state's role as a facilitator of growth and promoter of technological change and innovation (Chant 1989). Large-scale water engineering projects followed the introduction of germ theory. The Medical Officer of the Public Health Authorities in London, for example, recognized that the importance of engineers was on a par with that of the medical profession (Coley 1989). Water and sewerage networks were being laid everywhere and dam and water tower constructions were accompanying that development. While the 'back to nature' movement with its garden city utopias was providing an anti-urban 'run-away solution', the urban environment was saturated with networks and the machine developed its own aesthetic form and language. . . .

In sum, during early modernity, the technological dream of a universal justice under the equalizing and totalizing powers of technology was widely held. The urban networks and connections had to keep expanding in order to both sustain and, moreover, visualize the ideology of progress in everyday urban experience. The urban became saturated with pipelines, cables, tubes and ducts of various sizes and colours; things that celebrated the mythic images of early modernity, encapsulating and literally carrying the idea of progress into the urban domain. Their material existence provided the confirmation and lived experience that the road to a better society was under construction and paved with networks. . . .

It is not only networks of technology that were fetishized as the material expressions of the ideology of progress. Along with the networks, the elements of the built environment that supported the functional role of the networks (i.e. water towers, power stations, reservoirs, pumping stations etc.) – what we have called the 'urban dowry of networks' – were also fetishized. As Portaliou (1998: 287) argues: 'The proliferation of phantasmagorical forms in public spaces encloses the aura of art and the content of fantasy and senses within the commodity's pitiless power, hidden behind the phantasmagoria'.... The 'urban dowry' became prominently visible in the urban during early modernity. These concrete shrines embodying the networks were sticking out of the city landscape; they provided the best form of 'landmarks' in the image of the city (Lynch 1960) and became the 'stuff of artistic renditions of the cityscape' (Becher and Becher 1988). Their beauty and fascinating character was no longer achieved through ornamental display. Their beauty lay in the promise they were carrying for a better future and a more equal society....

The phantasmagoria of and fascination with technological networks and cathedrals in the urban experience, and their combined role as ideology supporters and objects of admiration and worship, suggests that they became fetishized products in a double sense. First, in a Marxist sense, these networks enshrined an instrumentality in terms of reifying social relations.... The fascination with technology and technological constructions in themselves made progress appear to be merely a matter of construction, of technological innovation and of connection. The fetish role of networks and the emphasis put on the new and the innovative masked the underlying relations of production and social power relations, which remained symptomatically the same. Second, in the way Walter Benjamin would define the fetish, they became objects of delight and desire in themselves, as signs and wish images of a better society that was yet to arrive (Buck-Morss 1995). Where Marx was using 'phantasmagoria' to describe the fetish character of commodities in the market, Benjamin was interested in the commodity on display, where the representational value of the commodity was emphasized.... In their fetish role, networks and their nodal infrastructures were not just carrying water, electricity etc. into the city, but also embodied

the promise and the dream of a good society. The cathedrals of progress represented, displayed and celebrated the aestheticized dreams of tomorrow's utopia.

The desire to connect to them was more than the desire to acquire the utility; it meant connecting to progress, 'betterment' and emancipation. Their display kept the dream alive, kept their phantasmagoric character vibrant; the city itself was the shop window for their display. It is the materiality of the fetish objects, infused with a utopian dream that permits the visualization of the dream itself. In this way, technology and networks, although failing to deliver the promise of a better society, became wish images for a better society that could be anticipated and desired.

But it is precisely this second element of fetishized desire that would eventually turn against itself, and erode or subvert the fetish character associated with their commodified reification. This subversion was expressed in material and visual terms within cities and transformed the very experience of urbanity in profound manners. The fetishized objects of desire, embodied in networks and enshrined in their 'urban dowry' became 'eidola', idols adored in themselves. Marvelling at networks, dams or water towers as embodiments of urban emancipation obscured seeing the exploitation of living labour and the socioecological transformation involved in the process of their production. Stripping those objects of their social meaning left them as just fetishes and idols, phantom-like material expressions of a myth of progress and an ideology of automatic emancipation.

MODERNITY RECAST (1918–60): THE SUBVERSION OF THE FETISH AND THE REINVENTION OF THE URBAN

[...]

[A]s modernity asserted itself with greater vengeance and shattered the experiences and practices of space and time (Kern 1983), the assumed emancipatory powers scripted into the urban began to fade away (Burchell 1970; Buck-Morss 1995). The contradictions and tensions of capitalist modernization increasingly revealed that technological innovation and progress were profiting at the end of the day the ones who had control over the means of production. At the same time, labour was turned into the appendage to the idea and practice of progress. It became abundantly clear that, although the networks did deliver the promised material in the form of commodified goods, they somehow failed to deliver in their wake a better society (Franklin 1990). The fetish character of the networks and technological artefacts collapsed under the weight of unfulfilled promises. Even the visual statements of the technological power could no longer feed the urban dream and function as an urban fetish, either in terms of reifying social relations or in terms of echoing desire and fascination. Of course, networks remained inevitably etched into the city. While the ideology of the power of technology faded, the networks and constructions were left behind in the cityscape, still prominently visible, sad material reminders of a promise that was never to be fulfilled. . . . The ruins of a now outlived urban dream revealed, more clearly than ever before, the phantasmagoric nature of the artefact and the hidden scripting of their making. While their initial social meaning and the subversion of their symbolic representations was hollowed out, they became re-inscribed with a different 'meaning', i.e. that of material embodiments of the failure of the emancipating project of modernity.

[. . .]

By the middle of the twentieth century, the cities of the industrialized world were left with the uncomfortable situation of being filled with material statements of an unfulfilled (and unfulfillable) promise, accentuated by two world wars and a period of depression. How was this crisis to be overcome? How could the urban regain its glitter, its appeal, and re-invent itself as material supporter of a new dream, a new way of vesting old power relations with new scriptings? The answer was as simple as it was cunning! Nourish monotony by the new! Urban technology networks and constructions, those witnesses of disillusionment with the patina of time added over them, rusting like the modern urban dream of emancipation and equality, had to be cleared away, literally swept underneath the carpet. They went underground, while a new form, ideology, and aesthetics, a new process of fetishization, became created and etched into the urbanization

process. The perpetuation of commodity production and exchange had to be vested in yet a new and innovative way; a new promise had to be made. High modernity emerged from the 1930s onwards, with its obsession with clarity of form, purity, functionalism and cleanliness, translating the myth of the machine from the distant future into everyday experience.

The 'key characteristics of the machine' (i.e. functionality and efficiency) were gradually translated into the cultural and the domestic sphere, into design and architecture, into a new way of living (Forty 1995; Lupton and Miller 1996). Le Corbusier's Citrohan house (1922), named and designed after the similarly named car, became the iconic example of the house as 'a machine for living in'. Schutte-Lihotzky's Frankfurt Kitchen translated the factory's functionality into the domestic space. The city itself is designed and planned after the machine. The factory assembly line – as practice and metaphor – permeates every aspect of people's lives: from their place of work, via the spaces for shopping and recreation, to their 'mechanized homes' (McLuhan 1967).

The phantasmagorial role of technical networks survives, of course, in new and revamped ways. While urban water networks become normalized and pushed to the status of the immanent and the invisible, a new form of networked spatiality emerges that links together the privatized spaces of high modernity by means of colonizing and erasing public space. Urban and regional highways become the new phantasmagoric networked constructions. Le Corbusier's planning vision, for example, brings forth a new world of sanitized urban mobility, while Robert Moses's emblematic reconstruction of New York City as a city of movement equally attempts to provide direct links between the modern, networked and private home to the places of consumption, leisure and work, often literally bypassing or overriding spaces of marginality and anomy (Stern et al. 1995).

The process of sanitizing and cleansing the city equally moves from the sphere of the public to the private and domestic. The rapid spread of private bathrooms by the 1920s, 'medicated' (chlorinated) domestic water supply, and the requirement for a water closet in Britain in 1936 (Coley 1989) signals this process of domestication. As Lahiji notes, '[m]odernity emerges from the belief that man is fundamentally a clean body' (Lahiji and Friedman 1997a). This coincided with efforts to strip away any visible connections between the urban and the domestic, further eroding any clue to the production process behind them. Networks started gradually disappearing underground and increasing efforts were made to render them invisible, to give way to a pure, clean and transparent new urban form (Lahiji and Friedman 1997b). In New York City's skyline or in North European cities, for example, one can no longer see any connection cables, antennas, pipes etc. Dams are no longer destinations for family trips and mass meetings. Urban technological cathedrals, now often scrap heaps, are turned into shopping centres or theme parks. The 'urban dowry' (pumping stations, purification plants, reservoirs) cannot even be located anymore in the city. The nature/city connection that was still present in the old forms and flows, demonstrating 'man's' control over nature, became totally severed, and, with it, the link between product and production process. The supply of water, electricity, information etc. now appeared to be 'miraculously' entering the domestic sphere, coming from nowhere in particular and from everywhere. The end of the flow became omnipresent, naturalized, inevitable, yet severed from any apparent connection to anywhere else.... The networks disappeared in the underground, materially and symbolically. Buildings, like people, are individualized, atomized, sanitized and seemingly disconnected. The u-topia of the home was finally achieved; the island of internal connections where everything arrived and from where everything left. It was no longer in the urban (public) sphere where emancipation potentially resided and could be enacted, but rather in the domestic sphere, at the individual rather than at the collective level. The domestic sphere took over the ideologies of the urban and reinterpreted them, creating a new ideology evolving around the belief that the myth, the desire and the wonder should be searched for in the domestic, the individual, the disconnected, the isolated, the suburban.... The promise of emancipation and freedom resided in the intimacy of the disconnected house. The house became the 'machine for living in'. The perfect house became individual, clear, pure, functional and safe for the inhabitant, protected from the anomie and the antinomies of the outside and the underneath, the urban.... The sanitized, piped, wired, plumbed house – the classic icon of 1950s and 1960s advertisements for home durables and home style – promised again the final delirious satisfaction of our dreams and desires.

STAIRWAY TO HEAVEN AND HIGHWAY TO HELL

[...]

However, no matter how sanitized and clean, both in symbolic and literary terms, our cities have become, the 'urban trash' in the form of networks, dirt, sewerage, pipes, homeless people etc. (Davis 1992) keeps lurking underneath the city, in the comers, at the outskirts, bursting out on occasion in the form of rats, disease, homelessness, garbage piles, polluted waters, floods, bursting pipes etc. They remain stubborn reminders of the materiality of the networked city, while threatening the city's existence. Despite the quest for clarity, purity and 'sanity' that was prominent throughout high modernity (or, rather, precisely because of this quest), the underlying contradictions of urban life, the ones that actually make it possible for clarity to exist, i.e. the urban 'trash' and underlying invisible networks, both inorganic (sewerage, water pipes) and organic (homeless people), become gradually more prominent (Davis 1992). The dystopian underbelly of the city that at times springs up in the form of accumulated waste, dirty water, pollution or social disintegration, produces a sharp contrast when set against the increasingly managed clarity of the urban environment. The contradictions are becoming difficult to be contained or displaced.

[...]

Walter Benjamin (Buck-Morss 1995: 93) identified the urban underworld with the urban Hell that exists underneath the urban splendour, underneath a supposedly heavenly urban environment. For him, everything about urban life was Hell disguised in Heaven. However, we should perhaps search instead for a dialectics whereby Heaven requires its Hell in order to exist (Merrifield 1993). Despite efforts to manage and control the city, it remains a realm carved out of the dialectics between clean and dirty, justice and injustice, underworld and high society, basements and lofts, Hell and Heaven. In fact, Heaven can establish itself as such only by contradistinction to a certain Hell. The urban paradise needs to exploit organic and inorganic, human and non-human urban trash in order to sustain itself, and urban trash permits the existence of the urban, dwelling at its margins or underneath its soil.

REFERENCES FROM THE READING

Bachelard, G. (1963) *L'eau et les reves; essai sur l'imagination de la matiere*, J. Corti, Paris.

Becher, B. and H. Becher (1988) *Water Towers*, MIT Press, Cambridge, MA.

Best, S. (1994) 'The commodification of reality and the reality of commodification: Baudrillard, Debord, and postmodern theory', In D. Kellner (ed.) *Baudrillard: A Critical Reader*, Blackwell, Oxford.

Buck-Morss, S. (1995) *The Dialectics of Seeing: Walter Benjamin and the Arcades Project*, MIT Press, Cambridge, MA.

Burchell, S.C. (1970) *The Age of Progress*, Time Life International, Amsterdam.

Castells, M. (1977) *The Urban Question: A Marxist Approach*, Edward Arnold, London.

Castells, M. (1985) *High Technology, Space, and Society*, Sage, Beverly Hills, CA.

Chant, C. (ed.) (1989) *Science, Technology and Everyday Life 1870–1950*, Routledge (in association with The Open University), London.

Coley, N. (1989) 'From sanitary reform to social welfare', in C. Chant (ed.) *Science, Technology and Everyday Life 1870–1950*, Routledge (in association with The Open University), London.

Cronon, W. (1991) *Nature's Metropolis: Chicago and the Great West*, W.W. Norton, New York.

Davis, M. (1990) *City of Quartz: Excavating the Future in Los Angeles*, Verso, London.

Davis, M. (1992) *Beyond Blade Runner; Urban Control – The Ecology of Fear*, Open Magazine pamphlet series, 23, Open Media, New Jersey.

Eagleton, T. (1991) *Ideology: An Introduction*, Verso, London.

Forty, A. (1995) *Objects of Desire: Design and Society since 1750*, Thames and Hudson, London.

Franklin, U. (1990) *The Real World of Technology*, CBC Enterprises, Toronto.

Gandy, M. (1999) 'The Paris sewers and the rationalisation of urban space', *Transactions of the Institute of British Geographers* 24: 23–44.

Goubert, J.P. (1989) *The Conquest of Water*, Polity Press, Cambridge.

Gympel, J. (1996) *The Story of Architecture: From Antiquity to the Present*, Konemann, Koln.

Harvey, D. and L. Chatterjee (1973) 'Absolute rent and the structuring of space by governmental and financial institutions', *Antipode* 6: 22–36.

Illich, I. (1986) H_2O *and the Waters of Forgetfulness*, Marion Boyars, London.

Kern, S. (1983) *The Culture of Time and Space, 1880–1918*, Harvard University Press, Cambridge, MA.

Lahiji, N. and D.S. Friedman (1997a) 'At the sink: architecture in abjection', in N. Lahiji and D.S. Friedman (eds) *Plumbing: Sounding Modern Architecture*, Princeton Architectural Press, New York.

Lahiji, N. and D.S. Friedman (1997b) *Plumbing: Sounding Modern Architecture*, Princeton Architectural Press, New York.

Latour, B. (1993) *We Have Never Been Modern*, Harvester Wheatsheaf, New York, London.

Latour, B. and E. Hermant (1998) *Paris – ville invisible*, La Decouverte, Paris.

Latour, B. and J.-P. Le Bourhis (1995) *Donnez-moi de la bonne politique et je vous donnerai de la bonne eau . . .*, Centre de Sociologie de l'Innovation, Ecole Nationale Superieure des Mines de Paris, Paris.

Lefebvre, H. (1991) *The Production of Space*, Blackwell, Oxford.

Lojkine, J. (1976) 'Contribution to a Marxist theory of urbanization', in C.G. Pickvance (ed.) *Urban Sociology: Critical Essays*, Methuen, London.

Lupton, E. and J.A. Miller (1996) *The Bathroom, the Kitchen and The Aesthetics of Waste: (A Process of Elimination)*, Kiosk (Princeton Architectural Press), New York.

Lynch, K. (1960) *The Image of the City*, Technology Press, Cambridge, MA.

Marx, K. (1976) *Capital Volume I*, Penguin, Harmondsworth.

McLuhan, M. (1967) *The Mechanical Bride: Folklore of Industrial Man*, Routledge and Kegan Paul, London.

Merrifield, A. (1993) 'Place and space: a Lefebvrian reconciliation', *Transactions of the Institute of British Geographers* 18: 516–31.

Mitchell, W.J.T. (1986) *Iconology: Image, Text, Ideology*, Chicago University Press, Chicago, IL.

Pietz, W. (1993) 'Fetishism and materialism: the limits of theory in Marx', in E. Apter (ed.) *Fetishism as Cultural Discourse*, Cornell University Press, Ithaca, NY.

Portaliou, E. (1998) *Alienation from urban space and the crisis of collective memory. The historical centres of cities and the spatial constraints. Space, inequality and difference*, Seminars of the Aegean Series, Aristotle University of Thessaloniki/National Technical University of Athens/University of the Aegean, Athens.

Reid, D. (1991) *Paris: Sewers and Sewermen: Realities and Representation*, Harvard University Press, Cambridge, MA.

Russell, N.W., L. McKnight and R.J. Solomon (1997) *The Gordian Knot*, MIT Press, Cambridge, MA.

Stern, R.A.M., T. Mellins and D. Fishman (1995) *New York 1960*, The Monacelli Press, New York.

Swyngedouw, E. (1996) 'The city as a hybrid: on nature, society and cyborg urbanisation', *Capitalism Nature Socialism* 7: 65–80.

Swyngedouw, E. (1997) 'Power, nature, and the city. The conquest of water and the political ecology of urbanization in Guayaquil, Ecuador: 1880–1990', *Environment and Planning A* 29: 311–32.

Swyngedouw, E. and M. Kaika (2000) 'The environment of the city or . . . the urbanisation of nature', in G. Bridge and S. Watson (eds) *Companion to Urban Studies*, Blackwell, Oxford.

Vigarello, G. (1988) *Concepts of Cleanliness*, Cambridge University Press, Cambridge.

Editors' references and suggestions for further reading

Gandy, M. (1999) "The Paris sewers and the rationalization of urban space," *Transactions of the Institute of British Geographers* 24: 23–44.

Gandy, M. (2002) *Concrete and Clay: Reworking Nature in New York City*, Cambridge, MA: MIT Press.

Graham, S. (2000) "Introduction: cities and infrastructure networks," *International Journal of Urban and Regional Research* 24, 1: 114–19.

Graham, S. and Marvin, S. (2001) *Splintering Urbanism: Networked Infrastructures, Technological Mobilities and the Urban Condition*, London: Routledge.

Harvey, D. (1996) *Nature, Justice and the Geography of Difference*, Oxford: Blackwell.

Swyngedouw, E. (1997) "Power, nature and the city: the conquest of water and the political ecology of urbanization in Guayaquil, Ecuador: 1880–1990," *Environment and Planning A* 29, 2: 311–22.

Swyngedouw, E. (1999) "Modernity and hybridity: nature, regeneracionismo and the production of the Spanish waterscape, 1890–1980," *Annals of the Association of American Geographers* 89, 30: 443–65.

Swyngedouw, E. and Kaika, M. (2000) "The environment of the city . . . or the urbanization of nature," in G. Bridge and S. Watson (eds) *A Companion to the City*, Oxford: Blackwell, pp. 567–80.

Warf, B. (2003) "Book review: *Splintering Urbanism*," *Annals of the Association of American Geographers* 39, 1: 246–7.

"Telecommunications and the Changing Geographies of Knowledge Transmission in the Late 20th Century"

from *Urban Studies* (1995)

Barney Warf

Editors' Introduction

Across the developed world, cities and the spaces between them are now criss-crossed by increasingly elaborate telecommunications grids, that include telephone networks, wireless and radio systems, cable networks, satellite systems, and Internet data and video networks. Few areas of economic, social and political life remain unaffected by this expanding telecommunications infrastructure, prompting observers to talk of a "digital age," and a "network society" (Castells 1996). But what do these developments mean for cities and the people who live and work in them? According to the "technological determinists," telecommunications spell "the death of distance and the end of cities" (Castells 2000: 18). Over thirty years ago, the geographer Ronald Abler (1970) predicted that advances in information transmission would not only "disperse information-gathering and decision-making activities away from metropolitan centers" but also that electronic communications would make "all kinds of information equally abundant everywhere in the nation, if not everywhere in the world." The inevitable consequence of these developments, it is claimed, is "the vanishing city" (Pascal 1987). As telecommunications technologies and "digital living" provide substitutes for face-to-face contact, and "the electronic cottage" replaces the need to physically go to the business meeting, library or concert hall, so proximity becomes increasingly redundant, and geographical dispersal and the dissolution of cities is the "logical" outcome. These bold claims have intriguing historical echoes. In the nineteenth century, the telegraph, wireless and telephone were imbued with similar "magical powers" in terms of liberating urban life from the frictional effects of distance. Then, as now, however, such thinking treats technology as something independent of society and rests on a simplistic understanding of technological "causes" and urban "effects" which radically overestimates the degree to which telecommunications can be a substitute for place-based, face-to-face interaction (Graham 1998).

As Barney Warf demonstrates in this reading, embedding telecommunications into the political, economic and social relations of capitalism provides a far more sophisticated understanding of the interplay between information technologies and the city. Focusing on the increasing reliance of financial and business services on telecommunications to relay information through international networks, Warf, Professor of Geography at Florida State University, argues that electronic data collection and transmission capabilities are now a crucial factor underpinning contemporary patterns of uneven urban, regional and international development. In relation to the Internet, for example, Warf demolishes the myth of "equal access for everyone" with evidence that its spatiality is "largely preconditioned by the legacy of colonialism." Warf also shows that far from leading to the dissolution of cities, telecommunication technologies permit the simultaneous centralization and decentralization of urban economic activity. Centralization may be seen in the continuing importance of a small number of "world cities," including London, New York and Tokyo, whose position has been strengthened not weakened by telecommunications (see Beaverstock et al., pp. 63–73). These cities have witnessed massive investment in the optic-fibre grids which provide the infrastructural foundations of the advanced telecommunications systems that allow global financial services industries and corporate headquarters to stay in contact with their operations around the world. World cities, then, are vital places of control and coordination where the demand for and production of high-order economic and political information requires the development of ever more sophisticated telecommunications systems. The decentralizing effects of telecommunications are evident in the expansion of offshore banking centres and the globalization of back offices. As the technological barriers to the movement of capital have declined, political factors, particularly favourable tax laws, have assumed a greater importance in the circulation of money, leading to a growth in offshore banking in such places as the Cayman Islands and Bahamas in the Caribbean, and Singapore and Hong Kong in Southeast Asia. In terms of the "back offices" that perform many of the routinized clerical functions in the producer services sector, the increased flexibility afforded by telecommunications technologies has allowed the relocation of these functions at intra- and international levels in order to exploit variations in labour and land costs. American Express, for example, has moved its back offices from New York City to Salt Lake City and Phoenix while some New York life insurance companies now have back offices in Ireland.

Warf's paper is therefore an important intervention in debates about the implications of telecommunications for urban geographies. Far from diminishing the importance of cities and space, Warf's urban political economy perspective shows how telecommunications systems "produce new rounds of unevenness, forming new geographies that are imposed upon the relics of the past"; and that rather than eliminating variations among places, such systems permit "the exploitation of differences between areas with renewed ferocity." If this perspective has a weakness, however, it is its tendency to overlook the people who interact daily with new information technologies in particular spaces and places around the globe. Studies of those working in the financial district of the City of London, however, have shown how the use of telecommunications systems actually increases the need for face-to-face contact in these environments: "The major task in the information spaces of telematic cities like the City of London becomes interpretation and, moreover, interpretation *in action* under the pressure of real-time events" (Thrift 1996: 1481). In contrast to the more extreme claims of technological determinists, then, information technologies can never be a substitute for social interaction in material spaces. As one business person wryly observed, "[Y]ou cannot look into someone's eyes and see they are trustworthy over the Internet" (quoted in Graham 1999: 933).

With James Wheeler and Yuko Aoyama, Warf is co-editor of *Cities in the Telecommunications Age: The Fracturing of Geographies* (London: Routledge, 2000) which provides a wide-ranging analysis of the importance of telecommunications in the context of urban planning, cyberspace and urban economic development. As this volume acknowledges, the academic context for studies of information technology and cities is dominated by the work of Manuel Castells and, in particular, *The Informational City: Information Technology, Economic Restructuring and the Urban-regional Process* (Oxford: Blackwell, 1989) which focuses on the crucial importance of information to the restructuring of capitalism; and his later trilogy *The Rise of the Network Society* (Oxford: Blackwell, 1996), *The Power of Identity* (Oxford: Blackwell, 1997) and *End of Millennium* (Oxford: Blackwell, 1998). Other important work in the field includes Stephen Graham and Simon Marvin's *Telecommunications and the City: Electronic Spaces, Urban Places* (London: Routledge, 1996), and their more recent *Splintering Urbanism: Networked Infrastructures, Technological Mobilities and the Urban Condition* (London: Routledge, 2001) which adopts a political economy perspective to highlight the growing inequalities that characterize the provision of and access to not only telecommunications networks but also a range of other urban infrastructures.

The late 20th century has witnessed an explosion of producer services on an historic scale, which forms a fundamental part of the much-heralded transition from Fordism to post-Fordism (Coffey and Bailly 1991; Wood 1991). Central to this transformation has been a wave of growth in financial and business services linked at the global level by telecommunications. The emergence of a global service economy has profoundly altered markets for, and flows of, information and capital, simultaneously initiating new experiences of space and time, generating a new round of what Harvey (1989, 1990) calls time-space convergence. More epistemologically, Poster (1990) notes that electronic systems change not only what we know, but how we know it.

The rapid escalation in the supply and demand of information services has been propelled by a convergence of several factors, including dramatic cost declines in information-processing technologies induced by the microelectronics revolution, national and worldwide deregulation of many service industries, including the Uruguay Round of GATT negotiations (which put services on the agenda for the first time), and the persistent vertical disintegration that constitutes a fundamental part of the emergence of post-Fordist production regimes around the world (Goddard and Gillespie 1986; Garnham 1990; Hepworth 1990). The growth of traditional financial and business services, and the emergence of new ones, has ushered in a profound – indeed, an historic – transformation of the ways in which information is collected, processed and circulates, forming what Castells (1989) labels the 'informational mode of production'.

[...]

THE GLOBAL SERVICE ECONOMY AND TELECOMMUNICATIONS INFRASTRUCTURE

There can be little doubt that trade in services has expanded rapidly on an international basis (Kakabadse 1987), comprising roughly one-quarter of total international trade. Internationally, the US is a net exporter of services (but runs major trade deficits in manufactured goods), which is one reason why services employment has expanded domestically. Indeed, it could be said that as the US has lost its comparative advantage in manufacturing, it has gained a new one in financial and business services (Noyelle and Dutka 1988; Walter 1989). The data on global services trade are poor, but some estimates are that services comprise roughly one-third of total US exports, including tourism, fees and royalties, sales of business services and profits from bank loans.

[...]

The increasing reliance of financial and business services as well as numerous multinational manufacturing firms upon telecommunications to relay massive volumes of information through international networks has made electronic data collection and transmission capabilities a fundamental part of regional and national attempts to generate a comparative advantage (Gillespie and Williams 1988). The rapid deployment of such technologies reflects a conjunction of factors, including: the increasingly information-intensive nature of commodity production in general (necessitating ever larger volumes of technical data and related inputs on financing, design and engineering, marketing and so forth); the spatial separation of production activities in different nations through globalised sub-contracting networks; decreases in price and the elastic demand for communications; the birth of new electronic information services (e.g. on-line databases, teletext and electronic mail); and the high levels of uncertainty that accompany the international markets of the late 20th century, to which the analysis of large volumes of data is a strategic response (Moss 1987b; Akwule 1992). The computer networks that have made such systems technologically and commercially feasible offer users scale and scope economies, allowing spatially isolated establishments to share centralised information resources such as research, marketing and advertising, and management (Hepworth 1986, 1990). Inevitably, such systems have profound spatial repercussions, reducing uncertainty for firms and lowering the marginal cost of existing plants, especially when they are separated from one another and their headquarters over long distances, as is increasingly the case.

Central to the explosion of information services has been the deployment of new telecommunications systems and their merger with computerised database management (Nicol 1985). This phenomenon can be seen in no small part as an aftershock of the microelectronics revolution and the concomitant switch from analogue to digital information formats: the digital format suffers less degradation over time and space, is much more compatible with the binary constraints of computers and allows greater privacy (Akwule 1992). As data have been converted from analogue to digital forms, computer services have merged with telecommunications. When the cost of computing capacity dropped rapidly, communications became the largest bottleneck for information-intensive firms such as banks, securities brokers and insurance companies. Numerous corporations, especially in financial services, invested in new communications technologies such as microwave and fibre optics. To meet the growing demand for high-volume telecommunications, telephone companies upgraded their copper-cable systems to include fibre-optics lines, which allow large quantities of data to be transmitted rapidly, securely and virtually error-free. By the early 1990s, the US fibre-optic network was already well in place. In response to the growing demand for international digital data flows beginning in the 1970s, the United Nation's International Telecommunications Union introduced Integrated Service Digital Network (ISDN) to harmonise technological constraints to data flow among its members (Akwule 1992). ISDN has since become the standard model of telecommunications in Europe, North America and elsewhere.

[. . .]

Telecommunications allowed not only new volumes of inter-regional trade in data services, but also in capital services. Banks and securities firms have been at the forefront of the construction of extensive leased telephone networks, giving rise to electronic funds transfer systems that have come to form the nerve centre of the international financial economy, allowing banks to move capital around at a moment's notice, arbitraging interest rate differentials, taking advantage of favourable exchange rates, and avoiding political unrest (Langdale 1985, 1989; Warf 1989). Citicorp, for example, erected its Global Telecommunications Network to allow it to trade $200bn daily in foreign exchange markets around the world. Such networks give banks an ability to move money – by some estimates, more than $1.5 trillion daily (*Insight*, 1988) – around the globe at stupendous rates. Subject to the process

of digitisation, information and capital become two sides of the same coin. In the securities markets, global telecommunications systems have also facilitated the emergence of the 24-hour trading day linking stock markets through the computerised trading of stocks. Reuters and the Chicago Mercantile Exchange announced the formation of Globex, an automated commodities trading system, while in 1993 the New York stock exchange began the move to a 24-hour day automated trading system.

Within the context of an expanding and ever more integrated global communications network, a central role in the formation of local competitive advantage has been attained by teleports, which are essentially office parks equipped with satellite earth stations and usually linked to local fibre-optics lines (Lipman *et al.* 1986; Hanneman 1987a, 1987b and 1987c). The World Teleport Association defines a teleport as:

> An access facility to a satellite or other long-haul telecommunications medium, incorporating a distribution network serving the greater regional community and associated with, including, or within a comprehensive real estate or other economic development. (Hanneman 1987a, p. 15)

Just as ports facilitate the transshipment of cargo and airports are necessary for the movement of people, so too do teleports serve as vital information transmission facilities in the age of global capital. Because telecommunications exhibit high fixed costs and low marginal costs, teleports offer significant economies of scale to small users unable to afford private systems (Burstyn 1986; Stephens 1987). Teleports apparently offer a continually declining average cost curve for the provision of telecommunications services. Such a cost curve raises important issues of pricing and regulation, including the tendency of industries with such cost structures to form natural monopolies. Government regulation is thus necessary to minimise inefficiencies, and the pricing of telecommunications services becomes complex (i.e. marginal revenues do not equal marginal costs, as in non-monopolistic, non-regulated sectors) (Rohlfs 1974; Saunders *et al.* 1983; Guldmann 1990).

In the late 1980s there were 54 teleports in the world, including 36 in the US (Hanneman 1987a).

Most of these are concentrated in the industrialised world, particularly in cities in which data-intensive financial and business services play a major economic role. In Europe, London's new teleport in the Docklands will ensure that city's status as the centre of the Euromarket for the near future; Hamburg, Cologne, Amsterdam and Rotterdam are extending telematic control across Europe.

Tokyo is currently building the world's largest teleport. In the 1980s, the Japanese government initiated a series of high-technology 'technopolises' that form part of a long-term 'teletopia' plan to encourage decentralisation of firms out of the Tokyo region to other parts of the nation (Rimmer 1991). In 1993 the city initiated the Tokyo Teleport on 98 ha of reclaimed land in Tokyo harbour (Tokyo Metropolitan Government Planning Department, 1993). The teleport's 'intelligent buildings' (those designed to accommodate fibre optics and advanced computational capacity), particularly its Telecom Centre, are designed to accommodate ISDN requirements. Wide Area Networks (WANs) provide local telecommunications services via microwave channels, as do Value Added Networks on fibre-optic routes. The site was originally projected to expand to 340 ha, including office, waterfront and recreational functions, and employ 100,000 people, but may be scaled back in the light of the recent recessionary climate there.

The world's first teleport is named, simply, The Teleport, located on Staten Island, New York, a project jointly operated by Merrill Lynch and the Port Authority of New York and New Jersey. Built in 1981, The Teleport consists of an 11-acre office site and 16 satellite earth stations, and is connected to 170 miles of fibre-optic cables throughout the New York region, which are, in turn, connected to the expanding national fibre-optic network. Japanese firms have taken a particularly strong interest in The Teleport, comprising 18 of its 21 tenants. For example, Recruit USA, a financial services firm, uses it to sell excess computer capacity between New York and Tokyo, taking advantage of differential day and night rates for supercomputers in each city by transmitting data via satellite and retrieving the results almost instantaneously (Warf 1989).

In addition to the US, European and Japanese teleports, some Third World nations have invested in them in order to secure a niche in the global information services economy. Jamaica, for example, built one at Montego Bay to attract American 'back office' functions there (Wilson 1991). Other examples include Hong Kong, Singapore, Bahrain and Lagos, Nigeria (Warf 1989).

THE INTERNET: POLITICAL ECONOMY AND SPATIALITY OF THE INFORMATION HIGHWAY

Of all the telecommunications systems that have emerged since the 1970s, none has received more public adulation than the Internet. . . . The Internet is the largest electronic network on the planet, connecting an estimated 20m people in 40 countries (Broad 1993). Further, the Internet has grown at rapid rates, doubling in networks and users every year; by mid-1992, it connected more than 12,000 individual networks worldwide. Originating as a series of public networks, it now includes a variety of private systems of access in the US including services such as Prodigy, CompuServe or America On-Line (Lewis 1994), which allow any individual with a microcomputer and modem to 'plug in', generating a variety of 'virtual communities'. By 1994, such services connected almost 5m people in the US alone (Lewis, 1994).

[. . .]

The Internet has become the world's single most important mechanism for the transmission of scientific and academic knowledge. Roughly one-half of all of its traffic is electronic mail, while the remainder consists of scientific documents, data, bibliographies, electronic journals and bulletin boards (Broad 1993). Newer additions include electronic versions of newspapers, such as the *Chicago Tribune* and *San Jose Mercury News*, as well as an electronic library, the World Wide Web. In contrast to the relatively slow and bureaucratically monitored systems of knowledge production and transmission found in most of the world, the Internet and related systems permit a thoroughly unfiltered, non-hierarchical flow of information best noted for its lack of overlords. Indeed, the Internet has spawned its own unregulated counter-culture of 'hackers' (Mungo and Clough 1993). However, the system finds itself facing the continuous threat of commercialisation as cyberspace is progressively encroached upon by corporations, giving rise, for example, to new forms of electronic shopping and 'junk mail' (Weis 1992). The combination of popular, scientific and commercial uses has led to an enormous surge in demand for Internet capacities, so much so that they frequently generate 'traffic jams on the information highway' as the transmission circuits become overloaded (Markoff 1993).

Despite the mythology of equal access for everyone, there are also vast discrepancies in access to the Internet at the global level (Cooke and Lehrer 1993; Schiller 1993). As measured by the number of access nodes in each country, it is evident that the greatest Internet access remains in the most economically developed parts of the world, notably North America, Europe and Japan. The hegemony of the US is particularly notable given that 90 per cent of Internet traffic is destined for or originates in that nation. Most of Africa, the Middle East and Asia (with the exceptions of India, Thailand and Malaysia), in contrast, have little or no access. There is, clearly, a reflection here of the long-standing bifurcation between the First and Third Worlds. To this extent, it is apparent that the geography of the Internet reflects previous rounds of capital accumulation – i.e. it exhibits a spatiality largely preconditioned by the legacy of colonialism.

[...]

GEOGRAPHICAL CONSEQUENCES OF THE MODE OF INFORMATION

As might be expected, the emergence of a global economy hinging upon producer services and telecommunications systems has led to new rounds of uneven development and spatial inequality. Three aspects of this phenomenon are worth noting here, including the growth of world cities, the expansion of offshore banking centres and the globalisation of back offices.

World cities

The most readily evident geographical repercussions of this process have been the growth of 'world cities', notably London, New York and Tokyo (Moss 1987a; Sassen 1991), each of which seems to be more closely attuned to the rhythms of the international economy than the nation-state in which it is located. In each metropolitan area, a large agglomeration of banks and ancillary firms generates pools of well-paying administrative and white-collar professional jobs; in each, the incomes of a wealthy stratum of traders and professionals have sent real estate prices soaring, unleashing rounds of gentrification and a corresponding impoverishment for disadvantaged populations. While such predicaments are not new historically – Amsterdam was the Wall Street of the 17th century (Rodriguez and Feagin 1986) – the magnitude and rapidity of change that global telecommunications have unleashed in such cities is without precedent.

London, for example, boomed under the impetus of the Euromarket in the 1980s, and has become detached from the rest of Britain (Thrift 1987; Budd and Whimster 1992). Long the centre of banking for the British Empire, and more recently the capital of the unregulated Euromarket, London seems to have severed its moorings to the rest of the UK and drifted off into the hyperspaces of global finance. State regulation in the City – always loose when compared to New York or Tokyo – was further diminished by the 'Big Bang' of 1986. Accordingly, the City's landscape has been reshaped by the growth of offices, most notably Canary Wharf and the Docklands. Still the premier financial centre of Europe, and one of the world's major centres of foreign banking, publishing and advertising, London finds its status challenged by the growth of Continental financial centres such as Amsterdam, Paris and Frankfurt.

Similarly, New York rebounded from the crisis of the mid-1970s with a massive influx of petrodollars and new investment funds (i.e. pension and mutual funds) that sustained a prolonged bull market on Wall Street in the 1980s (Scanlon 1989; Mollenkopf and Castells 1992; Shefter 1993). Today, 20 per cent of New York's banking employment is in foreign-owned firms, notably Japanese giants such as Dai Ichi Kangyo. Driven by the entrance of foreign firms and increasing international linkages, trade on the New York stock exchange exploded from 12m shares per day in the 1970s to 150m in the early 1990s (Warf 1991). New York also boasts of being the communications centre of the world, including one-half million jobs that involve the collection, production, processing, transmission or consumption of information in one capacity or another (Warf 1991). This complex, including 60 of the largest advertising and legal services firms in the US, is fuelled by more word-processing systems than in all of Europe combined. The demand for space in such a context has driven an enormous surge of office construction, housing 60 headquarters of US Fortune 500 firms. Currently, 20 per cent of New York's office space is foreign-owned, testimony to the need of large foreign financial firms to establish a presence there.

Tokyo, the epicentre of the gargantuan Japanese financial market, is likely the world's largest centre of capital accumulation, with one-third of the world's stocks by volume and 12 of its largest banks by assets (Masai 1989). The Tokyo region accounts for 25 per cent of Japan's population, but a disproportionate share of its economic activity, including 60 per cent of the nation's headquarters, 65 per cent of its stock transactions, 89 per cent of its foreign corporations, and 65 per cent of its foreign banks (Cybriwsky 1991). Tokyo's growth is clearly tied to its international linkages to the world economy, particularly in finance, a reflection of Japan's growth as a major world economic power (Masai 1989; Cybriwsky 1991). In the 1980s, Japan's status in the global financial markets was unparalleled as the world's largest creditor nation (Vogel 1986; *Far Eastern Economic Review*, 1987). Tokyo's role as a centre of information-intensive activities includes a state-of-the-art telecommunications infrastructure, including the CAPTAIN (Character and Pattern Telephone Access Information Network) system (Nakamura and White 1988).

Offshore banking

A second geographical manifestation of the new, hypermobile capital markets has been the growth of offshore banking, financial services outside the regulation of their national authorities. Traditionally, 'offshore' was synonymous with the Euromarket, which arose in the 1960s as trade in US dollars outside the US. Given the collapse of Bretton Woods and the instability of world financial markets, the Euromarket has since expanded to include other currencies as well as other parts of the world. The recent growth of offshore banking centres reflects the broader shift from traditional banking services (loans and deposits) to lucrative, fee-based non-traditional functions, including debt repackaging foreign exchange transactions and cash management (Walter 1989).

Today, the growth of offshore banking has occurred in response to favourable tax laws in hitherto marginal places that have attempted to take advantage of the world's uneven topography of regulation. As the technological barriers to capital have declined, the importance of political ones has thus risen concomitantly. Several distinct clusters of offshore banking may be noted, including, in the Caribbean, the Bahamas and Cayman Islands; in Europe, Switzerland, Luxembourg and Liechtenstein; in the Middle East, Cyprus and Bahrain; in southeast Asia, Singapore and Hong Kong; and in the Pacific Ocean, Vanuatu, Nauru and Western Samoa. Roberts (1994, p. 92) notes that such places "are all part of a worldwide network of essentially marginal places which have come to assume a crucial position in the global circuits of fungible, fast-moving, furtive money and fictitious capital." Given the extreme mobility of finance capital and its increasing separation from the geography of employment, offshore banking can be expected to yield relatively little for the nations in which it occurs; Roberts (1994), for example, illustrates the case of the Cayman Islands, now the world's fifth-largest banking centre in terms of gross assets, where 538 foreign banks employ only 1,000 people (less than two apiece). She also notes that such centres are often places in which 'hot money' from illegal drug sales or undeclared businesses may be laundered.

Offshore markets have also penetrated the global stock market, where telecommunications may threaten the agglomerative advantages of world cities even as they reinforce them. For example, the National Associated Automated Dealers Quotation System (NASDAQ) has emerged as the world's fourth-largest stock market; unlike the New York, London, or Tokyo exchanges, NASDAQ lacks a trading floor, connecting half a million traders worldwide through telephone and fibre-optic lines. Similarly, Paris, Belgium, Spain, Vancouver and Toronto all recently abolished their trading floors in favour of screen-based trading.

Global back offices

A third manifestation of telecommunications in the world service economy concerns the globalisation of clerical services, in particular back offices. Back offices perform many routinised clerical functions such as data entry of office records, telephone books or library catalogues, stock transfers, processing of payroll or billing information, bank cheques, insurance claims, magazine subscriptions and airline frequent-flyer coupons. These tasks involve unskilled or semi-skilled labour, primarily women, and frequently operate on a 24-hour-per-day basis (Moss and Dunau 1986). By the mid-1980s, with the conversion of office systems from analogue to digital form largely complete, many firms began to integrate their computer systems with telecommunications.

Historically, back offices have located adjacent to headquarters activities in downtown areas to ensure close management supervision and rapid turnaround of information. However, under the impetus of rising central-city rents and shortages of sufficiently qualified (i.e. computer-literate) labour, many service firms began to uncouple their headquarters and back office functions, moving the latter out of the downtown to cheaper locations on the urban periphery. Most back office relocations, therefore, have been to suburbs (Moss and Dunau 1986; Nelson 1986). Recently, given the increasing locational flexibility afforded by satellites and a growing web of inter-urban fibre-optics systems, back offices have begun to relocate on a much broader, continental scale. Under the impetus of new telecommunications systems, many clerical tasks have become increasingly footloose and susceptible to spatial variations in production costs. For example, several firms fled New York City in the 1980s: American Express moved its back offices to Salt Lake City, UT, and Phoenix, AZ; Citicorp shifted its Mastercard and Visa divisions to Tampa, FL, and Sioux Falls, SD, and moved its data-processing functions to Las Vegas, NV, Buffalo, NY, Hagerstown, MD, and Santa Monica, CA; Citibank moved its cash management services to New Castle, DE; Chase Manhattan housed its credit card operations in Wilmington; Hertz relocated its data entry division to Oklahoma City; Avis went to Tulsa. Dean Witter moved its data-processing facilities to Dallas, TX; Metropolitan Life repositioned its back offices to Greenville, SC, Scranton, PA, and Wichita, KS; Deloitte Haskins Sells relocated its back offices to Nashville, TN; and Eastern Airlines chose Miami, FL.

Internationally, this trend has taken the form of the offshore office (Wilson 1991). The primary motivation for offshore relocation is low labour costs, although other considerations include worker productivity, skills, turnover and benefits. Offshore offices are established not to serve foreign markets, but to generate cost savings for US firms by tapping cheap Third World labour pools. Notably, many firms with offshore back offices are in industries facing strong competitive pressures to enhance productivity, including insurance, publishing and airlines. Offshore back-office operations remained insignificant until the 1980s, when advances in telecommunications such as trans-oceanic fibre-optics lines made possible greater locational flexibility just when the demand for clerical and information-processing services grew rapidly (Warf 1993). Several New York-based life insurance companies, for example, have erected back-office facilities in Ireland, with the active encouragement of the Irish government (Lohr 1988). Often situated near Shannon Airport, they move documents in by Federal Express and the final product back via satellite or the TAT-8 fibre-optics line that connected New York and London in 1989. Despite the fact that back offices have been there only a few years, Irish development officials already fret, with good reason, about potential competition from Greece and Portugal. Likewise, the Caribbean has become a particularly important locus for American back offices, partly due to the Caribbean Basin Initiative instituted by the Reagan administration and the guaranteed access to the US market that it provides. Most back offices in the Caribbean have chosen Anglophonic nations, particularly Jamaica and Barbados. American Airlines has paved the way in the Caribbean through its subsidiary Caribbean Data Services (CDS), which began when a data-processing centre moved from Tulsa to Barbados in 1981. In 1987, CDS opened a second office near Santo Domingo, Dominican Republic, where wages are one-half as high as Barbados (Warf, forthcoming). Thus, the same flexibility that allowed back offices to move out of the US can be used against the nations to which they relocate.

[. . .]

CONCLUDING COMMENTS

[. . .]

[I]t is vital to note that, contrary to early, simplistic expectations that telecommunications would 'eliminate space', rendering geography meaningless through the effortless conquest of distance, such systems in fact produce new rounds of unevenness, forming new geographies that are imposed upon the relics of the past. Telecommunications simultaneously reflect and transform the topologies of capitalism, creating and rapidly recreating nested hierarchies of spaces technically articulated in the architecture of computer networks. Indeed, far from eliminating variations among places, such systems permit the exploitation of differences between areas with renewed ferocity. As Swyngedouw (1989) noted, the emergence of hyperspaces does not entail the obliteration of local uniqueness, only its reconfiguration. That the geography engendered by this process was unforeseen a decade ago hardly needs restating; that the future will hold an equally unexpected, even bizarre, set of outcomes is equally likely.

REFERENCES FROM THE READING

Akwule, R. (1992) *Global Telecommunications: The Technology, Administration, and Policies*, Focal Press, Boston, MA.

Blazar, W. (1985) 'Telecommunications: harnessing it for development', *Economic Development Commentary* 9: 8–11.

Broad, W. (1993) 'Doing science on the network: a long way from Gutenberg', *New York Times*, 18 May: B5.

Budd, L. and Whimster, S. (eds) (1992) *Global Finance and Urban Living: A Study of Metropolitan Change*, Pergamon, London.

Burstyn, H. (1986) 'Teleports: at the crossroads', *High Technology* 6, 5: 28–31.

Castells, M. (1989) *The Informational City*, Blackwell, Oxford.

Coffey, W. and Bailly, A. (1991) 'Producer services and flexible production: an exploratory analysis', *Growth and Change* 22: 95–117.

Cooke, K. and Lehrer, D. (1993) 'The Internet: the whole world is talking', *The Nation* 257: 60–3.

Cybriwsky, R. (1991) *Tokyo: The Changing Profile of an Urban Giant*, G.K. Hall and Co, Boston, MA.

Dicken, P. (1992) *Global Shift: The Internationalization of Economic Activity* (2nd edn), Guilford Press, New York.

Far Eastern Economic Review (1987) Japan banking and finance, 9 April: 47–110.

Garnham, N. (1990) *Capitalism and Communication: Global Culture and the Economics of Information*, Sage, Beverley Hills.

Gillespie, A. and Williams, H. (1988) 'Telecommunications and the reconstruction of comparative advantage', *Environment and Planning A* 20: 1311–21.

Goddard, J. and Gillespie, A. (1986) *Advanced Telecommunications and Regional Development*, Centre for Urban and Regional Development Studies, Newcastle-upon-Tyne.

Guldmann, J. (1990) 'Economies of scale and density in local telephone networks', *Regional Science and Urban Economics* 20: 521–33.

Hall, P. and Preston, P. (1988) *The Carrier Wave: New Information Technology and the Geography of Innovation, 1846–2003*, Unwin Hyman, London.

Hanneman, G. (1987a) 'The development of teleports', *Satellite Communications*, March: 14–22.

Hanneman, G. (1987b) 'Teleport business', *Satellite Communications*, April: 23–6.

Hanneman, G. (1987c) 'Teleports: the global outlook', *Satellite Communications*, May: 29–33.

Harvey, D. (1989) *The Condition of Post-modernity*, Blackwell, Oxford.

Harvey, D. (1990) 'Between space and time: reflections on the geographical imagination', *Annals of the Association of American Geographers* 80: 418–34.

Hepworth, M. (1986) 'The geography of technological change in the information economy', *Regional Studies* 20: 407–24.

Hepworth, M. (1990) *Geography of the Information Economy*, Guilford Press, London.

Insight (1988) 'Juggling trillions on a wire: is electronic money safe?', 15 February: 38–40.

Kakabadse, M. (1987) *International Trade in Services: Prospects for Liberalisation in the 1990s*, Croom Helm, London.

Langdale, J. (1985) 'Electronic funds transfer and the internationalisation of the banking and finance industry', *Geoforum* 16: 1–13.

Langdale, J. (1989) 'The geography of international business telecommunications: the role of leased networks', *Annals of the Association of American Geographers* 79: 501–22.

Lewis, P. (1994) 'A boom for on-line services', *New York Times*, 12 July: Cl.

Lipman, A., Sugarman, A. and Cushman, R. (1986) *Teleports and the Intelligent City*, Dow Jones, Homewood, IL.

Lohr, S. (1988) 'The growth of the global office', *New York Times*, 18 October: Dl.

Markoff, J. (1993) 'Traffic jams already on the information highway', *New York Times*, 3 November: I, C7.

Masai, Y. (1989) 'Greater Tokyo as a global city', in R. Knight and G. Gappert (eds) *Cities in a Global Society*, Sage, Newbury Park, CA.

Mollenkopf, J. and Castells, M. (eds) (1992) *Dual City: Restructuring New York*, Russell Sage Foundation, New York.

Moss, M. (1987a) 'Telecommunications, world cities and urban policy', *Urban Studies* 24: 534–46.

Moss, M. (1987b) 'Telecommunications and international financial centres', in J. Brotchie, P. Hall and P. Newton (eds) *The Spatial Impact of Technological Change*, Croom Helm, London.

Moss, M. and Dunau, A. (1986) 'Offices, information technology, and locational trends', in J. Black, K. Roark and L. Schwartz (eds) *The Changing Office Workplace*, Urban Land Institute, Washington, DC, pp. 171–82.

Mungo, P. and Clough, B. (1993) *Approaching Zero: The Extraordinary Underworld of Hackers, Phreakers, Virus Writers and Key Board Criminals*, Random House, New York.

Nakamura, H. and White, J. (1988) 'Tokyo', in M. Dogan and J. Kasarda (eds) *The Metropolitan Era*, Volume 2, Sage, Mega-Cities, Newbury Park, CA.

Nelson, K. (1986) 'Labor demand, labor supply and the suburbanization of low-wage office work', in A. Scott and M. Storper (eds) *Production, Work, Territory*, Allen and Unwin, Boston, MA.

Nicol, L. (1985) 'Communications technology: economic and spatial impacts', in M. Castells (ed.) *High Technology, Space, and Society*, Sage, Beverly Hills, CA, pp. 191–209.

Noyelle, T. and Dutka, A. (1988) *International Trade in Business Services*, Ballinger, Cambridge, MA.

Office of Technology Assessment (1993) *Automation of America's Offices*, US Government Printing Office, Washington, DC.

Poster, M. (1990) *The Mode of Information: Poststructuralism and Social Context*, University of Chicago Press, Chicago, IL.

Quinn, J., Baruch, J. and Paquette, P. (1987) 'Technology in services', *Scientific American* 257, 6: 50–8.

Rimmer, P. (1991) 'Exporting cities to the western Pacific Rim: the art of the Japanese package', in J. Brotchie, M. Baity, P. Hall and P. Newton (eds) *Cities of the 21st Century*, Longman Cheshire, Melbourne.

Roberts, S. (1994) 'Fictitious capital, fictitious spaces: the geography of offshore financial flows', in S. Corbridge, R. Martin and N. Thrift (eds) *Money Power Space*, Blackwell, Oxford.

Rodriguez, N. and Feagin, J. (1986) 'Urban specialization in the world-system', *Urban Affairs Quarterly* 22: 187–219.

Rohlfs, J. (1974) 'A theory of interdependent demand for a communications service', *Bell Journal of Economics and Management Science* 5: 13–37.

Sassen, S. (1991) *The Global City: New York, London, Tokyo*, Princeton University Press, Princeton, NJ.

Saunders, R., Warford, J. and Wellenius, B. (1983) *Telecommunications and Economic Development*, Johns Hopkins University Press, Baltimore, MD.

Scanlon, R. (1989) 'New York City as global capital in the 1980s', in R. Knight and G. Gappert (eds) *Cities in a Global Society*, Sage, Newbury Park, CA.

Schiller, H. (1993) ' "The information highway": public way or private road?', *The Nation* 257: 64–5.

Shefter, M. (1993) *Capital of the American Century: The National and International Influence of New York City*, Russell Sage Foundation, New York.

Stephens, G. (1987) 'What can business get from teleports?', *Satellite Communications*, March: 18–19.

Strange, S. (1986) *Casino Capitalism*, Blackwell, Oxford.

Swyngedouw, E. (1989) 'The heart of the place: the resurrection of locality in an age of hyperspace', *Geografiska Annaler* 71: 31–42.

Thrift, N. (1987) 'The fixers: the urban geography of international commercial capital', in J. Henderson and M. Castells (eds) *Global Restructuring and Territorial Development*, Sage, Beverly Hills, CA.

Tokyo Metropolitan Government Planning Department (1993) *Tokyo Teleport*, Tokyo Metropolitan Government Information Centre, Tokyo.

Vogel, E. (1986) 'Pax Nipponica?', *Foreign Affairs* 64: 752–67.

Walker, R. (1985) 'Is there a service economy? The changing capitalist division of labor', *Science and Society*, spring: 42–83.

Walter, I. (1989) *Secret Money*, Unwin Hyman, London.

Warf, B. (1989) 'Telecommunications and the globalization of financial services', *Professional Geographer* 41: 257–71.

Warf, B. (1991) 'The internationalization of New York services', in P. Daniels (ed.) *Services and Metropolitan Development: International Perspectives*, Routledge, London, pp. 245–64.

Warf, B. (1993) 'Back office dispersal: implications for urban development', *Economic Development Commentary* 16: 11–16.

Warf, B. (forthcoming) 'Information services in the Dominican Republic', *Yearbook of the Association of Latin American Geographers*.

Weis, A. (1992) 'Commercialization of the Internet', *Electronic Networking* 2, 3: 7–16.

Wilson, M. (1991) *Offshore relocation of producer services: the Irish back office*, Paper presented at the Annual Meeting of the Association of American Geographers, Miami.

Wood, P. (1991) 'Flexible accumulation and the rise of business services', *Transactions of the Institute of British Geographers* 16: 160–72.

Editors' references and suggestions for further reading

Abler, R. (1970) "What makes cities important," *Bell Telephone Magazine*.

Castells, M. (1996) *The Rise of the Network Society*, Oxford: Blackwell.

Castells, M. (2000) "Grassrooting the space of flows," in J. Wheeler, Y. Aoyama and B. Warf (eds) *Cities in the Telecommunications Age: The Fracturing of Geographies*, London: Routledge, pp. 18–27.

Graham, S. (1998) "The end of geography or the explosion of place? Conceptualizing space, place and information technology," *Progress in Human Geography* 22: 165–85.

Graham, S. (1999) "Global grids of glass: on global cities, telecommunications and planetary urban networks," *Urban Studies* 36: 929–49.

Pascal, A. (1987) "The vanishing city," *Urban Studies* 24: 597–603.

Thrift, N. (1996) "New urban eras and old technological fears: reconfiguring the goodwill of electronic things," *Urban Studies* 33: 1463–93.

"City Watching: Closed Circuit Television Surveillance in Public Spaces"

from *Area* (1995)

Nicholas Fyfe and Jon Bannister

Editors' Introduction

People are now rarely free from the electronic gaze of closed circuit television (CCTV) cameras. Routine activities, such as walking along streets in city centers or through shopping malls, traveling on trains and buses, or sitting in a sports stadium, are all captured by an elaborate network of cameras which now seem as much part of the urban infrastructure as traffic-lights or post-boxes. CCTV surveillance cameras are, of course, not new. This technology has been used as a crime prevention tool in many privately owned spaces, for example, banks and shops, since the late 1960s. The extension of CCTV into public spaces, such as city streets, is more recent. In Britain, the first public space CCTV system went "live" in the seaside town of Bournemouth in 1985 but by 1990 there were still only five towns that had public space CCTV. Then, largely as a result of financial incentives offered by the British government, the number of town center CCTV systems grew rapidly, reaching seventy-nine by 1994 and over 500 by 1999. While Britain now has the most intensive concentration of electronic "eyes on the street" of anywhere in the world, other cities in Europe and North America also have extensive public space CCTV systems (see Hempel and Topfer 2002; Sorkin 1992).

As Fyfe and Bannister argue in this reading, this rapid and spectacular diffusion of CCTV surveillance technology into urban public space is highly controversial. Symbolically, public spaces have considerable significance as a geographical area which is based on rights of universal access, so that any changes in the ways in which such spaces are controlled raises important political questions about the powers of inclusion and exclusion. Indeed, several writers have viewed CCTV from a Foucauldian perspective, seeing it as part of a disciplinary network for producing obedient individuals in public spaces. Much like Jeremy Bentham's Panopticon, a model prison in which inmates in cells on the periphery of a circular tower are always, potentially, under the gaze of an official in a central tower, CCTV surveillance could therefore be interpreted as an elaborate political technology for the exercise of state power. But, as Fyfe and Bannister also argue, the spread of CCTV into public space raises some more immediate social and political questions. How are such CCTV systems established? How effective are they in reducing crime? And what, if any, resistance is there to public space CCTV surveillance?

In terms of establishing a "modern Panopticon," CCTV systems in public spaces crucially depend on a strategic alliance between the local state and local private capital. Yet, as the example of Glasgow's City Watch shows, constructing a partnership between the local public and private sectors is fraught with tensions. In terms of effectiveness, Fyfe and Bannister question the dramatic claims made for CCTV in reducing crime. Although some "before" and "after" CCTV studies show sharp falls in crime, the many methodological difficulties involved in accurately measuring crime mean that the results of such studies need to be interpreted cautiously.

Indeed, far from being the "silver bullet" of crime control suggested by supporters of the technology, recent research indicates that CCTV systems are associated with only modest reductions in crime of around 4 percent, most of which is accounted for by falls in vehicle crime while violent crime levels are unaffected (Welsh and Farrington 2003). Finally, Fyfe and Bannister consider some of the civil liberties implications of CCTV systems. Despite the absence of "revolts against the gaze," the cameras clearly raise important questions about who is monitored in public space and how the operators of the systems are called to account. Fyfe and Bannister's conclusion is that the "democratic deficit" associated with local governance organizations is clearly a characteristic of city center CCTV systems where operational control and strategic management are typically the responsibility of non-elected bodies.

In the context of wider debates and developments in urban geography, Fyfe and Bannister's paper is important at two levels. First, CCTV is only one of a range of technologies that are increasingly being used in an attempt to monitor, control and guide social and economic processes in cities. Other examples include the use of Geographical Information Systems by retailing organizations for targeting investment by profiling census tracts in terms of consumption and spending potential; and the integration of CCTV with computerized tracking and charging devices by road transport bodies in order to allow the automatic charging of drivers for using particular sections of road (Graham 1998). Although many of these technologies are quite crude at present, recent innovations in CCTV have considerably enhanced the panoptic power of these surveillance systems. Digitalization means that cameras are now able to extract vehicle licence plate details and this information can be automatically checked against other databases containing the registration numbers of vehicles linked to suspected criminals or terrorists. Developments in facial recognition software mean that it is now also technically possible for computers to match a face from a city center surveillance camera with a computerized database of known and suspected offenders. Second, CCTV and related technological developments clearly have significant implications in terms of the restructuring and governance of cities. There are important connections, for example, between the spread of CCTV surveillance and the emergence of "interdictory spaces" highlighted by postmodern urbanists such as Michael Dear and Steven Flusty (see their reading in Part 3, Restructuring). Indeed, according to Flusty (1994: 37), video cameras are now part of "an infrastructure restructuring the city into electronically linked islands of privilege embedded in a police state matrix." Similarly, the use of CCTV appears to fit well with arguments about the kind of "revanchist" urban policy which is claimed to characterize neo-liberal cities (Smith 1996). According to this perspective, CCTV provides city authorities with a crucial instrument of social control, so that those perceived not to belong in commercial public spaces now risk being "monitored and harassed, losing rights as citizens just because they aren't seen to be lucrative enough as consumers" (Graham et al. 1996: 11). The result is a subtle privatization of public space as commercial imperatives come to define acceptable behavior, excluding those who detract from the consumption experience (Fyfe and Bannister 1998: 263; see also reading by Goss, p. 293).

These critical, almost dystopian, readings of CCTV, however, risk falling into the trap of technological determinism. There is an assumption that such is the impressive technological power of CCTV surveillance cameras that little or no human input is required for them to be effective. Yet, as the discussion of Warf's paper in this section highlights, it is important not to overlook the important interactions between people and technology. For CCTV, this includes those on the street aware of the gaze of the cameras, the control room staff who watch the screens for suspicious behavior, and the police officers who must decide whether to respond to incidents reported to them by the control room. By considering these "actor-networks" of people and technology it is possible to gain a much richer understanding of what CCTV means for urban life. The Finnish social geographer, Hille Koskela (2000), for example, has emphasized how being under surveillance is an ambivalent emotional event. A surveillance camera, she contends, can simultaneously represent safety and danger to those aware of its gaze and create a paradoxical emotional space in which people can feel both more secure and more fearful. Similarly, studies of local people's responses to the proliferation of CCTV cameras reveals a degree of ambivalence to this technology, combining "a grudging acceptance of the 'need' for video surveillance with a range of (often diffuse) worries about how CCTV might come to be used and about the kind of world it signifies" (Sparks et al. 2001: 894).

In addition to their paper here, Fyfe and Bannister's pioneering research on CCTV surveillance appears in chapters they have written for *Images of the Street: Planning, Identity and Control in Public Space* (London: Routledge 1998), edited by Nicholas Fyfe; and *Surveillance, Closed Circuit Television and Social Control* (Aldershot: Ashgate 1998), edited by Clive Norris, Jade Moran and Gary Armstrong. Other useful accounts of CCTV surveillance include Clive Norris and Gary Armstrong's *The Maximum Surveillance Society: The Rise of CCTV* (Oxford: Berg 1999) and Martin Gill's edited collection *CCTV* (Leicester: Perpetuity Press 2003). For more general discussions of crime control and city, see Nicholas Fyfe's chapter, "Zero tolerance, maximum surveillance: crime control and the late modern city," in Loretta Lees' book *The Emancipatory City: Paradoxes and Possibilities* (London: Sage 2004) and David Garland's *The Culture of Control: Crime and Social Order in Contemporary Society* (Chicago, IL: University of Chicago Press 2001).

Nicholas Fyfe is Reader in Human Geography at the University of Dundee and author of *Protecting Intimidated Witness* (London: Ashgate 2001), editor of *Images of the Street: Planning, Identity and Control in Public Space* (London: Routledge 1998) and co-editor with David Evans and David Herbert of *Crime, Policing and Place: Essays in Environmental Criminology* (London: Routledge 1992). Jon Bannister is Senior Lecturer in the Department of Urban Studies, University of Glasgow, and an editor of the journal *Urban Studies*.

INTRODUCTION

In the 1990s closed circuit television (CCTV) surveillance cameras have become an increasingly common feature in the public spaces of towns and cities across Britain. A survey of London boroughs, metropolitan authorities and a sample of district councils in England found that thirty-nine had CCTV cameras in public spaces[1] in 1993 compared with just two in 1987 (Bulos and Sarno 1994). The results of another survey . . . showed that by August 1994 seventy-nine towns and cities had CCTV, while by March 1995 the figure was over ninety.

CCTV surveillance cameras are, of course, not new. They have been operating in privately owned (but publicly accessible) spaces such as shopping malls . . . banks and football stadia for several years. But the extension of CCTV from these locations into *publicly owned* urban-space – the streets and squares of town and city centres – raises important questions about such intensive surveillance of spaces which, as Goheen notes, have particular importance as areas where people's 'collective rights to performance and speech are entrenched' (Goheen 1994: 431).

THE PANOPTIC DREAM?

The sight of an individual sitting at a console in front of a bank of TV monitors displaying pictures of the streets of a city centre, using the controls to make cameras pan across a crowded shopping area or zoom in on a group of youths gathered on a street corner, and dispatching police officers to the scene of anything that arouses suspicion, has prompted many comparisons with Orwell's dystopian vision of 'Big Brother' in *Nineteen Eighty-Four*. But it is 1791 rather than 1984 which provides a more instructive date for comparison. This was the year Jeremy Bentham published plans for a 'Panopticon', a model prison based on an optical-mechanical technique whereby inmates in cells on the periphery of a circular building are always, potentially, under the gaze of an official in a central tower. . . . Although Bentham's Panopticon was never built in Britain . . . it was a scheme which nevertheless had an 'imaginary intensity' that has given rise to many variations (Foucault 1977: 205) of which CCTV can be seen as one of the most recent.

Drawing on Foucault's discussion of Bentham's Panopticon there are several important parallels with CCTV. Like the Panopticon, CCTV schemes meet Bentham's principle that power should be 'visible and unverifiable'. Visibility is ensured by the fact that just as the inmate of Bentham's prison has constantly 'before his [*sic*] eyes the tall outline of the central tower from which he is spied upon' (Foucault 1977: 201) so too anybody in Glasgow city centre, for example, can see cameras on top of six-metre poles or jutting out from the sides of buildings, while street signs proclaim 'This area is protected by City Watch'. Unverifiability reflects the way in which, just as the inmate in Bentham's scheme never knows whether he is being looked at at any one moment 'but he must be sure that he may always be so' (op. cit.), so too anyone in the city centre never knows whether the control room operator is looking at them but always knows that they might be. This pressure of surveillance is particularly effective because like Bentham's Panopticon, CCTV is a mechanism which 'automatizes and disindividualizes power' (Foucault 1977: 202). Power becomes vested not in the surveillance by a particular person, like a police officer, but in the electronic eye of the camera, inducing a 'state of conscious and permanent visibility that assures the automatic functioning of power' (Foucault 1977: 201). The product of such intensive surveillance is, as both Bentham and the proponents of CCTV claim, the deterrent of deviant behaviour and the possibility of rapid intervention at any moment if something suspicious is detected.

[. . .]

Employing CCTV in public spaces raises important theoretical questions about the relationships between civil society and the state. CCTV could be seen from a Foucauldian perspective, for example, as a manifestation of a general expansion of power, as a new component of a disciplinary network, an elaborate political technology for producing obedient individuals in public spaces (Foucault 1977: 214). CCTV might also be seen, however, as reinforcing the infrastructural or administrative power of the state to penetrate and regulate the activities of civil society (see Mann 1984 . . .). Both perspectives potentially offer insights into the significance of CCTV but both tend to obscure some of the more immediate but nevertheless important questions concerning CCTV which we are interested in here. How, for example, is a panoptic scheme like CCTV *constructed*? What are the *limits* of CCTV in terms of its effectiveness? And what, if any, *resistance* is there to CCTV? These three themes are explored in more detail drawing upon research on city centre CCTV in general and on Glasgow's City Watch scheme in particular.

CONSTRUCTING A MODERN PANOPTICON: THE DEVELOPMENT OF CITY CENTRE CCTV

Construction of CCTV surveillance systems in public spaces depends crucially on a strategic alliance between the local state and local private capital. Local state involvement is necessary because of municipal responsibility for the areas that make up the public spaces of city centres in which cameras operate. The high financial costs of installing and running a system, however, mean that individual local councils are unable or unwilling to finance CCTV systems unilaterally. Although some central government finance for local CCTV schemes is now available . . . local private capital is a vitally important component of most (although not all) city centre CCTV-schemes. The construction of a partnership between the local public and private sectors is, however, fraught with tensions because of the way in which CCTV occupies an ambiguous position, both geographically and conceptually, on the boundary between the private and public domains. From the perspective of local councils there are anxieties about committing public funding to a project which may mainly appear to serve the needs of local private commercial interests and which raises sensitive civil libertarian questions about the invasion of privacy. From the perspective of the private sector, however, there is reluctance to contribute to a scheme which is viewed as part of the public urban infrastructure and which should therefore be funded from contributions businesses already make to the public domain through business rates.

Glasgow provides an instructive example of the role of local political and economic interests in the development of a city centre CCTV system and the attempted resolution of these public–private tensions. The idea for city centre CCTV originated with the Glasgow Development Agency (GDA), a government 'quango' with a remit for promoting the economic development of the city. Worried that business drift from the city in the early 1990s was partly the product of crime (actual and perceived), GDA presented proposals for CCTV to Glasgow District Council (GDC) and Strathclyde Regional Council (SRC), drawing attention to the positive economic and social impact of CCTV claimed by established schemes elsewhere in Britain. While keen to contribute to crime prevention in the city, both GDC and SRC were concerned about the civil liberties implications of CCTV in the city's public spaces. But these concerns were diffused when GDA demonstrated, first, widespread popular support for the scheme . . . and, secondly, the involvement of the Scottish

SEVEN

Council for Civil Liberties in producing operating guidelines for the system. Although the councils then agreed to give £200,000 to the project, with the GDA contributing a further £100,000, this was far short of the £1.1 million needed to set up and run the scheme for three years. GDA therefore approached city centre businesses for contributions, stressing the financial benefits of the scheme by using the slogan CCTV 'doesn't just make sense – it makes business sense'. On the basis of a public opinion survey, GDA calculated that introducing CCTV would encourage 225,000 more visits to the city a year, creating 1,500 jobs and an additional £40 million of additional income to city centre businesses. The private sector responded by contributing £270,000. While short of GDA's original target this was sufficient to decide to install and run the system for one year. In November 1994 Glasgow's CCTV scheme, City Watch, went 'live' with thirty-two cameras distributed across the city centre, monitored by civilian operators in a control room located in the central police station,

and providing twenty-four-hour surveillance of the city's main business, commercial, cultural and tourist areas.

'BIG BROTHER' IS PROTECTING YOU? THE POTENTIAL EFFECTIVENESS OF CITY CENTRE CCTV

[...]

Tackling crime

In planning Glasgow's CCTV system, GDA visited schemes in Birmingham and Airdrie (a market town to the east of Glasgow) both of which provided dramatic evidence of the impact of CCTV on crime. Table 1 shows the police recorded crime statistics for the areas covered by the cameras in Birmingham and Airdrie before and after the installation of CCTV.

Table 1 'Before' and 'After' crime statistics for areas covered by CCTV in Birmingham and Airdrie

	Birmingham	
	Before CCTV (3 months to 3/91)	After CCTV (3 months to 9/91)
Woundings	46	27
Robberies	79	55
Thefts from a person	89	63
Indecency	8	3
Damage	62	80

Source: police data quoted in Birmingham City Centre Development Group (1992: 8)

	Airdrie	
	Before CCTV (12 months to 8/82)	After CCTV (12 months to 8/93)
Car break-ins	480	20
Theft of cars	185	13
Serious assaults	39	22
Vandalism	207	36
Break-ins to commercial premises	263	15

Source: police data quoted in Wills (1993: 13)

While these statistics are impressive, the evaluations on which such dramatic claims are based have been called into question. . . . Several problems have been identified. First, the 'before' and 'after' time periods are often too short and not matched for time of year. Secondly, the data only relates to crimes reported to and recorded by the police which may not accurately reflect actual changes in crime. Thirdly, the possibility that CCTV has displaced crime to surrounding areas not in view of the cameras is rarely mentioned or studied, and nor are control areas identified to assess comparable changes in crime in places without cameras. Against this background the claims by City Watch that 'there has been a significant decrease in certain types of criminal activity' in Glasgow, particularly, counterfeit trading, bag-dipping, pickpocketing, robberies at automatic cash machines, shoplifting and break-ins to commercial premises need to be interpreted cautiously. But claims about decreasing criminal activity through CCTV surveillance need to be interpreted cautiously for theoretical as well as methodological reasons. CCTV is bound up with a 'master shift' in the discourse of social control from a concern with the mind (and issues of motivation, thought and intention) to a concern with the body (and issues of observable behaviour) (Cohen 1985). Rather than attempting to tackle crime by investing in the treatment and rehabilitation of offenders, the discourse of 'new behaviourism' of which CCTV surveillance is a part, is less interested in the causes of crime than with its prevention, and is less concerned with trying to change social conditions than with the more modest aim of 'changing behaviour sequences' (Cohen 1985: 150). From this perspective if CCTV surveillance 'works' by reducing crime, it works at the level of deterrence not at the level of causation.

Reviving the city centre

Whatever its impact on crime, city centre CCTV systems are not simply a piece of crime prevention technology. In Glasgow, as elsewhere, the hope is that CCTV will enhance what City Watch calls the 'feel good factor' by making people more confident about coming to the city and thus combat what Bianchini (1990) calls the crisis of urban public sociability. Of course, part of this crisis is the product of the fear of crime (as much as the objective risks of crime) and its negative impacts on urban life and culture (see Smith 1989: 279–81). Anecdotal evidence from Glasgow suggests anxiety about coming into the city is already diminishing as a result of City Watch, a representative of one of the city's shopping malls claiming that 'people today are demonstrating a positive response to a safer environment which offers additional security to their family and friends'. As with crime and fear of crime reduction, it's too early to assess these claims rigorously but it is important to recognise that the decline of city centres has a whole variety of causes. The proliferation of out-of-town shopping centres and leisure facilities and the trends towards the domestication and privatisation of social life, such that 'the cinema and the theatre have long since turned into the home video; the laundrette and the laundry into Ariston and Hotpoint' (Taylor quoted in Bianchini 1990: 4), have all conspired to reduce the significance of old urban centres. Moreover, at a conceptual level both Sennett (1970, 1977) and Giddens (1984) argue that people's search for 'ontological security' (a sense of well-being and identity) is increasingly focused on the residential neighbourhood, decreasing the importance of social encounters in the wider, public environment of the city. Against this background, it is unlikely that CCTV on its own can reverse the decline of urban public social life. Indeed, CCTV may have disbenefits in terms of urban public sociability by increasing bystander indifference and reducing people's propensity to report incidents to the police. . . .

'REVOLTS AGAINST THE GAZE'? CCTV AND CIVIL LIBERTIES

Whatever its actual impact, public support for CCTV in public spaces currently appears to be high. . . . A Home Office public attitude survey conducted in various sites with and without CCTV found that 85 per cent of those interviewed in shopping centres, 89 per cent of those in streets and 92 per cent of those in car parks said they welcomed (or would welcome) the installation of CCTV in those sites (Honess and Charman 1992: 12). . . . But such statistics need to be interpreted cautiously. As the Home Office researchers concluded 'public acceptance is based on limited, and partly inaccurate knowledge of the functions and capabilities of CCTV systems in public places' (Honess and Charman 1992: 25). Furthermore, although there appears to be little opposition to CCTV, it would be misleading to conclude that the public is not concerned about its development. Almost three-quarters of respondents in the Home Office survey believed that CCTV cameras could easily be abused and used by the 'wrong people' (who these were was not specified) and 38 per cent felt that the people in control of the cameras could not be completely trusted to use them only for the public good (Honess and Charman 1992: 9). In addition fears were expressed that the pressures to demonstrate the effectiveness of having a CCTV system may lead those monitoring the cameras into 'over-scrutinising particular groups – for example young black males, "scruffy people" – without due cause' (Honess and Charman 1992: 8). This fear is echoed by Bianchini (1990) in his concerns about the use of CCTV for the 'moral regulation' of city centres. 'In many city centres', he observes, 'the paramount need to create a safe and attractive environment for shoppers led to the virtual disenfranchisement from city life of young people with low spending power and of other – generally low-income – residents, whose appearances and conduct did not conform to the moral codes of well-ordered consumption enforced by shopping centre managers'. Indeed, as Mulgan (1989: 276) suggests, attempts to create a 'convivial milieu' for economic and socio-cultural life in the city using CCTV may become attempts to purify space of those 'troublesome others' – the underclass, the homeless, the unemployed – reducing exposure to what Sennett (1990) calls 'the presence of difference'. The commonly expressed view that the law-abiding have nothing to fear from CCTV has thus been dismissed by Britain's National Council for Civil Liberties as of little comfort to those who already experience discrimination and harassment (Liberty 1989).

Against this background, the operational control of CCTV is vitally important. . . . While most city centre CCTV schemes allow the public access to the control room so they can 'observe the observers', this is clearly not a sufficient condition for making schemes formally accountable to the local community. Nor are operational guidelines, however strict, a substitute for the local democratic accountability of CCTV systems, particularly where monitoring the implementation of such guidelines lies, as in Glasgow, with an employee of the public–private sector partnership responsible for running the system. The 'democratic deficit' associated with local governance organisations (agencies responsible for the provision and management of local public services . . .) is clearly a characteristic of city centre CCTV systems, including Glasgow, where both the operational control and strategic management of the system are the responsibility of non-elected bodies.

CONCLUSIONS

[S]upport for CCTV is partly a faith in its crime-reducing effects, but it also reflects the way CCTV fits neatly with both the neo-liberal and neo-conservative dimensions of the New Right law and order policy. CCTV shifts responsibility for controlling crime and undertaking policing from the central state and onto local civil society *and* it enhances the state's ability to penetrate and control civil society (see Fyfe 1995 . . .). This is of theoretical as well as political importance. Returning to our opening remarks, it suggests the relevance of CCTV to both a Foucauldian concern with an expanding disciplinary network and the interests of Michael Mann (and others) with the infrastructural power of the state. . . . Rather than focusing on narrow operational questions about effectiveness, the development of CCTV therefore needs to be set in its political context. But there are also key economic questions given our earlier remarks about the funding of city centre CCTV and the fact that CCTV is now big business with an estimated £300m spent on video surveillance a year. Finally, there are important social issues bound up with CCTV. Groombridge and Murji warn that the massive expansion of CCTV will yield public indifference to the world around them: no one will care what they see on the streets 'as they move about . . . head down' (1994: 289) because they will assume someone else is watching. . . . Resorting to cameras as a technocratic solution to a social problem may therefore have disturbing social consequences.

NOTE

1 In this study public space was defined as 'one to which normally people have unrestricted access and right of way' (Bulos and Sarno 1994: 7).

REFERENCES FROM THE READING

Bianchini, F. (1990) 'The crisis of urban public social life in Britain: origins of the problem and possible responses', *Planning. Policy and Research* 5, 3: 4–8.

Birmingham City Centre Development Group (1992) *City Watch*, Birmingham City Centre Development Group, Birmingham.

Bulos, M. and Sarno, C. (1994) *Closed Circuit Television and Local Authority Initiatives: The First National Survey*, School of Land Management and Urban Policy, South Bank University, London.

Cohen, S. (1985) *Visions of Social Control*, Polity Press, Cambridge.

Foucault, M. (1977) *Discipline and Punish: The Birth of the Prison*, Allen Lane, London.

Fyfe, N.R. (1995) 'Law and order policy and the spaces of citizenship in contemporary Britain', *Political Geography* 14, 2: 177–89.

Giddens, A. (1984) *The Constitution of Society*, Polity Press, Cambridge.

Goheen, P.G. (1994) 'Negotiating access to public space in mid-nineteenth century Toronto', *Journal of Historical Geography* 20, 4: 430–49.

Groombridge, N. and Murji, K. (1994) 'As easy as AB and CCTV', *Policing* 10, 4: 283–90.

Honess, T. and Charman, E. (1992) *Closed Circuit Television in Public Places*, Home Office, London.

Liberty (1989) *Who's Watching You? Video Surveillance in Public Places*, Liberty Briefing Paper No. 16, London.

Mann, M. (1984) 'The autonomous power of the state: its origins, mechanisms and results', *Archives Europeenes de Sociologie* 25: 185–213.

Mulgan, G. (1989) 'The changing shape of the city', in Hall, S. and Jacques, M. (eds) *New Times: The Changing Face of Politics in the 1990s*, Lawrence and Wishart, London, pp. 262–78.

Sennett, R. (1970) *The Uses of Disorder*, Penguin, Harmondsworth.

Sennett, R. (1977) *The Fall of Public Man*, Cambridge University Press, Cambridge.

Sennett, R. (1990) *The Conscience of the Eye: The Design and Social Life of Cities*, Faber, London.

Smith, S.J. (1989) 'The challenge of urban crime', in Herbert, D.T. and Smith, D.M. (eds) *Social Problems and the City: New Perspectives*, Oxford University Press, Oxford, pp. 271–88.

Wills, J. (1993) 'Candid cameras', *Local Government Chronicle*, 17 September.

Editors' references and suggestions for further reading

Flusty, S. (1994) *Building Paranoia: The Proliferation of Interdictory Space and the Erosion of Spatial Justice*, Los Angeles Forum for Architecture and Urban Design, West Hollywood, CA.

Fyfe, N. and Bannister, J. (1998) "The eyes upon the street: closed circuit television surveillance and the city," in N. Fyfe (ed.) *Images of the Street: Planning, Identity and Control in Public Space*, Routledge, London, pp. 254–67.

Graham, S. (1998) "Spaces of surveillant simulation: new technologies, digital representations, and material geographies," *Environment and Planning D: Society and Space* 6: 483–504.

Graham, S., Brooks, J. and Heery, D. (1996) "Towns on the television: closed circuit TV in British towns and cities," *Local Government Studies* 22, 3: 3–27.

Hempel, L. and Topfer, E. (2002) *Inception Report: On The Threshold to Urban Panopticon? Analysing the Employment of CCTV in European Cities and Assessing its Social and Political Implications*, Urban Eye Working Paper No. 1, Berlin Center for Technology and Society, Technical University Berlin, Berlin.

Koskela, H. (2000) "'The gaze without eyes': video-surveillance and the changing nature of urban space," *Progress in Human Geography* 24: 243–65.

Smith, N. (1996) *The New Urban Frontier: Gentrification and the Revanchist City*, Routledge, London.

Sorkin, M. (ed.) (1992) *Variations on a Theme Park: The New American City and the End of Public Space*, Hill and Wang, New York.

Sparks, R., Girling, E. and Loader, I. (2001) "Fear and everyday urban lives," *Urban Studies* 38: 885–98.

Welsh, B.C. and Farrington, D.P. (2003) "Effects of closed-circuit television on crime," *Annals of the American Academy of Political and Social Science* 587 (May).

"GIS Use in Community Planning: A Multidimensional Analysis of Empowerment"

from *Environment and Planning A* (2002)

Sarah A. Elwood

Editors' Introduction

Geographical Information Systems (GIS) are "organized collections of data processing methods which act on spatial data to enable patterns in that data to be understood and visualized" (Batty 2003: 409). GIS technology therefore treats different sets of attributes, such as land use or population, as map layers which can be examined in order to establish what, if any, relationships exist between these different layers. Although early examples of GIS date from the 1960s, it was in the 1980s that GIS began to grow in importance as practical applications for the technology were found increasingly in fields such as resource management and urban planning. The 1980s also saw a growing number of geographers exploring the potential of GIS, and by the 1990s it was firmly established as an important tool for conducting spatial analysis.

As the importance of GIS research grew, however, an increasing number of geographers began to voice concerns about the use of GIS technology (Schuurman 2000). Some critics focused on what they perceived to be flaws in the philosophical underpinnings of GIS, and particularly the positivist assumptions of objectivity and value neutrality of GIS analysis. According to Peter Taylor, for example, GIS was a collection of quantitative tools incapable of meaningful analysis, and therefore marked "a return of the very worst sort of positivism, a most naïve empiricism" (Taylor 1990; see also Lake 1993; Sui 1994). Other critics were uneasy about the broader social and political implications of GIS (Pickles 1995). There were concerns that GIS was advancing the trend towards a "surveillance society" in which the increasingly sophisticated ability to map people and their socio-economic characteristics constituted an erosion of privacy (Curry 1997). Others, for example, Neil Smith, were disturbed by the links between GIS and its military uses, claiming that "GIS and related technologies contribute[d] to the killing fields of the Iraqi desert" (Smith 1992: 257) during the 1991 Gulf War. Another line of criticism, however, focused on the failure to democratize GIS. The expense and expertise required to access the technology meant it was beyond the reach of many small organizations and communities, and therefore its use was confined largely to the corporate planning decisions of major public and private institutions. Concern that this perpetuated and reinforced significant inequalities in power relations within society prompted some human geographers and GIS researchers to focus on means of increasing public access to GIS technology. From this emerged PPGIS (Public Participation GIS), a field of study that investigates the use by and value of GIS for "marginalized peoples and communities engaged in social change" (Sieber 2003: 50), and it is this move towards greater community involvement with GIS which provides the immediate context for Sarah Elwood's paper.

A lecturer in geography at DePaul University in Chicago, Elwood focuses on a crucial area of ambivalence that surrounds the implications of GIS technology for the empowerment and disempowerment of individuals and communities. On the one hand, there is research that suggests GIS might be designed and used in ways which empower organizations and social groups that have been marginalized from decision-making; yet, on the other hand, there is evidence that GIS consolidates the power of existing dominant actors because of its high costs and skill requirements. To understand such potentially very different social implications of GIS, Elwood argues that it is vital to "unpack" what is meant by empowerment. Her account distinguishes three specific dimensions of this concept. It can refer to *distributive change* in the form of greater access to both goods and services and increased opportunities for political participation; *procedural change* whereby the views of citizens or community groups are given authority and legitimacy in the decision-making process; and *capacity building* in which the ability of citizens or communities to take action on their own behalf is enhanced. Armed with this conceptual framework, Elwood examines the implications of GIS use for a community organization in Minneapolis – the Powderhorn Park Neighborhood Association (PPNA) – which used GIS technology to oversee the implementation of a neighborhood revitalization plan (see also Elwood 2001, 2002). Drawing on ethnographic fieldwork, including a ten-month period working for PPNA, Elwood vividly reveals the varying impact of community-based GIS for different forms of empowerment. In particular her analysis shows that GIS use in community planning can "foster changes that are simultaneously empowering and disempowering," depending on the geographical scale and social focus of the analysis. Within the neighborhood, for example, the language and forms of information used in GIS have raised the level of expertise and knowledge needed by residents in order to participate in community planning discussions. While some, mainly upper-middle-class white homeowners, have mastered this new discourse, other residents, typically black, low income and old, have not, leading to their exclusion and disempowerment in relation to local planning debates. In terms of neighborhood interaction with the local state, however, PPNA's use of GIS has provided local representatives with a means of gaining greater participation, authority and legitimacy in city-level decision-making on planning issues, thus contributing to their empowerment via distributive and procedural change within this arena.

Elwood's case study therefore provides a fascinating neighborhood-level analysis of how GIS technology can be a tool for both empowerment *and* marginalization. In this respect, it stands out from much of the previous research in this field which tended to present a polarized picture of the links between GIS and society, contrasting community organizations successfully using GIS as a way into the policy-making process with organizations being overwhelmed by the resource requirements of GIS such that they became the "tools of corporate concerns" (Sieber 2003: 51). Elwood's analysis is more sophisticated. It shows not only that empowerment is a multidimensional concept, but also that the use of GIS in community planning can result in empowerment and disempowerment occurring simultaneously in different geographical contexts and with different social actors.

In cities across the United States, community organizations ranging from small resident groups active in single neighborhoods to large nonprofit organizations engaged in complicated tasks such as affordable housing development are becoming increasingly involved in urban planning and problem solving. This growing involvement is part of a shift in the roles and responsibilities of community organizations and, by extension, of the citizens who participate in them. Municipal government still plays a central role in planning, but community organizations are being charged with direct responsibility for planning and problem solving in their local communities. . . . In assuming responsibility for a greater range of urban planning and revitalization tasks, community organizations are incorporating new tools and practices. No longer, for instance, are information technologies such as geographic information systems (GIS) tools used solely by local state planners – they are increasingly being used by community organizations as part of their planning and revitalization efforts. The adoption of this technology by community organizations is particularly significant in light of ongoing debates in critical GIS research about the ways in which this technology may alter social and political relationships structuring participation and power of various actors in decision-making processes.

[. . .]

CRITICAL GIS RESEARCH

The critical GIS research agenda began with heated debates about the implications of this technology for the discipline of geography (compare Openshaw 1991; Taylor and Overton 1991), but the focus soon shifted to include an examination of the broader social and political impacts of this technology. Several propositions have been developed in the literature about ways that GIS may alter social and political processes and the power of individuals, institutions, and knowledge claims. These different arguments about the mechanisms mediating the impacts of GIS focus on different ways in which power is negotiated through GIS and the broader processes in which it is embedded.

A number of scholars have focused on the design of GIS to privilege information that can be displayed visually and quantitative techniques for spatial analysis, further arguing that these data storage and processing techniques are part of an empiricist and positivist logic (Lake 1993; Pickles 1995; Rundstrom 1995; Sheppard 1995). The empiricism and positivism of GIS are not inherently problematic, they contend, but become so because of the possibility that other forms of knowledge and logic (and by extension, the people and communities they represent) may be excluded from processes in which GIS is used. Other researchers have focused on the closely related role GIS may play in constituting expertise in decision-making. Specifically, they argue that GIS advances an instrumental rationalist approach to decision-making that is already dominant in decision-making processes where GIS is used, such as land-use planning. The validation by the state of this approach and of GIS as an acceptable tool for information analysis and decision-making thus mutually reinforces the hegemony of instrumental rationalism, at the expense of other approaches and knowledge systems (Aitken and Michel 1995; Harris *et al.* 1995; Lupton and Mather 1997; Yapa 1991). Both kinds of perspectives are rooted in a concern with the construction of power through the privileging of some knowledge claims or forms of logic over

others. In particular, there is an attempt to understand how GIS may reinforce an existing hierarchy in which some forms of knowledge and decision-making logic are given greater legitimacy, thereby excluding alternative forms of logic and information, marginalizing the knowledge and needs of communities represented by them.

A number of other studies in the critical GIS literature focus more on how access to GIS alters the knowledge production and discursive strategies available to different actors and, by extension, alters their relative power. For example, some suggest that GIS may enable less powerful actors to create alternative representations that contrast or challenge those of the state, potentially giving them a greater voice in policy debates (Barndt and Craig 1994; Ghose and Huxhold 2001; Sawicki and Craig 1996; Sieber 2002; Stonich 1998). A growing literature on 'public participation GIS' implicitly incorporates this understanding of how and why GIS might alter social and political relationships – it is argued that extending GIS access to grassroots groups and other nontraditional users is beneficial because it enables development of alternative knowledge and its inclusion in decision-making (Allen 1999; Ghose 2001; Ghose and Huxhold 2001; Jordan 2002; Kyem 2001; Obermeyer 1998). Recognizing concerns raised by other scholars about the exclusion of certain types of knowledge, a number of researchers in this area have sought ways of extending the representational capacities of traditional GIS to include, for instance, narratives or alternative cartographies (Harris and Weiner 2002; Krygier 2002).

[. . .]

Drawing on these different explanatory frameworks, the critical GIS literature has developed a number of perspectives that detail the contradictory nature of the technology. Clark (1998), for instance, contends that empowerment and marginalization are closely linked within GIS, arguing that any tool fostering information access, management, and analysis can be used in liberatory or repressive ways. Harris and Weiner (1998) and Stonich (1998) illustrate how GIS might be designed and used in ways that grant a wider range of actors a voice in decision-making and enable analysis of multiple forms of information. Simultaneously, however, they show that GIS has the potential to exclude and marginalize individuals and communities because of its high cost, technical skill requirements, and reliance on information that lends itself to cartographic and quantitative analysis.

In critical GIS research, it is increasingly clear that the impacts of this technology are contingent on and shaped by complex social and political relationships that constitute the power of different knowledge systems, decision-making processes, actors, and institutions. The question is less whether GIS is empowering or disempowering, but in what ways does it foster empowerment and disempowerment, and for whom? What is the basis of this empowerment or disempowerment for different actors and institutions? . . .

CONCEPTUALIZING EMPOWERMENT

[. . .]

The majority of definitions of empowerment developed in the social sciences have been singular – conceiving of empowerment (or disempowerment) as constituted by a single type of change. Although these conceptualizations of empowerment have been developed in a wide range of research contexts, it is possible to identify three groups of similar definitions. First, some scholars have defined empowerment as constituted by *distributive change*, such as greater access to goods and services (Jacobs 1992), or a greater number of opportunities for participation in political processes (McClendon 1993; Regalado and Martinez 1991). These definitions tend to be outcome oriented – conceiving of empowerment as a tangible or material change to be achieved. Second, some scholars use definitions that focus on *procedural changes* – arguing that empowerment occurs when social and political *processes* shift such that the contributions of citizens or community groups are granted greater legitimacy, or that their knowledge and needs are incorporated in decision-making processes (Allen 1993; Lake 1994; Tinker 1990; Young 1990). In contrast to distributive definitions that focus on material gain or on increasing opportunities for involvement, definitions focusing on procedural change are based on the premise that empowerment occurs when participation processes are structured not merely to include multiple views or ideas in decision-making dialogue, but to do so in a way that gives them authority and

legitimacy. Although this distinction between distributive and procedural change is subtle when applied to the issue of citizen participation, it is important in distinguishing between participation as simple inclusion and participation as accompanied by expanded legitimacy for participants and their priorities and arguments.

Third, a large number of studies conceive of empowerment as constituted by *capacity building*, generally framed as an expansion in the ability of citizens or communities to take action on their own behalf. Some scholars contend that such a capacity is a critical precursor to altering structures of oppression that have led to disempowerment (compare Fetterman *et al.* 1996). These definitions outline a variety of avenues through which that capacity building might occur. Some scholars focus on capacity building as the acquisition of new skills that help individuals and communities actively assert control over their circumstances (Barber 1984; Boyte 1989; Rappaport 1984; Zimmerman 1990). Others have argued that capacity building occurs through community-based knowledge production, in which members are engaged in gathering information about community conditions or existing resources. They contend that such knowledge production can inform action, serving as the basis for strengthening the capacity of a community to change its circumstances (Gaventa 1993; Heiman 1997; Perkins and Zimmerman 1995). Finally, a large number of scholars argue that the capacity of a community for effective action is increased through development of a politicized consciousness – an understanding of structural power inequities and how these affect them (Fals-Borda and Rahman 1991; Freire 1970). Much of Alinsky's writing on community organizing is informed by this assumption, as he asserts that a community must be politicized in order to mobilize effective collective action (Alinsky 1946, 1989).

[. . .]

In differentiating dimensions of empowerment along lines of distributive, procedural, or capacity-building change, scale remains an essential consideration. Efforts to change distribution of opportunities for involvement or distribution of material goods and services, to alter processes that determine the legitimacy or influence of different actors and institutions, or to expand their capacities for action all have inherently scalar dimensions, because these negotiations occur among actors and institutions positioned at (and working to affect) different spatial scales. . . . Such a 'politics of scale' must be taken into account in assessing empowerment. In the case of community organizations involved in urban planning and problem solving, key actors include citizens, community organizations, and the state. An assessment of empowerment in this context needs to examine distributive, procedural, and capacity-building changes in a way that explores multi-scalar implications of these changes for the relationships and interactions of these different actors engaged in urban planning and revitalization.

It is also crucial to consider the temporal scale of changes that might occur. Not only is empowerment a multifaceted process, it is shifting – very rarely constituting a fixed or permanent outcome. Thus, a crucial question to be asked of any kind of empowerment, at any scale of interaction, is the extent to which it is sustainable. Although many scholars writing on empowerment recognize its potential impermanence (Handler 1996; Hasson and Ley 1994; McClendon 1993; Perkins and Zimmerman 1995), there has been little extensive consideration of what types of empowerment might be most sustainable, or what strategies an organization or institution might pursue in trying to maintain advantages won.

[. . .]

APPLYING THE FRAMEWORK: GIS, NEIGHBORHOOD, REVITALIZATION AND EMPOWERMENT

Powderhorn Park neighborhood, located in south central Minneapolis, Minnesota, is a community of approximately 10,000 residents, diverse with respect to race and ethnicity, class and income, age, and housing tenure. Like many US inner-city neighborhoods, Powderhorn Park experienced decline during the 1970s and 1980s, through deterioration of its aging housing stock, loss of local employment opportunities and business activity, increasing poverty, and rising crime rates. Throughout its history, the neighborhood has had a high degree of citizen involvement in a number of neighborhood organizations. One of these, the Powderhorn Park

Neighborhood Association (PPNA), has become a key actor in neighborhood revitalization efforts since the 1970s. In particular, PPNA's role has been strengthened because of its responsibility for overseeing implementation of a comprehensive plan written by neighborhood residents in the early 1990s as part of the Minneapolis Neighborhood Revitalization Program (NRP).

[. . .]

In 1994, PPNA hired a local nonprofit software developer to build a complex database of neighborhood information, and purchased MapInfo GIS software to use in analysis and mapping of these data. Although staff and residents refer to this database as the 'housing database', it contains information on property and land use, as well as information about people and activities in the neighborhood. . . . The property and land-use variables include lot sizes, zoning, and age and condition of structures, and these data were obtained from city and county government offices. The 'people and activity' data were gathered by staff and residents and include information concerning past problems or changes in conditions at neighborhood properties, as well as histories of PPNA involvement in resolving problems or improving conditions at a site. Finally, the database includes information about how to contact occupants, owners, or managers of neighborhood properties.

[. . .]

DISTRIBUTIVE DIMENSIONS OF EMPOWERMENT

In the context of neighborhood planning and problem-solving efforts, distributive empowerment can be conceived of as individuals or organizations expanding their opportunities for involvement. PPNA's use of the GIS has fostered several changes that have distributive implications, both at the neighborhood level and at the city level. One of the most important of these changes in neighborhood-level processes has been a shift in the language and forms of information that dominate PPNA's deliberations about the neighborhood. Most of the information held in their GIS database is quantitative in form, and further, understanding these data requires experience with and knowledge of city land and property policies. Through their growing use of GIS and these data, residents have altered the terms and information they use in describing neighborhood conditions. For example, archival records show that, in the past, discussions of neighborhood housing conditions were couched in a language of visual description and relied heavily on residents' observations. Since the advent of GIS at PPNA, such discussions more commonly use numerical property-condition codes contained in the housing database. . . .

Such changes in the language and forms of information used in PPNA have contrasting implications with respect to distributive dimensions of empowerment. On [the] one hand, they raise the level of expertise and knowledge one must have in order to understand and contribute to dialogue in the organization, in effect restricting opportunities for involvement of some neighborhood residents. In the words of one resident,

> "I've been a member of this committee for years, and with the discussion of tonight's agenda items, I couldn't understand a thing. We've gotten so far into zoning code that the lay person can't figure out what is going on!" (Melissa, personal interview, 1999)

Though the new discourse of decision-making at PPNA and exclusions that have occurred because of it can be understood as barriers of expertise and knowledge, it is critical to recognize that they are intricately intertwined with exclusions along lines of race, class, ethnicity, gender, and other axes of difference. The individuals who have most readily incorporated this new discourse into their activities at PPNA have largely been upper-middle-class white homeowners. The emerging discourse has tended to impinge on participation of people of color, senior citizens, low-income individuals, and residents with limited English skills. These patterns of inclusion and exclusion are inscribed in PPNA's particular history and social relations, but of course are simultaneously an expression of broader structural opportunities and constraints.

Although PPNA's use of GIS has resulted in changes that raise barriers to participation of some residents, the organization has simultaneously been trying to use its GIS as part of a strategy to lower these barriers. One of the most common daily uses of PPNA's GIS is providing information to

neighborhood residents. For instance, residents experiencing problems with a nearby rental property could obtain information from the housing database about past strategies used in trying to resolve the problem, or information about how to contact the owner and manager. In an alternative scenario, a block club seeking to establish a community garden on a vacant lot could obtain zoning and ownership data needed for their project. Prior to the existence of PPNA's databases, acquisition of these type of data required calling or visiting several city government offices, as well as PPNA's own office. For many residents, this series of steps constituted a barrier to participation. . . . PPNA's choice to use its GIS to create an information clearinghouse within the neighborhood thus lowers the barriers to involvement experienced by some residents in neighborhood problem solving and revitalization, fostering distributive empowerment for these individuals.

Beyond the scale of neighborhood interactions, PPNA's use of GIS has also fostered changes that constitute distributive empowerment for PPNA in its relationship to local government. PPNA has used GIS to foster greater information access and analysis capabilities, which in turn it has used to leverage greater participation in city-level decision-making. One of PPNA's community organizers offered this example:

> "Say there is a dilapidated house and the City Inspector walks by it once on the sidewalk and says it is all cleaned up and there ain't no reason to put more energy into it. Well, we may still be getting a bunch of calls from residents who say the place is trashed again and it's noisy. By keeping this information in the database, we can start to see a pattern emerge. We can then show the City there might be a need for more monitoring, maybe checking in with the rental property owner. We use this information to justify and push for greater means to address a problem." (Anthony, personal interview, 1999)

In this instance, PPNA is using its database to monitor neighborhood conditions, and then using this information to insert its voice into city plans regarding housing improvement. Although the City of Minneapolis does have a history of willingness to involve neighborhood organizations in city-level decisions, this opportunity to make specific recommendations for city actions in the neighborhood constitutes an expansion of previous opportunities for community-organization participation. Such new opportunities for participation constitute a distributive form of empowerment.

[. . .]

PROCEDURAL DIMENSIONS OF EMPOWERMENT

PPNA's use of GIS has also fostered changes in its priorities, practices, and discourses that have altered the legitimacy or influence of actors and their knowledge claims. These changes in legitimacy and influence constitute procedural forms of empowerment for some residents and disempowerment for others. The procedural shifts outlined here are occurring largely because the use of digital databases and GIS at PPNA has been accompanied by a change in the type of information considered most important and appropriate for planning and problem solving, and in the relative authority of different claims to knowledge in these processes. Specifically, within PPNA's planning and problem-solving efforts, decisions and opinions that are grounded in the language and information of city policy and professional planning expertise are being granted increasing authority.

PPNA's GIS has enabled the organization for the first time to obtain, store, and analyze large volumes of information that directly draws on city policy and professional planning expertise. The shift in the type of information considered to be most important or legitimate is evident in the types of information and justifications given in the organization's decision-making. One resident offered this explanation of PPNA reasons for rejecting a variance to add more rental units to a property over ten years ago, prior to PPNA's use of GIS:

"We said no. It was an ugly property! There was tarps hanging off of it and pigeons making a home there and everything!" (Max, personal interview, 1999)

Explaining the information used in reviewing a variance request made in 1999, well after the advent of information technologies at PPNA, a staff member described the present decision-making process as one that relies much on different types of information:

"As a neighborhood, we want consistent zoning and land-use patterns, and we want to strengthen our commercial nodes. So we wanted to rip down commercial building like this [one which is] away from the nodes. And we didn't want to encourage a land use that wasn't [consistent] with the zoning." (Jeremy, personal interview, 1999)

The past choice, described in the first statement, is justified with site-specific conditions and visual observations of the property. The recent action is justified with the language and information of city policy and professional planning – "consistent zoning", "land-use patterns", and "commercial nodes". It is not the case that the first type of decision-making is no longer present at PPNA, just that the second has become increasingly common and is now actively defended by many participants as the best and most appropriate way of making decisions. This development has significant implications for the procedural empowerment of different groups of residents. Specifically, those residents who have the education, expertise, or experiences to be able to frame their contributions to neighborhood dialogue using the newly dominant forms of information can make a greater claim to authority. This shift has meant the empowerment of a particular social group in the neighborhood vis-à-vis all other residents – white, upper-middle-class homeowners with specialized education or experience in law, local government, or business.

The statements above also illustrate the way in which PPNA's revitalization efforts increasingly prioritize goals and plans framed in geographically comprehensive ways, referencing conditions or patterns across the neighborhood. This scalar shift contrasts with earlier decision-making priorities and practices that were more site-specific. The effect of this shift in terms of procedural empowerment has been roughly the same as the schism described above. It empowers individuals who have access to such comprehensive information, either through experience with PPNA's GIS and database or through other experiences such as professional experience in planning or local government. The contributions of individuals whose experiences or knowledge enable them to express their ideas and preferences with respect to only site-specific information are less influential.

[. . .]

Alongside these changes that foster procedural empowerment for some residents while disempowering others, PPNA's use of GIS is also altering the community's interactions with local-government actors in ways that constitute procedural empowerment. PPNA has used its GIS to generate information that these officials need, but do not have access to. This is particularly true because PPNA's database incorporates both local knowledge and government-collected information. Analysis of this information has afforded PPNA greater legitimacy and influence as an informed and knowledgable actor in local state processes – a procedural dimension of empowerment.

[. . .]

CAPACITY-BUILDING DIMENSIONS OF EMPOWERMENT

Capacity-building dimensions of empowerment may take several forms in the context of community organizations and their activities. Expansion of the capacity of individuals and communities to take action on their own behalf could occur through development of new skills, production of new knowledge to inform or guide their action, or development of a new understanding of community conditions that motivates further action. As with both the procedural and distributive dimensions of empowerment, PPNA's use of GIS has had varying impacts with respect to capacity building.

In terms of development of new skills, empowerment of the organization as a whole is clear. The use of GIS has given PPNA access to new data-storage and analysis capabilities – a significant new skill informing planning and problem-solving efforts. However, capacity building through expansion of the skills of individuals is unevenly distributed. Although a large number of involved residents use data and analysis generated from the GIS, PPNA staff members do most of the data acquisition, coding, and analysis. Thus, although development of new skills is occurring, this element of empowerment is restricted to the staff members. This uneven distribution in dimensions of

capacity-building empowerment has important potential consequences in the context of community groups. These organizations tend to have high staff turnover (Fisher 1994), and so GIS skills are unlikely to be retained as community resources if they are not also fostered in residents.

Capacity building through knowledge production is closely related to capacity building through new understanding of community conditions or needs, and in both areas PPNA's use of GIS fosters community-level empowerment. GIS enables PPNA to store community-collected observations (and local state data) for a much larger number of neighborhood properties, and to maintain these data over a longer period of time. Prior to the creation of its GIS, PPNA's local knowledge resources consisted of notes taken by a frequently changing array of staff members, and the memories of an equally shifting group of active residents. Their use of GIS to create a greatly expanded data resource for the community is a form of capacity building through knowledge production. As well, some residents who use data and maps from the database suggest that they experience capacity building through knowledge production and an enhanced personal sense of efficacy in creating change in the neighborhood. One resident who has frequently used information from the housing database argued:

"As you gain expertise in knowing what you can do, it makes you so much more powerful. And the reward is knowing you're getting something done, that you're making a difference to the place where you live!" (Jane, personal interview, 1999)

Further, by bringing analytical capabilities of their GIS to bear on this expanded community data resource, PPNA has developed new understandings of neighborhood conditions and needs. For example, PPNA's housing organizer explained that, prior to the creation of the housing database, the Housing and Land Use Committee felt that the area of the neighborhood in greatest need of revitalization was the northern portion, both rental and owner-occupied structures. However, after using the GIS to create a series of maps of condition, value, and tenure status, the committee changed its long-standing assessment of the pattern. They found a strong need for revitalization of rental properties on the three largest transportation arteries that cross through the neighborhood. This new interpretation of neighborhood conditions resulted in a significant change in the geographical pattern of their revitalization strategy, and a move toward assistance programs in those locations to target rental properties. These community-based analyses can be an important element of capacity building in terms of expanded political consciousness. GIS analysis may engage the community in critical re-examination of data and representations of their community being produced by local-government actors, potentially mobilizing and informing community efforts to present alternative views.

In understanding how GIS might foster capacity-building empowerment at the community level, existing patterns of participation and exclusion in an organization are crucial, because these determine which individual residents are directly involved in these dimensions of empowerment. The capacity building that is occurring at PPNA through its use of GIS is most immediately relevant for the organization as a whole, and for those individuals who are actively participating in the organization. Residents who are not involved in the organization are not directly experiencing this form of capacity-building empowerment through PPNA's use of GIS. Such differences illustrate the importance of assessing empowerment at multiple scales of interaction. Solely focusing at the level of the organization would overlook the uneven development of this capacity-building empowerment among Powderhorn Park residents.

CONCLUSION

[. . .]

As countless other scholars studying participatory democratic practice have observed, organizations engaged in community-based planning or neighborhood revitalization occupy a complicated position. They are simultaneously striving for involvement and influence within the state-level processes in which they are embedded, while at the same time working at the community level to include and represent the needs and priorities of a diverse range of social groups. This study illustrates the growing complexity of these tasks, in the face of new tools and practices that may alter participation and power relations differently in these multiple

spheres of interaction. Further, I have developed and highlighted the importance of a multi-scalar analysis of empowerment in assessing the implications of such new tools and practices. Analysis at a single scale may obscure the potential complexity of these impacts, whereas examination of multiple spheres of interaction may highlight the presence of trends toward both empowerment and disempowerment. A challenge for community-based organizations, and for those of us trying to understand their role and impact in urban neighborhoods, lies in further exploration of how these organizations might balance the oppositional tendencies of such changes – working to enhance empowering changes while ameliorating disempowering developments.

REFERENCES FROM THE READING

Aitken, S. and Michel, S. (1995) 'Who contrives the "real" in GIS? Geographic information, planning, and critical theory', *Cartography and Geographic Information Systems* 22: 17–29.

Alinsky, S. (1946) *Reveille for Radicals*, University of Chicago Press, Chicago, IL.

Alinksy, S. (1989) *Rules for Radicals: A Practical Primer for Realistic Radicals*, Vintage Books, New York.

Allen, J. (1993) 'Friends and neighbors: knowledge and campaigning in London', in R. Fisher and J. Kling (eds) *Mobilizing the Community: Local Politics in the Era of the Global City*, Sage, London, pp. 223–45.

Allen, M. (1999) 'A participatory model of information system design: a case study of the economic human rights documentation information-management system of the Kensington Welfare Rights Union', paper presented at the First International Conference on Geographic Information Society, Minneapolis, MN, 20–22 June.

Barber, B. (1984) *Strong Democracy: Participatory Politics for a New Age*, University of California Press, Berkeley, CA.

Barndt, M. and Craig, W. (1994) 'Data providers empower community GIS efforts', *GIS World* 7: 49–51.

Boyte, H. (1989) *CommonWealth: A Return to Citizen Politics*, Free Press, New York.

Clark, M. (1998) 'GIS – democracy or delusion?' *Environment and Planning A* 303-16.

SEVEN

Fals-Borda, O. and Rahman, M. (1991) *Action and Knowledge: Breaking the Monopoly with Participatory Action Research*, Apex, New York.

Fetterman, D., Katerarian, S. and Wandersman, A. (eds) (1996) *Empowerment Evaluation: Knowledge and Tools for Self-assessment*, Sage, Thousand Oaks, CA.

Fisher, R. (1994) *Let the People Decide: Neighborhood Organizing in America* (2nd edn), Maxwell Macmillan, New York.

Freire, P. (1970) *Pedagogy of the Oppressed*, Seabury, New York.

Gaventa, J. (1993) 'The powerful, the powerless, and the experts: knowledge struggles in an information age', in P. Park, M. Brydon-Miller, B. Hall and T. Jackson (eds) *Voices of Change: Participatory Research in the United States and Canada*, Bergin & Garvey, Westport, CT, pp. 21–40.

Ghose, R. (2001) 'Use of information technology for community empowerment: transforming geographic information systems into community information systems', *Transactions in GIS* 5, 2: 141–63.

Ghose, R. and Huxhold, W. (2001) 'The role of local contextual factors in building public participation GIS: the Milwaukee experience', *Cartography and Geographic Information Systems* 28, 3: 195–208.

Handler, J. (1996) *Down From Bureaucracy: The Ambiguity of Privatization and Empowerment*, Princeton University Press, Princeton, NJ.

Harris, T. and Weiner, D. (1998) 'Empowerment, marginalization, and community-oriented GIS', *Cartography and Geographic Information Systems* 25, 2: 67–76.

Harris, T. and Weiner, D. (2002) 'Implementing a community-integrated GIS: perspectives from South African fieldwork', in W. Craig, T. Harris and D. Weiner (eds) *Community Participation and Geographic Information Systems*, Taylor & Francis, London.

Harris, T., Weiner, D., Warner, T. and Levin, R. (1995) 'Pursuing social goals through participatory GIS: redressing South Africa's historical political ecology', in J. Pickles (ed.) *Ground Truth: The Social Implications of Geographic Information Systems*, Guilford Press, New York, pp. 196–221.

Hasson, S. and Ley, D. (1994) *Neighbourhood Organizations and the Welfare State*, University of Toronto Press, Toronto.

Heiman, M. (1997) 'Science by the people: grassroots environmental monitoring and the debate over scientific expertise', *The Journal of Planning Education and Research* 16, 4: 291–9.

Jacobs, B. (1992) *Fractured Cities: Capitalism. Community, and Empowerment in Britain and America*, Routledge, New York.

Jordan, G. (2002) 'GIS for community forestry user groups in Nepal: putting people before the technology', in W. Craig, T. Harris and D. Weiner (eds) *Community Participation and Geographic Information Systems*, Taylor & Francis, London.

Krygier, J. (2002) 'A praxis of public participation GIS and visualization', in W. Craig, T. Harris and D. Weiner (eds) *Community Participation and Geographic Information Systems*, Taylor & Francis, London.

Kyem, P. (2002) 'Promoting local community participation in forest management through a PPGIS application in Southern Ghana', in W. Craig, T. Harris and D. Weiner (eds) *Community Participation and Geographic Information Systems*, Taylor & Francis, London.

Lake, R. (1993) 'Planning and applied geography: positivism, ethics, and geographic information systems', *Progress in Human Geography* 17: 404–13.

Lake, R. (1994) 'Negotiating local autonomy', *Political Geography* 13: 423–42.

Leitner, H., Elwood, S., Sheppard, E., McMaster, S. and McMaster, R. (2000) 'Modes of GIS provision and their appropriateness for neighborhood organizations: examples from Minneapolis and St Paul, Minnesota', *The URISA Journal* 12, 4: 43–56.

Lupton, M. and Mather, C. (1997) ' "The anti-politics machine": GIS and the reconstruction of the Johannesburg local state', *Political Geography* 16: 565–80.

McClendon, B. (1993) 'The paradigm of empowerment', *Journal of the American Planning Association* 59: 145–7.

Obermeyer, N. (1998) 'The evolution of public participation GIS', *Cartography and Geographic Information Systems* 25, 2: 65–6.

Openshaw, S. (1991) 'A view on the GIS crisis in geography, or, using GIS to put Humpty-Dumpty back together again', *Environment and Planning A* 23: 621–8.

Perkins, D. and Zimmerman, M. (1995) 'Empowerment theory, research, and application', *American Journal of Community Psychology* 23: 569–79.

Pickles, J. (1995) 'Representations in an electronic age: geography, GIS, and democracy', in J. Pickles (ed.) *Ground Truth: The Social Implications of Geographic Information Systems*, Guilford Press, New York, pp. 1–30.

Rappaport, J. (1984) 'Studies in empowerment: introduction to the issue', *Prevention in Human Services* 3, 2/3: 1–7.

Regalado, J. and Martinez, G. (1991) 'Reapportionment and coalition building: a case study of informal barriers to Latino empowerment in Los Angeles County', in R. Villareal and N. Hernandez (eds) *Latinos and Political Coalitions: Political Empowerment for the 1990s*, Greenwood Press, New York, pp. 126–43.

Rundstrom, R. (1995) 'GIS, indigenous peoples, and epistemological diversity', *Cartography and Geographic Information Systems* 22: 45–57.

Sawicki, D. and Craig, W. (1996) 'The democratization of data: bridging the gap for community groups', *Journal of the American Planning Association* 62: 512–23.

Sheppard, E. (1995) 'GIS and society: towards a research agenda', *Cartography and Geographic Information Systems* 22: 5–16.

Sieber, R. (2002) 'Geographic information systems in the environmental movement', in W. Craig, T. Harris and D. Weiner (eds) *Community Participation and Geographic Information Systems*, Taylor & Francis, London.

Stonich, S. (1998) 'Information technologies, advocacy, and development: resistance and backlash to industrial shrimp farming', *Cartography and Geographic Information Systems* 25, 2: 113–22.

Taylor, P. and Overton, M. (1991) 'Commentary: further thoughts on geography and GIS', *Environment and Planning A* 23: 1087–94.

Tinker, I. (1990) *Persistent Inequalities*, Oxford University Press, Oxford.

Yapa, L. (1991) 'Is GIS appropriate technology?', *International Journal of Geographical Information Systems* 5, 1: 41–58.

Young, I. (1990) *Justice and the Politics of Difference*, Princeton University Press, Princeton, NJ.

Zimmerman, M. (1990) 'Taking aim on empowerment research: on the distinction between individual and psychological conceptions', *American Journal of Community Psychology* 18, 1: 169–77.

Editors' suggestions for further reading

Batty, M. (2003) "Using Geographical Information Systems," in N.J. Clifford and G. Valentine (eds) *Key Methods in Geography*, Sage, London, pp. 409–24.

Batty, M. and Kim, K.S. (1992) "Form follows function: reformulating urban population density functions," *Urban Studies* 29: 1043–70.

Curry, M. (1997) "The digital individual and the private realm," *Annals of the Association of American Geographers* 87: 681–99.

Elwood, S.A. (2001) "GIS and collaborative urban governance: understanding their implications for community action and power," *Urban Geography* 22, 8: 737–59.

Elwood, S.A. (2002) "The impacts of GIS use for neighbourhood revitalization in Minneapolis," in W.J. Craig, T.M. Harris and D. Weiner (eds) *Community Participation and Geographical Information Systems*, Routledge, London.

Lake, R.W. (1993) "Planning and applied geography: positivism, ethics, and geographic information systems," *Progress in Human Geography* 17: 404–13.

Pickles, J. (1995) *Ground Truth: The Social Implications of Geographic Information Systems*, Guilford Press, New York.

Schuurman, N. (2000) "Trouble in the heartland: GIS and its critics in the 1990s," *Progress in Human Geography* 24: 569–90.

Shaw, S.L. (1993) "Hub structures of major U.S. passenger airlines," *Journal of Transport Geography* 1: 47–58.

Sieber, R.E. (2003) "Public participation geographic information systems across borders," *Canadian Geographer* 47: 50–61.

Smith, N. (1992) "History and philosophy of geography: real wars, theory wars," *Progress in Human Geography* 16: 257–71.

Sui, D.Z. (1994) "GIS and urban studies: positivism, post-positivism, and beyond," *Urban Geography* 15: 258–78.

Taylor, P.J. (1990) "GKS," *Political Geography Quarterly* 9: 211–12.

Taylor, R.W. and McDonald, D. (1989) "GIS techniques applied to spatial analysis of crime in the city of Dallas: preliminary findings," *GIS/LIS'89* 1: 171–80.

Waddell, P. and Shukla, V. (1993) "Employment dynamics, spatial structuring, and the business cycle," *Geographical Analysis* 25: 35–52.

ILLUSTRATION CREDITS

Every effort has been made to contact copyright holders for their permission to reprint plates and figures in this book. The publishers would be grateful to hear from any copyright holder who is not here acknowledged and will undertake to rectify any errors or omissions in future editions of this book. The following is copyright information for the figures and plates that appear in this book.

PLATES

Part 1 German immigrants in front of their boarding house in Milwaukee, Wisconsin (late 1880s). © Judith Kenny. Reproduced with permission.

Part 2 The towers of Paris' La Defense office center. © Harold Mayer Collection. From the American Geographical Society Library, University of Wisconsin-Milwaukee Libraries. Reproduced with permission.

Part 3 Detroit's Renaissance Center with an old manufacturing site in the foreground. © Elvin Wyly. Reproduced with permission.

Part 4 Police and anti-war protesters meet on the streets of Cleveland. © Elvin Wyly. Reproduced with permission.

Part 5 Anti-racist public message. © Elvin Wyly. Reproduced with permission.

Part 6 Concord Pacific Place on the north shore of False Creek, Vancouver, BC. © Elvin Wyly. Reproduced with permission.

Part 7 Surveillance camera in Glasgow, Scotland. © Elvin Wyly. Reproduced with permission.

1 "Typical" Polish housing as illustrated in a 1906 housing study of Milwaukee, Wisconsin. Public domain.

2 Piazza Tolomei in Siena, Italy. © Thomas Harvey. Reproduced with permission.

3 Looking through the window of the car showroom on to a street in Calcutta. © Scott Douglas Purl. Reproduced with permission.

4 The law school of West Bengal National University, Calcutta. © Rina Ghose. Reproduced with permission.

5 Loft housing along the Brooklyn shore of New York City as seen from the East River. © Harold Mayer Collection. From the American Geographical Society Library, University of Wisconsin-Milwaukee Libraries. Reproduced with permission.

6 One of the towers of the Cabrini-Green public housing development stands in contrast to the skyline of downtown Chicago. © Jeffrey Zimmerman. Reproduced with permission.

7 High-rise public housing in Harlem. © Elvin Wyly. Reproduced with permission.

8 A Prairie Crossing advertisement. © Jeffrey Zimmerman. Reproduced with permission.

9 Neo-traditional architectural styles and new urbanist design guidelines. © Jeffrey Zimmerman. Reproduced with permission.

10 The Arcade, Letchworth Garden City, a covered pedestrian alley built in 1922. © Judith Kenny. Reproduced with permission.

11 South Street Seaport, New York City. © Harold Mayer Collection. From the American Geographical Society Library, University of Wisconsin-Milwaukee Libraries. Reproduced with permission.

12 The Watergate complex in Washington, DC, built in the late 1960s. © Harold Mayer Collection. From the American Geographical Society Library, University of Wisconsin-Milwaukee Libraries. Reproduced with permission.

FIGURES

Burgess Figure 1. The growth of the city. Burgess, Ernest W., "The Growth of the City: An Introduction to a Research Project." From R. Park, E. Burgess *et al.* (eds) *The City*, 47–62. Copyright © 1925 The University of Chicago Press. Reprinted by permission.

Burgess Figure 2. Urban areas. Burgess, Ernest W., "The Growth of the City: An Introduction to a Research Project." From R. Park, E. Burgess *et al.* (eds) *The City*, 47–62. Copyright © 1925 The University of Chicago Press. Reprinted by permission.

Hoyt Figure 1. Shift in location of fashionable residential areas in six US cities, 1900 to 1936. Hoyt H., "The structure of American cities in the post-war era," *American Journal of Sociology* 48, 4: 475–481. Public domain.

Ullman Figure 1. Theoretical shapes of tributary areas. Ullman, Edward L., "A Theory of Location for Cities," *American Journal of Sociology,* XLVI, 853–64. Copyright © 1941 The University of Chicago Press. Reprinted by permission.

Harris and Ullman Figure 1. The pattern of growth of the city. Harris, Chauncy D. and Ullman, Edward L., "The Nature of Cities," *Annals of the Association of Political and Social Science*, 1–17. Copyright © 1945 The American Academy of Political and Social Science. Reprinted by permission.

Beaverstock *et al.* Figure 1. World-city network. Beaverstock, Jonathan V., Smith, Richard G. and Taylor, Peter J., "World-City Network: A New Metageography?," *Annals of the Association of Political and Social Science*, 90, 123–34. Copyright © 2000 Blackwell Publishing. Reprinted by permission.

Chakravorty Figure 1. The spatial structure of colonial Calcutta. Chakravorty, "From Colonial City to Globalizing City? The Far-from-complete Spatial Transformation of Calcutta." From P. Marcuse and R. van Kempen (eds), *Globalizing Cities: A New Spatial Order?*, 56–77. Copyright © 2000 Blackwell Publishing. Reprinted by permission.

Chakravorty Figure 2. The spatial structure of post-colonial Calcutta. Chakravorty, "From Colonial City to Globalizing City? The Far-from-complete Spatial Transformation of Calcutta." From P. Marcuse and R. van Kempen (eds), *Globalizing Cities: A New Spatial Order?*, 56–77. Copyright © 2000 Blackwell Publishing. Reprinted by permission.

Chakravorty Figure 3. The spatial structure of post-reform Calcutta. Chakravorty, "From Colonial City to Globalizing City? The Far-from-complete Spatial Transformation of Calcutta." From P. Marcuse and R. van Kempen (eds), *Globalizing Cities: A New Spatial Order?*, 56–77. Copyright © 2000 Blackwell Publishing. Reprinted by permission.

Harvey Figure 1. The relations considered for 'reproduction on an expanded scale'. Harvey, David, "The Urban Process Under Capitalism: A Framework for Analysis," *International Journal of Urban and Regional Research*, 2, 101–31. Copyright © 1978 Blackwell Publishing. Reprinted by permission.

Harvey Figure 2. The structure of relations between the primary, secondary and tertiary circuits of capital. Harvey, David, "The Urban Process Under Capitalism: A Framework for Analysis," *International Journal of Urban and Regional Research*, 2, 101–31. Copyright © 1978 Blackwell Publishing. Reprinted by permission.

Dear and Flusty Figure 1. A concept of protopostmodern urbanism. Dear, Michael and Flusty, Steven, "Postmodern Urbanism," *Annals of the Association of American Geographers*, 88, 50–72. Copyright © 1998 Blackwell Publishing. Reprinted by permission.

COPYRIGHT INFORMATION

1 FOUNDATIONS

BURGESS Burgess, Ernest W., "The Growth of the City: An Introduction to a Research Project." From R. Park, E. Burgess *et al.* (eds) *The City*, 47–62. Copyright © 1925 The University of Chicago Press. Reprinted by permission.

HOYT Hoyt, H., "The structure of American cities in the post-war era," *American Journal of Sociology* 48, 4: 475–481. Public domain.

ULLMAN Ullman, Edward L., "A Theory of Location for Cities," *American Journal of Sociology,* XLVI, 853–64. Copyright © 1941 The University of Chicago Press. Reprinted by permission.

HARRIS and ULLMAN Harris, Chauncy D. and Ullman, Edward L., "The Nature of Cities," *Annals of the Association of Political and Social Science*, 1–17. Copyright © 1945 The American Academy of Political and Social Science. Reprinted by permission.

2 GLOBALIZATION

BEAVERSTOCK, SMITH and TAYLOR Beaverstock, Jonathan V., Smith, Richard G. and Taylor, Peter J., "World-City Network: A New Metageography?," *Annals of the Association of Political and Social Science*, 90, 123–34. Copyright © 2000 Blackwell Publishing. Reprinted by permission.

HAMNETT Hamnett, Chris, "Social Polarisation in Global Cities: Theory and Evidence," *Urban Studies*, 30, 401–24. Copyright © 1994 Taylor and Francis. Reprinted by permission. Journal website http://www.tandf.co.uk.

CHAKRAVORTY Chakravorty, Sanjoy, "From Colonial City to Globalizing City? The Far-from-complete Spatial Transformation of Calcutta." From P. Marcuse and R. van Kempen (eds), *Globalizing Cities: A New Spatial Order?*, 56–77. Copyright © 2000 Blackwell Publishing. Reprinted by permission.

NIJMAN Nijman, Jan, "Cultural Globalization and the Identity of Place: The Reconstruction of Amsterdam," *Cultural Geographies*, 6, 146–64. Copyright © 1999 Hodder Arnold. Reprinted by permission.

Index

Figures and Tables in *Italic*
PL before a number indicates the plate number in plate section
"n" after page number indicates material in notes